KINESIOLOGY

Fifth Printing

KINESIOLOGY

OF THE HUMAN BODY

Under Normal and
Pathological Conditions

By

ARTHUR STEINDLER, M.D., (Hon) F.R.C.S.
Eng. F.A.C.S., F.I.C.S.

Professor of Orthopedic Surgery, Emeritus
State University of Iowa
Head of Orthopaedic Department, Mercy Hospital
Iowa City, Iowa

CHARLES C THOMAS · PUBLISHER

Springfield · Illinois · U.S.A.

Published and Distributed Throughout the World by

CHARLES C THOMAS • PUBLISHER

BANNERSTONE HOUSE

301-327 East Lawrence Avenue, Springfield, Illinois, U.S.A.

© 1955, *by* CHARLES C THOMAS • PUBLISHER

ISBN 0-398-01846-4

Library of Congress Catalog Card Number: 55-6121

First Printing, 1955
Second Printing, 1964
Third Printing, 1970
Fourth Printing, 1973
Fifth Printing, 1977

With THOMAS BOOKS careful attention is given to all details of manufacturing and design. It is the Publisher's desire to present books that are satisfactory as to their physical qualities and artistic possibilities and appropriate for their particular use. THOMAS BOOKS will be true to those laws of quality that assure a good name and good will.

Printed in the United States of America

00-2

To my good Wife and Helpmate
Louise Steindler

PREFACE

A PREVIOUS publication on the subject of the *Mechanics of Normal and Pathological Locomotion*,[1] has met with what the author believes to be an indifferent reception. There seemed to be two reasons for it. The first problem was that not enough consideration had been accorded to the practical application of rehabilitation and re-education of locomotor deficiencies. With this in mind, the author has endeavored to give the theoretical facts a much wider interpretation in connection with diagnostic and therapeutic questions, fully realizing that the value of the book lies in what help it can offer to clinician and physical educator.

The other criticism was that the analysis of locomotion has ventured too much into the field of general mechanics which made the reading of the book uncomfortable and required much needless concentration. It was argued that practical conclusions could be attained more directly by observation and empiricism without any tedious theoretical preambles. It is with this second argument that the author has to disagree. He is neither willing nor able to reconcile himself to a short cut method based entirely on impressions and to summersault over the hard facts of basic sciences just in order to arrive at quick and usually superficial practical conclusions.

He is convinced that Kinesiology is an indispensable background for the prevention, the treatment, and rehabilitation of locomotor disorders.

On the other hand, it is conceded that mathematical accuracy in the calculation of locomotor events is only too often swallowed up by the stream of physiological and especially of pathological fluctuations; that certain allowances must be made because no two situations, be they normal or abnormal, are so alike that they can be expressed accurately by the same mathematical formula.

After all the analysis of human motion is primarily the job of the biologist, and only secondarily that of the physicist or engineer. The purely mechanical aspect of locomotion, which is the province of kinesiology, is merely a part of the physiology of motion in general.

All motor events are the end result of a long chain of cause and effect, the beginning of which, as in all things, is steeped in mystery. The study of kinetics taps this chain of events at a point just behind the clinical manifestation while the biologist tries to penetrate much farther into the background. In other words, kinetics explains only the physical events which lead immediately to the locomotor performance and it is not concerned with the more remote causes of human motion.

[1] Steindler, A.: Charles C Thomas, Publisher, Springfield, Illinois, 1935.

With these considerations in mind the author has tried to reduce all mathematical calculations to a minimum and has advanced on every occasion the clinical significance and the practical implication of kinetic findings. This, he hopes, should make for more palatable reading.

A. STEINDLER

Iowa City

ACKNOWLEDGMENTS

T HE author has received most valuable assistance in the preparation of this book. He is particularly grateful for the help extended to him by his associate, Dr. Webster B. Gelman. He not only furnished the illustrative drawings but he also helped with the arrangement of the book and with reading of the text, for which he made many useful suggestions. For the clerical help in typing the text the author is indebted to Mrs. June Clendenin, Mrs. Marian Heilbrun, and Mrs. Margaret Washburn. The photographic work was in the able hands of the Messrs. Kent. Mrs. Thea Wiegand compiled the table of contents and author and subject indices. The publisher, Mr. Charles C Thomas, has bestowed upon this work his highly appreciated painstaking efforts in the examination of the text, its editing and printing. All of this assistance is hereby thankfully acknowledged by the author.

A. STEINDLER

CONTENTS

PART II

THE TRUNK

LECTURE IX. THE MECHANICS OF THE SPINAL COLUMN

LECTURE X. THE DYNAMICS OF THE NORMAL SPINE

LECTURE XI. THE PATHOMECHANICS OF THE LUMBOSACRAL JUNCTION

LECTURE XIV. THE MECHANICS OF POSTURE

LECTURE XV. THE PATHOMECHANICS OF SCOLIOSIS

PART III

THE EXTREMITIES

LECTURE XVI. MECHANICS OF THE HIP JOINT

LECTURE XXV. THE LOWER EXTREMITY AS A WHOLE

LECTURE XXVI. MECHANICS OF SHOULDER-ARM COMPLEX

LECTURE XXVII. THE PATHOMECHANICS OF PARALYSIS OF THE SHOULDER

KINESIOLOGY

Part I

GENERAL KINETICS

Lecture I

THE AIMS AND PURPOSES OF KINESIOLOGY

KINESIOLOGY is that part of physiology of motion which describes and analyzes locomotor events so far as they reflect the action of mechanical forces. In other words, it presents bodily motion as a special case in mechanics.

The question is, can it be so represented; can human motion, with its boundless variability and being the product of a multitude of only partially known factors, be forced into the narrow frame of precise and austere mathematical and physical laws?

This is the very point which ultimately decides whether or not there is enough practical value in the undertaking to warrant this line of investigation. It must be admitted that locomotion involves certain unknown and incalculable factors which preclude a strictly mathematical presentation.

Nevertheless, the operation of known physical laws can be clearly recognized and evaluated in properly observed locomotor events. The Newtonian laws are accepted as being fundamental for terrestrial mechanics. We intend to interpret the operation of these laws in terms of human motion. We do not attempt to formulate general mechanical laws from our observation as Newton formulated his law of gravitation from observing the falling apple. In other words, our reasoning is *deductive* from the accepted mechanical laws to their interpretation in terms of our observations; it is not *inductive* from the observation to the formulation of the laws.

If we are in error the quarrel is with the observation and its interpretation and not with the law itself. The danger then lies not in failing to recognize that a mechanical law is operating but rather in overlooking the existence of other laws, not mechanical in nature, which have their share in human locomotion.

A good example is the interpretation of the functional adaptation of bone as formulated by Julius Wolff in 1876. When it was recognized by Culman that the internal structure of the upper end of the femur presents a perfectly calculated construction of a loaded crane, it was believed that the physical laws had absolute and complete control over the form and texture of the human bone and that the lamellar arrangements or trajectories, as they are called, fall strictly in line with tension and pressure stresses. This is no doubt true to a large extent. The only mistake was that the operation of other laws biological in nature had been overlooked because their influence on the structure of the bone was less evident. As a matter of fact, the architectural response of bone to the mechanical forces of stress and strain merely approaches but never completely conforms with the accurate engineering calculation. The reason is that there are other biological laws in operation which, under certain conditions, may become more conspicuous than the mechanical laws themselves. For instance, the organization of bone after fracture may proceed under the laws of transformation on stress and strain resistance. But it would be more difficult to demonstrate the same law of functional adaptation in cases of severe osteomalacia or in severe rickets.

This is merely one of the many situations in which biological factors may take precedence over mechanical ones.

The apparent discrepancy which exists between the biomechanical laws of Hueter-Volkmann and those of Wolff and Roux is another illustration. The former stipulates that pressure produces bone atrophy, the latter that it leads to bone formation. The fact is that under pressure both bone production and bone atrophy may occur as the degree and duration of stress varies and as the resistance to pressure fluctuates under the biological influence of growth and maturity. When Scaglietti in 1930 examined the effect of pressure on the growing femur, he found that the epiphyseal cartilage responded with arrest of bone growth despite a marked proliferation of cartilage cells. Such an arrest must be interpreted as the result of excessive and continuous pressure at the time when the growing bone shows marked biological activity and when it is therefore much more vulnerable than resting bone. Such instances reveal how fallacious it would be to explain morphological phenomena of the locomotor system purely in terms of mechanical laws.

With all these complicating influences at work, how can it still be maintained that locomotion is a special case of mechanics? Is it possible to analyze and calculate human motion on similar lines as we calculate the movement of an inert mechanical device?

Our contention is that this is indeed possible under certain reservations; and the question now arises whether the concessions which must be made to mathematical accuracy are so far reaching as to make the value of calculations illusory or whether they are still within acceptable limits of accuracy.

Here are some of the principal concessions to be made.

1. In human locomotion we are dealing with bodies and parts of irregular morphological shapes composed of tissue of varying density and specific gravity for which data regarding weight, volume, center of gravity and inertia must be established. To do that one must approximate the human body or its part to standard geometrical forms of homogeneous structure. The arm is considered as a cylinder, as is the trunk, the whole extremity as a truncated cone, the head as a sphere and so forth. The inertia of the body or its parts which resists rotatory motion is then calculated on this basis.

2. Although we know that the different systems of the bones, muscles, and so forth, are not of uniform density, we accept an average density without great loss of accuracy.

3. We know that the rotatory moment produced by muscle contraction depends upon the angle of application of the muscle to the bone and that this angle changes with the position of the joint; we further know that with increasing contraction the tension which the muscle can develop rapidly diminishes; consequently, for both situations we must accept mean values provided it can be done without too great damage to accuracy.

4. Some motor events are so complicated and the result of so many individual factors that to compute mechanical events, factors of lesser importance must be eliminated. This is particularly the case in computing the mechanics of the gait. It is a license which admittedly detracts from the desirable accuracy of our computation. On the other hand it is useful and practical for purposes of drawing comparisons between normal and pathological situations.

The argument is advanced that all these concessions add up to a degree of inaccuracy which makes any mechanical analysis of locomotor events a futile effort. Against this stands the fact that the effect of mechanical forces on the human locomotion is so strong and dominant that even the aggregate inaccuracies do not vitiate the practical value of mathematical computations.

In comparing the relative efficiency of normal and pathological motor performances, another source of information comes from the science of physiochemistry. It consists in the determination by gasometric methods of the expenditure of energy and its relation to the output of visible motion, thereby establishing the efficiency quotient of the specific motor act. Any movement which requires a greater amount of energy for production of the same mechanical effect can be said to be performed with less efficiency and with less skill. The normal efficiency for certain movements such as walking, bicycling and so forth have been determined. For instance, normal walking has an efficiency index of 35%, which is a high rate for a combustion engine such as the human body represents. We also had occasion to compare efficiency indexes in certain pathological conditions, for example, paralysis, dislocation of the hip, etc., where the efficiency index was found considerably lowered. High efficiency can be established against clumsiness; economical against uneconomical technique of manual work; physical alertness can be compared with lassitude and sluggishness. Important as such observations are in industry, they are equally helpful to the physical trainer and still more essential to the surgeon engaged in the rehabilitation of locomotor function.

It is hardly necessary to adduce further evidence to show that the study of the mechanics of the human body is a practical even though a formidable undertaking. Its approach involves a number of fundamental principles.

The first of these is unbiased observation and truthful recording. In the last decades the methods of observation and records have made great advances: from simple photography to cinematography; from palpation and faradization to electromyography; from the kymograph to the oscillograph. We are now able to observe and measure motion as it occurs in the joints under normal and pathological conditions and to determine its effect on different parts of the body with much greater accuracy than before.

The second principle is that of analyzing the forces which produce motion. This makes it unavoidable to invade to some extent the field of general mechanics. Kinesiology differs from systematic anatomy in that it analyzes motion as it occurs under actual living conditions when carried out against some extrinsic force such as gravity or any other external resistance. We call such an arrangement a kinetic chain, meaning that a situation exists where muscle action is used primarily to establish stabilization and equilibrium rather than free motion. These stabilizing and equilibrating efforts which do not manifest themselves in visible motion play the most important part in our locomotor performances.

Part of this analysis is to investigate the effect of muscle action on the proximal lever arm and even beyond this upon remote articulations of the body, which calls in calculations of body mass and inertia.

These and other considerations are essential to anyone dealing with locomotor activity of the human body: the athletic trainer, the time and motion expert, the physical educator, the pediatrist and the osteopath; but above all others, it is the orthopedist who should be interested because he is the one upon whose shoulders falls most of the responsibility for dealing with locomotion disorders.

Knowledge of the conditions under which the body maintains its equilibrium is the best way to become aware of approaching pathological situations. On the other hand, it is difficult to imagine that an efficient prophylactic or therapeutic regime against any deformity or disability can be instituted by one who is not acquainted with the normal kinetics of the human body.

ON THE PHYSICAL PROPERTIES OF BONE

I. HISTORICAL

IN THE preceding lecture we have referred to the fact that the inner architecture of bone reflects to a high degree the strains and stresses produced by
external mechanical forces. It required a long period of observation before this
concept of interrelationship between form and function of bone was finally
accepted. The old idea of A. Fick[2] was that the role of bone was entirely a
passive one and that the surrounding musculature determined and restricted
its growth. This was strongly opposed by R. Virchow who emphasized that
bone plays an active role in developing its form and structure and that for this
reason constant internal changes must take place even in mature bone. The
theories of Hueter[4] and Volkmann[16] simply stipulated that pressure inhibits
bone growth and that release of pressure promotes it.

It is due to Hermann von Meyer[10] that the relationship between the architecture of bone and its function was definitely recognized. The occasion was
a meeting of scientists held in 1867, where sections of the upper end of the
femur were shown. Culman, a mathematician, called attention to the similarity
of the trabecular arrangements to that of a crane constructed upon the principle of highest efficiency and the greatest economy of material.

In 1870, Julius Wolff[17] gave expression to the idea that the spongiosa of bone
is able to re-orient itself by a process of rearrangement of its trabecular system
to new mechanical tension or pressure stresses. It was the first formulation
of his well-known law of functional transformation of bone which recognized
this interdependence between form and function in the living. His final observations are masterfully recorded in his classical work which appeared in 1892.[18]

Identical conclusions were arrived at by Roux[13] who examined the trajectory
systems of other bones, especially of the ankylosed knee. Here he also found
that the reorientation of the trabecular systems corresponded to the direction
of tension and pressure stresses and that it developed with maximum economy
of material. True to good engineering principles, bone reshapes itself both
under normal and pathological conditions so as to sustain a maximum of stress
with a minimum expense of bone tissue, and this reorientation of the archi-
tectures is being carried out to a degree of almost mathematical perfection by a
process of absorption and apposition.

II. THE MODIFICATION OF THE LAW OF FUNCTIONAL ADAPTATION OF BONE

Wolff and Culman based their contention that the construction of bone was
mathematically correct upon the so-called orthogonality, that is, the crossing
at right angles of the different lamellar systems. Although this statement of

orthogonality of lamellar crossing was challenged later by Murk Jansen,[5] it is generally considered that the structural plan is in keeping with the best engineering principles.

Of much greater importance to Wolff's law are the observations of Jores[6] to the effect that pressure is not always a growth stimulus but often has an inhibitory effect causing absorption of bone. Such atrophy occurs where pressure is excessive or constant or where periods of pressure exceed those of release. On the other hand, bone reacts to concussion or intermittent pressure by proliferation which is shown by the formation of spurs and bridges in spondyloarthritis when the buffer action of the degenerated intervertebral disc has become inadequate.

Obviously there is a certain threshold beyond which the bone reacts by atrophy and below which it reacts by proliferation (Lange[8]). We have already made reference to the growth inhibiting effect of pressure on growing epiphyseal cartilage as it has been established by Scaglietti.[14]

Experiences of this kind make it necessary to restate the law of functional adaptation of bone. It can no longer be considered as the supreme and exclusive factor which determines form and structure. Definite concessions must be made to other biological laws. The biological factors which determine the pattern of growth is one of these. It is transmitted by heredity, although physical stresses ultimately underlie the philogenesis of the architectural arrangement. We have already indicated that in the growing bone the turgor of growth is a very potent factor, more powerful in determining the form than is the functional response to mechanical strains and stresses. Under the effect of enchondral growth, bone deformation may occur directly in opposition to the laws of strain and stress. The deformity which we see in achondroplasias is a case in point.

Yet with all these limitations and exceptions, the effect of mechanical stresses can always be recognized. Deformities such as osteomalacia or rickets or malunited fractures will still reveal a pattern of static reorganization even if it sometimes is very incomplete and remains only vestigial. All criticism which can be leveled against Wolff's law is based upon the fact that there are other factors which determine growth and formation of bone and which must be duly considered; one must not fall into the error of ascribing to the functional adaptation exclusive control over the formation of bone.

Under these restrictions it is fair to state that the functional adaptation to static stress in the sense of Wolff's law still remains the major factor which determines the postnatal form and texture of the skeletal system.

III. ON ELASTICITY AND STRESS RESISTANCE OF BONE

The two physical properties of bone which determine its reaction to stresses are the elasticity and the unit stress resistance. Both qualities are interrelated; elasticity in general means deformation without breaking; unit stress resistance indicates the force by which a break or failure is produced per unit of cross section.

A. THE ELASTICITY OF BONE

On general principles one must assume that all tissues are elastic within a certain limit. If subjected to stress they change their form but return to their original shape when the stress is removed, and if each unit of stress produces a corresponding and constant unit of elongation, we say that the body is perfectly elastic in the sense of Hooke's law (1660). This law states that there is a constant arithmetical relationship between force and elongation, one unit of force, one unit of elongation, two units of force, two units of elongation, and so forth.

In as much as different structures vary greatly in their ability to become elongated without breaking, the physicist, in order to give this ability a numerical value, has set up a definite point of reference relative to the resistance of a body against stretching. It is represented by a theoretically assumed force which, per cross section unit, elongates the body by its own length, or doubles the original length of the body. This force is called Young's module of elasticity. For most bodies it is an imaginary term because they break long before such a degree of elongation is reached. The practical value of this concept lies in the fact that the point of break (or failure) can be expressed as a fraction of this elasticity module.

For example, steel has an elasticity module of 29,000,000 pounds per square inch cross section. It breaks, of course, long before this point is reached. Theoretically, if there were no previous breaking point the amount of force necessary to double the original length would be 29 million pounds per square inch. For practical purposes it is assumed then that in a steel rod one inch in diameter one pound will produce an elongation 1/29,000,000 part of its length.

In the ordinary sense bone is a very inelastic structure. Its elasticity module for tension resistance as determined by Rauber[12] and by Huelsen[3] is 2000 kg. per mm.[2]. The breaking point of bone of this cross section area, however, is only 10 kg. per mm.[2]. This means that the elongation up to the breaking point is only the 200th part of the length of the rod.

Even within the narrow field up to the breaking point there is only a small portion in which bone follows strictly Hooke's law; that is, where it returns to its original length after the stress is released and where, according to this law, there is an arithmetical proportion between distending force and elongation.

Beyond this limit deformation and force are no longer arithmetically proportioned, and when the force is removed the bone does not return to its former dimensions but the deformity remains. The practical point is that under certain pathological conditions bone becomes much more extensible, i.e., its elasticity module is lowered; but at the same time the elasticity has become imperfect; it remains distended and the deformation is permanent. This is strikingly demonstrated in such conditions as osteomalacia or rickets.

B. THE UNIT RESISTANCE OF BONE

The breaking point of bone determined at 10 kg. per mm.[2] applies only to the resistance of bone to a distending force. But the force to which bone is exposed is usually pressure, bend or shear. The breaking points for these forces have been established likewise.

1. Compression (Fig. 1)

Here the bone is exposed to two forces, the compressing force and its reaction. In pure compression these forces are co-planar and co-linear, that is, they lie in the same straight line and they are directed toward each other. Koch[7] examining the compression resistance found it to be 12.56 to 16.85 kg. per mm.[2], values which come close to those of the unit resistance of bone against distention.

Many years ago Messerer[9] found the breaking point of the patella on compression to be 192 kg.; that of the humerus 600 kg.; that of the femur 756 kg.;

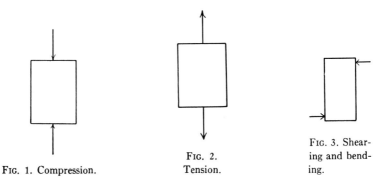

FIG. 1. Compression.   FIG. 2. Tension.   FIG. 3. Shearing and bending.

that of the tibia 450 kg. or more. These values are all above the usual fracture producing forces, but fractures are seldom due to compression alone. Usually there is a combination with bending and shearing stresses. On the other hand, much of the compression force may be absorbed by intervening elastic tissue. In cadaver experiments, Messerer[9] found that greater forces were required to produce a fracture of the vertebrae if the intervertebral discs were left intact and so could absorb much of the impact. Against the increased fragility of the vertebrae due to degenerated discs, nature usually sets up a safeguard in the form of arthritic spurs and bridges.

Schanz[15] found that when a skull rested on a hard surface the compression force necessary to produce a fracture was less than one-half of what was required when the skull rested on an elastic block made of wood and cork.

Static pressure stresses are produced by a resting load; dynamic compression stresses are produced by a falling body. In dynamic compressions the energy of the falling body equals the weight times the distance, and the living force is the mass of the falling body times the square of the average velocity which is acquired during the fall $\frac{(mv^2)}{2}$. Hence the greater effect of dynamic compression. Most fractures occur by dynamic rather than static compression forces.

It has been mentioned that Rauber[12] found the resistance of bone against this distention to be 10 kg. per mm.^{2} cross section in dry bone. For fresh compact bone in man he determined the tension resistance at 12.41 kg. per mm.^{2} and the compression resistance at 16.85 kg. Thus it will be seen that the resistance to compression is considerably greater than that to distension, the ratio being 1:0.73. In bending, tension (Fig. 2) develops on the convexity and compression on the concavity of the bent bone. Failure, therefore, occurs first at the side of the convexity. Proof of this can be seen in some of the fatigue fractures of bones which are normally curved such as the femur or the tibia. Here the absorption of bone starts regularly at the convexity (the so called Looser's zones of absorption).

2. The shearing and bending stresses (Fig. 3)

Two forces are operating here; they are not in the same line, co-linear, but they are in the same plane, co-planar; parallel to each other and directed toward each other.

The situation can be illustrated by a horizontal beam which is supported at both ends and loaded in the middle by a superincumbent weight. The load from above and the reaction from the supports represent two co-planar forces. The load produces a bending stress in the beam which is greatest in the middle where the load applies; at the same time it develops a shear which has its maximum at the ends. Consequently bending and shearing stresses cannot be separated as they are interdependent. The maximum of one corresponds to the minimum of the other. To produce a pure shear and a pure bending stress in the beam, the forces must not only be parallel among themselves but they also must apply perpendicularly to the beam.

The bending resistance of bone also has been determined by Rauber[12] for a concentrically loaded beam. He found that a rod of bone 80 mm. long and 2-3.5 mm.^{2} thick will break under a load of 2.1 kg. If the rod has bent 4 mm. before it breaks, then the visible work to produce the break would be 0.0084 kgm., which is the product of the deformation of 0.004 m. times the force of 2.1 kg.

Assuming by the same token that the deformation or bend of the whole tibia before the break is 4 cm., then the work to produce its bending would amount to 2 to 3 kgm. This means that it would take a static load of 400 to 600 kg. before a break is accomplished. We have found experimentally that for bending, the breaking point of the fibula was 216 inch pounds, of the humerus 885 to 1500 inch pounds, of the ulna 340 inch pounds, of the radius 391 inch pounds of bending moment.

The bending stress is a composite of two elementary stresses: a tension stress on the convexity and a pressure stress on the concavity of the bending beam. Between these is a central neutral axis. It was stated before that the resistance of bone to tension is less than that to pressure and that bone under stress is therefore failing on the convex side first (Looser's zones of absorption).

The effect of dynamic loads on tension resistance was studied on the femur by Evans, Lissner and Pedersen[1] by the so-called stress coat method. This consisted of a lacquer coating applied to the bone and the deformation was

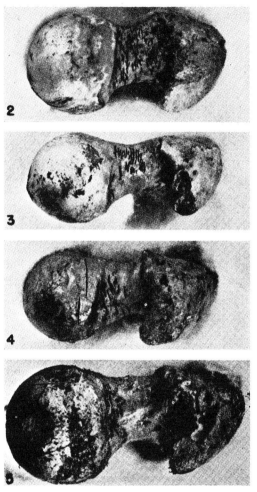

FIG. 4. "Stresscoat" deformation pattern produced under dynamic vertical loading at the superior aspect of the neck. (Evans, Lissner, & Pederson)

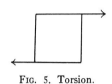

FIG. 5. Torsion.

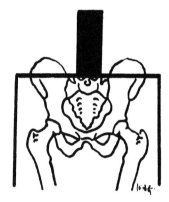

FIG. 6. Human pelvis as a beam resting upon two femoral heads.

measured by the cracks appearing in this coat which was sensitive to 0.000085 inch deformation per inch.

By dropping a weight of 7.9 pounds a distance of two inches, an energy of 15.8 inch pounds was developed. This produced under dynamic vertical loading a definite pattern at the superior aspect of the neck and also at the antero-lateral convex aspect of the proximal or middle third of the shaft (Fig. 4). It was found that this relatively small load of 15.8 inch pounds of energy dynamically applied could produce similar deformation patterns in the same part of the femur as 400 to 750 pounds of static load.

3. The torsion

Torsion is produced by two couples of forces acting in parallel planes at right angle to the axis and working in opposite directions (Fig. 5). If the forces are not in parallel planes and not at right angles to the axis of the bone, the torsion effect is combined with other mechanical stresses such as tension and compression. The effect of the torsion stress on the skeleton can be demonstrated in numerous instances: the forward twist of the upper end of the femur; the outward twist of the lower end of the tibia; the torsion of the vertebral bodies in scoliosis are examples of such torsion effects. The moment of the torque depends upon the force itself and the perpendicular distance of its application from the axis of the bone. A mutual relationship exists between torque and shear.

Torsion stresses often lead to failure. Let us assume a violent torsion of the back is transmitted to the lower extremity. If the hip joint is ankylosed and locked against rotation it forms with the pelvis a mechanical unit and as the rotatory momentum is transmitted to the sacro-lumbar junction, the ankylosed hip prevents it from resolving itself in visible motion and strain occurs at the junction. If this junction does not absorb the rotatory stress, its momentum is transmitted to the femur or the tibia below and may then produce a torsion failure of these bones. Many of the sport fractures, especially in skiing, come about in this manner.

Under ordinary static conditions the danger of bone being unable to resist mechanical stresses is very small; dynamic conditions, however, such as a fall or jump or violent rotation of the trunk, produce a deforming force so great that failure of bone occurs frequently. It is therefore of more practical value in fractures of bones to compute the moment of the force rather than the deformation it produces. The two situations which should be especially considered in this connection are the bending failure and the failure on torsion.

IV. APPLICATION TO THE THEORIES OF BEAM AND COLUMN

We are now in a position to apply some of the facts relating to bone resistance to conditions of weight bearing and other stresses as they appear under living conditions. From the engineering point of view the situations can be represented by two mechanical systems: the loaded beam and the excentrically loaded column.

A. THE THEORY OF THE BEAM APPLIED TO THE SKELETON

The human pelvis can be compared to a beam resting upon the two femoral heads (Fig. 6). A load applied in the midline of the pelvic ring produces a bending stress which has its maximum at the site of the concentric load. At both ends of the beam the shearing stress has its maximum. Similarly also the human foot can be represented as a beam loaded somewhat excentrically at the ankle joint which produces bending and shearing stresses distributed in a comparable manner. Here, also, the maxima and minima occur at the corresponding sites.

In both situations we consider for the present only those gravitational stresses which transmit themselves in a translatory direction and not the rotatory effect which these stresses produce in the articulations joining the pelvis to the femur or the tibia to the astragalus. A consideration of these rotatory components and the condition under which they are held in balance will be the subject of later discussions. For the moment we keep in mind merely the bending stress produced by gravity upon the pelvis or upon the foot.

It is conceivable that the superincumbent load becomes so great that failure results. While this does not happen under normal static conditions it may well occur under the effect of dynamic forces, as for instance by fall from a height. The living force developed in the fall on both feet would equal the mass of the superincumbent body times the square of the average of the velocity which the body acquires at the end of the fall $\dfrac{mv^2}{2}$ (Koch[7]). The reactions at the hip joint must neutralize this force if the body is to be in equilibrium. In the case of the pelvis the interposition of the sacroiliac junction at the ends of the beam lessens the shearing resistance of the structure and failure occurs by shear. Fall upon the foot also causes failure by shearing stress because the shearing resistance is lowered by the interposed tarsal articulations. At the same time the superincumbent load produces a bending stress with its maximum at the point of loading.

Now let us suppose that the beam is supported only at one end, as for instance in standing or falling on one foot or standing on the ball of the foot. The situation is comparable to a cantilever. The static load applied to the end of the bar or, in a fall, the living force, produces a rotatory moment which is equal to this force times its distance from the unilateral support. If we again consider pelvis and femur, or foot and leg, as one mechanical unit not interrupted by any articulation which absorbs the bending stress, then the bending moment may become of sufficient magnitude to cause a bending failure. Such a situation arises, for instance, by a fall upon an ankylosed hip or an ankle ankylosed in equinus position (Fig. 7).

B. THE THEORY OF COLUMNS APPLIED
TO THE HUMAN SKELETON

The theory of columns provides that if the length of the column greatly exceeds the diameter, any load placed upon it is not strictly concentric but that under the effect of vertical stresses certain bending deformation occurs. The weight bearing pipe bone can be considered as such an excentrically loaded column. Concentricity of the load merely transmits weight pressure; excentricity has a bending effect upon the column. The pressure at the side of the load shift is increased and decreased on the opposite side (Pauwel[11]).

No doubt the curvations of the long bones such as the femur and the tibia are predetermined by the prenatal growth pattern, but the post-natal development of curves or angulations such as the decrease of the cervicofemoral angle or the torsion of the tibia are the product of gravitational stresses and of muscle action.

To what degree do these curves which are seen in the long bones conform with the theory of excentrically loaded columns?

The type of curve depends largely upon the anchorage of the column at one or the other end. If we compare the femur with a crane firmly implanted, then the bending moment produces pressure stresses on the concave side and tension stresses on the convex side. These are normally absorbed by the resistance of the structures. But if the column is movable at the lower end, then neutralizing forces are necessary to maintain the mechanical axis of the column strictly vertical, i.e., in equilibrium. Thus there are two aspects to the proposition.

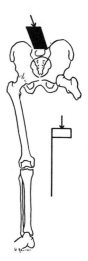

FIG. 7. Standing on one ankylosed hip reproduces the principle of cantilever beam.

One is the passive development of curves under the excentric load; the other is the neutralizing muscular forces and their effect on the torsion and compression stresses on the column. The changes in these stresses as they occur with

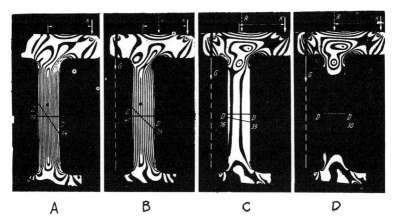

A   B   C   D

FIG. 8. The reduction of tension and pressure by a counterweight against the load K. Z = Tension. D = Pressure. A. No counter pressure. B. Counter pressure and resultant moves close to mechanical axis, therefore stress diminishes. C. Resultant moves closer to mechanical axis and stresses diminish further. D. Resultant goes through the mechanical axis and consequently no compression or tension stress. (Pauwels)

the shifting of the load can be demonstrated by the so-called tension optical picture (Pauwel[11]) (Fig. 8). It is of further significance that the concave side pressure and the convex side tension increases with the distance from the neutral axis which means that economy can be achieved by making the peripheral portion of the bone stronger and the central portion weaker. In other words, in a tubular or pipe bone (Pauwel[11]), the bending resistance against excentric load is also greatly strengthened by the internal struts which are represented by the trajectory system (Pauwel[11]).

The femur as seen in the frontal plane represents such a column anchored at its base and excentrically loaded. It is weighted above by the superincumbent weight and it is anchored below because in the frontal plane no movement is possible between femur and tibia. The column theory postulates that in a post so anchored the deformation of the upper part forms a cosine curve, the length

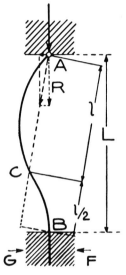

FIG. 9. Column theory, one end fixed, the other turning about one or more horizontal axes, but not laterally. L (height of column) $= 1\frac{1}{2}l$ (length of curve).

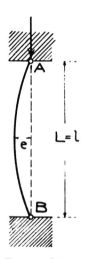

FIG. 10. Column theory, both ends hinged. l (length of curve) $= L$ (straight height of column).

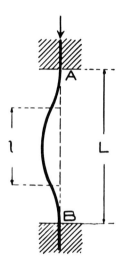

FIG. 11. Column theory, both ends fixed and anchored. L (height of column) $= 2l$ (length of curve).

of which (1) is about two-thirds the length of the rod (L). The deflection is maximal at the junction of the upper and middle third (Fig. 9).

Because of the increased pressure at the concavity of the curve we notice that the weight sustaining trajectories arise from the medial contour of the shaft.

In the sagittal plane the situation is different. Here the femur represents a column hinged at both ends, since movement is possible both in the hip and in the knee joint in this plane. According to the column theory the bend should be a cosine curve which is equal in length to the rod. The deflection now has its maximum at the middle (Fig. 10). With the concavity being directed backward the trabeculae are more densely arranged at the posterior wall.

The tibia is also movable at both ends in the sagittal plane: in the knee and ankle joint. The bend likewise represents a cosine curve, the length of which corresponds to the whole column with the maximum at the middle. Again the greater condensation of the cortical trabeculae appears posteriorly at the side of the concavity.

In the frontal plane both upper and lower ends of the tibia are fixed because there is no movement in either knee or hip joint in this plane. The cosine curve is half the length of the bone, the maximum deflection being at the middle (Fig. 11).

Many objections can be raised against applying the theory of column to the long weight bearing bone. There is no uniformity in mass distribution and the cross section areas are by no means equal at all levels. Furthermore, aside from the gravitational stresses, muscular action has a great influence upon the molding of bones which is particularly evident in rhachitic and paralytic deformities.

Bone deformation also results from aberrations of enchondral growth which has little relation to gravitational or other mechanical stresses. The changes from genu varum to genu valgum in the second year of life represent a growth pattern rather than a functional response to gravitational stresses. Yet with all these reservations one can hardly deny the similarity which exists between the normal curves of the weight bearing bones and the deformation of excentrically loaded columns.

C. THE EFFECT OF MUSCULAR TENSION ON GRAVITATIONAL STRESSES IN BONE

To illustrate the effect which the tension of the muscles has on weight bearing bones Pauwel[11] compares the pull of the muscles with a system of guy wires counteracting gravitational stresses. The bending effect of gravity upon bone may be met by muscle forces which are able to neutralize the pressure and tension stresses produced by the bending. In the case of the human femur, when the individual stands on one leg, the role of these guy wires or ropes is taken over by the ilio-tibial tract which holds the pelvis against gravity. The vastus lateralis, in contracting, pulls the ilio-tibial tract away from the bone thereby giving it an increase of its rotatory moment.

These changes of stress on bone can be seen in what Pauwell[11] calls the tension optical picture (Fig. 12a). It shows that pressure and tension of the bone becomes diminished by the effect of muscle function, a fact which is reflected by the architecture of the upper end of the femur (Pauwel[11]) (Fig. 12b).

Here is another proof that the construction of bone is thoroughly rational from the engineering point of view. The same principle can be demonstrated in the humerus, although it is not a weight bearing bone (Pauwel[11]) (Fig. 13a). Here the tension optical picture also shows a diminishing of the bending stresses by the action of the brachialis or deltoid or both (Pauwel[11]) (Fig. 13b). A similar effect is produced by the bi-articular muscles; since they are spanned over the full length of the bone their guy wire action uniformally reduces the tension stresses of the whole bone (Pauwel[11]). In the elbow, the combination of bi- and uniarticular muscles results in release of tension stress of the humerus on contraction of the biceps, whereas in contrast, the contraction of the brachio radialis relieves only the tension stress on the upper radius and ulna and on the lower humerus.

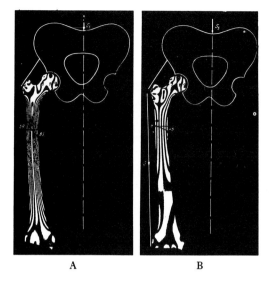

A B

FIG. 12. Tension optical picture. A. Under weight bearing alone. B. Under pressure plus countertension of tensor fascia, showing pressure and tension of bone diminished. (Pauwels)

The lower the insertion of the muscle, i.e., the farther it reaches down on the bone the greater is its effect on lowering the tension stresses on bone. Also, a greater amount of release is produced by muscles with broad insertions.

In addition, gravity itself may diminish tension stresses which are produced in curved bones such as in the saggital curve of the femur. In the second half of the standing period when the line of gravity passes in front of the ankle joint, the body weight, in addition to the extensor musculature, acts as a sort of guy wire which reduces the tension stresses at the anterior convexity of the curve.

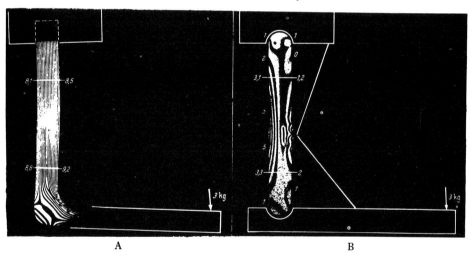

A B

FIG. 13. A. Pressure tension stress of humerus without muscular countertension. B. Diminution of bending stresses by action of branchialis and deltoid. (Pauwels)

The meaning of these construction principles for the locomotor performances is obvious. They are installed to counteract the bending stresses produced by gravitational forces on the long bone. So far as these counteracting forces are furnished by the tension of the surrounding muscles, they are designed to make the danger of the bone failing under gravitational and dynamic stresses as remote as possible. In pathological conditions which affect the musculature, bending fractures of the long bones occur more easily because the tension releasing effect of the muscles is missing.

V. SUMMARY

1. The inner structure of bone reflects to a high degree the physical forces acting upon it. Nevertheless, the law of functional adaptation of bone must be modified to make allowance for other biological laws influencing bone growth, which under certain circumstances may act contrary to purely mechanical expectations.

2. The two physical qualities of bone to be considered are the elasticity and the unit stress resistance. The elasticity module of bone in regard to distention has been determined. So have also the unit stress resistances for tension and compression. The force necessary to produce a break by bending, although varying in wide limits, has been established for most of the long bones experimentally.

3. The principle of the loaded beam can be applied to the pelvis and the foot. The gravitational strain on an ankylosed hip or ankle joint is comparable to that of a cantilever beam.

4. The column theory applies with reservations to such bones as the femur and the tibia. One proposition is the passive development of cosine curves in the long bones under the column theory. This is represented by the femur and tibia in both the frontal and sagittal planes.

5. The other is the effect muscular tension has upon diminishing the tension effect of gravitational stresses on the convexity of curved bones. In this respect Pauwel's studies are particularly illuminating. He showed that certain muscles, for instance, the tensor fasciae, act as guy wires which release tension stresses on the bone, and he demonstrated in his tension optical pictures the actual decrease of tension lines in the bone as the result of the reduction of tension by muscle action. This is shown both for uni- and biarticular muscles.

BIBLIOGRAPHY

1. EVANS, E. G., LISSNER, H. R., and PEDERSON, H. E.: Deformation Studies of the Femur under Dynamic Vertical Loading. *Anat. Rec., 101:2*, June, 1948.
2. FICK, A.: *Medizinishe Physik*. Braunschweig, 1856.
3. HUELSEN: Spezifisches Gewicht, Elastizität und Festigkeit des Knochengewebes. *Anz. des Biol. Laboratoriums, St. Petersburg*, 1898 (russisch).
4. HUETER, C.: Anatomische Studien an den Extremitätengelenken Neugeborener und Erwachsener. *Virchows Arch. Path. Anat., 25:572*, 1862; and *26:484*, 1863.
5. JANSEN, MURK: *On Bone Formation*. Manchester, Manchester University Press, 1920.
6. JORES: Experim. Untersuchungen über die Einwirkung Mechanischen Druckes auf Knochen. *Beitr. path. Anat. u. allg. Path. 66:453*, 1919.
7. KOCH, JOHN C.: The Laws of Bone Architecture. *Am. J. Anat., 21:177*, March 1917.
8. LANGE, CHRISTIAN: Über Elasticitätswerte in Rückenwirbeln und über Osteomalacia Traumatica. *Verhandl deutsch. Orthop. Gesellsch*, 1920.
9. MESSERER: Über Elastizität und Festigkeit des menschlichen Knochens. Stuttgart, 1880.
10. MEYER, H. V.: Die Architekturen der Spongiosen. *Arch. f. anat. physiol. und wissensch. Med.*, 1867.
11. PAUWEL, F.: The Significance of the Architectural Principles of the Supporting and Locomotor Systems in Regard to Stresses of the Long Bones. *Ztschr. f. Anat. und Entwicklungsgesch., 114:1* and 2, 1948.
12. RAUBER, A.: *Elastizität und Festigkeit der Knochen*. Leipzig, 1876.

13. ROUX, W.: *Theorie der funktionellen Anpassung.* 1883.
14. SCAGLIETTI, O.: "Mikroskopische Untersuchungen über die Folgen dauernden örtlichen
    Durckes auf die Gelenkflächen." *Ztschr. f. Orthop. Chir., 52:*577, 1930.
15. SCHANZ, A.: Untersuchungen über das Entstehen von Schädelbrüchen. *Mediz. Jahr-
    bücher: Wien,* 1881, p. 291.
16. VOLKMANN, R. v.: Chirurgische Erfahrungen über Knochenverbiegungen und Knochen-
    wachstum. *Virchows Arch. path. Anat., 24:*152, 1862.
17. WOLFF, J.: Über die innere Architektur des Knochens. *Virchows Arch. path. Anat., 50:*
    389, 1870.
18. ———: *Gesetz der Transformation der Knochen.* Berlin; Aug. Hirschwald, 1892.

Lecture III

ON FUNCTIONAL ADAPTATION OF BONE UNDER PATHOLOGICAL CONDITIONS

IN A general way it has been intimated that the great natural law under which bone adapts its structure to mechanical demands operates in the morbid as well as in the normal state. However, the pattern of this functional orientation varies greatly with the underlying pathological conditions of the bone, all differing from the normal response both in kind and degree.

For the sake of clearness these various reactions are better classified on the basis of patho-physiology rather than on a purely morphological basis. From this point of view we are proposing the following grouping:

I. *Static Conditions.*

a. Functional transformation and adaptation of deformations arising from static conditions in otherwise normal bone.

b. Functional transformation and adaptation of deformations arising from lack of bone apposition.

c. Functional transformation and adaptation of deformations which arise from the deficiency of enchondral or endosteal bone formations on endocrine or metabolic basis.

II. *Traumatic Conditions.*

a. Reorganization of form following fractures.

b. Readaptive changes following dislocation or other disalignments.

III. *Inflammatory Conditions.*

a. Static orientation of reactive bone formation in osteomyelitis and tuberculosis.

b. In arthritis.

IV. *Static Reorganization in Tumors.*

V. *Static Reorganization in Congenital Deformities.*

I. STATIC CONDITIONS

a. An example of the purely static transformation of bone is the thickening of the second metatarsal in cases of so called Morton's syndrome. The shortening of the first metatarsal makes it necessary for the second to bear more weight than usual and it responds to it by increasing the thickness of the shaft (Fig. 1).

Another instance is seen in what is called the Looser transformation zones. These are areas of absorption developing at the convexity of a curved bone, for instance, the tibia. They are caused by excess of tension stresses ultimately leading to discontinuity of the bone (Hartley[7]). It is believed that the hair line fracture of the march foot, which also develops gradually, starts with a stress absorption of one of the metatarsals.

Miller[10] recently observed a case of bilateral stress fracture of the neck of the femur. Such fractures have been reported for all bones of the lower extremity as well as in the pelvis, ribs, humerus and vertebrae.

External pressure upon normal bone may produce absorption if the pressure

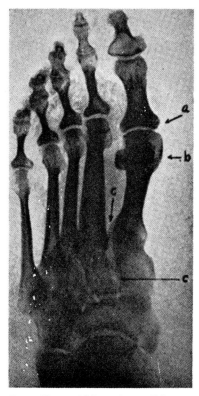

FIG. 1. X-ray of Morton's toe. (Morton)

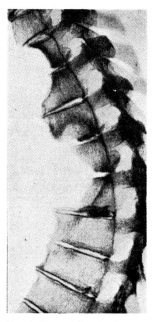

FIG. 2. Deep defect in the vertebral bodies due to constant pressure in a case of aneurysm. (Schmorl)

is constant or of long duration. An example is the hollowing out of the vertebral body by an overlying aortic aneurysm (Fig. 2).

b. The osteoporotic bone is deficient in the normal process of apposition while the normal bone resorption continues.

Changes occur in form of lacunar absorption. The law of functional adaptation manifests itself by the preservation of the statically more important trabecular systems. For instance, in the osteoporotic vertebra, the perpendicular weight bearing trajectories remain while the oblique systems which serve the dynamic requirements of motion become atrophic; at the same time bone proliferations uniting vertebral bodies restrict the intervertebral movement. Ultimately an equilibrium is established between the atrophic vertebral bodies and the turgor of the intervertebral disc which results in a flattening of the vertebrae. They assume the biconcave shape of a fish tail and the disc protrudes into the concavities (Fig. 3).

The atrophy which we encounter in paralytic conditions is an adjustment to the decreased functional demands in connection with the muscular deficiency. When the static functions are taken up again, a reorganization takes place. The degree of reorganization is not influenced by weight bearing alone but also by the amount

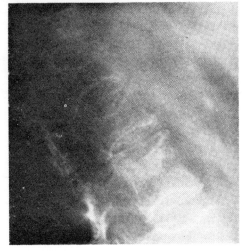

FIG. 3. Vertebrae with biconcave shape.

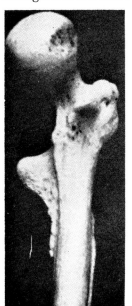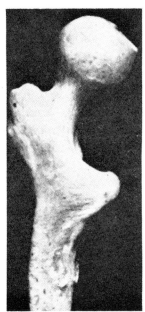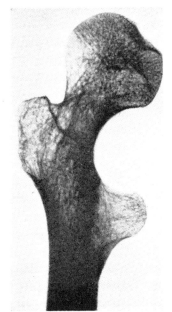

FIG. 4. Coxa valga. (Putschar)

of musculature remaining, which again shows that in functional adaptation both static stresses and muscular tension play a role.

In this connection it is of interest to study the mechanogenesis of coxa valga (Putschar[12]) (Fig. 4) where the inclination angle of the femoral neck is more than 140 degrees (Storck[15]).

It develops, for example, in young amputees (Reich[13]). Here the trajectory system into the greater trochanter becomes atrophic and the massive trajectories from the head to the medial cortex run in almost straight lines, showing that a greater part of the architecture is engaged in sustaining perpendicular pressure while the trajectories subserving traction disappear.

Likewise a femur which is eliminated from weight bearing because of paralysis shows in younger individuals an increase of the angle of inclination. The stimulation of weight bearing being absent, there is not only atrophy of the weight bearing trabecular system, but the weight of the hanging limb actually produces a stretching of the angle between neck and shaft.

c. Metabolic and endocrine deficiencies lessen the stress resistance of bone. This resistance depends largely upon its mineral content and any condition which causes demineralization of bone will lower it. In rickets the seat of the deficiency is the metaphysis; it is due to the formation of osteoid bone of much reduced unit stress resistance. Characteristic deformations develop: the rachitic coxa vara, the tibia vara, the genu valgum and varum.

The deformations seen in osteomalacia are quite different from the rachitic deformities being due to deficient endosteal ossification. Here also osteoid instead of calcified bone is laid down, but the entire bone is involved, especially in the long bones and the pelvis and the resulting deformations are therefore much more formidable.

Both gravitational and muscular stresses are engaged in producing the deformities in osteomalacia as well as in rickets. But the functional adaptation to static and dynamic demands is much more effective in rickets than it is in osteomalacia which is often called adult rickets.

Weir, Bell and Chambers[16] examined the changes in Young's module and the breaking point of bone in rats in whom rickets was produced by rachitogenetic diet. They found the module reduced to one-tenth of normal and the breaking stress resistance also was greatly diminished though the elasticity limit of the bone showed no significant change.

It is of clinical interest to examine the limit of spontaneous correction of some of the more common rachitic deformities. The juvenile tibia vara often represents merely a persistence of the genu varum of the first year of life. In other cases it is of definitely rachitic nature. The tibia may show an arcuar deformity, i.e., one in which the bent is evenly distributed over the whole length, or it may be a sharp angulation at the upper or more often at the lower end. The greater pressure strain at the concavity stimulates bone apposition and thickening of the corticalis. In the arcuar deformity this may proceed to complete straightening of the curve provided early measures are taken to eliminate gravitational stresses. In the angular deformity the outlook for spontaneous correction is much less promising.

The so-called idiopathic genu valgum appears particularly in two periods of growth, namely between the second and fifth year and after the twelfth to fourteenth year; the infantile and the juvenile type. Usually the upper metaphysis of the tibia is the principal seat of the deformity. In milder cases there is only an underdevelopment of the lateral condyles, but in more severe ones an actual kink exists above the epiphysial lines of both femur and tibia.

It is definitely related to the static functions of the limb even if it develops on a rachitic basis. Bragard[2] finds it frequently associated with genu recurvatum, and Hueter[8] emphasized its association with abduction and outward rotation of the tibia. Whether the deformity is concentrated at the upper tibial metaphysis or is distributed along the entire bone, the lateral concave contour shows the thickening and the adaptive bone proliferation (Putschar[12]).

Spontaneous rectification of these deformities seems to go farther in the genu valgum than in the opposite deformity. The reason may be that the deformity falls in more with the normal growth pattern, being an accentuation of the normal valgity which follows the varity of the limb in the second year. From the purely static standpoint, the weight pressure transmitted through the hip joint counteracts rather than promotes the valgity in the knee so long as the neck of the femur does not yield and becomes deformed itself. The femur acts on the knee joint as an angulated lever in a varus direction (Putschar[12]).

II. TRAUMATIC CONDITIONS

Functional adaptation reaches its highest degree in purely traumatic conditions where the bone is essentially sound and had its normal stress resistance before the injury occurred.

In fractures the amount of spontaneous correction depends on several factors: the age of the patient, the type and location of the fracture, the degree of disalignment and displacement, the constitution of the patient, the degree of damage done to the soft structures, the circulatory supply of the bone, and the presence or absence of infection.

The function of the provisory callus is to establish and to maintain continuity of the bone; the static orientation of the inner architecture occurs later by transformation of the endosteal callus.

Experiments on animals on whom a portion of the bone is resected (Zondeck[18]) show that the functional orientation of the callus begins as soon as the extremity is put to use. The massive callus formation is always at the point of the greatest functional stimulus. In malaligned fractures it is on the concave side of the deformity where the pressure is the greatest.

This raises the question of the effect compression has on the healing of fractures. It has been found experimentally (Eggers, Schindler, and Pomerat[5]) that the contact compression factor, i.e., the broad contact area and the compression force have a favorable influence on the healing of fractures. This is a principle opposite to that of traction which is so universally applied in the alignment of fractures; an excess of traction causes diastasis and non-union. The favorable effect of compression has been demonstrated in fusion of the knee by Charnley[4] (Fig. 5) and by Michele and Kruger,[9] and again for the ankle and shoulder joint by Charnley[3] (Fig. 6).

It may be taken for granted that as soon as the callus is formed it undergoes some functional transformation, even if the static orientation of the trabeculae occurs much later in the endosteal callus. In many fractures which are healed with angulation a natural spontaneous straightening can be expected through functional transformation.

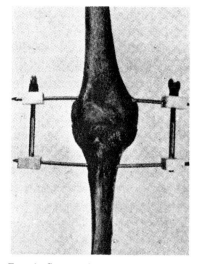

FIG. 5. Compression arthrodesis of the knee. (Charnley)

Age is the most potent factor which determines the limitations of functional transformation after fracture. For instance, in the supracondylar fracture of the humerus, even a considerable anterior kink is likely to straighten out in the young and restore the flexory range while the same situation in an adult would leave a noticeable restriction of motion.

Other factors are the type, site and degree of the deformation. In all situations the factor which stimulates functional reorganization of bone is the unevenly distributed tension and pressure stress. Bone apposition and bone absorption under this stimulus is able, within certain limits, to correct deformities in the

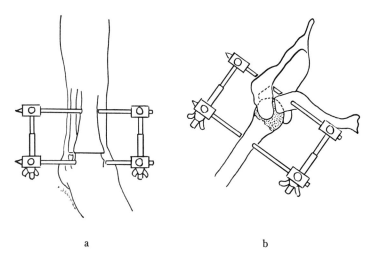

a   b

FIG. 6. Compression arthrodesis of ankle (a) and shoulder (b). (After Charnley)

sagittal or in the frontal planes, i.e., lateral or antero-posterior angulations. What cannot be accomplished is the correction of disalignments in the transverse plane of the bone, the so-called axial deviations which are caused by the fragments rotating against each other about the length axis of the bone. Here all stresses are equal and all are pressure stresses.

This is of great clinical importance. These axial disalignments in fractures are produced by muscle disequilibrium. The functional disabilities they cause are far greater than those which are due to angulation. A fracture of the femur healed with a strongly outward rotated leg is much more disabling than one with considerable angulation. Yet the functional transformation of bone can do little for these axial deviations. In the end it is the position of the neighboring joints which determines the degree of disability. While in simple angulation there is some obliquity of the joint axis, some correction of this deflection can be achieved by functional adaptation. The axial deviation, on the other hand, however great the functional damage may be, cannot be changed by a process of adaptive transformation.

III. FUNCTIONAL ADAPTATION IN INFLAMMATORY CONDITIONS

Only as the destructive phase of the inflammation subsides and reparatory processes set in can one speak of functional adaptation. The state of the joint at this point determines how far this natural process can go.

a. The early periosteal reaction one sees in osteomyelitis is nature's first measure to establish bone continuity, but the real process of reorganization of the

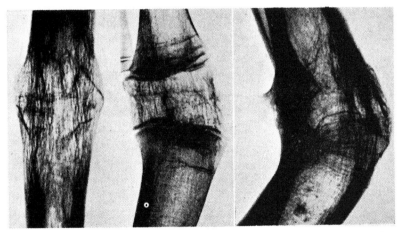

Fig. 7. Bony ankylosis. Complete transformation of trabecular pattern forming a continuous system. (Putschar)

new formed bone is a matter of the quiescent stages. The transformation which follows resection or inflammatory ankylosis of a joint was studied by Roux[14] and Wolff[17] from the viewpoint of the reorientation of internal architecture. This undergoes a complete transformation to the effect that the trabecular systems of the adjoining bones become continuous (Fig. 7).

The rearrangement reflects the effect of the stresses to which the ankylosed joint is exposed. When complete fusion occurs with angulation, then the cortex on the concave side becomes enormously thickened while the convex cortex becomes absorbed; even some degree of streightening may be observed in younger patients. This ultimate reorganization indicates both in tuberculous and pyogenic joints that the pathological process has come to a state of complete quiescence.

In the spinal column a similar event occurs, though not as often and as complete as it is seen in the peripheral joints. When the x-ray shows that two or

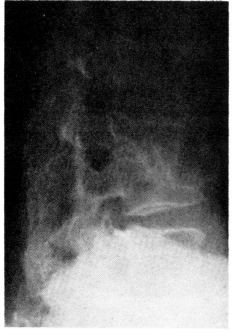

FIG. 8. Tuberculosis of spine with fusion of vertebral bodies.

more vertebral bodies are fused together so completely that their trajectory systems are one without interruption, one can be sure of cure and one may assume that the structures have become biologically sound (Fig. 8).

On the other hand there are instances where one finds clinically that the joint is quiescent but the x-ray fails to show trabecular continuity across the joint space. In these cases one is not so sure of the biological cure. A case in point is the tuberculous hip joint. The clinical examination may be entirely negative for signs of inflammation, but the x-ray fails to show uninterrupted trabecular systems. These are the cases in which contractures develop even after corrective osteotomies and sometimes after years, an event hardly conceivable in biologically

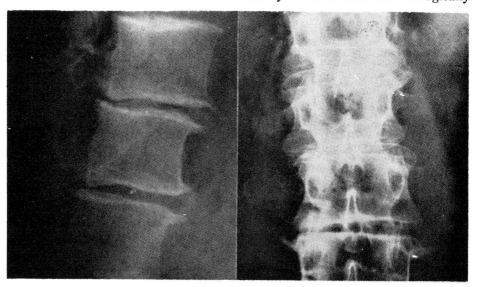

FIG. 9. Osteophytes seen in osteoarthrosis of the spine.

sound ankylosis. The finding of complete static rearrangement of the trabecular system is especially important to the surgeon when he intends to perform corrective operations without incurring the danger of reactivating the inflammatory process.

b. The so-called arthrosis of the spine is another illustration of functional adaptation. Nature restricts intervertebral motion because the degenerative process of the discs and ligaments no longer permits it without strain to these structures. This is the meaning of osteophytes found in arthrosis of the spine. They are not strictly morbid but rather curative phenomena. The morbidity lies in the degeneration of discs, ligaments, and of capsular reinforcements of the intervertebral articulations (Fig. 9).

In the degenerative arthritis of the hip joint, similar principles prevail. The pathological process precludes obliteration of the joint and the production of trabecular continuity such as is the case in the more destructive inflammatory lesion. Instead, bone proliferates in forms of spurs and bridges, protecting the joint against movements which are no longer tolerated.

The degenerative osteoarthritis starts with degeneration of the joint cartilage; the osteoarthrosis of the spine starts with degeneration of the disc which can be considered a universal joint, and with cartilage degeneration in the intervertebral articulations. Since the disc can no longer function as a shock absorber and the spine has lost its elasticity, the only way to protect it is to restrict its motion, and nature accomplishes this by the checking effect of the bridges and spurs.

That the purpose of nature to salvage some of the static function of the part by restricting its motion is not always achieved and that contracture in faulty position or painful phenomena appear does not disprove nature's intentions. In the stricter sense these proliferative phenomena are definitely adaptive and protective measures. Nor do these measures always keep in step with the degree of degenerative changes which require protection. We see articulations with extensive osteophytes and little subjective complaints, and in other cases, the subjective symptoms prevail while the x-ray changes are inconspicuous. Yet there is no mistaking the tendency of nature to protect the joint against poorly tolerated motion by the blocking effect of the exostoses, which may be, and often are, inadequate, but still represent in a measure a functional adaptation.

IV. THE STATIC REORGANIZATION IN NEOPLASM

Where functional adaptation of bone meets with defeat is when facing the irresistible forces of neoplastic growth. Still, even in malignant bone tumors functional adaptation does not lie idle; pathological fractures heal, sometimes with surprising promptness. Naturally the callus will not stand up against the invading force of the tumor.

In benign tumors, on the other hand, adaptive changes may persist. An instance is the periosteal reaction in cystic tumors where the destruction of the cortex from the turgor of growth causes the periosteum to lay down layer upon layer of cortical bone. Furthermore, after the removal of such a cyst or of a chondroma, a textural rearrangement takes place by the formation of lamellae which fill the cavity and align themselves with the lamellar systems of the rest of the bone. It is only in the rapidly growing malignant tumors where the attempt at static reorganization fails completely.

V. CONGENITAL DEFORMITIES

One can hardly speak of a spontaneous correction of congenital deformities by means of a static reorientation of the architecture of bone because there is no such thing as "outgrowing" a congenital skeletal malformation. On the other hand, adaptive changes do occur for the purpose of improving the function, and there are many indications that such changes are effective. A case in point is the congenital coxa vara. Coxa vara has many causes: it may be rachitic, it may be due to a slipped epiphysis, or it may be of true congenital

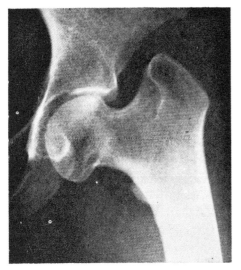 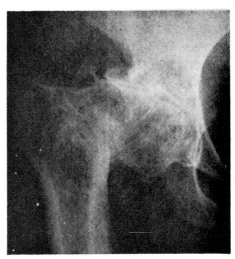

Fig. 10. (Left) Congenital coxa vara with adaptive changes. Thickening of the femoral neck is apparent and the trabecular system has been rearranged conforming with the changed static conditions.
Fig. 11. (Right) Acquired coxa vara occurring after maturity. Functional readaptation is poor.

nature, often occurring familially (Nilsonne;[11] Barr;[1] Fairbank[6]). Whatever the cause is, it gives rise to static stresses which are unphysiological and which elicit certain adaptive morphological changes.

How far a congenital coxa vara can reach a state of natural repair is difficult to state; there is scant possibility of some spontaneous repair because the deformity develops at an intensely formative embryonic period and therefore may attain a high degree. Yet adaptive changes do occur in form of thickening of the femoral neck and in form of rearrangement of the trabecular system conforming with the changed static conditions (Fig. 10). In the acquired coxa vara the outlook for functional readaptation is better if the deformity appears before maturity (Fig. 11).

In most cases of congenital club foot the structural changes of the skeleton develop after the static functions are taken up. Weight bearing increases the deformity. Since weight is now borne largely on the lateral side of the foot, the

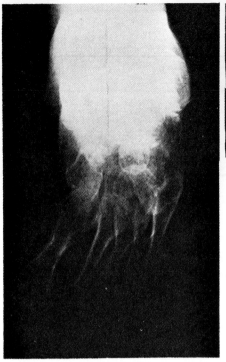

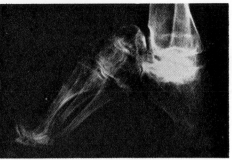

Fig. 12. Structural changes in an inveterate congenital clubfoot due to the abnormal static conditions.

os calcis, the cuboid and the astragalar neck adapt themselves to the mechanical situation by overdevelopment and elongation, and their trajectories assume an arrangement which is appropriate to the abnormal static conditions (Fig. 12).

VI. SUMMARY

In summing up the situation, one can say that the entire phenomenon of functional adaptation of bone which plays such a prominent part in the patho-kinetics of the locomotor system of deformities represents three stages.

1. In what may be called functional rehabilitation, a static disalignment or disarrangement of structure is gradually corrected, partially or completely, by a process of corrective growth in response to the influence of static stresses; here belong the self-correction of static deformities, of bends, and of fracture disalignments.

2. As a protective functional rearrangement which restrains dynamic functions for the purpose of preserving the static functions. Examples are the osteoarthritic bridges or the obliteration of a joint followed by establishment of a continuous trabecular system.

3. A static rearrangement which is purely compensatory is seen in pathological conditions which are irreversible; examples are the structural changes in coxa vara or in congenital club foot.

The few situations mentioned above will suffice to convey to the reader the meaning and the operation of Wolff's great law of functional adaptation as it manifests itself in pathological situations. How much can be expected of it in the sense of alleviating or minimizing the mechanical effect of deformities we cannot state in precise terms; we can only illustrate it by example. In the given case it must be left to the experience of the surgeon to decide how much these efforts of nature can be relied upon to correct deformity and to restore function.

BIBLIOGRAPHY

1. BARR, J. S.: Congenital Coxa Vara. *Arch. Surg., 18:*1909: 1929.
2. BRAGARD, K.: Das Genu Valgum. *Ztschr. Orthop. Chir., 57,* Beilage Hefte, 1932.
3. CHARNEY, J. C.: Compression Arthrodesis of the Ankle and Shoulder. *J. Bone & Joint Surg., 33b:2,* 180, May 1951.
4. ———: Positive Pressure in Arthrodesis of the Knee Joint. *J. Bone & Joint Surg., 30B:3,* 478, August 1948.
5. EGGERS, G. W. W., SCHINDLER, T. O., and POMERAT, C. M.: The Influence of the Contact Compression Factor on Osteogenesis in Surgical Fractures. *J. Bone & Joint Surg., 31A:4,* 693, October 1949.
6. FAIRBANK, T.: Coxa Vara Due to Congenital Defect of the Neck of the Femur. *Am. J. Anat., 62:232:* 1928.
7. HARTLEY, J. B.: Fatigue Fracture of the Tibia. *Brit. J. Surg., 30:*9, 1942.
8. HUETER, C.: Ein Beitrag zur Anatomie des Genu Valgum. *Arch. f. Klin. Chir., 2:*622, 1862.
9. MICHELE, A. A., and KRUGER, F. J.: Compression Fixation Arthrodesis of the Knee. *J. Internat. Coll. Surg., XVI:*1, 93, July, 1951.
10. MILLER, L. F.: Bilateral Stress Fractures of the Neck of the Femur, *J. Bone & Joint Surg., 32A:3,* 695, July 1950.
11. NILSONNE, H.: Beitrag zur Kenntniss der Kongenitalen Form der Coxa Vara. *Acta. radiol. scandinav., 3:*383, 1924.
12. PUTSCHAR, W.: Der Funktionelle Skelettumbau und die Sogenannten Belastungsdeformitaeten. O. Lubarsch und F. Henke, *Handbuch der Speziellen Pathologischen Anatomie und Histologie.* Vol. IX. 3. 617. Berlin, Julius Springer, 1937.
13. REICH: Die Amputationen im Kindesalter und ihre Folgen für das Knochenwachstum, *Beitr. klin. Chir., 68:*260: 1910.
14. ROUX, W.: Theorie der Funktionellen Anpassung. 1883.
15. STORCK, H.: Coxa Valga, ein Beitrag zur Frage der den Knochen bildenden Kräfte. *Arch. f. Orthop. Chir., 32:*133. 1933.
16. WEIR, J. B. DE V., BELL, G. H., and CHAMBERS, J. W.: The Strength and Elasticity of Bone in Rats on a Rachitogenic Diet. *J. Bone & Joint Surg., 31B:3,* 454, August 1949.
17. WOLFF, J.: *Gesetz der Transformation der Knochen.* Berlin, Aug. Hirschwald, 1892.
18. ZONDECK: Experimentelle Studie zur Struktur des Knochenkallus. *Zentrlbl. Chir.,* p. 210, 1914.

Lecture IV

ON THE PHYSICAL PROPERTIES OF CARTILAGE, MUSCLES, FASCIA AND TENDONS

A. ON THE PHYSICAL PROPERTIES OF NORMAL CARTILAGE

I. STRESS AND STRUCTURE

OF ALL the structures composing the locomotor system there is none which is so consistently and severely exposed to stress as is the joint cartilage. Its primary function is to provide for the joint a smooth surface throughout the full range of motion. During this function the cartilage is continuously subjected to stresses; to gravitational in weight bearing joints and to stresses produced by muscle tension in all joints.

When the muscle contracts, the tension it develops can be resolved in two components: one vertical to the axis of the lever arm produces the rotation in the joint; the other, parallel to the axis of the lever arm, stabilizes the joint by pressing the articular surfaces together. When one considers that this stabilizing component of the contracting muscle is considerably greater than the mobilizing one which carries out a visible motion, it becomes clear that the joint cartilage has to sustain a compression stress of considerable degree.

During motion, nature minimizes the wear and tear upon the cartilage by reducing the pressure stress. It does so by changing the contact areas so that stress is not concentrated in one small field. For instance in the knee joint where the ligamentous and muscular tension as well as the gravitational pressure holds the joint surfaces together, we find that in extension the joint contact is broadened by the semi-lunar cartilages. At the beginning of flexion the joint still sustains the superincumbent weight; for this reason flexion is initiated by a rocking type of motion in which equidistant points of the femur come in contact successively with equidistant points of the tibia, and so the pressure is distributed more evenly. In higher degrees of flexion, however, the motion must change into a gliding one where a circumscribed area of the tibia slides over the convexity of the femoral condyle. But at this point the weight pressure has already been released. The foot has left the ground with the take-off and besides the relaxation of the collateral ligaments further diminishes the pressure. This again facilitates the length rotatory movement of the knee joint.

In flexion position the metacarpo-phalangeal joints are held tightly by the lateral ligaments; but here also the ligamentous tension eases when the joint goes in extension. This then permits lateral movements in this articulation.

In the elbow joint the ulno-humeral portion represents the most perfectly adjusted hinge joint in the body. Therefore pressure remains uniformly distributed while the articular surfaces of both bones glide over each other. However in the radio-humeral part, in flexion and extension as well as in rotation, the pressure is concentrated upon a small area of the radial head. Here also the stress is alleviated by the comparative looseness of the joint in contrast with the tightly knit ulno-humeral portion.

By these and other similar provisions nature tries to prevent the joint cartilage from degenerating prematurely by wear and tear.

The fact that with all these precautions, wear and tear of joint cartilage still occurs readily and sometimes prematurely, constitutes a major orthopedic problem, and it makes the study of the physical properties of the cartilage more important. One must, above all, get an insight into the durability of normal joint cartilage and the danger of degeneration if exposed to excessive and prolonged stresses.

When we think of the liberties we take with joints in surgical operations ranging from the simple arthrotomies to all kinds of reconstructive procedures, it would seem that a knowledge of the physical properties should be an essential prerequisite before entering into joint surgery.

According to MacConaill,[23] the stress put upon the cartilage is "deployed" across it in the form of a pressure wave which has a constant intensity. Shear and tension stresses develop which must be met by adequate resistance of the histological makeup. These stresses are taken up by a tension resisting network. It consists of fibers which are part of a laminal system tangential to the wave front and which form a virtual galea aponeurotica beneath the free surface (Fig. 1). On the other hand, Benninghoff[4] argues on the basis of his experiments for a more vertical disposition of the fibers which is objected to by MacConaill on the ground that it cannot be demonstrated histologically nor is it in accord with the mechanical principles of the type of stress involved. Some information can be gained from the structure of the menisci of the knee. According to MacConaill, here the tension is more marked in the peripheral position which sustains the greater stress. Under the combined muscular and gravitational forces a deformation of the cartilage occurs with vertical compression and lateral expansion (MacConaill[23] (Fig. 2).

Benninghoff[4] pointed out that the structure of cartilage consists of cells, collagen fibrils and hyaline substance. In their superficial or tangential zone the cartilage cells are narrow, elliptical and flattened. This is followed then by the oblique or transitional zone, while in the deepest zone adjacent to the bone, the cell groups are arranged perpendicularly.

Studying the fracture lines of cartilage he[4] found them perpendicular to the surface in the deeper layers and then, bending sharply in the transitional zone, they run parallel to the surface in the superficial zone where the fibrils are particularly abundant and where this collagen framework acts as a sort of protecting net allowing the fibers to move apart without loss of tensile strength. The fibrils of the perpendicular zone weave around the cartilage cell forming the so called chondromes. Hultzkranz[19] studied the splitting line of the cartilage and the arcuar arrangement of the fibrils (Hirsch[17]) (Fig. 3) and also observed that these fibers separate vertically in the perpendicular zone toward the zone of calcification into which they finally become anchored.

Of vital importance for the study of the physical properties of cartilage is the hyaline ground substance.

It was Morner in 1887 who discovered a substance with a high sulfuric content which he called chondroit, identical with the chondroitin sulfuric acid of Schmiedeberg (1890). This intercellular substance contains collagen and chondro-mucoid, the latter being a product of chondroitin sulfuric acid and protein. It is a polymer of the acetylated disaccharide chondresine. Miyazaki (1934) found the percentage of chondroitin sulfuric acid to be between 28-42% in dry cartilage. Hirsch[17] estimated it at 19-25%. This chondroitin can be visualized by a so-called metachromatic staining method, a term Ehrlich applied to coloring by certain pure and chemically defined stains which come

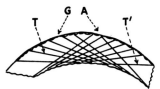

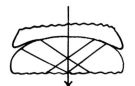

FIG. 1. The disposition of some of the collagenous fibers that take tensile stresses consequent upon the deployment of pressure. *T* and *T'*—two lines of fiber, each tangential to a potential curved isobar. *G.A.*—the *galea aponeurotica* of the cartilage formed by conjunction of the juxta-synovial ends of the collagenous fibers. (MacConaill)

FIG. 2. The effect of combined muscular and gravitational forces acting in the direction of the arrow. The effect of these forces is to cause vertical compression and lateral dilation, as shown by the network of lines. (MacConaill)

FIG. 3. Schematic representation (after Benninghoff) showing the main course of the collagenous fibrils in articular cartilage and arcuar arrangement of fibrils. The chondromes are drawn as black ovals. (Hirsch)

out in different color than the stain itself. If the stain is tuloidin blue, the chondroitin appears blue-red.

Of especial interest is the distribution of this substance. The surface zone always contains a smaller amount of chondroitin due to the fact that this zone is the result of adaptation to mechanical stresses. In this zone also the collagen fibrils surrounding the intercellular substance form a much denser network than they do in the deeper zones. The collagen skeleton thickens in contact with synovial membrane and periosteum. This is a phenomenon of adaptation to tension stresses. The important question is, then, have these variations in intercellular substance anything to do with the development of pathological conditions?

Hirsch[17] finds that in healthy cartilage the chondroitin sulfuric acid is always retained in the perpendicular zone and varies only in the tangential and transitional zones.

It was the observation of Weber, Sappey and others that the thickness of the articular cartilage is in proportion to the pressure it has to sustain. This is to be expected. But Braune and Fischer[7] found that the thickness of the joint cartilage is also inversely proportional to the congruency of the joint surfaces. H. Virchow observed, for example, that the patellar cartilage is thickest at the sagittal crest where it faces the intercondyloid notch; here its thickness amounts to as much as 6 mm.

II. ELASTICITY, DEFORMATION AND PRESSURE

a. Normal cartilage

Under normal conditions, cartilage is extensible and compressible to a considerable degree (Fick[12]). While the elasticity is high it is not uniform in all

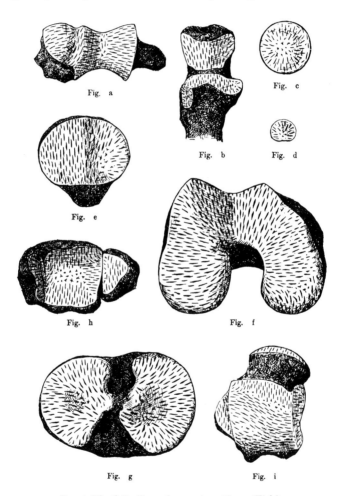

FIG. 4. Elasticity lines of normal cartilage. (Fick)

parts. Hultzkranz'[19] fissure lines after puncturing the cartilage showed that the joint cartilage is under less tension in transverse than in longitudinal direction. This applied to the elbow, knee and ankle joints. In the lower end of the femur and the upper end of the tibia the extensibility lines are more radially distributed (Fick[12]) (Fig. 4). In short, the greater elasticity develops always in the direction of joint motion and where the joint pressure is concentrated the tension lines always run radially from the point of greatest pressure.

The compressibility of cartilage has been investigated by Braune and Fischer[7] who in specimens succeeded in compressing cartilage from 5 to 2.5 mm. and from 2.5 to 1 mm. in thickness.

Another point of interest is to investigate how cartilage adapts itself to the joint contours under pressure. A certain deformability is necessary if the greatest possible contact is to be achieved. Fischer[14] found that these deformations not only increase the contact area but they also increase the range of motion. An example of perfect contact is the ulno-humeral portion of the elbow joint which shows almost complete congruency. The motion is conducted or "track bound" similar to what we see, for instance, in the ankle joint. In saddle joints such as the carpo-metacarpal joint of the thumb, the cartilage is deformable in different directions corresponding to the different planes of motion. Within the limits of its normal function, that is while it is still perfectly elastic, cartilage can fully regain its normal shape after release of pressure (Fischer[14]).

The deformation of the joint contours of the femur after removal of the menisci was studied by Fairbank.[11] He noticed the formation of an antero-posterior ridge from the margin of the femoral condyle over the side of the removed meniscus; a generalized flattening of the marginal half of the femoral articular surface and a narrowing of the joint space on the side of operation with a corresponding widening on the other side. Meniscectomy results in overloading the articular surface with increasing compression of the cartilage and an increase in friction of about 20%.

Again the requirement of the cartilage to change its form varies with the size of contact areas. Joints with small contact areas combine gliding with rocking or rolling motion and therefore the strain on the cartilage changes constantly. In gliding the surface of one body contacts only a point of the other, while in rocking, as mentioned above, equidistant points touch each other consecutively. All this makes a difference in the degree of deformation the cartilage must undergo for the purpose of adapting itself to the contact areas.

1. In gliding motion (for example the radius contacting the capitellum) the contact is not really a point as one would expect, but it is a small contact area. It is obvious that this can occur only by a flattening of the contact surfaces either of the moving joint constituent (gliding of the first order (Fig. 5) or at the resting part (gliding of the second order) (Fig. 6) (Fischer[14]).

2. In the rocking motion the same adaptation of contours occurs by virtue of the compressibility of the cartilage but the contacting surfaces are small and they change their position on both joint constituents consecutively (Fig. 7).

3. The elastic adaptability is most important in length rotatory motion of incongruent joints such as the radio-humeral or the knee joint where the development of contact areas depends primarily on compressibility and the flattening of the convex joint constituent.

According to Rauber[29] the elasticity module of cartilage is 0.9 kg. per mm.^{2}

and the tension resistance is 0.17 kg. per mm.². For pressure the resistance is much greater, namely 1.57 kg. per mm.² while the shearing resistance is 0.35 kg. and the torsion resistance 0.24 kg. per mm.².

Experiments show that cartilage is perfectly elastic for a small load only if the load acts for a short time; for example, a load from 50 to 120 gm. per mm.² applying for not more than one hour.

Braune and Fischer[7, 8] found that cartilages can be greatly deformed for the purpose of contact pressure distribution. The radius of the contact surface of the elastic body depends upon the force (P), the elasticity module (E) and the

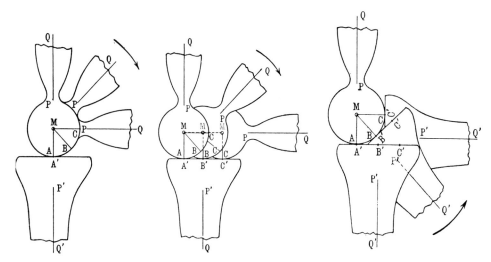

Fig. 5. Joint motion—gliding of first order. (Fischer)

Fig. 6. Joint motion—gliding of second order. (Fischer)

Fig. 7. Rocking motion—contacting surfaces change position of both joints consecutively. (Fischer)

diameter of the contacting convexity (Hirsch[17]) (Fig. 8). The important point is that when the deformity increases the compression modules also increase; hence the cartilage is more resistant. Baer, in 1926, estimated that total deformation in healthy hyaline articular cartilage could be most suitably produced by a pelotte of 3 mm. diameter with a load of 500 gm. He estimated that pressure on the condyles of the knee joint could be 5.2 kg. per square cm. which corresponds to body weight of 70 kg. Older persons are less elastic and the reapplication of loads leads to loss of elasticity.

Hirsch[17] uses a special elasto-meter with a plane surface of about 10 mm.² and a load of 500 g. Applied for five minutes this gives a healthy elasticity curve (Fig. 9). When pressure is concentrated the size of the deformation depends not only on the vertical force but also on lateral and oblique forces of the neighboring cartilaginous regions (Fig. 10).

Cartilage is probably always in a certain state of deformation, being exposed to constant stresses. It resumes its shape during rest, and persons are taller in the morning.

The fact that the elasticity module increases with increasing load is proof of the capacity of cartilage to adapt itself functionally. When the load is increased the cartilage is compressed while chondromes are used to bear it and to act as a shock absorber.

As far as the duration of load is concerned, Hirsch[17] finds that a load of

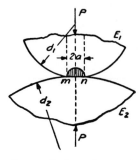

Fig. 8. Radius of contact surface of the elastic body depends on the force (P), elasticity module (E), and diameter of the contacting convexity (Hirsch). (From Timoschenko's "Strength of Materials" 1941.)

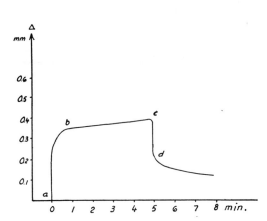

Fig. 9. Typical elasticity curve from healthy patellar cartilage. (Hirsch)

100-1200 g. causes immediate deformation during the next sixty seconds and then the deformation increases slowly. Then when the load is removed the cartilage returns to a normal state, being perfectly elastic within the limits of load and time. Beyond one minute the deformation gradually increases, though

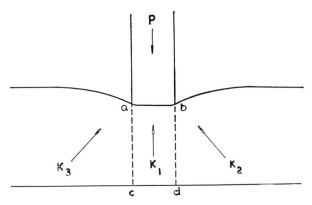

Fig. 10. Drawing showing the forces which may be thought to prevail on loading of a limited area of articular cartilage. (Hirsch)

more slowly, but no stage of absolute equilibrium is seen (Fig. 11). The longer the cartilage is loaded the greater is the impairment of its elasticity.

In other words, the longer the load continues the less complete is the rebound after the compression ceases. Consequently Hooke's law applies to normal cartilage only up to a certain load and up to a certain time of load bearing. While pressure controls the size of the contact area, i.e., the deformation as well as the elastic rebound, the time factor limits the physiological elasticity of cartilage. In practical terms it means that it matters not only how much load the cartilage carries, but also how long it carries it. It is in the latter

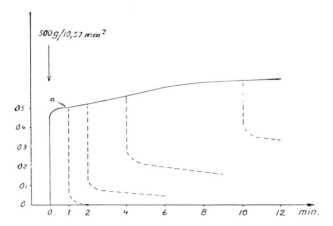

FIG. 11. Schematic drawing showing the influence of the duration of the loading on the size of the deformation. (Hirsch)

aspect that most abuses of the joints occur such as prolonged standing or prolonged maintenance of the joint in one position.

b. The pathokinetics of cartilage

In hypertrophic arthritis the degenerative changes of cartilage constitute morphologically the starting point of the disease. The pathologist finds the first signs in the formation of fissures in the ground substance and the accumulation of cartilage cells in the so-called Weichselbaum lacunae. In the more superficial, tangential layers, cells of the cartilage are arranged horizontally. In the underlying transitional layer the cells are placed obliquely and in the deep layers they are arranged perpendicularly.

The latter represent the bulk of the cartilage and end in the zone of provisory calcification which is lying immediately above the subchondral layer of bone. For the mechanics of the cartilage it is significant that the intercellular fibrils are arranged in perpendicular and tangential systems. When degeneration occurs it manifests itself in a fibrillary peeling and in fissuring of the ground substance. However this is preceded by histological signs of degeneration in form of changes in the staining quality of the hyaline ground substance. The essential constituent of this substance is chondroitin sulfuric acid which, as we have mentioned above, forms from 19-25% of the hyaline cartilage. By metachromatic differential staining with tuloidin blue one observes in early stages of degeneration an increasing loss of this substance and parallel with it goes a corresponding loss of elasticity. The relation of these pathological changes to physical properties has been⋅the object of intensive studies. As early as 1897, Beneke[3] found that the cartilaginous degeneration is anteceded by impairment of the main physical properties of cartilage, i.e., elasticity and unit strength.

Since then, many investigators have emphasized the role which this early change of physical properties plays in cartilage degeneration. Axhausen's theory of 1919 stipulates that trauma leads to cartilaginous lesions and that physical defects of the cartilage are a pre-stage of arthrosis deformans. Preiser's static theory is based upon the incongruency of the joint surfaces, and Wiberg[35] has pointed to contact pressure between the patella and condyle, especially at the

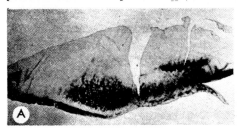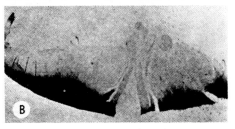

Fig. 12. A. Specimen of deeply fissured malacic cartilage taken at operation, stained with tuloidine blue. Microphotograph enlarged 8 times. Great reduction of the content of chondroitin-sulphuric acid. Preservation of metachromatically staining substance around the bottom of the cracks. B. Operation specimen of tufted malacic cartilage with deep fissures in the center of the softened region, showing large reduction of the content of chondroitin-sulphuric acid. ×8. (Hirsch)

median facet, as one of the earlier causes of the chondromalacia of the patella. In this connection it is of interest to note the report of Aleman.[1] In reviewing his arthrotomies of the knee, he found that one-third of 220 operative cases showed chondromalacia, especially of the medial part of the patellar cartilage, in some cases accompanied by an involvement of the synovial membrane. He showed that recovery from this stage with freedom from symptoms is quite a possibility which has a definite bearing upon the conservative management of the condition, whereas a second trauma may reactivate latent symptoms.

Hirsch[17] demonstrated that all lesions with actively impaired elasticity showed poor metachromatic staining. The chromosome substance was always more reduced in the superficial layers of the cartilage. Wherever such cartilage shows a loss of pressure elasticity, one will find the chondroitin sulfuric acid reduced, and this in particular pertains to the perpendicular zone of the cartilage. What is of special significance is that long before the fissure state the cartilage shows a reduced content of chondroitin sulfuric acid indicating a loss of its normal physical properties (Fig. 12).

If fissures develop in the borderline between soft and firm tissue, they may reach deep enough to bare the underlying bone, though as a rule there is healthy epi-ossal cartilage in the depth (Fig. 13).

In the knee joint the incongruency of the cartilaginous surfaces appears preeminently in position of flexion (Wiberg[35]), causing the deformation produced by pressure to become increased. As the cartilage loses its elas-

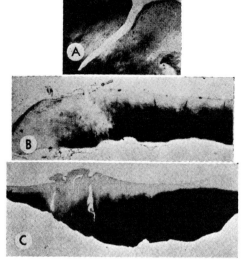

FIG. 13. A. Operation specimen of malacic patellar cartilage with a deep fissure lying between a soft and a firm region. Staining with tuloidine blue shows how the content of chondroitin-sulphuric acid is more reduced on one side of the crack than the other. The chromotropic substance is well preserved at the bottom of the fissure. ×26. B. Malacic cartilage with a relatively deep fissure on the borderline between firm and soft tissue. Poorer metachromatic staining on one side of the crack than the other. ×8. C. Tufted malacic cartilage with a main crack on the borderline between the relatively normal and the malacic tissue. Tuloidine blue stain. ×8. (Hirsch)

ticity its circulation becomes impaired, which under repeated or prolonged stress may become so poor as to cause softening and defibrillation, while, even before, the chondroitin sulfuric acid content shows a marked decrease.

c. Summary

The physical properties of cartilage under normal and pathological conditions can be summarized as follows (Hirsch[17]):

1. NORMAL CARTILAGE

Healthy cartilage has a typical elasticity curve. Cartilage can adapt itself to a certain extent to increasing loads of short duration. Healthy cartilage shows variation in the metachromatic staining of the superficial layers only. Loads of short duration within certain limits show no deterioration, while the same load repeated or prolonged may cause degeneration.

2. MALACIC CARTILAGE

There is a distinct metachromatic effect in all zones but the reduction in chondroitin sulfuric acid is always greater in the superficial layers. This lessening of the chondroitin sulfuric acid may exist without visible disintegration of the tissues, but disintegration always exists in the softened cartilage with reduced chondroitin sulfuric acid.

B. ON THE PHYSICAL PROPERTIES OF MUSCLES

I. INTRODUCTION

As introduction into the mechanics of muscle action, we refer briefly to the immediate biochemical changes which accompany it. They represent a complex phenomenon, the details of which are by no means fully understood. We know that contraction is initiated by choline, and the restraining action of cholinesterase under the influence of the central nervous system controls the degree of muscular contraction.

In the course of these biochemical changes the glycogen furnished by the liver is oxydized into lactacidogen and then into lactic acid. The lactic acid is then carried off by the muscle blood stream where part of it is resynthetized by the liver into glycogen and part is oxydized into CO_2 and H_2O. Intensive muscular action produces a surplus of lactic acid. Part of it is taken up by the blood stream, but a surplus of lactic acid lingers in the muscle and is one of the essential features of fatigue and exhaustion. The final removal of lactic acid from the muscle depends upon its circulatory system. One can appreciate the importance of the blood supply by the fact that, according to Krogh,[22] no less than 2,000 capillaries can be found in 1 mm.[2] of cross section of the human muscle, while in the same territory only 200 single muscle fibers are counted.

In the end these chemical changes are transformed into mechanical energy; from this point on, the analysis of muscle function must proceed on physical lines which is the prime purpose of this lecture.

II. THE ELASTICITY AND CONTRACTILITY OF MUSCLE

The principal physical properties of the muscle are:

1. The elasticity which is a property common to all structures taken as inert living bodies.

2. The contractility which occurs under nervous or electric stimulus.

According to Sherrington's[31] law of all or none reaction, the single muscle fiber, once innervated, goes into maximum contraction. There is no gradation. The individual fiber is either totally relaxed or it is in maximum contraction. However the tension of the muscle must be controlled according to the mechanical requirement which determines the degree of contraction necessary. Therefore this degree can only be a quantitative phenomenon, that is, one that depends upon the number of individual muscle fibers which contract at the same time. It is this number of muscle fibers which determines the degree of tension the muscle will display.

a. The elasticity

The only stress to which muscle, as a passive structure, is exposed is tension. When such tension is applied a passive elongation occurs accompanied by a decrease in the cross section of the muscle.

The histological structure upon which this elasticity depends is the sarcolemma and the connective tissue sheath enveloping the muscle fiber. The amount of elastic fibers in this connective tissue is a decisive factor (Proebster[27]). They take care of the shortening when distension ceases, while the collagen fibers protect the muscle against over-distension.

If a weight is attached to the relaxed muscle, an elongation (Δ) occurs which is directly proportional to the original length (l) to the pulling force (P) and to a certain constant (k) varying with each body, and is inversely proportional to the cross section area (A).

$$(\Delta = \frac{P \ l \ k}{A})$$

This means that the greater the original length and the greater the distending force and the smaller the cross section area, the greater will be the elongation.

The elasticity coefficient, or Young's module (E) is the force necessary to produce for the unit of length and cross section area an elongation equal to its original length. This module (E) has not been definitely determined for the muscle, but we know that a normal muscle fiber can be stretched to 1.6 times its original length before it ruptures.

The relaxed muscle obeys Hooke's law of the arithmetical proportion between force and elongation only to a certain point. Beyond this point the same amount of force produces an increasingly lesser amount of elongation, or the same amount of elongation requires an increasingly greater amount of force. From

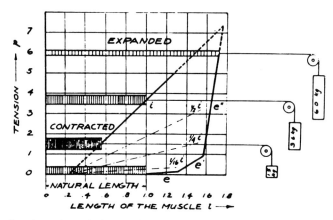

FIG. 14. The length tension relation of the relaxed muscle represented by a hyperbolic curve.

there on, the length-tension relation is no longer represented by a straight line but by a hyperbolical curve (Fig. 14). The breaking point of the muscle fiber lies considerably before the doubling of its length has been reached.

The unit resistance of the muscle to tear is much less than that of its tendon. The latter is a much more slender structure with a very much smaller cross section, while at the same time it must sustain the full stress transmitted to it from the muscle. The ratio of cross section of the tendon to that of its muscle is in some instances 1:60.

Continuous passive tension produces permanent structural changes in the muscle in the form of interstitial fibrosis, and this occurs long before the breaking point is reached. The muscle fiber undergoes degenerative changes from overstretching which vary from simple atrophy to albuminous or waxy degeneration (Zenker) or cloudy swelling, all of which greatly interferes with the natural elasticity of the muscles.

b. The contractility

The second physical property of the muscle is its contractility. It is the ability of the muscle to shorten under innervational impulses. While elasticity is an entirely passive phenomenon, contractility is an active process. A contracted muscle also retains a certain elasticity in its state of contraction, but from the physical point of view this elasticity differs from that which obtains when the muscle is in the relaxed state. In order to determine the elasticity of a contracted muscle it is first necessary to ascertain to what degree such a muscle resists elongation. Here we note that the muscle approaches much more Hooke's law of arithmetical proportionality between force and elongation than it does in the relaxed state. In other words, when a muscle is at maximum contraction, a force necessary to pull it out to its original length is more proportional, unit by unit, to the elongation which is thereby produced. If a certain amount of force causes a certain amount of elongation, then twice the force will produce twice the amount of elongation, etc. Graphically presented, this relationship between force and elongation, approaches more a straight line (Fig. 15) while in the case of the relaxed muscle the curve is more hyperbolic.

This relationship continues until the maximally contracted muscle has been pulled out to its normal resting length. To accomplish this, a certain constant force is required for each muscle. This force is equal to the greatest tension the muscle can display on contraction. Any further elongation beyond the point of natural length would then be resisted by the inert elasticity of the muscle as well as by whatever tension the muscle retains at this point of elongation. The amount of tension which will stretch out a maximally contracted muscle to its original length is called the absolute muscle power. The magnitude of this muscle power can be ascertained for each muscle provided that the physiological cross section area of the muscle is known. It has been determined at 3.6 kg. per cm.^{2} of physiological cross section by Recklinghausen,[30] whose findings coincide with our own (Arkin[2]) rather than the value of 10 kg. per cm.^{2} as established by Fick.[13]

However it must be understood that the figures on maximal muscle strength apply only to the so called isometric condition of the muscle where the muscle is prevented from actual shortening during innervation. In other words the maximally innervated muscle is held to its original length by an external force and an equilibrium is thus established between this external force and the maximally innervated muscle distended to its natural length. In this state of unchanged length, the tension of the muscle remains at its highest level.

If, on the other hand, the muscle is permitted to shorten, then the tension rapidly decreases until it finally becomes unable to produce any further visible motion. One can easily satisfy oneself of this fact when one flexes the wrist strongly and then tries to flex the fingers at the same time. The flexion of the wrist shortens the finger flexors to such a degree that in this position they can no longer display any power. We call this a state of active insufficiency of the muscle.

The maximum tension displayed by a shortened muscle is therefore a very inconstant factor while in contrast that of the muscle maintained at its natural length is constant (Bethe[5]). The difference between the tensile strength of the muscle in the isotonic state, i.e., when the muscle is allowed to shorten, from that in isometric state where shortening is prevented, varies from 20-80%.

Computed on the isometric basis, the agregate strength of the musculature of the body is surprisingly great. Martin and Rich[25] estimated that for the adult the combined strength of the muscles of the extremities alone, not counting those of the back muscles, amounts to 3,900 pounds, the ratio to the body weight being 26.6 to 1.

c. Electrophysiology of muscle

1. THE ACTION CURRENT

So far the behavior of muscle has been considered in two situations only: namely, in complete relaxation and in maximum innervation. In the ordinary activities of everyday life, however, complete relaxation as well as maximal innervation are the exceptions. It is therefore most important to know what occurs within the field bounded by these two extremes. In everyday life there is a constant gradation of muscular effort regulated by innervation which adapts itself to the mechanical requirements of the specific situation. The question is how this field of partial innervations of the muscle can be explored. It should be theoretically possible to determine any degree of muscular tension from moment to moment by simply equilibrating it against an external resistance strong enough to prevent the muscle from shortening. It is evident, however, that such a method would not only be cumbersome and laborious but also highly inaccurate because it would be difficult to maintain the muscle under the constant innervational load. In other words, stabilization by external resistance would require a continuous adjustment of the external forces as the tension of the opposing muscle fluctuates.

In the last decades the study of muscle tension has been greatly advanced by the introduction of the action current methods. The basic fact is that a muscle in contracting develops an idiomuscular current, the action current (Einthoven[10]).

It was found that this electric phenomenon was inseparable from muscle contraction and that it disappeared when the muscle became entirely relaxed. No action current is produced by the metabolism of the muscle itself so far as we know. This action current can be led off from the muscles by electrodes but it must be magnified several hundred thousand times before it becomes strong enough to be recorded by an oscillograph.

The action current studies show that the intrinsic tension in the muscle during its physiological activity is a very finely graded phenomenon. It reveals all the fluctuations of muscular tension as it varies with the number of muscle fibers which are engaged at one time in accord with the mechanical demands of the situation.

The action current picture also changes according to whether the motion is free or is resisted, whether it is forceful or slight, and whether it is rapid or slow.

Taking, for instance, a free pendulum movement such as for and backward swinging of the arm, the action current picture would show a simple and short impulse, enough to overcome the inertia and to set the arm in a swinging motion (Fig. 16). The curve denotes a rhythmic rise and fall of the motor impulse.

If one current is led off from the anterior and another from the posterior portion of the deltoid, and at the same time the motion is slightly accelerated, then we will find that the action currents occur in the anterior deltoid at the forward and the posterior deltoid at the backward swing. The motor impulse does not cover the entire swinging phase because it is only necessary in this situation to give the limb sufficient impulse to complete its swing.

On the other hand, the swing may become faster than is provided for by the natural swinging time of the pendulum which is about 0.45 seconds for a swinging leg and somewhat less for a swinging arm. In this case the waves of the action current become higher, although still apart (Fig. 16). The explanation

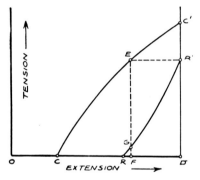

FIG. 15. Maximum innervated muscle. Force—elongation ratio represented by a relatively straight line.

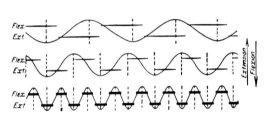

FIG. 16. Action current diagram. Relation of antagonistic action: upper, slow; middle, moderate; lower, fast motion. (Wachholder)

is that the greater mechanical impulse produces a greater momentum which is capable of carrying on motion to the end of the swinging phase. However in contrast to the first situation where only the natural swinging velocity was mentioned, now the accelerated swing of the muscle must be checked by its antagonist. As the anterior deltoid swings the arm forward, the posterior deltoid is set in motion and exercises its antagonistic effect before the forward swing of the anterior deltoid is completed. Here the dynamic explanation is that at this speed gravity alone is insufficient to check the forward swing and therefore the antagonistic restraint has to be called in; the earlier, the more rapid the motion (Fig. 17).

The action current also shows how nature uses gravity as an adjuvant in motion. If the elevated arm is allowed to drop, the deltoid is relaxed and the drop is produced entirely by gravity (Duchenne[9]). Similarly, if the arm is slowly abducted by the deltoid, then the antagonistic pectoralis and latissimus remain relaxed.

On the other hand, if this elevation is rapid and forceful, then an antagonistic action of these muscles sets in to check motion, or if the elevated arm is adducted rapidly to the side of the trunk by the action of the adductors, the deltoid exercises a restraining influence which is expressed in its action current.

The state of the muscle working under isometric conditions is reflected by the action current picture with the same accuracy as is the isotonic state when the muscle carries out visible motion. The positioning of the limb or of parts of it in space and the maintenance of position for the purpose of carrying out certain directed and well poised movements with calculated precision and speed calls for great stabilizing efforts. For instance, directing the fingers to a certain point is an effort which involves maintaining of the entire extremity in a suitable position and this can only be obtained by a complex scheme of balance in

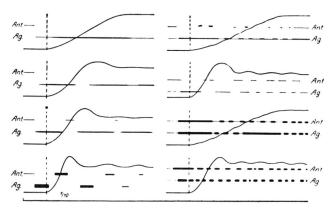

FIG. 17. Action current diagram. Agonist and antagonist. 1. Slow to fast (1.—4. row downward). 2. Free or rigid (left and right column). (Wachholder)

the different articulations and the cooperation of a number of muscle groups. These postural positions which furnish the base for directed motion therefore include a complicated innervational scheme of the stabilizers of all joints involved. In this situation the action current no longer shows the synchronicity or periodicity of the free motion. The spikes are now bunched together and they are more or less continuous.

Let us suppose now that external resistance is added to this postural rigidity; that body or the limbs are not only poised for a certain effort but also have to carry it out against external opposition, lifting a load, for instance, or pushing a cart. Here the action current picture is still more complicated. The curves are more crowded and their peaks are higher because the greater effort produces action currents of higher voltage.

All in all the action current reveals the quality as well as the quantity and duration of the motor activity of the muscle (Wachholder[33]). It has its different characteristic features in the free swing, in accelerated motion and in fast or rigid motion.

One feature of clinical interest is that the action current picture reveals the action of the antagonist as checker or moderator of motion. Such a simple act as abducting the hip, for instance, requires the cooperation of the adductors to make the movement purposeful. Another value lies in the study of the muscle synergism which operates in maintaining posture. Inman[20] points out that muscles like the tensor fasciae which had not been credited before with immobilizing the pelvis can be shown to hold the pelvis firmly when standing on the other leg and that its abductory action is merely secondary. While the abductory function of this muscle was known before, this investigator established its function as stabilizer of the pelvis by action current studies.

If a number of the muscles engaged in a motor act could be recorded simultaneously, the action current picture would be a visual presentation of what might be called the symphonic score of complex motor performances. For instance, throwing a ball or even picking up an object with the fingers are visible

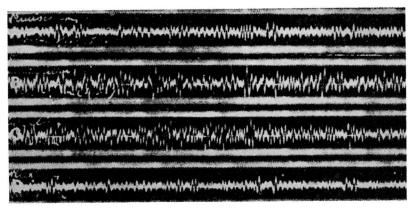

FIG. 18. Action current from the spastic muscle. Frequent and large oscillations, even in state of apparent rest. (Steindler and Lindemann)

acts superimposed upon a chain of motor combinations necessary to furnish the base upon which the final and visible motor act is built up.

Under pathological conditions the action current shows definite variations. In more recent years observations have accumulated on such conditions as anterior poliomyelitis, muscular dystrophies and atrophies. These methods have been refined to the point of picking up single muscle fibers and of establishing motor unit potentials which give a closer insight into the changes of the action current patterns as they occur in degenerative conditions.

Proebster,[28] investigating the action currents under pathological conditions observed in the severely paralyzed poliomyelitic muscle that the action current curve was of low intensity and the waves were small compared with the normal. The same is true of muscular dystrophy. In more recent electromyographic studies, Hirschberg and Abramson[18] make a distinction of:

(1) A muscle fiber or *fibrillation* potential with diphasic or monophasic spikes, with an amplitude of 50-100 µV, and a twitch duration up to 2 Σ, which is pathognomonic of a denervated muscle and which increases progressively after denervation.

(2) A *fasciculation* which is a spontaneous contracture of a motor unit,

sporadic with more or less irregular intervals, with a duration of spikes of 5-10 Σ and with an amplitude of 100-500 μV. The usual duration of the fasciculation itself is from 10 seconds to two minutes. Faster ones from eight to 10 seconds are observed in poliomyelitis.

(3) The potential of *myogenic dystrophy* is of medium or low voltage which is also seen in a muscle in a state of reinnervation.

(4) An *interfering potential* from multi-channel recordings at various areas of a simple normal muscle is observed, but it is also seen in spasms, in tremor, and in rigidity. Here the amplitude is increased, and the frequency is three to five per second.

We have been able to observe (Steindler and Lindemann[32]) in the spastic muscle an increase of amplitude commensurate with the increase of the innervational stimulus. After ramisection a change in the action current was observed and it appeared more normal (Fig. 18).

2. THE CHRONAXIA

Another method to determine muscle response to electric stimulation is the so-called chronaxia. This is the shortest duration which an electric current requires to elicit muscle contraction. A standard current is used twice the strength of the minimum contraction producing current (rheobasic current). The time necessary to produce contraction with such a current is called the chronaxia, and it varies with the different muscles of the body. It averages about 0.5 sigma or 0.0005 seconds. It also varies with age. In the newborn child the chronaxia is about ten times that of the adult which is in accord with the slower movements of the infant.

A change in chronaxia is an important feature in pathological conditions. For instance in the paralytic muscle it may be considerably prolonged, sometimes 100 times of normal, indicating degeneration of the muscle and the scarcity of still functioning muscle fibers. In the spastic muscle the chronaxia is normal or even accelerated.

d. The contraction length

Having discussed in general the physical properties of the muscles and the ways and means to determine them, the next step is to investigate the relationship of their contractility to the production of visible rotatory motion.

If the muscle tension represents the force, the changes in length during contraction represent the distance covered by the operation of this force and their product is the visible work thus accomplished.

To what length can a muscle contract? Borelli[6] knew that the lifting height of a muscle is the greater the greater the length of the fiber. Fick and Weber[34] were more specific. The so-called Fick-Weber law specifies that the fiber grows in length until in relaxed position it is twice as long as in maximum contraction, which means that it can contract about 50%.

Marey[24] even estimated the shortening up to 75%. It is probably nearer to the truth to follow the estimate of Recklinghausen,[30] namely, that the maximum contracted muscle is about 25% shorter than in the relaxed state, while the greatest amount of passive stretching will add to the normal length of the muscle about 60%, or 1.6 times its relaxed length. Jansen,[21] checking the muscles of the lower extremity, found that the so-called distensors, that is the lengtheners of the leg, differ in fiber length from the proximators, i.e., the shorteners. He stated that the former have an average fiber length of 7.6 cm., the latter of 13.9 cm.

A close relationship exists between the contractile and the natural length of the muscle so far as the range of motion and the efficiency of motion are concerned. Since the maximally distended muscle is 1.6 times its natural length and the maximally contracted about 0.5 its natural length, all motor events must occur between these limits. The difference in length between the completely distended and the maximum contracted muscle determines the range. The greater this difference in relation to the natural length $\dfrac{\text{Lmax} - \text{Lmin}}{\text{Lnormal}}$, the greater is the amplitude the muscle controls.

On the other hand there are included in this amplitude the terminal ranges of maximal shortening and of maximal lengthening which are not very suitable for the display of muscle tension, neither the extreme distension nor the extreme shortening. The closer the muscle remains to its natural length, the more strength it can display or the more efficient it is. Hence we can say that the reciprocal of the formula for the amplitude, that is $\dfrac{\text{Lnormal}}{\text{Lmax} - \text{Lmin}}$ is the formula for the efficiency of the muscles (Recklinghausen[30]).

The muscle is especially adapted in its construction for the specific mechanical function it has to perform. Some have a greater contraction length and less power, and those are built for speed. Others have a lesser contraction length and therefore a larger index of efficiency, and these muscles are built for strength.

This explains why so many articulations are endowed with pairs of muscles: one muscle for speed, the other for strength. For instance in the knee joint, muscles like the inner hamstrings have a greater contractile length which makes them more suitable for speed, while the biceps serves more prominently for the display of strength.

A knowledge of the contractile length of the different muscles is particularly necessary for muscle or tendon transference. Many operative plans have failed because the surgeon was not aware of the insufficient contractile length of the transplanted muscle.

SUMMARY

1. Elasticity of the muscle fiber is its resistance to stretching and its ability to assume passively its natural length when the distending force has ceased.

In terms of Hooke's law, muscle is very incompletely elastic. The breaking point is reached long before Young's module of elasticity where elongation equals original length.

2. Contractility is, in contrast, an active process. The maximally contracted muscle yields to a distending force more in arithmetical proportion than does the relaxed muscle. The amount of tension which holds the equilibrium against a force trying to distend the maximally contracted muscle beyond its natural length is the absolute muscle power. It has been determined at 3.6 kg. per cm.^{2} physiological cross section. Under Sherrington's law of all or none reaction of the individual muscle fiber, the voluntary gradation of tension of the muscle is a quantitative phenomenon, i.e., depending upon the number of individual muscle fibers engaged at one time.

3. The action current is an idio-muscular current produced by contraction. It can be presented oscillographically and its picture is characteristic for slow, free, rapid or resisted motion, respectively. Action current pictures are especially valuable for the study of combined, resistant motion, of coordinate and synergistic motor pattern and of antagonistic control. It shows characteristic changes under pathological conditions such as poliomyelitis, muscular dystrophies and atrophies and in spastic conditions.

4. Chronaxia is a phenomenon of the neuromuscular conduction. It represents the time necessary for a standard (rheobasic) current to produce contraction. It is found greatly increased in paralytic conditions.

5. The contraction length of a muscle is the difference between its extreme distended and its extreme contracted length. It varies with individual muscles. The degree of contractility of the muscle fiber is variously given between 25 and 75% of its natural length. The greater this contraction length the greater is the amplitude (in degrees) which a muscle can cover in contracting.

On the other hand, the extreme positions of extension and of contraction occupy a larger portion of the total amplitude in a muscle with a wide contraction range than in a muscle with more restricted contractility. These extreme positions are unfavorable for the display of muscle tension; the most suitable positions are those closest to the natural length of the muscle.

Consequently, the efficiency of a muscle is just the reciprocal of the amplitude; the greater the difference between terminal positions in relation to natural length, the greater the amplitude; the lesser this difference, i.e., the closer the terminal positions lie together (i.e., the closer they come to the natural length), the greater is the efficiency.

C. THE PHYSICAL PROPERTIES OF LIGAMENTS AND TENDONS

There are two reasons why these tissues should be included in the discussion of the physical properties of locomotor structures. First, they are susceptible to strain, and it behooves the clinician as well as the physical educator to know something about the safety limits of these structures.

Secondly, they are extensively used in surgical procedures to supply or reinforce ligaments or tendons. Gallie and Le Mesurier[15] introduced the use of ligaments and fascia for suture material as well as for stabilizing joints.

A rough test of the tensile strength of fascia taken at intervals from two days to two years led these two observers to the conclusion that the implanted fascia did not stretch and that its length remained the same as it did when it was implanted. The question then arises, what is the tensile strength of normal fascial tissue? Gratz[16] in a very painstaking and accurate experimental study found that the ultimate strength of fascia lata amounts to about 7,000 pounds to the square inch. It has a considerable degree of elasticity which, up to the maximum safety stress, was found as high as 91%. That means that within certain limits one can depend on fascia not to become elongated and relaxed after the stress has ceased. The safety stress of the fascia should be about 2,000 pounds per square inch cross section, or if one takes a strip of 0.5 inch wide and 0.02 inch thick, that is 0.01 inch^2 cross section area, the safety stress for such a fascial strip would be about 20 pounds.

However it should be kept in mind that it is not only the amount of stress but also its duration which have an effect upon the fascial structures. They will withstand certain momentary stresses up to the mentioned safety limit, but a protracted stress will result in a permanent elongation, similarly as cartilage reacts with a permanent deformation to prolonged pressure stresses. In this event a fascial structure does not resume its former shape but remains elongated or relaxed and consequently it no longer complies with Hooke's law.

However the comparative strength of fascial tissues varies so much with the individual that it is difficult to set up standards of stress resistance; but all have in common that duration of the stress for a longer time is liable to damage these tissues permanently even when the amount of stress is still within safety limits.

Ligaments also vary in their histological makeup. Those which are especially exposed to tension stresses during normal motion are composed largely of elastic fibers. This gives them a greater resistance to tension than the non-elastic fibers can provide. The ligamentum flavum is a good example. Its frequent thickening shows that it is particularly taxed for stresses on movements of the spine, and this hypertrophy is more marked in spines which are unstable. But even this ligament can become relaxed and lose its elasticity if the stress is excessive or is too continuous. An instance of what happens to a ligament when persistently overstressed is the deltoid ligament of the ankle joint or the internal collateral ligament of the knee.

The physical property of the *tendon tissue* is similar to that of fascia. If one considers that the tendon has to sustain all the stresses transmitted to it by the muscle and that its cross section area is so much smaller than that of the muscle, one will realize that the tensile strength of the tendon is very considerable. This problem has been investigated by McMaster[26] on the tendo-Achilles of the rabbit. In nine tendon-bone preparations he found that the load which produced rupture of the tendon varied from 10.4 to 23 kg. The rupture occurred at the tendon insertion or at the musculo tendinous junction. He found the unit tensile stress resistance of the tendon to be about 5,000-6,000 pounds per square inch cross section, or, for a small tendon of 0.01 square inch about 50-60 pounds. The safety stress which is about one-third of the breaking stress would then be about 2,000 pounds per square inch cross section or 20 pounds for 0.01 inch². These values come very close to the figures which we mentioned for fascial tissue. Figuring the cross section of the tendo-Achilles at about 0.25 square inch, the safety margin of this tendon would be about 500 pounds.

We know, however, that tendons undergo early degenerative changes. Its central artery disappears as early as the third decade so that the tendon becomes entirely dependent on circulation from the surrounding area. So called "spontaneous" tendon ruptures are therefore nothing unusual. We see it most often in the long extensor tendon of the thumb, in the tendo-Achilles, in the patellar tendon, and especially in the rotator cuff of the shoulder joint. Unless the trauma is excessive, one is likely to find in all these instances degeneration of the tendons as the cause of the rupture.

The tension resistance of the tendon often exceeds that of the bone to which it is attached so that avulsions of the tendon rather than tears of the tendon itself will result. This is particularly true of younger individuals. In older people the rupture occurs more easily in the tendon itself and avulsion by tension stress is less common.

SUMMARY

1. The tension resistance of normal fascia has been determined by Gratz[16] at 7,000 pounds per inch², giving a safety margin of about 2,000 pounds (one-third of the breaking point) or about 20 pounds for a fascial strip 0.01 inch².

2. Permanent elongation of fascial tissue results not only from excessive stress but also from its excessive duration.

3. Ligaments endowed heavily with elastic fibers such as the ligamentum flavum show a greater stress resistance, but even these become permanently elongated with excessive duration of the stress.

4. The physical properties of tendons are similar to those of fascia and ligaments. Their tensile stress resistance was found by McMaster to be 5,000-6,000 pounds per inch², with a safety margin of 20 pounds per 0.01 inch².

5. However, tendon undergoes early degenerative changes mainly on circulatory basis, and so called "spontaneous" ruptures are not infrequent after the third decade.

6. In younger individuals with healthy tendons avulsions are more frequent; in older people rupture occurs more easily in the tendon itself.

BIBLIOGRAPHY

1. ALEMAN, O.: Chondromalacia Posttraumatica Patellae. *Acta chir. scandinav., 63:*149, 1928.
2. ARKIN, A. M.: Absoute Muscle Power. The Internal Kinesiology of Muscle. Research Seminar Notes. *Dpt. Orth. Surg., State Univ. of Iowa, 12D:*123, 1938.
3. BENEKE, R.: Zur Lehre von Spondylosis Deformans. *Beitr. Z. Wissensch. Med.; Festschrift d. 69.* Vers. Dtscher Naturforscher und Aerzte, 1897.
4. BENNINGHOFF, A.: *Lehrbuch der Anatomie des Menschen.* Vol. I, Munich, J. F. Lehmann, 1939.
5. BETHE, A.: Active und Passive Kraft Menschlicher Muskeln. *Erg. d. Physiol., 24:*71, 1925.
6. BORELLI: *De Motu Animalium.* Lugduni Batavorum. 1685.
7. BRAUNE, W., and FISCHER, O.: *Die Bewegungen des Kniegelenkes.* Leipzig, S. Hirzel, 1911.
8. ———: *Die Bewegungen des Kniegelenkes.* Leipzig, 1891.
9. DUCHENNE DE BOULOGNE: *Physiologie des Mouvements.* Paris, Baillière, 1867.
10. Einthoven: Sur Les Phenomènes Electriques du Tonus Musculaire. *Arch. Neerland. de Physiol., 2:*489, 1918.
11. FAIRBANK, T. J.: Knee Joint Changes after Meniscectomy. *J. Bone & Joint Surg., 30B-4:*664, November 1948.
12. Fick, R.: *Spezielle Gelenk und Muskelmechanic, Vol. I.* Jena Gustav Fischer, 1911.
13. ———: Tätigkeitsanpassung der Gelenke und Muskeln. *Akademie d. Wissenschaften,* Berlin, 1922, 353-383.
14. FISCHER, O.: *Kinematik Organischer Gelenke.* F. Vierveg & Son, Braunschweig, 1907.
15. GALLIE, W. E., and LE MESURIER, A. B.: The Transplantation of Fibrous Tissue in the Repair of Anatomical Defects. *Brit. J. Surg., 12:*289, 1924-25.
16. GRATZ, C. M.: Tensile Strength and Elasticity Tests on Human Fascia Lata. *J. Bone & Joint Surg., 13:*2, April, 1931.
17. HIRSCH, C.: A Contribution to the Pathogenesis of Chondromalacia of the Patella. *Acta. chir. scandinav.*
18. HIRSCHBERG and ABRAMSEN: *Arch. Phys. Med., 31:*576, September, 1950.
19. HULTZKRANZ, J. W.: Uber die Spaltrichtungen der Gelenkknorpel. *Verh. Anat. Ges. Kiel.* 1898.
20. INMANN, V. T.: Functional Aspect of the Abductor Muscles of the Hip. *J. Bone and Joint Surg., XXIX:3* 607, July, 1947.
21. JANSEN, MURK: Die Länge der Muskelbündel. *Ztschr. f. Orthop. Chir., 36:*1, 1917.
22. KROGH: Anatomie und Physiologie der Capillaren. Berlin, 1924.
23. MACCONAILL, M.A.: The Movements of Bone and Joints; The Mechanical Structure of Articulating Cartilage. *J. Bone & Joint Surg., 33-B:*2, 251, May 1951.
24. MAREY, J.: *La Machine Animale.* Paris, F. Alcan, 1891.
25. MARTIN, E. G., and RICH, W. H.: Muscular Strength and Muscular Symmetry in Human Beings. *Am. J. Physiol., 47:*29, 1918.
26. McMASTER, P. E.: Tendon and Muscle Rupture. *J. Bone & Joint Surg., 15:*3 July, 1933.
27. PROEBSTER: Ueber Muskeltonus. *Ztschr. f. Orthop. Chir., 48:*541, 1928.
28. ———: *Ueber Muskelaktionsstrôme am Gesunden und Kranken Menschen.* Stuttgart, F. Enke, 1928.
29. RAUBER, A. A.: *Elastizität und Festigkeit der Knochen.* Leipzig, 1876.
30. RECKLINGHAUSEN: *Gliedermechanik und Lähmungsprothesen.* Berlin, J. Springer, 1920.
31. SHERRINGTON, C.: *Proc. Royal Soc. London, 52:*81B.
32. STEINDLER, A., and LINDEMANN, E.: Alteration of the Action Current of Skeletal Muscles Following Sympathetic Ramisection. *J. Bone & Joint Surg., 11:*1, January, 1929.
33. WACHHOLDER: *Willkürliche Haltung und Bewegung.* München, J. F., Bergmann, 1928.
34. WEBER, E. F.: Ueber die Längenverhältnisse der Muskeln im Allgemeinen. *Verh. Kgl. Sachs. Ges. der Wissensch,* Leipzig, 1851.
35. WIBERG, G.: Arthritis Deformans; Orsaker, Symtom och Behandlung. *Med. Forening Tidskrift,* 1942, 9.

ON THE MECHANICS OF JOINT AND
MUSCLE ACTION

A. THE JOINTS

I. GENERAL MECHANICAL PRINCIPLES

THE function of a joint is determined primarily by the shape and contours of the contacting surfaces, and particularly by the degree of congruency of the articulating bodies and by the manner in which they are fitted together.

The joint cartilage is subject to a great deal of pressure produced both by gravitational forces and by muscular tension. In order that the pressure bearing surfaces may be as large and pressure distribution as even as possible, great demands are made on the elastic properties of cartilage. How much of this adaptation is necessary depends upon the construction of the joint. There are those in which extensive contact areas glide upon each other such as the elbow joint. There are others of equally broad contact in which the constituents describe a peripheral gliding around an extraarticular center: a conducted or trackbound (zwangsläufig) motion, such as is seen in the intervertebral articulations. Others again have closely fitted cylindrical contact surfaces such as the ankle joint or they are ball and socket joints, as for example the hip joint. All these articulations have in common a high degree of congruency of their surface and because of their broad contact do not require of the cartilage to undergo adaptive changes under pressure.

On the other hand there are joints with great incongruency of their surfaces. These exact under pressure adaptive deformations of the joint cartilage. To this group belongs the knee joint, the subastragalar and the midtarsal joint.

Under pressure the contact area increases in some joints more, in others less, according to the thickness and elasticity of the joint cartilage. In that sense Fischer[6] makes a distinction between deformable and non-deformable joints.

II. THE SHAPE OF THE ARTICULAR SURFACES

Simple and composite joint surfaces

We can imagine all joint surfaces to be the product of a generating line which rotates about an axis. According to the shape of this generating line and its relation to the axis around which it revolves we distinguish (Fick[5]):

1. Simple cylindrical or conical surfaces. A straight line rotating about a parallel axis describes the surface of a cylinder: example, the ankle joint (Fig. 1). If the generating straight line is not parallel but inclined to the

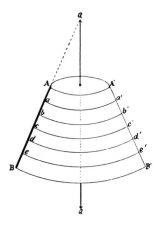

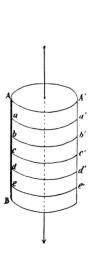

FIG. 1. A straight line rotating about a parallel axis describes the surface of a cylinder. (Fick)

FIG. 2. Conical surface developed by a generating straight line not parallel but inclined to the axis. (Fick)

 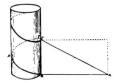

FIG. 3. Screw surface generated by a point moving both in circular and progressive direction. (Fick)

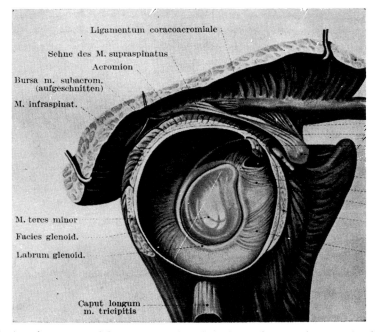

FIG. 4. Oval surface generated by curve rotating about its tendon as axis: example, the glenoid.

axis (Fig. 2), a conical surface is developed; there is no example of this in the human body. If the rotating line is a semicircle with the concavity facing the axis, the surface produced by rotating about this axis will be a sphere: examples, the capitellum, the head of the femur. If the convexity of the line faces the axis the surface generated by its rotation will be concave: example, the head of the radius.

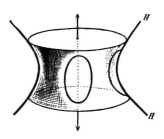

FIG. 5. Saddle joint generated by a hyperbola with symmetrical ends rotating about an imaginary axis. (Fick)

2. Screw surfaces are composite in origin. They are generated by a point moving both in circular and in progressive direction (Fig. 3).

3. There are oval joint surfaces which are generated by a curve of less than a semicircle rotating about its tendon as axis. The characteristic point is that the angular value differs in the different axes. The least angular value is that of the length axis of the oval. This type is represented by the glenoid or by the radio-carpal articulation (Fig. 4).

4. A saddle joint is generated by a hyperbola with symmetrical ends rotating about an imaginary axis. The surface thereby created is inversely curved in two perpendicular directions, i.e., in one concavely and in the other convexly. This is represented by the carpo-metacarpal joint of the thumb and also by the lumbar intervertebral articulations (Fig. 5).

III. THE JOINT CONTACT

The cohesion of the joint surfaces

Cohesion of the joints is produced:

1. By the joint ligaments, especially in hinge joints, such as the lateral ligaments of the elbows and knees or of the metacarpo-phalangeal joints. The ligaments are relaxed in mid position where they exert the minimum of pressure.

2. By muscle tension through the stabilizing components of the muscle force. This component is considerably greater than the rotatory component of the muscle.

3. By fascial structures which cover the surrounding muscles, for instance, the fascia lata.

4. By atmospheric pressure (760 mm. mercury). This pressure is especially effective in the hip joint. According to the brothers Weber, all muscles, the whole capsule and all ligaments can be severed without the head falling out of the acetabulum. But it does so promptly when a hole is bored into the socket. Atmospheric pressure at the hip joint amounts to 25 kg., depending somewhat on the position of the limb (Fick[4]). It seems that the incisura acetabuli does not interfere with the carrying ability of the atmospheric pressure.

In other joints which are less hermetically fitted, the atmospheric pressure causes the surrounding soft tissues to penetrate into the joint, as for example, in the knee joint. Under normal conditions the joint cavity is filled with synovial fluid, synovial folds, fringes, and the menisci filling the joint spaces under atmospheric pressure. A reactive joint effusion produces a positive pressure which is higher than the opposing atmospheric pressure; hence the latter cannot prevent subluxations.

An interesting phenomenon is that under traction there appears in the Xray picture a translucency between the joint ends which indicates a vacuum. This has been recently emphasized by D. Fuiks.[8] According to A. Fick[4] and Gürter,[9] this separation begins at the hip joint when traction reaches the atmospheric pressure of 1 kg. per cm.[2].

The cracking which is heard in joints such as the knee joint, the mandible, the wrist or the finger joints is more likely due to the gliding of tendons or ligaments over bony prominences, or it is produced by the gliding of the tightened capsule over the bone.

In high altitudes there is a deficiency of atmospheric pressure. At 1,300 m. it becomes insufficient to bear the weight of a leg weighing 10 kg., at 3,000 m. a leg of 7.5 kg. cannot be born by the atmospheric pressure and the deficiency must then be made up by muscle tension. It is also said that at 8,800 m., the height of the Gaurisankar in the Himalayas, the atmospheric pressure amounts to only 4 kg. for the hip joint.

IV. THE TYPE OF JOINT MOVEMENT

a. The gliding

1. SURFACE GLIDING

This motion can be carried out in purely translatory direction, as in the shoulder which glides over the thoracic wall up or down, backward or forward, or it can be a rotational gliding over a rather flat surface. It can also be a combination of both, such as the movement carried out by the shoulder blade in ab and adduction of the arm. A gliding type of motion is also that of the intervertebral articulations. In this case the area described by the articulation is not a flat surface but that of a cone or cylinder.

2. LINEAR GLIDING

In linear gliding a point of a shallow concave gliding body sweeps in linear direction over a larger surface of the other convex joint body. This moving concave body is generally the peripheral part of the joint, for instance the basal phalanx moving over the metacarpal. The free movement of the peripheral part represents what is called an open kinetic chain (see below). On the other hand, the central joint constituent may be the movable part while the other is stationary as in the so called closed kinetic chain (see below); or both joint constituents may be in motion against each other.

b. Rocking or rolling motion

In rocking or rolling motion equidistant points touch each other in the course of motion. The one body is concave and the other one is convex since there is no instance of a bi-convex contacting of the joint surfaces. Here the best example is the initial range of knee motion. Due to the fact that the earlier ranges of flexion are associated with inward rotation, the movement is not a true rocking one in the strict mathematical sense. However it comes so close to the true rocking motion (see lecture on the mechanics of the knee joint), that one would have to be very dogmatic to take issue on that point.

c. Rocking combined with gliding joints

In joints in which rocking combines with gliding, the equidistance of the contact points gradually disappears with the transition of the one type of motion into the other. For instance in the knee joint the first 15 or 20 degrees are rocking motion modified as indicated above and then gradually the gliding element prevails.

d. Axial rotation

Axial rotation differs from the former motion in that the contacting surfaces of both bones are small. They may be restricted to a small point as in the radio-humeral articulation. In the axial rotation of the knee joint, the bodies describe circular movement against each other. This occurs because the axis of length rotatory motion falls into the intercondyloid notch so that the contacting points of the moving body describe circles.

V. THE DEGREES OF FREEDOM OF MOTION

a. We designate a joint as having one degree of freedom of motion if it can be moved in one plane only; two degrees of freedom of motion means that motion is possible in two planes perpendicular to each other, and three degrees of freedom of motion is one in which motion is possible in three planes all perpendicular to each other. Accordingly, in joints with one degree, the motion is either that of a point moving against a convex line, as in the interphalangeal articulation, or of two lines moving against each other as in the ulno-humeral articulation.

b. Two degrees of freedom of rotatory motion. Examples are represented by the knee joint where motion occurs in two planes perpendicular to each other; flexion and extension in the sagittal, and internal-external rotation in the transverse plane; or the radio-humeral joint where flexion-extension as well as pro and supination is carried out. In the knee joint the contact areas differ with the position and are always small because of the incongruity of the joint constituents. For the same reason the contact area in the radio-humeral joint is restricted to almost a point for pro and supination while that for the flexion-extension in the ulno-humeral joint is broad because the joint constituents are congruent.

c. Three degrees of freedom of motion is represented by the amphi-arthrotic joints such as the shoulder joint, the hip joint or the metacarpo-phalangeal joint.

It will be seen that no joint can have more than three degrees of purely rotatory motion because the three planes represent the tridimensional space, but a joint can participate in the translatory motion of the proximal part against the trunk which again has three degrees of freedom of motion that is in three perpendicular planes.

The only example is the shoulder joint where the scapulo-humeral joint follows the translatory movement of the scapula against the thorax; and the temporo mandibular joint in which straight forward and backward movement is associated with rotatory movement about three perpendicular axes.

VI. A KINETIC CHAIN

A kinetic chain is a combination of several successively arranged joints constituting a complex motor unit. We designate as *open* kinetic chain a combination in which the terminal joint is free. The waving of the hand is an open kinetic chain in which the action of the shoulder joint, the elbow joint, and the wrist joint are successively involved.

A *closed* kinetic chain, on the other hand, is one in which the terminal joint meets with some considerable external resistance which prohibits or restrains its free motion. Eventually, the external resistance may be overcome and the peripheral portion of the joint may move against this resistance, for instance, in pushing a cart or lifting a load; or the external resistance is absolute, in which case the proximal part moves against the peripheral, as for instance, in chinning oneself on a horizontal bar; or the limitations of the muscular effort may assert itself both peripherally and proximally and may be unsurmountable, in which case no visible motion is produced. Only in the latter instance is the kinetic chain strictly and absolutely closed. However in common use we apply the term to all situations in which the peripheral joint of the chain meets with overwhelming external resistance.

The importance of the summation of all the freedoms of motion of the different successive joints for precision performances becomes at once evident.

For the *upper extremity* the summation amounts to as follows:

The first interphalangeal joint 1 degree, the second interphalangeal joint 1 degree, the metacarpo-phalangeal joint (voluntary motion) 2 degrees, the carpus 2 degrees, the elbow 2 degrees, the shoulder joint 3 degrees, a total of 11 degrees of freedom of motion from the fingers against the shoulder girdle.

To this one still add the 3 degrees of freedom of the acromioclavicular and 3 degrees of the sterno-clavicular joint, making a total of 17 degrees of freedom of motion of the fingertips against the trunk, not counting the 2 degrees of translatory motion which the scapula can carry out against the thoracic wall.

The sweep of the extremity against the trunk alone requires the 3 degrees for the shoulder and the 2 for the elbow joint, a total of 5 degrees of freedom of motion. Adding then to it the motions of the wrist with 2 additonal degrees, it sums up to a total of 7 degrees of motion between the wrist and the trunk.

Thus, of the 11 degrees of freedom of motion between fingers and shoulder girdle or of the 17 between fingers and trunk, only a certain number is made use of in some of the motor performances while others may require movements in all degrees. For instance, picking up a pencil calls for abduction, forward flexion and inward rotation of the humerus; for flexion and pronation in the elbow; for extension and adduction in the wrist; for flexion adduction and pronation of the thumb; and for flexion in the metacarpo-phalangeal joint of the thumb and the interphalangeal joint of the thumb and index finger, a total of 11 degrees of freedom of motion, or even more if one adds the elevation of the shoulder blade.

For the *lower extremity* which is less movable the kinetic chain in respect to the trunk can be analyzed as follows:

The hip joint has 3 degrees of freedom of motion: ab and adduction in the frontal, flexion-extension in the sagittal, and internal and external rotation in the transverse plane. Aside from these 3 degrees in the hip joint there are, between femur and tibia, 2 degrees of motion, flexion-extension and axial rotation. Consequently between tibia and trunk there are 5 degrees of freedom of motion. The tibio-astragalar joint is "trackbound" in the mortice and therefore has only 1 degree of freedom of motion making a total of 6 degrees for the foot as a whole against the trunk (Fischer[7]).

The 3 degrees of motion in the hip joint and the 3 degrees of motion in the shoulder joint serve to change the position in space of the full length limb in all directions, without, however, changing the form of the link itself. Consequently all motions which occur in the hip joint alone or in the shoulder joint alone place the end of the extremity on the surface of a globe, the radius of which equals the entire length of the extremity.

However, more often than not the motor performance requires that the end of the extremity be placed not on the surface of the globe which represents the excursion field of shoulder or hip joint, but at a point in the interior of this globe. Therefore there must be provisions for shortening and lengthening the extremity, that is for changes of its form. For this purpose the extremities are endowed with joints which provide lengthening and shortening. For the lower extremity it is the knee joint and for the upper extremity it is the elbow joint which performs this function.

VII. THE TRANSFORMATION OF ROTATORY INTO TRANS-LATORY MOTION IN A KINETIC CHAIN

All movements in the joints are rotatory motion about the centers located within the body. The question how these rotatory motions can be transformed into a translatory end effect is of considerable interest.

Let us imagine a rod rotating around one end. After having completed this rotation about one end, let us next assume that the rod now rotates around the other end the same amount but in the opposite direction of the clock, that is, counterclockwise if the first rotation has been clockwise (Steindler[10]) (Fig. 6). The end effect of these two motions is that the rod is now in the same position it would have assumed if it had been moved forward in a

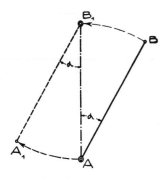

FIG. 6. Translatory end effect of two consecutive rotations of a rod about opposite ends: AB to AB₁ to A₁B₁.

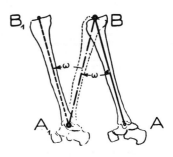

FIG. 7. Transformation of two rotations (about ankle and knee) into a translatory forward motion of the leg: AB to A₁B to A₁B₁.

parallel direction, that is, if all points of the rod had moved forward in a straight line. Therefore one can state that the condition for a translatory effect resulting from two rotatory motions about two different centers is the first angle of rotation (w) is equal and opposite to the second (— w). The general formulation would then be that whenever rotatory motion about different joints is carried out in such a way that the end effect is a straight translatory progression, then the sum total of all rotating angles is zero. (w) + (— w) = 0.

Now let us apply this to the human body. The tibia represents a rotating rod flexed against the femur at the start. The first movement is a forward swing of the tibia at the knee joint (Fig. 7). Next the tibia rotates forward the same amount at its lower end about the ankle joint; the swinging angle of w is the same but in opposite direction from the former swinging angle —w. We therefore find that the algebraic sum of the two swinging angles (w plus — w) is zero and consequently we find at the end of the second rotation that the tibia has advanced forward in strictly parallel or translatory direction.

Now if, by contrast, we combine two or more rotatory motions about the centers of rotation which do not run in opposite but move in the same direction, we find as follows (Fig. 8): The first rotation of the tibia is about the upper end (counterclockwise), and then the next rotation is carried out about its lower end, also counterclockwise. The end effect, now is not a parallel forward shift but it is a summation of the forward swing, presenting the sum of the angles w and w₁ which now have the same, and not opposite signs (plus or minus respectively). It appears, then, that combinations of two or more rotatory motions can be arranged in such a way that they result either in a translatory progression if the angles are equal and

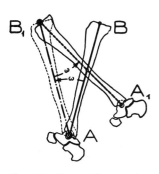

FIG. 8. Consecutive motion about opposite ends in the same direction.

opposite or in a summation of the rotatory effect, the extensory or flexory, if the angles are not in the opposite but in the same direction.

In a kinetic chain such as represented by the extremities, the practical significance of this combination for the function of the extremity can be easily demonstrated. In the gait the lower extremity alternates between shortening and lengthening on one hand, and for and backward swing on the other (Fig. 9). The centers of these motions are the hip joint, the knee joint, and the ankle joint. The purpose of elongating the limb during the supporting period of the gait is to press the foot against the ground and thereby give it

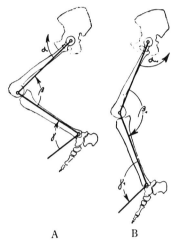

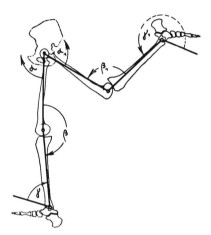

FIG. 9. A. Shortening, B. lengthening of limb. Rotation in three joints in oppoite direction neutralizing angles of rotation. A—Translatory movement of heel in flexion. B—Translatory movement of heel in extension.

FIG. 10. Forward and backward swing of limb. Motions in different joints are all in the same direction: counterclockwise in backward swing, clockwise in forward swing.

an upward acceleration. In the next phase the leg becomes shortened. This is accomplished by a similar mechanism, this time by a combined flexion in all joints, the rotation occurring alternatingly in opposite direction of the clock so that the sum total of all the angles of rotation again adds up to zero. This, as pointed out above, is the condition for a straight translatory movement.

But the gait is not only an alternation of shortening and lengthening; it is also a periodic back and forward swing of the extremity. In the case of a full backward swing of the hip, knee, and ankle joint, all angles of rotation point in the same direction, for instance, all are counterclockwise, while in full forward swing all joints should move opposite to that of the backward swing, that is, all clockwise (Fig. 10).

A third situation arises when the angles of rotation alternate in direction, plus or minus, but still do not check each other out so that the total sum is not zero but there is a certain plus or minus value left. This means that the movement is a shortening or lengthening effect combined with a for or backward swing. If the size and direction of each angle of rotation and the size of the links which constitute the kinetic change are known, we can determine accurately what point in space is reached by the terminal end point of the limb. It makes no difference in which succession the motions of the several joints occur so far as the location of this point in the space is concerned. If there is a certain sequence maintained in the scheme of a motor performance its purpose is not static, not for allocation of the movement in space itself, but it is founded on dynamic reasons, namely the development of force and speed.

B. THE MECHANICS OF MUSCLE ACTION

I. STABILIZING AND ROTATORY COMPONENTS

As the muscle moves the lever arm to which it is attached around the articulation as a center of motion, it develops two types of acceleration:

1. There is a tangential acceleration which produces motion in linear direction and continues in its path in a straight line. The acceleration is equal to the velocity over the time $(a_t = \dfrac{v}{t})$.

2. Then there is a normal or perpendicular acceleration which is directed away from the center in a radial direction (a_n). This is equal to the square of velocity (v^2) over the radius (r) $(a_n = \dfrac{v^2}{r})$ (Fig. 11). It is this radial or centrifugal acceleration which has the tendency to separate the moving parts away from the center of motion. Let us imagine a violent swing of the arm. The hand will receive an acceleration which carries the hand tangentially forward and another centrifugal acceleration which tries to tear it away from the center of motion in a radial direction. To take care of this situation, the muscle must develop one component which performs the visible rotatory motion and another component which counteracts the disruptive tendency of the centrifugal force. In other words, it must develop a rotatory as well as a stabilizing component. The respective proportions of these two components of the total muscle force depend upon the angle at which the tendon

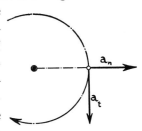

FIG. 11. Diagram of tangential and centrifugal force in rotatory motion.

applies to the bone at its insertion, or more accurately, the angle formed by the mechanical axis of muscle with the lever arm into which it is inserted. The rotating component is proportional to the sine and the stabilizing component is proportional to the cosine of this angle.*

---

* If the muscle is applied to the bone at an angle of 90 degrees, the sine of this angle would be 1, which means that all muscular force developed would be rotatory. If the angle of the muscle is applied to the lever arm at 0 degrees, that is, when the muscle axis is parallel to the bone, the cosine of this angle is 1, which means that all muscular force is used for stabilization.
$$(F_r = F. \text{sine } \alpha, \; F_s = F. \text{cosine } \alpha)$$

It is easy, therefore, to compute the relative size of these two components. When the angle of application is 45 degrees they are equal because the sine of 45 equals the cosine ($= 0.707$).

There are, however, very few muscles which have that large an angle of application. The pectoralis major attaches itself at almost right angles; therefore its rotatory component comes close to the entire muscle force; but most of the muscles are so arranged around the limb that the axis runs almost parallel to the axis of their lever arm so that the angle of application is small. Hence the stabilizing component of the muscle is, as a rule, much larger than the rotatory component.

The practical significance of this arrangement is that the maintenance of position, balance and stabilization are really the prime functions of the muscle while the visible rotatory motion almost appears to be second in importance.

II. LEVERAGE AND EQUILIBRIUM

Although a single muscle develops in contracting a relatively large stabilizing component, it alone cannot stabilize the joint. In order to accomplish this, its own rotatory component must be suppressed. This calls the antagonist of the muscle into action. For example, to stabilize the ankle joint, the tension of both the extensors and the flexors of the joint is required. In this case, the joint is stabilized by the antagonistic action of both muscle groups, which mutually neutralize their respective rotatory components. However this antagonistic action does not necessarily come from the muscles. It may be furnished by external forces such as gravity. The pressure of the floor forces the ball of the foot upward into dorsiflexion against which the tension of the calf muscles holds the equilibrium. In neither instance is the stabilizing effect accomplished by one muscle group alone.

Stabilization of a joint is a case of equilibrium between two forces. They may be arranged on opposite sides of the joint in which case the situation is that of a lever of the first order (Fig. 12). Or, the two opposing forces may be attached to the same lever arm, but they hold the equilibrium by being directed oppositely, in which case the situation is that of a lever of the second or third order according to what one considers the primary force and what one considers the resisting force (Fig. 12).

Examples of all three kinds of lever arms are very numerous. In standing on the foot the reaction from the floor applying to the heel and ball of the foot, respectively, represents an arrangement of a lever of the *first* order. Rising on the ball of the foot the force is provided by the tension of the tendo Achilles and the resistance by the superincumbent body weight. Here is an example of a lever of the *third* order. In flexing the elbow against a weight held in the hand, the force and resistance are also on the same lever arm but the force is closer to the center of motion than is the resistance. This is an example of a lever of the *second* order.

The *angle of application* of the muscle is not constant. The more flexion proceeds the greater the angle becomes. For instance the biceps has a very small angle of attachment at the elbow when the latter is in full extension. As the flexion increases the angle increases. This would indicate that the rotatory strength is increased in full flexion while the stabilizing force of the muscle is greatly diminished (Fig. 13). However the contrary is the case. The tension which a muscle is capable of developing decreases rapidly as the muscle becomes

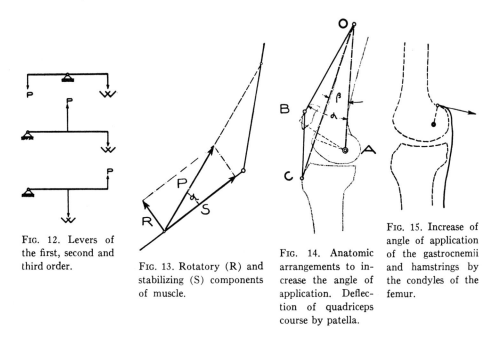

FIG. 12. Levers of the first, second and third order.

FIG. 13. Rotatory (R) and stabilizing (S) components of muscle.

FIG. 14. Anatomic arrangements to increase the angle of application. Deflection of quadriceps course by patella.

FIG. 15. Increase of angle of application of the gastrocnemii and hamstrings by the condyles of the femur.

shorter by contraction. This loss far outweighs the advantage which the increase of the angle of application gives to the leverage.

Nature makes certain provisions to increase the angle of application without decreasing the muscle length. The most effective of these is deflecting the course of the tendon of insertion. An example of it is the patellar tendon. Here the interposition of the patella into the quadriceps tendon changes the direction of its tendon favorably by increasing its obliquity.

Such a deflection of the course of the tendon for the purpose of increasing its leverage angle is of advantage only if it is carried out on the pulley principle, that is if the tendon is lifted off away from the bone without change in the plane of the pull. This is the case in the quadriceps tendon (Fig. 14). Similarly are the heads of the gastrocnemius deflected by the protruding posterior mass of the femoral condyles (Fig. 15).

There is another way of increasing the leverage of a tendon, namely by using muscle pull to separate the tendon from the underlying bone. In the case of the flexor digitorum sublimis, the tendon runs almost parallel to the metacarpals and basal phalanges. Before it attaches itself into the bases of the middle phalanges, it is perforated by the profundus tendon which passes between the two terminal slips of the superficial flexor so that the latter straddles, so to speak, the deep tendon. When the deep flexor tendon then contracts it lifts the superficial flexor away from the bone and thereby increases its angle of application (Fig. 16).

Another instance of deflection of tendons is seen in the volar carpal groove. Here the walls of the groove act as pullies which deflect the finger tendons to give them proper divergence for their insertion into the phalanges.

Deflection of tendons around projecting bones by restricting fibro-cartilaginous sheaths is found in many places. Examples are the fibrous sheaths enveloping the tendons of the peronei as they wind around the lateral malleolus or the fibro-osseous canal which contains the long head of the biceps giving the tendon a direction parallel to the humerus.

The result of these deflections is that a great portion of the muscle pull is absorbed in pressure against the retaining vincula. Sometimes this strain causes

Fig. 16. Increase of angle of application of the superficial finger flexor by the deep flexor, which lifts the former off the bone.

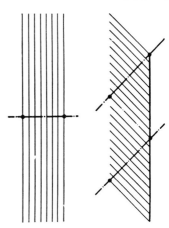

Fig. 17. Parallel fibered and pennated muscles.

the tendons to dislocate laterally over their pulleys or restraining prominences. The peronei, for example, dislocate forward over the external malleolus if for some reason the restraining action of the peroneal sheath is relaxed. At the shoulder the biceps tendon may dislocate medially on outward rotation of the humerus, leaving the bicipital groove.

III. MORPHOLOGICAL ADAPTATION OF THE MUSCLE

a. Anatomic construction

The anatomic makeup of a muscle indicates whether it is built primarily for power or for speed, while the total work it can develop depends primarily on its mass. The total tension can be measured by the physiological cross section area of the muscle and a constant for the cross section unit which, according to Recklinghausen, amounts to 3.6 kg. per cm.[2]. The difficulty consists in determining the physiological cross section which is the section that cuts all fibers transversely. It therefore has to be determined differently in the parallel fiber and in the pennated muscle (Fig. 17). In the parallel fibers the effective tension equals the anatomical cross section area times the tension per unit. In the pennated muscle an anatomical cross section would not cut all fibers. Consequently a series of sections must be made, all perpendicular to the direction of the fibers in such a manner that all fibers are cut.

The pull which each individual fiber exerts in the direction of the axis of the muscle depends on the angle the fiber forms with the muscle axis, being equal to its tension times the cosine of that angle (t $\times$ cos). Should the obliquity of the fiber approach 90 degrees (cos 90 $=$ 0), all of the tension would be spent in lateral pull. An example is the digastric muscle which tightens the hyoid bone. In the pennated or bipennated muscle a good deal of the tensile strength is expanded in a lateral pull or tug on the central tendon of the muscle.

b. Speed and power

Taking two muscles of the same volume, the one consisting of comparatively few and long fibers (such as the sartorius), and the other of more numerous but shorter fibers (such as the gastrocnemius), both when contracting will display a difference in the lifting power and a difference in the lifting height. The one with many short fibers has the lesser lifting height (h). However both will do the same amount of work provided they have the same volume, the work being the product of force times distance. The difference is only that parallel fibered muscle lifts a lesser load a greater distance; the short fibered or pennated muscle will lift a greater load a lesser distance. The power of the first one is smaller but its lifting height is greater. The power of the second one is greater but its lifting height is smaller. The first, therefore, is built for speed and the second is built for power.

The point arises what arrangement can be made in the lever system of the muscle which would make use of this difference in power and lifting height of the individual muscles. It is done by varying the length of the lever arm. Everything being equal, the muscle attached closer to the center of the joint would require the greater power to lift a given load a certain distance, but to do so it will contract only one-half of what another muscle must contract, which applies at twice the distance from the center of motion (Fig. 18).

The conditions suitable for the respective display of speed or power can be summarized as follows:

1. ANATOMIC BUILD OF MUSCLE

   a' For power: the fibers are short, more numerous, in pennated arrangement.
   b' For speed: the fibers are long and less numerous, parallel arrangement.

2. LEVERAGE OF THE MUSCLE

   a' For power: the muscle is attached at a greater distance from the center of motion.

b′ For speed: the muscle is attached at a lesser distance from the center of motion.

In the larger joints of the body, especially those subject to weight bearing, both speed and strength is required, and for this reason they are provided with both types of muscle. The elbow has its biceps and brachialis, the knee has its biceps and hamstrings (Fig. 19).

3. The Natural Length, Contractile Distance and Tension of the Muscle.

A constant relation exists between the natural length of the muscle (L), its variable length in contraction (l), and the degree of rotation. This means that for each unit of shortening there is a constant angular value of rotation; shortening and rotation angle are arithmetically proportional.

There is also a certain relation between the contraction length (l) and the

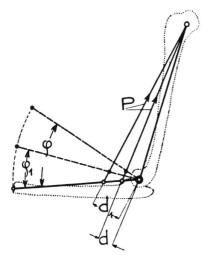

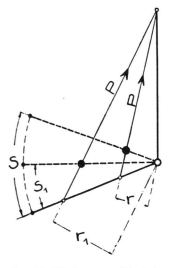

Fig. 18. Relation of length of lever arm to speed and power. Lever arm at a short distance from the center of motion, greater excursion per unit of contraction. Longer lever arm, lesser excursion per unit of contraction.

Fig. 19. Work accomplished by speed and power muscle. Relation of rotatory angle S and S_1 to rotation moment (Pr and Pr_1). Shorter lever arm for speed; longer lever arm for strength.

muscle tension (t). The absolute muscle power coefficient of 3.6 kg. per cm.[2] applies only to the natural length of the muscle. The relation of tension to muscle length is not in arithmetical proportion as is length and angle of rotation, but as contraction proceeds the contractile power decreases much more rapidly.

The question is, what is the natural length? There is a certain known difference in muscle length between a muscle which is under some innervational tone and another which is completely relaxed. In the living we see, however, that the muscle will retract even under deep anesthesia when its tendon is sectioned. This makes it hardly possible to ascertain, under normal living conditions, the actual length of the relaxed muscle.

For practical purposes we may accept the definition of Fick[5] that the length of the completely relaxed muscle is one in which there is neither innervation, nor stretching by the antagonist.

Can a muscle contract actively beyond its terminal position?

We know that the fully contracted muscle still responds to electrical stimulation. The question is, can it display voluntary contractile power at terminal positions or is its power then completely exhausted? From an experiment of the Weber brothers, it seems likely that even at terminal positions some voluntary contractile power persists. He found that a dissected frog muscle still contracts to even one-sixth of its natural length, that is about 85%, which is certainly much beyond the maximum contraction shortening in the living.

One may then distinguish the following conditions regarding muscle length:

1. Length of the muscle in extreme stretching.
2. Length of the muscle when its antagonist assumes terminal position.
3. Natural length of the non-innervated muscle.
4. Length of the habitually contracted muscle.
5. Length of maximum contraction.

The natural length of a non-innervated muscle is used as reference both for distension and contraction length. It is estimated that in maximum contraction, that is in the isotonic state, the muscle loses up to 80% of the contractile power which it can display in isometric contraction.

IV. COORDINATION OF SKELETAL MUSCLE ACTION

The comparison of muscle action as discussed above only serves as an elementary introduction to the analysis of muscle activity. It does not reach far enough to analyze the intricate and complicated muscle movements as they occur in actual life. One reason for this is that the muscle contraction takes effect both at the points of origin and its insertion. In other words, it acts both peripherally and centrally. Another is that the effect of muscle contraction asserts itself not only upon the joint which it serves directly but also upon the neighboring and even upon the remote articulations. Thirdly, there are pluri-articular muscles which stretch across more than one joint and which have a distinct motor effect on all joints they cross. And finally gravity and the inertia of the moving parts must be included in the calculations of motor performance.

Beginning with a single uniarticular muscle the simple concept of leverage can be expressed in terms of the three coordinate system, which means that their leverage effect must be projected into three orientation planes to arrive at the resultant leverage effect produced by the muscle.

So far we have considered the muscle action only as occurring in a plane perpendicular to the axis of motion. For instance in flexion and extension of the elbow joint the analysis of the muscle action with its rotatory and stabilizing components was based on the proposition that all movement occurs in one plane, namely the sagittal plane perpendicular to the axis of the elbow joint and that force and lever arm both occupy the same sagittal plane. However, on examining joints with three degrees of freedom of motion such as shoulder and hip, we soon find that most muscles have their axes oblique to the orientation planes of the joint; in fact, most of these muscles are oblique in all three planes. Consequently, only a component of their strength is operative in any one of these planes. The problem is to calculate these components trigonometrically under the three coordinate system.

To make matters more complex, the respective coordinates for the different planes change with the position of the joint, and muscles may even reverse their action. For instance, in the case of the adductors of the hip, the flexion components of these muscles reverse themselves as the hip goes from extension into flexion. The exact points of these reversals differ for the respective adductor muscles. The adductor longus which starts as a secondary flexor becomes an extensor at 70 degrees flexion; the magnus reverses its action as secondary flexor into an extensor at 50 degrees.

A second prerequisite for analyzing muscle action is to calculate the resistance which the muscles must overcome in performing their task. We have so far only dealt with the lever arm as a purely mathematical concept; but the moving portion has a mass and as such offers an inert resistance to rotatory motion. The value of the mass of the body can be ascertained from its weight, the mass being weight or force of gravity over the gravitational acceleration $(m = \dfrac{F}{g})$ or to express it in more general terms, for any acceleration the mass is, according to Newton's second law, equal to the force over the acceleration $(m = \dfrac{F}{a})$, the latter (a) then equalling the force (F) over the mass (m) or $a = \dfrac{F}{m}$.

In translating this general law into rotatory motion which is merely a special case of motion, the force is replaced by the moment of rotation (R) and the mass is replaced by inertia (I), the acceleration (a) then being $a = \dfrac{R}{I}$.

The inertia which resists rotatory motion produced by a certain force can be computed. It is the sum of all mass points (m) and all their distances (r) from the center of motion, or, $\Sigma m\rho$ times this distance, ρ being the average distance of all mass points. The formula for inertia is therefore expressed as $m\rho^2$.

We can thus compute the acceleration generated by any rotating force $(a = \dfrac{R}{I})$ from the following data:

a. To determine the moment of rotation (R), we must know:
   1. The size of the rotating muscle, its force and direction.
   2. The distance of this force from the center of the axis of rotation.
b. For the inertia (I) we must know:

1. The average distance of all mass points of a body from the center of gravity which gives us the radius of gyration (ρ).
2. The mass of the body (m) which is its weight over the gravitational acceleration (g): $(m = \dfrac{w}{g})$.

The resistance of the body to rotation or the inertia (I) is then given by the product of the mass and the square of the radius of gyration $(\rho)^2$: $(I = m\rho^2)$.

To calculate such values as the radius of gyration, the location of the center of gravity or determining the inertia for the parts of the human body would be a difficult task if it were not for the expedient, mentioned above, of reducing the human body or its parts to regular geometrical bodies; for instance, the trunk to a cylinder, and the limb to a truncated cone. For the lower limb of the adult, O. Fischer[6] computed the center of gravity to lie 0.38 m. below the hip joint and the mass of the whole limb at 12 kg./g. The inertia of the lower limb was computed by Amar[1] to be 1,460 cm.[2] kg. For an adult weighing 65 kg., Amar found as follows: The trunk with a mass of $\dfrac{32.5 \text{ kg.}}{g}$ has a moment of inertia, I = 8,600 cm.[2] kg.

The whole upper limb with a mass of $\dfrac{4.2 \text{ kg.}}{g}$ has a moment of inertia, I = 300 cm.[2] kg.

These data apply only for the case that the angle which the axis of rotation forms with the length axis of the limb is a right angle or is perpendicular to the length axis, in which case the radii of gyration for cylindrical and conical bodies are greatest. They are smallest when the body rotates about its own length axis. For instance; for the thigh rotating at the angle of 0 with the length axis the radius of gyration is 4.56 cm., at 45 degrees it is 8.56 cm., at 90 degrees, that is perpendicular to the length axis of the limb, it rises to 11.22 cm. For the calf: at an angle of 0 the radius is 3.09, at 45 degrees it is 7.00 cm., and at 90 degrees it is 9.41 cm. For the upper arm: at angle 0 the radius is 2.77 cm., at 45 degrees it is 5.90 cm., and at 90 degrees, it is 7.87 cm. For the lower arm: at angle 0 the radius of gyration is 2.73 cm., at 45 degrees it is 7.9 cm., and at 90 degrees it is 10.84 cm. (Fischer[7]).

For the extremity as a whole, therefore, the moment of inertia for movement about axes going through the center of gravity is greatest for the axis which is vertical to the length axis of the limb because the radius of gyration is the largest; and it is smallest for rotation about the length axis of the limb because the radius of gyration now is smallest. To apply these data to the joint movements in the living, a further difficulty lies in the fact that in most instances the axis of rotation does not contain the center of gravity but the center of motion is situated at the end of the limb which naturally increases the moment of inertia. This means that another factor must be added to it, which consists of the mass of the limb (m) times the square of the distance (e) between the center of gravity and the axis of rotation (me²).

The equation for the inertia in this case is then $I = M\rho^2 + me^2$ where I is the inertia, m is the mass, ρ is the radii of gyration, and e is the distance of the joint center from the center of gravity.

No such increment exists in case the rotation occurs about a length axis since the latter already contains the center of gravity. Therefore e, the distance of the axis of rotation from the center of gravity equals 0, and the increment me² is 0. It is obvious, therefore, that the rotation of the limb about its length axis meets with the least resistance (Fig. 20).

We have a simple expedient to arrive at an approximate formula for the radii of gyration of cylindrical bodies, thereby determining the inertia resistance which these bodies offer against rotatory motion. There is a constant relation

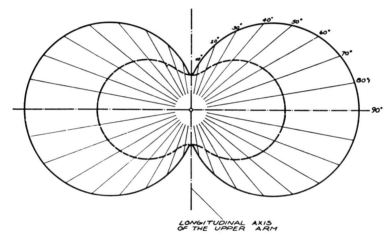

LONGITUDINAL AXIS
OF THE UPPER ARM

Fig. 20. Radii of giration of the upper arm for all axes going through the center of the humeral head, at different angles to the longitudinal axes, all lying in the same plane. (Baune and Fischer) The dotted curve shows the radii of giration of the upper arm for all axes going through its center of gravity.

existing between the length of the limb, the diameter of the limb, and the values of the radius of gyration for motion about transverse and longitudinal axes. For all motion about transverse axes going through the center of gravity the radius of gyration is about 0.3 times the length (L) of the limb. For all motion about the longitudinal axis the radius of gyration is about 0.35 times the diameter (D) of the limb.

Thus the radius of gyration (ρ) can be simply expressed in terms of length (L) of the limb for the transverse ($\rho = 0.3$ L) and in terms of diameter (D) of the limb for the longitudinal rotation ($\rho = 0.35$ D).

In the analysis of motion the inertia is essential in determining the external resistance. It is of particular importance to keep in mind that the inertia resistance is the smallest in all movements which go about a longitudinal axis. While this seems to be a purely theoretical consideration it has, nevertheless, an enormous practical application for all problems of locomotion of the human body. Thousands of details characterizing various phases and types of locomotion are clarified thereby. It explains why one can rotate a limb so very much easier and with much less expenditure of muscle power about a longitudinal axis than one can about a transverse axis of the same articulation. Length rotation of humerus or femur is easier than the ab or adduction of flexion or extension. Since such length rotatory movement can be carried out with so much greater facility, it is used extensively in the accumulation of momentum. Herein lies the answer to the numerous rotatory twists the body assumes during the gait. Most of all it explains the length rotatory movements, the "wind ups" which are practiced when a great accumulated momentum is to be imparted to the upper extremity, for instance, in baseball pitching, in hammer throwing, in golf and numerous other motor performances in athletic and sport events.

V. THE COORDINATE FUNCTION OF THE BIARTICULAR MUSCLES

It has been stated that a muscle in contraction exerts a rotatory effect on both ends. This effect is measured by the acceleration which is imparted to the movement. According to what has been explained above, this acceleration is inversely proportional to the mass of the moved part and hence to its inertia which is the product of the mass times the square of the radius of giration.

We recall that the formula for the rotatory acceleration is $(a = \dfrac{R}{I})$ where R is the moment of rotation (force times lever arm) and I the inertia. Let us, for instance, compare the values of the accelerations which the iliacus muscle develops in respect to the trunk on the one hand and the thigh on the other. Given an equal angle of application on both ends these rotatory moments should be equal and opposite but the acceleration produced by them differs greatly because of the inequality of the masses. The rotatory effect upon the greater mass will be commensurately smaller than that upon the smaller mass.

To visualize the situation one can mark on the peripheral lever arm the respective masses by the distance from the axis of the joint in inverse proportion to the masses. If mass A is ten times mass B, mass B is marked at a distance ten times that for mass A (Fig. 21). Thus the lever is divided by the two mass point in inverse ratio to the magnitude of the masses.

Let us assume a biarticular muscle such as the hamstrings is at work. The movement occurs: trunk against the thigh at the upper end and leg against the thigh at the lower end. Taking the mass of the leg and thigh on one hand and the mass of the rest of the body on the other, then the ratio between the two is 0.122 to 0.878. If we now take the length of the thigh as a unit and mark on it the points representing the lever arms in inverse ratio to the masses, the point representing leg and thigh would be 0.878 from the hip joint, and the one representing the rest of the body would be 0.122 (Fig. 21).

In practical terms this means that the hamstring produces the same rotatory moment at knee and pelvis providing that their lever arms are the same but the acceleration they develop on both ends is inversely proportional to the masses presented by the two lever arms.

In this case the total mass of the body being 1, that of the lower lever arm

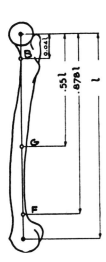

Fig. 21. The three cardinal points of the thigh. *G.* Center of gravity of the thigh. *B.* Common mass point representing rest of body. *F.* Common mass point representing foot and leg.

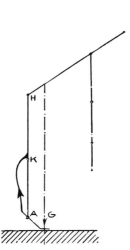

Fig. 22. Biarticular muscles. Gastrocnemius. Rotation imparted to knee and ankle by gravity neutralized by tension of the gastrocnemius.

Fig. 23. Biarticular muscles. The tension of the gastrocnemius neutralizes the lesser gravitational rotation moment at the knee (k), while the tension of the soleus neutralizes the greater gravitational rotation moment at the ankle (A). (Baeyer's analysis)

consisting of foot, leg and thigh, would be 0.122; that of the upper lever arm consisting of the rest of the body, including the other lower extremity would be 0.878 (W. Braune and O. Fischer[3]) (Fig. 21).

In short, the effect of the hamstrings on moving the entire body against the limb would be hardly 1/10 of its effect on moving the limb against the body, measured by the acceleration they produce at the two ends.

THE ECONOMY OF THE BIARTICULAR MUSCLES

In general, muscles which are spanned over two joints come into action when gravity rotates the two joints in opposite direction. For example, if the line of gravity falls in front of both knee and ankle joint, then gravity will extend the knee in one direction (say clockwise) and at the same time dorsiflex the ankle (say counter clockwise). Thus gravity imparts rotation to the knee and ankle joint in opposite direction. In this situation tension is imparted to the gastrocnemius at both ends as the muscle holds the equilibrium to gravitational stresses (Fig. 22).

The action of a biarticular muscle is distributed between the two ends in such a manner that the tension suffices to neutralize the lesser antagonistic gravitational moment at one end. In the knee the gastrocnemius at the upper end neutralizes the lesser rotatory moment of gravity thereby stabilizing the knee; at the lower end, however, the gastrocnemius together with the soleus overcome the greater rotational moment of gravity in respect to the ankle joint; the result is, generally, plantar flexion of the foot, provided always that the motion is not inhibited by some external resistance (Fig. 23).

The effect which contraction of a pluriarticular muscle has upon two or more joints naturally depends upon the relative rotation moments and the relative inertias, as has been shown above. There are many instances where visible motion is produced on both ends at the same time, although always on the principle of distribution of masses.

For the viewpoint of biarticular motion, one can therefore distinguish two situations. Either the two ends of the muscle approach each other in contraction in which case motion occurs in both joints, or else the muscle is held in check at one end by its antagonist and the joint at this end is stabilized, while motion is carried out in the other joint. In the first case the ends move together toward each other. This is called a countercurrent or syntactic movement (Baeyer[2]). In the second case the ends move parallel since as one moves up or down in contraction the other moves up or down in the same direction. This is called a concurrent or parallel shift (Baeyer[2]).

Here are some illustrations:

1. Simultaneous movement of the rectus femoris and the hamstrings: the rectus flexes the hip joint at the pelvis pulling the hamstrings upward, the hamstrings flex the knee pulling the rectus downward. This is an example of a parallel shift of the rectus against the hamstrings (Fig. 24a).

Or: the hamstrings extend the hip joint and pull the upper end of the rectus upward, the rectus extends the knee joint and thereby pulls the lower end of the hamstrings downward (Fig. 24b).

The effect is obvious. Both muscles retain their length; their action is isometric which, as explained above, is most favorable for display of tension. In fact this parallel or concurrent shift results in shortening and lengthening of the extremity where the force is most necessary. That does not mean that the lengthening and shortening operations of the extremities are carrid out by biarticular muscles alone; in fact the extremes of either flexory or extensory ranges are entrusted to monoarticular muscles (the vasti, the short head of the biceps femoris, the brachialis and so forth). But so far as biarticular muscles are concerned, the shortening and lengthening of the limb is procured by the con-current shift.

On the other hand we assume that the hamstrings and the long head of the biceps work at both ends and extend the hip and flex the knee at the same time; then, passively the upper end of the rectus is pulled up and the lower is pulled down. Thus the ends of the hamstrings move together by active contraction while the ends of the rectus move apart by passive stretching (Fig. 25).

Or the rectus flexes the hip and extends the knee simultaneously; in doing so it pulls the upper end of the hamstrings upward and the lower downward; in this case the ends of the rectus move actively together while the ends of the ham-

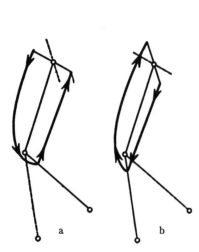

FIG. 24. Parallel shift of rectus down-ward, hamstrings upward and vice versa.

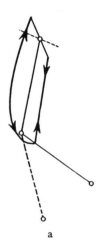

FIG. 25. Counter current shift; hamstrings con-tracted at both ends, rectus dis-tended at both ends.

strings are passively pulled apart. These are examples of countercurrent motion.

The effect of this type of muscle action is backward or forward swing. The contracture is now isotonic and not isometric; the tension displayed therefore is much less. Here again, forward and backward swing is not carried out by biarticular muscles alone; but so far as they are involved their mode of operation is countercurrently.

Instances like the above should give an insight into the economic plan which nature has laid out for the combinations of motion. The significance of these intricate schemes becomes still more evident under pathological, and especially under paralytic conditions.

One remaining question to be considered is what effect external resistance has on the muscle effort. But here again the situations are so variable and unpredictable for each single joint and even for every individual muscle that the purpose is served better by dealing with specific situations.

Gravity is a universal external resistance which affects not only the actions of a single muscle in producing motion in its proper joint against gravity, but it also has an indirect effect through the changes in position which the partial centers of gravity undergo when the shape of the limb changes. For instance, as the brachialis flexes the elbow, it shifts the common center of gravity of the upper arm-forearm system posteriorly and thereby produces an extensory effect upon the shoulder joint (Fig. 26).

Another effect of external resistance in a closed kinetic chain is to stabilize

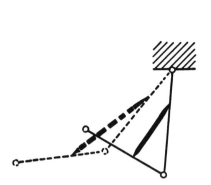

FIG. 26. Uniarticular muscle. External resistance (gravity). Brachialis anticus flexing the elbow and extending the shoulder.

FIG. 27. Uniarticular muscle; soleus. Closed kinetic chain (Baeyer). Plantar flexion resisted; extension of the knee.

the peripheral portions of the extremities while the central portions become the moving part of the system. This arrangement also has its effect on remote joints which are not within the compass of the contracting muscle.

For example if the soleus contracts but the plantar flexion of the foot is resisted by the floor (Fig. 27), the effect is extension of the knee, whereas if the plantar flexion is free so that the individual may rise on his toes, the effect is plantar flexion in the ankle joint (Fig. 28).

Again, if plantar flexion is free but rising on the toes is inhibited by the weight on the pelvis, then contraction of the soleus produces plantar flexion of the foot; at the same time it will flex the knee, corresponding with the increased plantar flexion (Fig. 29).

In a similar manner the hamstrings may produce unexpected locomotor effects. Let us assume the knee is flexed and prevented from extending by a super-incumbent load and that at the same time the ankle is in plantar flexion with the ball of the foot fixed by external resistance. This is the situation of a cyclist. The contracting hamstrings now will extend the knee at the same time forcing the heel down and the ankle in passive dorsiflexion (Fig. 30).

In the upper extremity similar situations arise. To mention only one: the biceps first flexes the elbow and brings the shoulder forward at the same time; with increased elbow flexion it adjusts itself to the backward displaced center of

FIG. 28. Uniarticular muscle; soleus. Closed kinetic chain (Baeyer), ball fixed: elevation and adduction of hip, extension of knee.

FIG. 29. Uniarticular muscle: soleus. Closed kinetic chain (Baeyer), ball fixed, pelvis fixed against rising: flexion of knee and hip.

FIG. 30. Biarticular muscle: hamstrings. Closed k i n e t i c chain (Baeyer), b all fixed, pelvis fixed against rising: extension of knee and dorsiflexion of foot.

gravity of the whole arm unit and the shoulder joint goes into backward extension.

However, if the motion of the forearm is fixed by external resistance or by a heavy load so that it cannot be flexed against the humerus, then the upper arm moves in flexion against the forearm and at the same time pulls the shoulder strongly forward.

These are only sporadic instances to indicate the intricacy of the effect of external resistances upon the function of the muscles. As the above named examples suggest, they bring about some surprising departures from the conventional concepts of muscle activity and yet they represent muscular performances in real life in ordinary static or dynamic situations.

VI. SUMMARY

a. Joints

1. Congruent and incongruent joints.
2. Generating joints surfaces by a line rotating about an extraneous axis:

a′ A straight rotating line generates cylindrical or conical joint surfaces.

b′ A helical surface is generated by a point moving both in circular and progressive direction.

c′ Oval surfaces are generated by a curve of less than a hemicircle rotating about its tendon.

d′ A saddle joint is generated by a hyperbola rotating about an imaginary axis.

3. Joint contacts.

a′ Ligament and muscle tension.

b′ Atmospheric pressure.

Action of sustaining weight of limb diminishes with altitude; vacuum translucencies appear in joints under traction.

4. Type of joint movements.

a′ Surface gliding: translatory: scapula: rotatory: intervertebral articulations.

b′ Linear gliding: metacarpo-phalangeal joints.

c′ Rocking motion: equidistant points touch each other (earlier ranges of knee flexion).

d′ Combinations of rocking and gliding (knee joint).

e′ Axial rotation: small contacting surfaces: radio-humeral articulation.

5. The degrees of freedom of motion are 1-3 according to the number of perpendicular planes in which motion can be carried out: ulno-humeral joint, 1 degree; knee and radio-humeral joint, 2 degrees; hip and shoulder joint, 3 degrees.

6. Kinetic chains: *open*—the terminal link is freely movable: waving hand; *closed*—the terminal link is fixed by external resistance and, if absolute, proximal part moves against distal.

7. Summation of freedoms of motion for the joints of the extremities:

a′ *Upper:* 11 degrees from fingers to shoulder girdle, 17 degrees from finger tips to trunk.

b′ *Lower:* 6 degrees from foot as a whole against the trunk.

8. The transformation of rotatory into translatory motion in a kinetic chain.

a′ A true translatory effect (shortening or lengthening of the extremity) is achieved by alternations in direction of rotation (clockwise and counter clockwise) in such a fashion that the algebraic sum of rotating angles in the different joints (plus and minus) adds up to zero.

b′ A true rotatory effect (flexion or extension) is achieved by the different joints rotating in the same sense (clockwise or counter clockwise) so that all angles of rotation have the same sign (plus or minus), the total degree of rotation then being the algebraic sum of these angles.

c′ An alternation of the direction of motion (clockwise and counter clockwise) in which the algebraic sum of the rotating angles (with alternating signs) does not sum up to zero but to an algebraic sum (plus or minus) represents a combination of shortening or lengthening with for or backward swing.

b. Muscles

1. The contracting muscle develops a tangential $(a_{tg} = \dfrac{v}{t})$ and a centrifugal $(a_n = \dfrac{v^2}{r})$ acceleration of the moved part: the first produces accelerated rotatory, the second accelerated centrifugal motion; the rotatory acceleration is checked by the antagonist of the rotating muscle; the centrifugal acceleration is checked by the anatomical cohesion of the system and by the stabilizing component of the rotating muscle. The latter is almost always greater than the rotating component, both being equal only when the angle of application of the muscle is 45° $(F_r = F \sin 45°; \; F_s = F \cos 45°; \; \sin 45° = \cosine 45° = 0.707)$.

2. Antagonistic suppression or control of rotation may also be furnished by gravity.

3. Rotatory equilibrium or stabilization of the joint may be on the principle of a lever of the first order (standing on foot, equilibrium between flexors and extensors of the ankle joint) or on the principle of a lever of the 2nd or 3rd order (standing on the ball of the foot, 3rd order, or flexing elbow against weight held in hand, 2nd order).

4. Angle of application increases with increased flexion; but tensile strength of the muscle decreases more rapidly with its shortening; hence extended position close to natural length remains most favorable for display of muscle power.

5. Natural provisions to increase the angle of application without decreasing muscle length are for example: insertion of patella into the quadriceps; the posterior projection of the femoral condyles for the hamstrings; the deflection of flexor tendons of the fingers in the carpal groove, etc.

6. Morphological adaptation of muscle: built for power—short, numerous fibers, pennated arrangement; built for speed—long and less numerous fibers, parallel arrangement.
   Mechanical adaptation of muscle: for power—attached at a greater distance from center to motion, greater lever arm; for speed—attached at a lesser distance, shorter lever arm.
   Muscle of same volume produces the same amount of work $(W = $ Force $\times$ distance). Those built for speed will cover a greater angular distance with less force; those built for power will cover a lesser distance with greater force.

7. Natural length, contractile distance and tension. There is a constant relation between natural length, contractile length and degree of rotation; each unit of shortening corresponds to a constant angular value of ro-

tation. There is no such arithmetical proportionality between contractile length and muscle tension because tension decreases more rapidly with increasing contraction. The absolute muscle power coefficient (3.6 kg. per cm.^{2} physiological cross section) applies only to natural length.

8. Coordination of muscle action.

a′ Obliquity of the muscle axis to the planes of the body (frontal, sagittal and transverse) results in the muscle acting upon the joint frequently in more than one plane; a muscle whose principal action is in one plane, therefore develops secondary action in other planes. In addition these secondary effects often change with the position of the joint. The adductor magnus is in normal position a secondary flexor of the hip but changes to an extensor when 50° of flexion is reached.

b′ Inertia as a factor resisting rotatory motion. Quite aside from gravity, the inertia of a body resist rotation. Inertia is related to gravity being the product of the mass of the body (mass being gravitational force or weight over terrestrial gravitational acceleration or $m = F/g$). For rotatory motion the acceleration produced by a rotating force equals the moment of rotation of this force R (force times lever arm) over the inertia I (I equals mass times square of the radius of gyration, which is the average distance of all mass points from the center of gravity, $I = m\rho^2$). The inertia of the different parts of the body has been determined: for the lower limb it is 1,460 cm.^{2} kg., for the trunk 8,600 cm.^{2} kg. The radius of gyration (ρ) is greatest for any body rotating about an axis going through the center of gravity and being at right angle to its length axis. It is smallest for rotation about the length axis. Hence also, the inertia (I) is smallest for length rotatory movements and the acceleration produced by it is, consequently, the greatest ($a = R/I$). This explains the great value of rotatory twists of the body in accumulating momentum, for example the "winding up" of the baseball pitcher, the disc thrower, etc. The computation of the radius of gyration for cylindrical bodies is greatly facilitated by the fact that for transverse rotation (through the center of gravity) it is computed to equal 0.3 times the length, and for length rotation 0.35 times the diameter of the cylindrical body.

9. Coordinate function of biarticular muscles. Muscles in contracting act on both ends producing accelerations inversely proportional to the masses to be moved. The hamstrings, for example, produce the same rotatory moment at knee and hip but the acceleration is in inverse proportion to the masses presented by both lever arms.

Muscles spanned over two joints come into action when gravitational stresses rotate the joints in opposite direction (line of gravity falling in front of knee and ankle joint; gravity extends the knee, for instance clockwise, and dorsiflexes the ankle joint counter clockwise).

In biarticular muscles, two dynamic situations are to be considered.

a' The two ends of the muscle approach each other in contraction: example, rectus contracting flexes hip and extends knee, while the hamstrings are distended at both ends; countercurrent shift.

b' The muscle is immobilized at one end by its antagonist. The muscle retains its original length. The contraction is isometric. Example: rectus flexes hip joint while hamstrings flex the knee; parallel or concurrent shift.

In lengthening and shortening, the biarticular antagonists undergo a parallel or concurrent; in for and backward swing a countercurrent shift.

10. The effect of external resistance on muscle action, closed kinetic chain: Examples:

a' Soleus contracting, plantar flexion resisted: extension of the knee; plantar flexion free: rising on toes.

b' Hamstrings contracting, knee flexed, ankle fixed prevented from plantar flexion: extension of the knee.

c' Forearm free, biceps first flexes elbow, then brings shoulder forward; with increased flexion, center of gravity of the unit is displaced backward and shoulder goes into backward extension.

d' Forearm fixed, upper arm flexes against forearm pulling shoulder strongly forward.

BIBLIOGRAPHY

1. AMAR, J.: *The Human Motor*. London, George Rutledge & Sons, 1920.
2. BAEYER, H. v.: Die Synapsis in der Algemeinen Gliedermechanik. *Report, 2nd Internat. Orth. Congress, London*, 1933.
3. BRAUNE, W., and FISCHER, O.: Bestimmung der Trägheitsmomente des Menschlichen Körpers. Vol. XVIII. Abhandl. d. Mathem. *Phys. Klasse der Königl. Sächs. Gesellschaft d. Wissench., VIII:*465, Leipzig, S. Hirzel, 1892.
4. FICK, A.: *Gesammelte Schriften, Vol. I*. Würzburg, Stahel, 1903.
5. FICK, R.: *Handb. d. Anatomie und Mechanik der Gelenke, Vol. II*. Jena, G. Fischer, 1911.
6. FISCHER, O.: *Kinematik Organischer Gelenke*. Braunschweig, F. Vierweg & Son, 1907.
7. ———: *Theoretische Grundlagen der Mechanik des lebenden Körpers*. Leipzig, Berlin, B. G. Teubner, 1906.
8. FUIKS, D.: Personal Communication.
9. GÜRTER, A.: Statische Betrachtungen der Muskulatur des Oberschenkels. A. Fick: *ges. Schriften, Vol. I,* Würzburg, Stahel, 1903.
10. STEINDLER, A.: *Mechanics of Normal and Pathological Motion in Man*. Springfield, Illinois, Charles C Thomas, Publisher, 1935.

Lecture VI

ON THE PATHOMECHANICS OF MUSCLE FUNCTION

I. FATIGUE AND RECOVERY

THE function of muscles may be impaired in two opposite directions. They may become *hypoactive* by changes interfering with the innervation or with the metabolic processes necessary to produce muscle tension; or they may become *hyperactive* due to hyperinnervation or, as it is called, innervational overload.

Representatives of the first group are the myopathies or neuropathic paralyses; of the second group, the cerebral palsy.

These conditions have their physiological counterpart: a state of hyperexcitability exists in the muscles from nervous tension; or a hypotonic state of the muscles results from overexertion.

Both of these physiological states represent essentially reversible conditions; but with proper intensity and duration they may ultimately assume pathological and irreversible changes; the hypertonus may develop contractures, while the hypotonus may end in a state of paralysis.

Fatigue can be defined from several angles: biochemically in terms of changes in the muscle metabolism; physiologically in terms of changes in individual muscle function; and physically in terms of changes in the performance of work.

a. Biochemically

Biochemically, fatigue means the accumulation of lactic acid in the muscles which has reached the point of seriously impeding muscular contractility. The removal of lactic acid from the muscle by the blood stream is inadequate. Acceleration of the circulation therefore retards fatigue. It has been found experimentally that the blood-fed frog muscles can be stimulated indefinitely if given a recovery period of six seconds between contractions. The accumulation of lactic acid in the muscle may reach a level of 10-15 times the normal resting value. Moderate exercise gradually increases the level of lactic acid. This is met by an appropriate increase of oxygen intake during and after exercise. This oxygen intake falls again to a normal level relatively soon after the exercise has ceased. In violent exercise the accumulation of lactic acid is more rapid. The circulation becomes more inadequate to carry off the surplus lactic acid and even the increased oxygen intake by rapid respiration may not be sufficient. Then oxygen is drawn from the circulatory system in the tissues and this so-called oxygen debt is slowly paid off by a long continued rise of the respiratory exchange.

The anatomic seat of fatigue are the motor end plates in the muscle fibers, but the fatigue products soon diffuse through the whole muscle. During muscle contraction lactic acid is the essential part in the cycle of oxidation from glycogen to its end products. From the thermodynamic point of view, 1 gram of lactic acid, when completely oxidized, represents 370 gram calories. But only a smaller part of lactic acid is completely oxydized; a greater part is resynthetized in the liver from where it originally came, to oxyglycogen (Hill[4]).

The efficiency of the muscle can be measured by the ratio between the lactic acid which is removed for resynthetization to the lactic acid which is completely oxidized. This index is normally as 5:1 (Meyerhoff[5]).

It has been mentioned that in severe exercise lactic acid accumulates rapidly (up to 100-200 mg. per 100 cc. of blood) and that a sharply increasing oxygen debt is incurred. The removal of highly concentrated lactic acid is a compara-

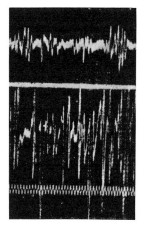

FIG. 1. Action current in fatigue. "Peloton fire" type of motor impulses. (Wachholder)

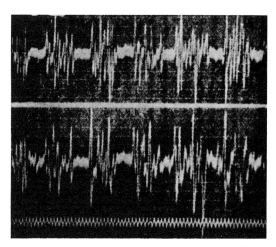

FIG. 2. Action current in approaching exhaustion. "Salvo type" of impulses. (Wachholder)

tively slow process. For instance, it may take several hours to remove a concentration of 100 mg.% of lactic acid which may have been stored up by a few minutes of violent exercise.

b. The physiological definition of fatigue

We recall that prolonged contraction requires a sort of rotational arrangement between the muscle fibers, by which one set of fibers relieves the others in turn, each fiber when innervated contracting at maximum according to Sherrington's law[10] of all or none reaction.

Only by alternating the different groups of muscle fibers in rotation is it possible to effect a sustained contraction. In this connection it is of interest to recall that the plurisegmental innervation of the muscles by the spinal nerves has an effect on the control of fatigue. If one spinal root is stimulated to exhaustion, the muscle still responds to stimulation of the other roots. The product of fatigue remains confined for a time to the group of end plates supplied by the same root. However, when the circulation of the muscle is impeded or stagnates, then a diffusion of the fatigue products takes place and complete exhaustion results.

The response of muscle to fatigue can be seen clearly in the action current picture. As motion becomes more difficult, the action current waves not only become more irregular but also more crowded (Fig. 1). The discharges of motor impulses resemble the peloton firing of a machine gun. This higher frequency and voltage can be noted in chronic muscle spasm which is comparable to a muscle entering a state of fatigue. Finally the limit of contractile capacity is reached. Then we note a peculiar grouping of the impulses with accompanying clonicity of contraction; this is the tremor one sees in muscles as they become exhausted. In addition to this grouping the waves assume higher amplitudes. The action current picture now resembles more a "salvo" fire (Fig. 2). In general the action current picture shows that, as external resistances increase, the muscle tension increases and the waves assume higher amplitudes.

The grouping of the waves and their high voltage indicate that a majority of the available muscle fibers participate in the contraction while the intermissions indicate periods of rest for these muscle fibers (Wachholder[11]).

One notices also that the tension of the muscles increases much more rapidly than the external resistance; and correspondingly also the metabolism of the muscle increases out of proportion with the resistance.

Finally as the stage of exhaustion is approached, the amplitudes of the waves again fall. The action current becomes more feeble until there is no longer any response to stimulation. The muscle appears paralyzed.

c. The physical definition

In terms of work and efficiency fatigue can be defined as a condition in which the performance of a certain amount of work meets with increasing difficulties and is carried out with a decreasing effect. Subjectively there is a feeling of locomotor inhibition which finally leads to complete muscular impotence. It takes greater and greater will power to perform a certain task; the precision of the locomotor act gradually decreases.

It is not the locomotor function alone which falls victim to fatigue. All systems are affected. The pulse rate is increased, the blood pressure rises with the increase of exertion, respiration becomes more frequent. All this means an increase of gas exchange which is first in proportion to the dynamic value of the muscular effects, but very soon increases out of proportion with the external resistance, the same as muscle tension increases out of proportion to the external load.

The motor performance becomes more and more uneconomical. Bernstein and Poher[1] express the situation as follows: "The expenditure in static work does not increase in proportion to the work (that is arithmetically) but it increases more rapidly."

Those engaged in physical education may draw from this the lesson that nothing is gained for the development of muscle strength if exercises are pushed to or even beyond the point of fatigue.

d. Recovery from fatigue

Hill and Long[4] divide fatigue recovery into two stages:

1. An early stage where the rapid fall of oxygen intake marks the quick removal of lactic acid from the muscle.

2. A second stage in which the carbohydrate stores which had been depleted by the previous exercise are built up. This part of recovery is the slowest and may require several hours. During severe exercise the lactic acid concentration becomes high and the removal of lactic acid is likewise slow and recovery prolonged.

There is evidence that the sympathetic nervous system has something to do with fatigue. When the muscle is stimulated electrically, the fatigue produced is less if the sympathetic system is stimulated at the same time; the action of the sympathetic system therefore retards fatigue probably by its adrenolergic effect (Orbeli[6]).

e. The prevention and treatment of fatigue

1. PREVENTION

Of particular interest for the physical trainer is the effect of diet on fatigue. A diet of fat reduces the amount of lactic acid produced by muscle contraction; also if by electric stimulation a maximum of fatigue is obtained, the blood contains only half the amount of lactic acid compared with carbohydrate feeding. The conclusion is that carbohydrates are the principal source of muscle energy and subjects fed on fat are much more easily fatigued because of the difficulty of producing, storing and oxidizing lactic acid.

It follows that efforts to prevent fatigue, especially in athletes, must first be concerned with an appropriate diet suitable to supply with the greatest ease the necessary amount of lactic acid to keep the threshold of fatigue high.

2. TREATMENT OF FATIGUE

The management of fatigue revolves around three points: rest, relaxation, and the reestablishment of normal blood circulation in the muscle.

The effect of fatigue on motor responses is best shown in the poliomyelitic muscles in which, because of loss of a large amount of muscle fibers, the fatigue threshold may be extremely low. If such a muscle is subject to stretching or contraction it easily becomes exhausted (Proebster[8]) under the influence of the nervous system. But, from the viewpoint of circulation, movement enhances lactic acid removal. Apparently the postulates of rest for the paralytic muscle from the nervous point of view and that of movement in the interest of circulation are contradictory. Inability to appreciate the proper course between these two apparently contradictory requirements has led to much confusion in the treatment of anterior poliomyelitis.

The principal requirement for the fatigued muscle is rest. The healthy muscle shows a remarkable ability to recover, particularly if the circulation is enhanced by massage and graded exercise, though the latter should be reserved until the acute stage of fatigue has passed.

In certain muscular affections such as muscular dystrophy, fatigue occurs early but recovery also occurs relatively soon.

In other conditions, particularly in peripheral paralysis and in poliomyelitis, the threshold of fatigue is also low and it occurs early, but recovery is likewise slow and protracted.

There is a great difference in the energy expenditure between the isometric and isotonic contraction. In the isotonic contraction the muscle shortens and in doing so performs visible work. However, because of the constantly decreasing tension the work accomplished by isotonic contraction amounts to only one-third of the theoretical maximum based on energy consumption.

In the isometric contraction the muscle does not shorten and no visible work is performed. Its contraction is balanced by external resistance. In this case the muscle develops its maximum tension. Chauveau[2] sums up the situation as follows:

Length being constant (that is isometric contraction), the energy expended by sustaining weight grows proportionally with the weight; weight being constant (that is isotonic contraction) the energy expended grows out of proportion with the degree of shortening.

These facts have a great bearing on the technique of the reeducation and development of muscle in physical education and in physical therapy. Exercises are most useful under isometric conditions where the muscle is held to its optimum length by an external resistance and is not allowed to shorten. Here it displays its maximum strength.

But this applies only to the point to which the muscle can resist an external force. When the force goes beyond it, then the muscle becomes passively stretched and is no longer in optimum position. This is an important point to remember when applying either passive or active resistance exercises.

One should also be aware of the fact that when a muscle contracts, either free or against resistance, its efficiency increases the slower and the more prolonged the contraction is. The best time for a single active contraction is about 1½ seconds.

It has been found that the efficiency coefficient which indicates the ratio of external work to energy consumption (gasometric studies) for the normal muscle is bout 26%.

Its maximum contractile power is somewhat beyond the resting length, usually in extreme joint position; for instance for the flexors of the elbow in full extension and vice versa for the extensors.

Early signs of fatigue manifest themselves in the action current picture by the decrease of amplitude of the waves; they also manifest themselves clinically both by loss of precision and by increase of effort. They thus show a mounting efficiency loss which is reflected in the increase of energy expenditure.

f. Constitutional and general pathological conditions facilitating fatigue

1. *Respiratory difficulties,* bronchial catarrh, bronchiectasis, asthma and tuberculosis, or any condition which impairs respiratory exchange facilitate fatigue. In fact fatigue may be one of the early signs of a beginning pulmonary tuberculosis and it is regularly found in cases of pneumothorax.

To this group also belong the respiratory difficulties of purely mechanical character, especially those due to weakness of the anterior abdominal wall. Here it is the expiratory back stroke which is particularly impaired. The vital capacity of the lungs is lowered and breathing is shallow. People with visceroptosis fatigue easily and complain of a constant feeling of tiredness which is not fully relieved by a night's rest.

2. Among the *circulatory conditions* which enhance fatigue, such conditions as arteriosclerotic disease, valvular heart lesions or myocarditis should be mentioned. The fatigue encountered in anemia and chlorosis is of particular diagnostic significance.

3. That deficiency or dysfunction of *endocrine nature,* especially hypothyroidism is accompanied by fatigue, is well known. Perhaps the most striking example is the extreme fatigue observed in Addison's disease or tuberculosis of the adrenal glands which is associated by a drastic fall of blood pressure.

4. The outstanding example of *metabolic diseases* where early fatigue is a prominent clinical sign is diabetes.

High altitudes produce in normal individuals temporary fatigue until an adjustment has taken place principally by increase of the red blood cells. However, in people with the above-mentioned general conditions, such an adaptation may not take place and the load upon the heart may be too great for safety.

The reduced oxygen intake associated with high altitudes has some effect upon vital processes; it causes a decrease of oxygen tension in the lung alveoli and with it goes a decrease of oxygen tension in the blood and tissues. This means early fatigue and a decrease of working efficiency. There are people, aside from those afflicted with disturbances of the respiratory or circulatory nature, who never make a complete adjustment; older people or people with a delicate constitution. In general, however, the adjustment of well people is prompt and fatigue is only a temporary phenomenon.

II. THE THEORY OF CONTRACTURES

a. Myogenic contractures

INTRODUCTION

Definition. Contractures, bizarre as they may seem, do not occur haphazardly. They have a certain regularity and consistency in their development which indicates that they are governed by certain biological and physical laws.

These laws are largely vested in the physical properties of the muscle itself. First of all, one should distinguish between contraction and contracture. Contraction is a voluntary change in muscle tone and length; contracture is a permanent state of muscle length different from the normal. Contraction is an active process; contracture is a passive state.

The physiology and mechanics of contraction have been dealt with in previous lectures; we are now engaged in the analysis of contracture which is entirely a pathological condition.

Even under the physiological conditions of contraction is it not easy to set up numerical standards for such factors as elasticity, viscosity and contractile power. It is much more difficult, if not impossible, to do so in the pathological conditions of contracture where the fluctuations of the physical properties of the muscle are infinitely greater.

Classification. From the standpoint of the physical properties of the muscle, contractures can be divided into four classes.

1. The contracture represents merely a *dynamic imbalance*. This imbalance is not due to a diseased state of the contracted muscle itself, but it is merely the result of rearranged distribution of muscle power about the joint, necessitated by the fact that a portion of the normal muscles about the joint has lost its contractile power due to paralysis. The normal equilibrium is therefore overthrown and a new and pathological equilibrium is taking its place. This type of contracture is seen in paralytic conditions. The contracture merely represents the new equilibrium; those muscles which no longer are opposed by antagonists because the latter are paralyzed, contract until the passive tension of the paralyzed antagonist sets a halt to further contracture. The contracted muscle itself is not changed either in its histological or its physical properties. Its tone is still within the physiological limits and the malposition of the joints arises solely from the fact that the muscles are no longer opposed by their antagonists.

2. In a second group it is not the muscle but its innervational element which is pathological. The normal distribution of muscle power is destroyed because part of the muscles are under an innervational overload. Inhibitions which normally regulate muscle tone and which safeguard the equilibrium have ceased to function, at least for a portion of the musculature surrounding the joint. This is called spastic contracture. If only a portion of the muscle around the joint is involved while the others are normal, the developing contracture represents the new muscle equilibrium. Here again the antagonist sets a halt to the contraction of the spastic muscle. The contracture position is more conspicuous because the spastic muscle is overinnervated and hence in a state of excessive tension; if all muscles about the joint are spastic one group of muscles usually shows a higher degree of spasticity than the other and the normal muscle equilibrium is overthrown. Or it may occur that one group of muscles is spastic while their antagonists are in a state of flaccid paralysis.

What is commonly called spastic contracture is by no means a pathological entity. In one case it may be a symptom of a cerebro-spinal lesion, in another a reflex contracture following injury such as fracture, in a third it may be a reflex to an inflammatory condition of the joint. The underlying pathological condition ultimately determines the course of the contracture.

3. In a third group of cases the contracture is not due to either muscular imbalance or to spasticity, but to intrinsic changes in the structure of the muscle itself: degenerative changes with myositic infiltration and fibrosis of the muscle. In such a case it is impossible to state even in approximate terms what these histological changes mean for the elasticity and contractility of such a muscle. All one can say is that these qualities are largely lost and one has to depend upon empirical knowledge to determine how much of the function of the muscle can be salvaged. This group is presented by the so-called myositic contractures of which the ischemic contracture is an example.

4. There is a fourth type which is really a combination of spastic and myositic contracture inasmuch as it starts with a spasticity and ends with myositic fibrosis. We refer here particularly to the reflex contracture as seen in arthritic joints or in other inflammatory joint conditions. In the earlier stages the muscle may be entirely spastic, but in the course of time structural shortening develops. Since we know that the spastic contracture can be managed by mechanical means while the myositic contracture cannot, early recognition and promptness of treatment is of utmost importance.

III. THE PROPENSITY AND ADAPTATION TO CONTRACTURE

There is a constitutional factor in the ability of an individual muscle to adapt itself to the changed mechanical conditions.

We recall that normally there is an individual difference in the tension at which the various muscles are spanned across the joints. Some individuals are loosely knit and in these the antagonist relaxes easily as the joint moves in opposite directions. Some others are tightly knit, and in these individuals the antagonist relaxes slowly.

Normally, also, there is a difference between the various joints of the body. The ankle joint, for instance, relaxes its antagonist slower than the wrist joint. On the whole, the loose joints have a greater amplitude of motion while the tighter joints display greater strength within their reduced amplitude. This length-amplitude and amplitude-strength relation is expressed mathematically:

1. The equation: $\dfrac{Lmax - Lmin}{Lnatural}$ indicates the range. The greater this difference between the terminal positions in relation to the natural length, the greater is the amplitude of the range.

2. On the other hand, the strength displayed by a muscle within this range is the efficiency expressed by the reciprocal of this equation, namely, $\dfrac{Lnatural}{Lmax - Lmin}$ = efficiency. As one can see, the lesser the difference in relation to the natural length, the greater is the efficiency. This becomes clear as one considers that the lesser the difference, the closer the muscle remains to its natural length at the end of contracture.

Now in contracture this process of adaptation sets in gradually to the degree in which the spastic tonus of the earlier stages disappears. The adaptation is necessary in the interest of the decreased range of motion. If the muscle were to retain its maximum length in spite of the decreased range, its control over this range would not be as forceful. The muscle would be too long and its optimum range for contracture would be consumed in taking up the slack.

In the paralytic and spastic contracture the normal shortened muscle can be brought out to natural length by an extending force because it has not suffered any change in its elasticity. The latter depends upon the sarcolemma and the connective tissue sheath which envelops the muscle fiber (Proebster[7, 9]).

In spasticity it is the contractural element which yields to stretching more in accordance to Hooke's law, the same as the voluntarily contracted muscle does. Once the spastic element is overcome and the muscle is again pulled out to its natural length, it offers the same elastic resistance as the non-innervated muscle. We have learned that the paralyzed antagonist of the contracted muscle can be stretched passively to 1.6 times its natural length at which time the breaking point is reached. This is the same as in a normal relaxed muscle. However distension beyond the natural length requires an increasing extension force. So long as the pathological change is a simple atrophy of the muscle fibers, this elasticity is not changed.

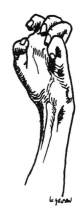

Fig. 3. Deformity of the hand following ischemic contracture.

The adaptation of a muscle to its state of contracture is a physiological process. For instance when the tuber of the os calcis is shortened in an animal so that the calf muscles control the same arc of motion by a lesser contractile distance, Marley found that the muscle belly of the gastrocnemius becomes shorter and the tendo Achillis corresponding longer.

Nature installs this process of physiological shortening of the muscle in adaptation to the reduced range of motion or the reduced contraction length in order to make the muscle more efficient in its reduced range, that is, to make it work with greater force.

Several practical lessons can be drawn from these facts:

1. That purely contractional (versus contractural) deformity such as the spastic are thoroughly amendable to correction by mechanical means, that is, by elastic traction, weight and pulley, turnbuckle cast, and that the force necessary to accomplish it is less than the absolute muscle power of these muscles.

2. That the extending force should be continuous since it becomes more effective as the contracted muscle fatigues and gives more readily. This can best be demonstrated in arthritic contractions which straighten out gradually under persistent traction.

3. In the myositic contracture, active resistance of the muscle is no longer a factor since the muscle is inert. Whatever is left of its natural elasticity

may be used, but it is very little. Mechanical means are ineffective. In most instances surgical measures of muscle lengthening have to be resorted to. However prevention of these contractures by early splinting may accomplish much. A good example is the ischemic contracture (Fig. 3).

IV. CONTRACTURE OF THE NON-CONTRACTILE SOFT TISSUES

a. Tendon resistance

One cannot speak of a contracted tendon because the tendon is a non-contractile tissue. However in muscular contracture the resistance to stretching is transmitted to the tendon and it is therefore important to know to what extent the tendon as such will resist stretching.

We recall that the tendon sustains a great deal more tension per unit of cross section than the muscle it serves because the stress of the muscle is becoming transmitted to a structure of much smaller cross section area. In some instances the ratio has been computed as being 1:60.

Therefore, if the muscle has its maximum strength computed at 3.6 kg. cm.^{2} or 0.036 kg. per mm.^{2} the tendon of this muscle has to sustain a tension sometimes as high as 2.18 kg. per mm.^{2} when the muscle goes into maximal contraction.

Why is it then that tendon ruptures are not uncommon? It is because degenerative changes occur early in the tendon and greatly reduce the stress resistance. In treating contracture by mechanical means one must keep in mind that the tendon may have degenerated and may have lost a great deal of its natural tensile strength.

The tendon rupture is usually situated at the musculotendinous junction or at the insertion and not in the tendon itself. The latter plays no part in the muscle contracture which follows rupture.

b. The role of capsular, fascial and ligamentous structures in contracture

Studying the elastic properties of ligaments, Fessler[3] found that the ligaments are strongest in the middle and weakest at their insertion. He determined the tensile strength of the fibular collateral ligament at 6.5 kg. per mm.^{2} which is close to Triepel's values of 5 kg. per mm.^{2} for the tendon. The bone resistance to tension in transverse direction is only 4 kg. per mm.^{2} It follows that the ligaments, at least in young individuals, offer stronger resistance with the result that deformation and even avulsion of bone may occur before the ligament tears. The congenital clubfoot is an example. Forcible manipulation may produce deformation of bone without breaking the ligaments.

It is of great clinical interest that the elasticity of ligaments is incomplete in the sense that after forced stretching, but before breaking, a certain elongation remains. According to Fessler[3] only about one-fourth to one-half of the elongation disappears after forceful stretching.

It may be that under immobilization the ligament recovers its elasticity. Yet it is certain that repeated and strong strains of the reinforcing ligaments which occur during corrective manipulations of contractures are apt to result in permanent relaxation and to loss of strength of these ligaments, which is added reason why correction of contracture should be gentle and gradual.

Forceful stretching results often in tears. When these are repaired by scar tissue the elastic yield of the ligament is greatly impaired. Fascial structures also share in the formation of contractures and the shrinkage of ligaments, and fascia constitutes a considerable obstacle to correction which is equally difficult to overcome as is the resistance of the fibrosed muscle itself.

Similarly as in the case of the myositic contractures, surgical measures become necessary to release the ligamentous and fascial component of contractures.

V. SUMMARY

a. Fatigue

1. Biochemicaly fatigue is the accumulation of lactic acid in the muscles, impeding muscular contractility. It may reach 10-15 times the resting value. The anatomical seat of fatigue are the motor end plates. The efficiency of the muscle is measured by the ratio between lactic acid removed for synthetization in the liver (to glycogen) and lactic acid completely synthesized, which is normally 5:1.

2. Physiologically, fatigue expresses itself in a change in the action current picture in that voltage and frequency become higher and the pattern changes from the "peloton" to the "salvo" fire type. This indicates a more massive grouping of muscle fibers contracting at the same time and is clinically expressed by the fatigue tremor. As the stage of exhaustion is approached the amplitudes again become lower.

3. Physically fatigue offers increasing difficulties in performance of work, requiring more and more effort and being carried out with less and less precision. The motor performance becomes less economical.

4. Recovery from fatigue shows an early stage with rapid fall of oxygen intake and a second stage in which the depleted carbohydrate stores are built up and which is much slower, requiring several hours. The stimulation of the sympathetic system retards fatigue.

5. Aside from the avoidance of excessive exercises, a good means to elevate the threshold of fatigue in the normal is proper diet. This should consist largely of carbohydrates. The management of fatigue centers around rest, relaxation and reestablishment of normal blood circulation in the muscle. Healthy muscle shows a remarkable ability to recover under such regime. In muscular dystrophies fatigue occurs early but recovery also sets in comparatively soon. In peripheral paralysis and in poliomyelitis fatigue comes early and recovery is slow.

6. There is a great difference in energy expenditure between isotonic and isometric contraction, the work accomplished by the former amounting only

to one-third of the theoretical maximum based on energy consumption.

7. Constitutional and general pathological conditions facilitate fatigue. Among the latter are: respiratory difficulties and obstructions, circulatory conditions, endocrine and metabolic disturbances. High altitudes cause fatigue because of the diminished oxygen intake.

b. Contractures

1. Dynamic imbalance contractures represent the overthrow of the normal muscular equilibrium in paralytic condition when a portion of the musculature is eliminated and the joint assumes a new, pathological position of equilibrium.

2. Spastic contractures are due to an innervational overload of part of the joint masculature. Spastic contractures may be of cerebral (as in cerebral palsy) or peripheral origin (as in fractures).

3. Myositic contractures are due to interstitial myositic changes. While imbalance and spastic contractures are highly amenable to correction by passive stretching, the myositic contractures are not.

4. A combination of spastic and myositic contractures are the reflex contractures of inflamed joints. They are spastic in the beginning but become structural in time.

5. There is a constitutional propensity to contractures. It has to do with the natural tightness which varies with the joints of different individuals and also with the different joints of the body. This tightness depends on how quickly the antagonist relaxes when the joint goes into opposite movement.

On the whole the loose joints have the greater amplitude, the tight joints the greater strength. The difference between maximum and minimum length of the muscle in relation to its natural length $\dfrac{\text{Lmax—Lmin}}{\text{Lnatural}}$ indicates the range. The index for strength and efficiency of the muscle is the reciprocal of the above equation, namely $\dfrac{\text{Lnatural}}{\text{Lmax—Lmin}}$. It indicates that the lesser the difference the greater is the efficiency, because the muscle remains closer to the natural length throughout its contraction. The adaptation of a muscle to its state of contracture is a physiological process in which the muscle reduces its contraction range to conform with the reduced range of the joint.

6. Practical conclusions are as follows:

  a′ purely contractional (versus contractural) deformities are amenable to correction by mechanical forces lesser than the absolute muscle power of these muscles.

  b′ The correcting force should be continuous as the fatigued contracted muscle yields more readily.

  c′ In the myositic contracture much less can be expected, if anything at all, by correcting forces.

7. Tendon does not contract, but its unit resistance to stretching is many times that of the muscle.

8. Fascial and ligamentous structures undergo inflammatory shrinkage which offers great obstacles to the correction of contractures and usually operative measures are necessary to release them. The unit tensile strength of ligaments often exceeds the tension resistance of bone and deformation of the latter may be produced without breaking the ligaments.

9. Repeated forceful stretching of ligaments impairs their elasticity and permanent lengthening and relaxation results.

BIBLIOGRAPHY

1. BERNSTEIN and POHER: *Pflüger's Arch. Physiol., 95:*146, 1903.
2. CHAUVEAU, A.: *Compt. rend. Acad. sc.,* Paris, 1904.
3. FESSLER, J.: Festigkeit des Menschlichen Gelenkes mit Besonderer Berücksichtigung des Bandapparates. Munich, 1893.
4. HILL, A. V., and LONG, C. N. H.: Muscular Exercise, Lactic Acid and the Supply and Utilization of Oxygen. *Ergebn. d. Physiol., 24:*43, 1925.
5. MEYERHOFF, O.: Die Energieumwandlung im Muskel. *Pflüger's Arch. Physiol., 191:*128.
6. ORBELI, L. A.: Die Sympathische Innervation der Skelettmuskeln. *J. Petrograd Med. Instit.,* 6:8. (See also abstract in *Med. Sc. Abstr. & Rev., 10:*486.)
7. PROEBSTER, R.: *Verhandl. d. Deutsch. Gesellsch. f. Orth.* 23rd Congress, 1929.
8. ———: Ueber Muskelaktionsströme am gesunden und kranken Menschen. *Ztschr. f. Orth. Chir.,* 50.
9. ———: *Ztschr. f. Orthop. Chir., 48:*451, 1928.
10. SHERRINGTON, C.: *Proc. Roy. Soc. London,* 52-81B.
11. WACHHOLDER: *Willkürliche Haltung und Bewegung.* Muenchen, J. F. Bergmann, 1928.

Lecture VII

ON BODY BALANCE AND BODY EQUILIBRIUM

I. INTRODUCTION

O UR previous contention that human locomotion is a special case of general mechanics subject to the same laws should now be followed through by a statement of these laws, and by demonstrating how they apply to body balance and body equilibrium.

1. Under the first law of Newton[6] a body is in equilibrium if no free force is acting upon it. If it is at rest it remains so; if it is in uniform motion it persists in uniform motion.

2. Under the third Newtonian law, a body is in equilibrium if all forces acting upon it mutually neutralize each other both in intensity and in direction so that the resulting force is zero.

3. Under the second Newtonian law, a force (F) acting upon a body imparts to it an acceleration (a) which is proportional to the force both in intensity and direction and is also inversely proportional to the mass (m) of the body $(a = \dfrac{F}{m})$.

Thus the mass of the body appears to be a purely quantitative term $(m = \dfrac{F}{a})$. The mass of any body under terrestial conditions does not change. It is the acceleration the force gives the mass as it sets it in motion, which determines the degree of force $(F = am)$. To apply this to terrestial conditions, the force is given by the weight of the body. This force is the product of the terrestial acceleration (or 980 cm. per second) times the mass of the body. The celestial bodies, moon, planets, etc. each have their own gravitational acceleration which is different from that of the earth. Consequently a body of the same mass will have a different weight on each of these celestial bodies.

The determination of the weight is an application of the third Newtonian law by which a body is in equilibrium if a force applied to it is met by a counter force of equal size and opposite direction. The weight on the scale supplies the counter force; the point at which it equals the gravitational force indicates the weight of the body.

For purposes of computation we may conceive the mass of the body being concentrated in one point which is called the center of gravity. A line passing through the center to the center of the earth would be the line of gravity of this body.

All forces applying to a body impart to it an acceleration not only in the translatory direction, that is, in the direction of the applying force but also transmit to it a rotatory acceleration which tries to rotate the body around its center of gravity. The rotatory movement of such a motion is proportional to the amount of force which applies and to the distance of the direction of this force from the center of gravity of the body. Hence, if the direction of the force goes through the center of gravity, no rotary moment can be created by it because the distance would be zero.

II. APPLYING OF THE ABOVE-MENTIONED GENERAL LAWS TO THE HUMAN BODY

A. TRANSLATORY EFFECT OF THE FORCE OF GRAVITY

The balance of the human body or its parts requires that all gravitational forces be completely neutralized by counter forces.

These counter forces are supplied by the resistance of the supporting surface of the body. As long as all gravitational forces fall into, and are intercepted by such supporting surfaces, the translatory force of gravity is neutralized. In order to neutralize also the rotatory forces it is further necessary that the line of gravity goes through the center of gravity which represents the gravitational forces applying to all mass points of the body, and that the line of gravity strikes the supporting surface and does not fall outside of it.

A body may be inclined and still be in gravitational equilibrium so long as the supporting surface is large enough to intercept the line of gravity of this body. The Leaning Tower of Pisa is an example (Fig. 1).

It has been possible to determine experimentally the center as well as the line of gravity of the human body and of all its parts.

B. THE ROTATORY EFFECT OF GRAVITY

If the applying force is not in line with the line of gravity it displays, as we mentioned above, a rotatory effect on the mass points of the body around the center of gravity. Pushing a table forward by one edge not only causes it to move forward but it also rotates the table around its middle where the center of gravity is located. In a solid mass this rotation movement is resisted by the inertia (I) of the mass which, as described above, equals the mass (m) times the square of the average distance (ρ) of all mass points from the center of gravity ($I = m\rho^2$). It has been possible, also, to determine this inertia for the human body and its parts by approximating their shapes to geometrical bodies (spheres, cylinders, truncated cones), an expedience which we have alluded to before as being permissible within the required limits of accuracy.

C. THE CENTER OF GRAVITY OF THE HUMAN BODY AS A WHOLE

For obvious reasons the determination of the center of gravity has engaged interest of anatomists and kinesiologists for centuries. There is a manifest difference in its position between biped and quadruped. But even among the quadrupeds there are great variations which depend upon their build and their locomotor activities.

Following it from the quadruped to the *biped* we notice that the change in the site of the weight center runs parallel with the evolution of the upright position. The ultimate acquisition of the orthograde type of locomotion involved the lifting of the weight center until it came to lie perpendicularly above the small supporting surface which is formed by the contact area of the lower extremities with the floor.

In contrast, the equilibrium of the *quadruped* is stable. Its weight center

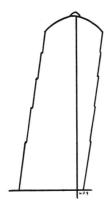

FIG. 1. An inclined body in gravitational equilibrium as the supporting surface is large enough to intercept line of gravity of body.

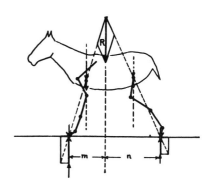

FIG. 2. Balance diagram of the horse. Center and line of gravity. Rotatory forces active at the joints of the extremities (redrawn from Strasser).

rests on the four extremities. However it is not exactly an automatic equilibrium because the extremities are articulated and because the line of gravity does not go to the center of the joints. It therefore develops rotatory moments which must be neutralized by muscle action. The equilibrium is therefore still an active one. Yet it is infinitely easier for the quadruped to maintain the body equilibrium in upright position than it is for man (Fig. 2). As the latter attained the bipedal stage the center of gravity had to become more elevated until it rested over the comparatively small area of the bipedal support.

There are two factors which influence stability. First, the broader the supporting area the greater is the force necessary to destroy the balance by throwing the line of gravity beyond the supporting surface (Fig. 3).

Secondly, the lower the center of gravity is located, the greater is the arc which an unbalancing force must describe before it can bring the center to fall outside the supporting surface (Fig. 4).

The "stand up" toys are based on the proposition that their center of gravity never falls outside of the supporting surface even if they are inclined 90° and the figure is flush with the ground.

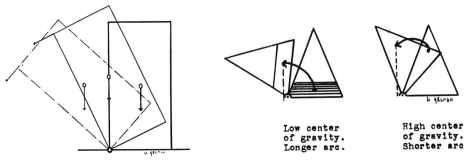

FIG. 3. Greater force necessary to throw line of gravity beyond supporting surface.

FIG. 4. The lower the center of gravity, the greater is the arc which an unbalancing force must describe.

D. THE LOCATION OF THE CENTER OF GRAVITY

When a dog stands with his head extended the weight center is close to the front legs and is almost in line with the shoulder blade. When he is trained to rear himself on his hind legs, he must perform the respectable feat of lifting

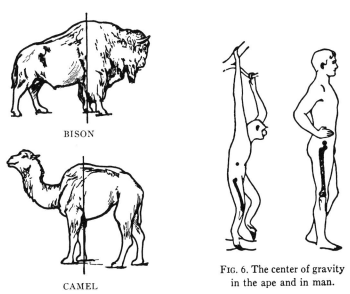

FIG. 5. Line of gravity of Bison and Camel.

FIG. 6. The center of gravity in the ape and in man.

his weight center to a considerable height and to maintain it perpendicularly over the extremely small supporting surface of his hind legs. Some animals such as the camel or bison (Fig. 5) have their weight center located very far forward so that the forelegs sustain much more of the body weight than the hind legs. Other animals, as for instance the kangaroo or some of the prehistoric reptiles such as the brontosaurus have their weight centers located farther backward close to the posterior extremities in accord with their ability to raise themselves on their hind legs and to use the tail as a sort of tripod support.

The philogenetic development of the upright position shows that when the animal strove to raise up on its hind legs, the first problem was to develop sufficient muscle power to hold the thigh extended against the pelvis. Even in the higher apes we find the extensors of the hip joint only incompletely developed compared to man. This insufficiency is particularly marked in some of the primates. In these the center of gravity wandered upward with the greater development of chest and arm when they were forced to acquire arboreal habits. Then again, when forced to the ground they were unable to balance their enormous trunks and their powerful arms upon the comparatively poorly developed and poorly extended hind legs and they consequently reverted again to the quadrupedal stage (Fig. 6).

Applying this to the upright position of man we find that what was most needed to acquire and maintain such a position was the increase in strength of the extensory apparatus of the back and of the extensors of the lower extremity, especially the hips. In the end this has resulted in the even and harmonious development of both upper and lower extremities which is the distinction and beauty peculiar to the human body. Inasmuch as the special development of the extensory muscles exacted a change in the distribution of the body mass compared with the primates, the weight center, as expected, had moved again to a lower level.

Borelli[1] determined the weight center of the human body by balancing the body over a prismatic wedge in three perpendicular planes. More accurate were the investigations of W. and E. Weber.[7] They found in their subject which measured 1,669 mm. that the center of gravity was located 721 mm. below the vortex and 948 mm. above the heel, and that it was 88 mm. above the common hip joint axis and 8.8 mm. below the promontory. Hermann von Meyer[5] found that in upright position the weight center was located at the level of the second sacral vertebra 5 cm. behind the common hip joint axis and the line of gravity touched the floor at 3 cm. in front of the ankle joint. It strikes the spine about the middle of the anterior border of the 10th dorsal vertebra and it touches the head a little in front of the atlanto-occipital articulation.

Thus the line of gravity has two intersections with the spine between occiput and sacrum, namely at the cervico-dorsal and approximately at the lower dorsal or the dorso-lumbar level.

Harless[3] repeated the experiments of the Weber brothers and located the center of gravity at $432°/_{00}$ from above and $568°/_{00}$ from below, corresponding approximately to the earlier figures of the Weber brothers.

The freezing method which Braune and Fischer[2] used on cadavers, suspending them in three perpendicular planes, confirmed in the main the data given by the previous observers.

The position of the line of gravity in relation to the supporting surface can also be determined in the living. Lovett and Reynolds'[4] method is based on the fact that the weight born by ball and heel of the foot respectively is inversely proportional to the perpendicular distance of these points from the line of gravity and that the sum of the two reactions acting at the ball and heel equals the total weight of the body (Fig. 7).

A board is placed on a scale which carries a prismatic ledge, the board resting upon the ledge with its anterior end while the posterior edge of the board rests on another fixed prismatic ledge outside the scale. The subject is then placed upon the board, the heel touching a small bar placed at the back end of the board in line of the posterior ledge while the free end of the board rests on the prismatic ledge of the scales (Fig. 8).

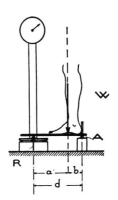

FIG. 7. Relation of reactions R_1 and R_2 to their respective distances (a and b) from the line of gravity.

FIG. 8. Lovett and Reynold's method of determining the situation of the line of gravity between ball and heel.

The distance between the two prismatic ledges and the total weight of the individual are first determined. Then as the patient steps on the board one reads the scale which gives the anterior weight component. As stated, this reading is inversely proportional to the distance of the line of gravity from the anterior prism. The portion of the body weight resting on the posterior prism is also known, being the difference between the known total weight and the amount of weight read from the scale. If a and b are the respective distances of the prism from the line of gravity, then R_1 (that is the amount read off the scale): R_2 (that is, the difference from the total weight) = b:a. Since a + b is known by measuring the length of two boards between the two supporting prisms, their respective distance from the line of gravity can easily be calculated.*

* Wt is the total weight of the individual and Wp is the partial weight as read from the scale. Then Wp: Wt = X:d (d being the total distance of the prismatic support), or $X = \dfrac{Wp \cdot d}{Wt}$.

If the total weight is 100 kg., the partial weight from the scale 40 kg., and the distance between the prism 50 cm., then $X = \dfrac{40 \times 50}{100}$, or 20 cm. from the posterior edge would then be the point of intersection of the line of gravity.

The difficulty is that even in strict upright standing the situation of the line of gravity is not constant but there is always some forward and backward swaying. However on the whole, there is an obvious tendency of the individual to maintain the line of gravity at a fairly stable distance from the ankle joint which is about 4 cm. in front of it. The reason for this constancy is that any strong forward displacement of the line of gravity would involve a considerable increase in the tension of the calf muscles.

The farther the line of gravity moves forward toward the ball of the foot, the greater will be this tension. When finally the line of gravity falls into the supporting surface of the ball of the foot, the latter bears all the translatory gravital stress which means all the body weight. But at the same time the gravity displays a rotatory force pressing the forefoot upward in the ankle joint; this rotatory force now must be resisted by the calf muscles. Since the latter are only half the distance from the center of motion of the ankle joint compared with the distance between the joint and the ball of the foot, it is obvious that to provide a rotatory moment adequate to neutralize the rotatory moment at the forefoot, the tension of the calf muscle must equal twice the body weight.

In certain morphological changes, for instance in pregnancy or in obesity, one might expect that the line of gravity would be thrown forward. However we see instead that the individual simply increases the plantar flexion of the foot and thus throws his body backwards so as to maintain the line of gravity closer to the center of motion at the ankle joint.

The same can be noticed in persons who wear high heels. The foot accommodates itself to it by an increased plantar flexion, but at the same time the body is thrown backwards and the knees are held straight and not flexed as one might expect.

III. THE EQUILIBRIUM OF THE HUMAN BODY AS AN ARTICULATED SYSTEM

So far we have treated the equilibrium of the human body as though the latter were a long rigid structure standing upon a comparatively small supporting surface. But the body is a complicated mechanical unit composed of many links which are joined together by articulations. For each link the gravitational forces must be equilibrated. As each link rests upon the one underneath it, the connecting articulation is the supporting surface bearing it. This surface being necessarily extremely small, the balancing of the supported link becomes difficult.

Can an automatic equilibrium be established for all articulations?

The condition of equilibrium for both translatory and rotatory components of the gravitational force requires that:

1. All respective centers of gravity of the individual links, and
2. All centers of motion fall in the line of gravity.
3. That this common line of gravity of the whole system falls into the area

of support. Only when these conditions are met can the system be kept in equilibrium; only then all translatory components of the gravitational forces are neutralized (Fig. 9); and only then all rotatory components of the gravitational force become zero because they all pass through the centers of motion. Such, for instance, would be the feat of a juggler when he skillfully balances a number of objects one upon the other over a small supporting surface.

A glance at the construction of the human body shows that such a passive equilibrium is impossible. Neither all the centers of gravity of the different

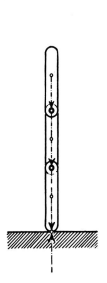

Fig. 9. Condition of balance in an articulating chain. The line of gravity goes through the centers of all joints and through all partial centers of gravity.

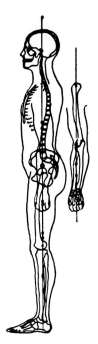

Fig. 10. Relation of center and line of gravity to the articulations of the body. (Braune and Fischer)

parts nor all the centers of motion between the different parts can be brought to coincide with the common line of gravity. Most of the joint centers are, in fact, at a considerable distance from the weight line (Fig. 10). Hence in the upright standing position the force of gravity develops active rotatory components in many of the joints.

These rotatory components of gravity must be neutralized by active opposing muscle forces, if an equilibrium is to be established and the parts are to be maintained in their relative positions.

In order to appreciate what these rotational components of gravity amount to and what counterforces are necessary to neutralize them, it is above all necessary to know where the partial centers of gravity and the centers of rotation of the different parts which are movable against each other are located in relation to the common line and center of gravity of the entire system. This is the factor which determines what resistance the opposing muscular forces will have to generate against gravity. Meyer[5] and Harless[3] undertook to locate the partial centers of gravity of the different parts relative to the sagittal and frontal planes of the body. Later Braune and Fischer[2] determined the situation of not only the partial centers of gravity but also of the centers of the connecting articulations in relation to the cardinal planes of the body which contain the line and center of gravity.

This makes it possible to calculate rotatory moments which gravity develops. Having ascertained, furthermore, the mass resistance or inertia (I) which each part offers to rotation (see above) they were able to determine the acceleration (a) which such a rotatory moment (M) imparts $(a = \dfrac{M}{I})$.

With this concept we are furnished at least with a theoretical basis for the computation of the rotational effect of gravity upon the large joints of the body and as the result it becomes possible to estimate the muscle effort required to hold the body and its parts in balance.

IV. SUMMARY

1. The Newtonian laws state the conditions for motion and equilibrium in a general way as follows:

a′ A body is at rest or continues in uniform motion if no force applies to it.

b′ If a force applies, it imparts to the body an accelerated motion, in which the acceleration is proportional to the force and inversely proportional to the mass of the body.

c′ A body is in equilibrium if all forces acting upon it mutually neutralize each other both in direction and intensity so that the resulting force is zero.

2. Applied under terrestrial conditions, any body is under influence of gravity and equilibrium requires neutralization of the gravitational force. Likewise any secondary force applying on the body must be neutralized under the provision of the second Newtonian law. Furthermore, any force which does not go to the center of gravity of the body imparts to it a rotatory as well as a translatory acceleration, the rotatory moment being the product of this force times its perpendicular distance from the center of gravity. The body resists this rotation by its inertia which is the product of its mass times the square of the average distance of all mass points (radius of gyration, ρ) from the center of gravity.

3. Applied to an articulated structure such as the human body under terrestrial conditions the condition for equilibrium of gravitational stresses is that the line of gravity falls:

a′ Through the center of gravity of all connecting links.

b′ Through the center of all connecting articulations.

c′ Through the supporting surface of the whole system.

4. From the quadrupedal to the bipedal stage the center of gravity of the human body has been gradually raised by the development of the extensory apparatus of back, hip and knee.

5. Its location was determined by Borelli, W. and E. Weber, H. V. Meyer, Harless and in the living by Lovett and Reynolds.

6. The line of gravity strikes the floor about 4 cm. in front of the ankle joint in upright standing, but there is a constant fluctuation both in antero posterior and lateral direction.

7. The farther it moves forward toward the ball of the foot, the greater is the tension placed on the muscles of the heel cord. There is therefore a tendency to keep this line within bounds.

8. Because of the situation of the articulations of the human body in relation to the line of gravity, the conditions for an automatic equilibrium of the gravitational forces (see 3) can never be fulfilled; therefore such an equilibrium is always active, i.e., requires muscular forces to counteract the rotatory effect of gravitational stresses.

9. Inasmuch as these rotatory moments depend upon the distance of the line of gravity from the centers of gravity of the different parts, the location of these partial centers were established for head, trunk and parts of the extremities by Meyer, Harless and Braune and Fischer.

10. And inasmuch as the resistance of a body to rotatory motion is expressed by its inertia ($I = m\rho^2$), the inertias for each part of the body were likewise determined by the latter investigators.

11. This makes it possible to calculate the accelerations which are imparted to the body parts in the different joints by the rotatory components of the gravitational stress ($a = \dfrac{R \text{ (moment of rotation)}}{I \text{ (inertia)}}$

BIBLIOGRAPHY

1. Borelli: *De Motu Animalium*. Lugduni Batavorum, 1679.
2. Braune, W., and Fischer, O.: Über die Lage des Schwerpunktes des Menschlichen Körpers, Vol. XV. *Abh. Math. Phys. Klasse d. Kgl. Sächs. Ges. d. Wiss.* Leipzig; S. Hirzel, 1889.
3. Harless: *Die Statischen Momente der Menschlichen Gliedmassen*. München Abh., 1857 and 1860.
4. Lovett, R. W., and Reynold, E. S.: Method of Determining the Position of the Center of Gravity in its Relation to Certain Bony Landmarks in Erect Position. *Am. J. Physiol.*, May, 1909
5. Meyer, Hermann von: *Die Wechselnde Lage des Schwerpunktes im Menschlichen Körper*. Leipzig, Engelmann, 1868.
6. Newton, I.: *Principia Mathematica*. London, 1687.
7. Weber, W. and E.: *Mechanik der Menschlichen Gehwerkzeuge*. Dieterich, Göttingen, 1836.

Lecture VIII

ON MEASUREMENT AND COMPUTATION
OF BODILY MOTION

I. ORIENTATION PLANES

IN THE preceding lecture we have become acquainted with the condition of balance of the body and with the effect of muscular forces in sustaining it. In addition it was shown how by muscular contraction the state of balance is overthrown and yields to visible motion which continues until some force, voluntary or otherwise, sets a halt to motion by establishing a new state of balance.

What lies ahead of us now is to follow the moving parts and to describe in more accurate terms the paths they take while in motion and the velocity and acceleration which they acquire by moving. The immediate need is to be able to define a point in space unequivocally and in mathematical terms. We can only do so by establishing its relation to points or planes which lie outside of the moving system. We then can follow the moving point and ascertain how this relation changes during motion. This would give us the pathway which a moving point describes. We can further determine the time necessary for such a change, and this gives us the velocity. And finally we can ascertain whether or not this velocity was uniform or whether it had changed in the different units of time; this would give us the acceleration and with it the moving force.

The position of a point in space can be expressed mathematically by stating its distance from another fixed point or from three planes perpendicular to each other which contain this point.

This would then be the reference point and the planes will be the reference planes. It is most advantageous to choose for this reference point the center of gravity of the body.

The great advantage in doing so is that we now can express the rotatory forces and their directions for the body as a whole so far as they are produced by gravity in terms of their distance from the center of gravity.

However the center of gravity is inside of the pelvis and is inaccessible for any direct measurement. It is for this reason that we must select for reference three planes which can all be established by accessible landmarks of the body and which are all perpendicular to each other. The three planes are so chosen that their intersection point is the center of gravity. Thus we make this center the origin of the three coordinate system. The point of the body, the position of which we want to ascertain, can then be projected perpendicularly upon each of these three planes. This, then, would give us the three coordinates of this point in respect of the three chosen planes. The resultant of these three coordinates is the distance of the point from the center of gravity. We shall call these planes which contain the center of gravity the cardinal reference planes of the body.

How can they be established?

We lay the first orientation plane horizontally through the level of the center of gravity when the body is in upright position and we call this the cardinal transverse plane of the body (Fig. 1).

The second plane is laid in antero-posterior direction so that it divides the body in two perfectly symmetrical halves, a right and left half. This we call the cardinal sagittal plane. This plane is perpendicular to the above-mentioned transverse plane.

Finally we lay a third plane from side to side through the center of gravity which is perpendicular to both the transverse and the sagittal planes. This plane divides the body into unequal anterior and posterior halves.

These three planes, then intersecting, produce three lines which are all perpendicular to each other, and which serve now as coordinates of the three coordinate system.

The intersection of the transverse with the sagittal plane generates the X coordinate; it is a line directed from forward to backward. By conventional usage the distances which lie in front of the center of gravity upon this line are marked positive and those which lie behind it are marked negative.

The intersection of the transverse plane with the frontal plane is the Y coordinate.

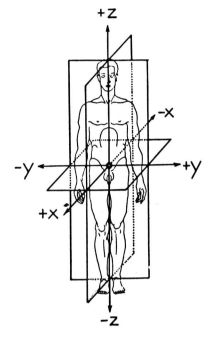

Fig. 1. General orientation planes of the human body.
  YZ frontal plane
  XZ sagittal plane
  XY transverse plane
The origin O is the common center of gravity.

It is directed from left to right. By conventional usage the distances to the right of the center of gravity are marked positive and those to the left negative.

Finally the intersection of the frontal and sagittal planes produces a line running perpendicularly from above downward. This is the Z coordinate. By conventional usage also we name the distances which lie above the center of gravity positive and those below negative.

Any point in space can thus be determined by these three coordinates.

II. THE RELATION OF THE PARTIAL CENTERS OF GRAVITY TO THE ORIENTATION PLANES OF THE BODY

The system of coordinates formed by the orientation planes makes it possible to express mathematically the relations which the partial centers of gravity of

COORDINATES OF THE CENTERS OF GRAVITY FOR THE COMFORTABLE POSITION
(Braune and Fischer)

For		X	Y	Z*
Head		−1	0	154.5
Trunk		−.6	0	111.8
Thigh	right	−.4	+7.8	69.4
	left	−.4	−7.8	69.4
Calf	right	−2.7	+6	29.8
	left	−2.7	−6	29.8
Foot	right	+1.3	+6.3	3
	left	+1.3	−6.3	3
Upper Arm	right	−1.9	+18	118.4
	left	−1.9	−18	118.4
Lower Arm	right	−1.3	+18	90.9
	left	+5.4	−15.9	102.2
Hand	right	0	+18	70.5
	left	+20.5	−10	102.0

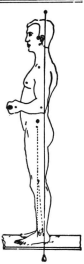

Location of center of gravity in comfortable posture (R. Fick).

FIG. 2.

THE COORDINATES OF THE CENTERS OF THE JOINTS IN THE
NORMAL POSITION IN CENTIMETERS

For		X	Y	Z*
Atlanto-Occipital Joint		0	0	152
Hip Joint	right	0	+8.5	87
	left	0	−8.5	87
Knee Joint	right	0	+8.5	47
	left	0	−8.5	47
Tibio-Astrag. Joint	right	0	+8.5	6
	left	0	−8.5	6
Posterior Rim of Foot	right	−4	+6	4
	left	−4	−6	4
Tip of Foot	right	+20	+16	1.5
	left	+20	−16	1.5
Shoulder Joint	right	0	+18	134
	left	0	−18	134
Elbow Joint	right	0	+18	103
	left	0	−18	103
Wrist Joint	right	0	+18	76
	left	+19	−11	108

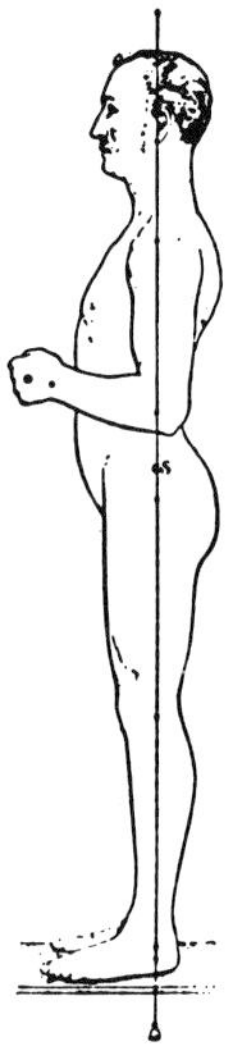

* Measured from floor.

FIG. 3.

the body parts bear to the common center of gravity.

a. Table I shows the coordinates of the *partial centers of gravity* relative to the common center of gravity in comfortable position (Fig. 2) (Braune and Fischer[1]).

b. Table II shows the coordinates of the *centers of joints* in relation to the common center of gravity in *normal* position (Fig. 3).

c. Table III shows the coordinates of the *centers of joints* in relation to the common center of gravity as they appear in *military* position (Fig. 4).

The different relationships of the joint in relaxed and military positions can be noted by comparing Table II with Table III.

1. In the sagittal plane (X coordinate).

In military position the head is stretched forward, the X coordinate of the atlanto-occipital joint being +5 cm. against 0 in normal position. The knee joint in military position is backward (the X coordinate being —4.5 against 0 in normal position).

The tibio-astragalar joint is also displaced farther backward (the X coordinate being —7 against 0 in normal position). The shoulder joint is carried forward, the X coordinate being +4.5 in military against 0 in the normal position.

THE MILITARY POSITION
THE COORDINATES OF THE CENTERS OF THE JOINTS
(Braune and Fischer)

For		X	Y	Z*
Atlanto-Occipital Joint		+5	0	152
Hip Joint	right	0	+8.5	87
	left	0	−8.5	87
Knee Joint	right	−4.5	+6	47
	left	−4.5	−6	47
Tibio-Astragaloid Joint	right	−7	+4.5	6
	left	−7	−4.5	6
Posterior Rim of Foot	right	−11	+2	4
	left	−11	−2	4
Tip of Foot	right	+13	+12	1.5
	left	+13	−12	1.5
Shoulder Joint	right	+4.5	+18	133
	left	+4.5	−18	133
Elbow Joint	right	+.5	+18	102.5
	left	+.5	−18	102.5
Wrist Joint	right	0	+18	96.0
	left	+7.5	−18	76.5

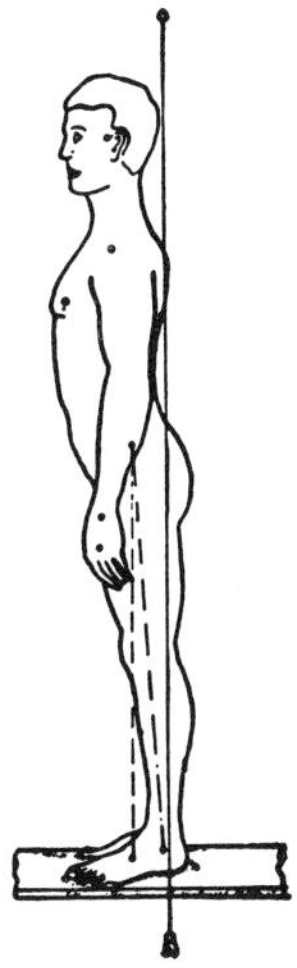

Line of gravity in "military" posture. (redrawn from Strasser).

* Measured from floor instead from center of gravity.

FIG. 4.

2. In the frontal plane (Y coordinate).

There is no difference, as long as the stance is symmetrical, between normal, relaxed or military position. For instance, the hip joints in both positions have the Y coordinate of $+8.5$ and -8.5, right and left respectively.

The shoulder joints have the same Y coordinates, namely $+18$ and -18 cm. in both normal and military positions, which shows that the symmetry of the body is little influenced by relaxation. The Y coordinates for the hip joint are also equal, namely, $+8.5$ and -8.5 in both positions.

These coordinates become important in the calculation of the rotatory stresses which gravity produces on the different articulations. For instance, seen in the frontal plane, the pelvis can be considered as a horizontal beam which is concentrically loaded by the superincumbent body weight and which is supported at its end by the hip joints.

Now the distance of this joint from the line of gravity, as determined by the above-mentioned coordinates, gives the lever arm about which the gravitational forces act upon the joints and hence the rotatory moment which gravity develops in respect to this joint (Fig. 5)*.

III. THE RELATION OF MUSCLE ACTION TO THE ORIENTATION PLANES

Most of the muscles of the body do not lie in any of the cardinal orientation planes, but their axes are oblique to one or all of them. They consequently develop rotatory components in respect to these planes. These rotatory components are commensurate with the sine of the angle of obliquity.

The action of a muscle can thus be determined by the three coordinates which its point of insertion forms with these three planes. For instance the sternocleidomastoid is oblique to all three planes, and therefore develops a component in the sagittal plane which flexes the head forward, then a second component in the frontal plane which turns the head toward the side of the muscle, and a third component in the transverse plane which rotates the head toward the opposite side.

Or, let us take the action of the external oblique which also has an axis oblique to all three planes. It bends the trunk forward in the sagittal, sideways in the frontal, and rotates it to the opposite side in the transverse plane. Its action is very much similar to that of the sternocleidomastoid.

IV. THE SECONDARY COORDINATION PLANES

When movement occurs in the joints of the extremities against the trunk as in ab and adduction of the arm or between one part of the extremity against

---

*In symmetrical standing the reaction R_1 and R_2 are equal, that is, one-half of the superincumbent weight is carried by R_1 and the other half by R_2. The bending moment at the point of the load (W) then would be $\frac{W + w}{2}$ (w = the weight of the beam) times half of the length of the beam (d) or $\frac{(W + w) d}{4}$.

the other as in flexion-extension of the elbow, the point of reference chosen for the determination of the rotatory moments produced by the muscle action is the center of the joint. The center thus becomes the origin of a three coordinate system similar to that which is used for the body as a whole.

We again lay three planes through the center of the joint perpendicular to each other. Each of these planes is parallel to the respective cardinal planes which go through the center of gravity of the whole body. An antero-posterior plane parallel to the cardinal sagittal plane is laid through the center of the joint; a horizontal plane parallel to the cardinal transverse plane and a plane parallel to the cardinal frontal plane are likewise laid through the center of the joint. It is here where all these planes intersect. The intersections of these three planes produce the same X, Y, and Z coordinates as did the intersection of the cardinal planes at the center of gravity (Fig. 6).

Braune and Fischer[1] made use of the system of the cardinal planes to express

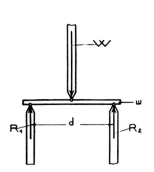

FIG. 5. Superincumbent weight and reactions. $W = R_1 + R_2$.

$$M(Moment) = \frac{(W + w)}{2} \frac{d}{2}$$

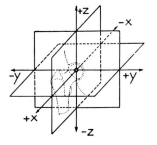

FIG. 6. Special orientation planes for the hip joint.
YZ frontal plane
XZ sagittal plane
XY transverse plane
The origin O is the center of the hip joint.

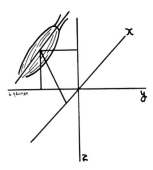

FIG. 7. Muscle axis oblique to all three planes. Lines perpendicular to axis are coordinates of a point of the axis.

in exact measurements the location of the partial center of gravity of the parts in relation to the common center of gravity. By using the special orientation planes of the joint as planes of reference, it is likewise possible to measure the leverage value of the partial masses in relation to the adjacent joints. This is essential if one wants to compute the muscular effort necessary to keep a joint in balance against gravitational stresses.

Furthermore the location of the insertion of a muscle and the direction of its axis in relation to the center of the joint can be similarly expressed in coordinates (X, Y, Z) of this special reference system. We have then arrived at a point where we can formulate both the rotary movement of the muscle (R) and the inertia resistance (I) of the part moved. The acceleration produced is the quotient ($a = R/I$) of these values.

To calculate the rotatory effect of any muscle in relation to a certain joint its force must be resolved into three coordinates relative to the center of the joint as origin of the system. The projection of the muscle axis on the three coordinate planes, sagittal, frontal, and transverse, then give the coordinates of this muscle (X, Y, Z) for each of these planes (Fig. 7).

If one wishes to find out the moment of rotation of a certain muscle of the hip joint which forms an angle with all three planes, then one must project the axis of the muscle upon these three planes and thus obtain the components which act in these planes, indicating the relative strength by their numerical value.

Let us consider, for example, the gluteus medius. It is inserted into the greater trochanter at an angle of, let us say, 30°. Its full rotatory component, therefore, is only its force (F) times the sine of 30. The rest of its force acts as stabilizers (S) (F times the cosine of 30). Let us further assume that the middle fibers of this muscle are strictly in the frontal plane. Consequently the angle (β_1) between these fibers and the frontal plane is zero. Its full abducting force would then be expressed by F $\cdot$ sin 30 $\times$ cos β_1; β_1 being 0, its cosine is 1; therefore the abducting strength of this muscle is the full rotatory component F $\times$ sin 30. The anterior fibers, on the other hand, form an angle (β_2) with the frontal plane. Hence the abducting power of the anterior fibers is F sin α cos β_2, which means the greater the angle β_2 is, the less is its abductory power.

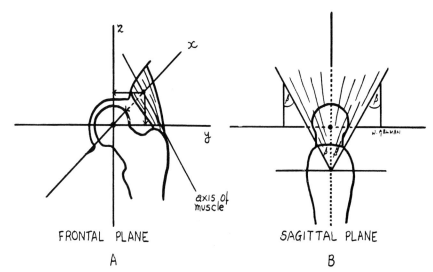

FRONTAL PLANE   SAGITTAL PLANE

A   B

FIG. 8. A. The axis of the gluteus medius is oblique to all 3 coordinates. It therefore develops rotatory components in respect to the sagittal (x) axis in the sense of abduction; in respect to the frontal (y) axis in the sense of flexion and extension; and in respect to the vertical (z) axis in the sense of inward and outward rotation. B. The midline fibers of the muscle fall in the frontal plane. Therefore, the angle it forms with this plane is 0. The angle of attachment (α) of this muscle is, for instance, 30°. Consequently, the rotary component of this muscle is F $\times$ sin 30, while the stabilizing component is F $\times$ cos 30.

Because of the fan-shape expansion of this muscle the rotatory action of the different fibers is different in the different planes. In the frontal plan the mid-fibers form an angle of 0 with this plane. Since the rotatory effect depends on the cosine of the angle to this plane and since the cosine of 0 is 1, it follows that the full rotatory component of this muscle (F sin α) is in force as abduction. The marginal fibers form an angle of 30° with the frontal plane ($\beta = 30$), therefore, in respect to this plane an abductory force of F sin 30 $\times$ cos 30. If the anterior fibers were at 90° to the frontal plane (a theoretical assumption), then the abductory power would be 0 because the cosine of 90 is 0.

Should this angle β_2 reach $90°$, that is, should the anterior fibers run straight forward, then the cosine β_2 would be zero. Consequently F sin α cos β_1 would be zero, i.e., no abductory power woud be left in this position (Fig. 8).

Let us next consider what happens in the transverse plane. The anterior fibers rotate the hip forward, the posterior fibers of the muscle rotate it backward. Again, expressed mathematically the middle fibers form with the transverse plane an angle of $90°$; with the cos of 90 being zero, they consequently have no rotatory effect. Let us assume the anteriormost fibers run straight forward, which is of course an entirely theoretical proposition, then the angle with the transverse plane would be zero, the cos of zero being 1; consequently all of their rotatory component (F sin α) would rotate the femur forward (Fig. 9).

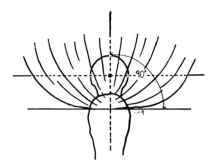

FIG. 9. Assuming that the marginal fibers run horizontally forward and backward (a purely theoretical case), then the angle (S) formed with the frontal plane would be $90°$, the cosine 0; therefore, there exists no component moving in the frontal plane. On the other hand, the angle (S) formed with the sagittal plane is 0, its cosine is 1; therefore, all rotatory force of the marginal fibers is in the direction of flexion or extension.

V. ORIENTATION AND MEASUREMENT OF JOINT MOBILITY

The simple recording of joint movement in a single plane can be accomplished by a goniometer which marks the angular value upon a sheet of paper. This applies, for instance, for the flexion extension ranges of elbow or knee.

To measure motion of a joint which has more than one degree of freedom of motion, that is which moves in two or three planes which are all perpendicular

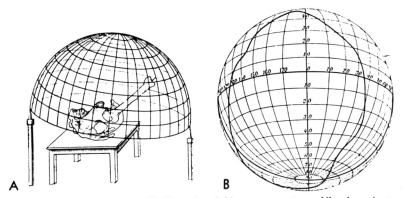

A   **B**

FIG. 10. A. Albert's perimeter. B. Excursion field measurement on Albert's perimeter.

to each other, the tracing must be made on the surface of a globe. Albert's perimeter is such an instrument (Fig. 10A).

For example, by placing the hip joint in the center of this globe, abduction and adduction can be measured by the meridians while flexion and extension can be measured by the parallels of the globe. In this way the excursion field can be traced upon the surface of the globe.

Or by attaching the instrument to the joint, a self-recording system can be used. A mercury drop in a hollow sphere makes connections with numerous electrical contacts which are arranged in the inside of the globe and which record on contact the angles of motion in all directions. Dann's[2] global joint perigraph is an instrument of this type.

VI. THE RELATION OF JOINT POSITIONS TO THE ACTION OF MUSCLES

As a result of these perimeter measurements an excursion field can be constructed which registers accurately the degree of motion possible in all planes (Fig. 10B).

The part which the individual muscle plays in the excursion is not revealed

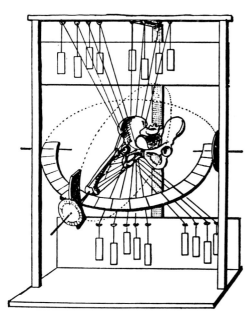

FIG. 11. Strasser and Gassmann's skeletal muscle phantom. (Strasser)

except for the fact that we know the principal and the secondary action of the muscles in a certain position of the joint, usually in the so-called neutral or normal position.

It must be remembered, however, that with the motion of the joint the relative position of the muscle to the orientation planes changes constantly; and with it also changes the relative components of the muscle force active in different planes, because the angle between muscle axis and the reference planes changes with the joint position.

In addition there is also a change in the angle of application which again means a fluctuation between the stabilizing and the rotatory components of the muscle.

Each joint position, therefore, postulates a different distribution in strength and leverage of the several components of muscle action with the result that the graphic picture of muscle action changes in kaleidoscopic fashion as the joint sweeps through its ranges of motion.

To visualize the direction and the amount of the force which is exerted by the individual muscle about a joint, a so-called muscle phantom as described by Strasser and Gassman[3] (Fig. 11) is useful.

But even these muscle phantoms present the movement of the different muscle about a joint only in a certain position and they do not reveal the changes which the different components of the muscular forces undergo with the changing positions of the joint.

A good example of such a change are the adductors of the hip joint. These muscles have secondary flexory and extensory components. When the hip joint is brought into a certain position of flexion a flexory component turns into an extensory component and vice versa. As a matter of fact the adductor longus changes from a flexor into an extensor when the hip joint reaches 70° and the adductor magnus when it reaches 50° of flexion.

VII. LINEAR EXPRESSION OF ANGULAR VALUES OF ROTATION

There remains only one more point to consider for the calculation of rotatory events. It is the fact that it is necessary to express the distance or length of the arc covered by rotation in linear terms. So far we have only spoken of an angle of a certain degree described on rotation.

How can this angular value of rotation be expressed in linear terms of, for instance, mm. or cm.?

So long as the problem was to determine the condition under which the body or parts of it remained in equilibrium, i.e., at rest, no rotatory motion occurred and consequently angular values of pathways or velocity were not a factor. For instance when the pelvis was presented as a concentrically loaded beam supported by the two hip joints, the conditions of equilibrium were analyzed merely in terms of superincumbent weight, and the moments of stresses were calculated on this basis. Consequently they were entirely expressible in metric terms. When it comes to calculating actual and visible motion, however, the degree of rotation in angular terms and the velocity and acceleration in such terms becomes a factor. The problem is now to fit these angular terms into metrical ones by expressing them in meters or fractions.

The linear value can be found for any kind of rotatory motion and for any point of the moving part rather simply. We suppose that a certain point in the rotating lever arm has a distance from the center of motion r expressed in linear values. For every degree of rotation the space covered by this point in linear values must be $\dfrac{2r\,\pi}{360}$ because the full circle in linear values equals $2r\,\pi$ and consequently one degree equals $\dfrac{2r\,\pi}{360}$.

Now for ω degrees of rotation the space covered by a rotating point would be $\dfrac{2r\pi\omega}{360}$ or $\dfrac{\omega r\pi}{180}$ or, since $\dfrac{\pi}{180}=\dfrac{1}{57}$, it would equal $\dfrac{\omega r}{57}$. This gives the pathway covered by the rotating point. If we want to express the velocity in similar terms, we use the formula $v=\dfrac{s}{t}$ or, substituting for s (Pathway) $\dfrac{\omega r}{57}$ would equal $\dfrac{\omega r}{57t}$

Next, to find the acceleration we again use the general formula $a = \dfrac{v}{t}$ or $\dfrac{s}{t^2}$; substituting for v, $\dfrac{\omega r}{57t}$ we arrive at the formula of acceleration $a = \dfrac{\omega r}{57t^2}$.

And finally in order to calculate the force which is creating this acceleration we can again refer to the formula which expresses the second Newtonian law, that the force equals the mass times acceleration ($F = am$). Substituting for acceleration, $a = \dfrac{\omega r}{57t^2}$, we then arrive at the formula for the rotating force,

$$F = \dfrac{m\omega r}{57t^2}.$$

There is therefore no difficulty in determining for a rotating point the space which it covers in rotation, the velocity with which it rotates, the acceleration with which it increases its speed per unit of time and finally the force itself which produces this acceleration.

The resistance to rotatory motion or the inertia (I) has already been discussed in a previous lecture. It represents the product of the mass times the square of the radius of gyration ($m\rho^2$). The rotatory acceleration is expressed by the equation $a = \dfrac{R}{I}$, R being the rotatory moment. This formula is the rotatory equivalent for the general formula $a = \dfrac{F}{m}$.

With these values established all pertinent data for the computation of rotatory motion are at hand.

VIII. SUMMARY

1. To follow a moving part in its pathway the determination of its place in space as it changes from moment to moment is essential.

2. This determination is carried out by means of a three coordinate system by which the distance of the point from each of the coordinates can be measured in linear values.

3. For the body as a whole this three coordinate system is created by the intersection of three planes, all at right angles to each other and all containing the center of gravity of the body. This center thus becomes the origin of the three coordinate system. The coordinates are named X for the one running in sagittal, Y for the one in frontal, and Z for the one in perpendicular direction.

4. The location of the partial centers of gravity of the parts of the body as well as that of the articulations in relation to the common center of gravity can thus be determined and the differences in the coordinates of the different joints, i.e., their varying distances from the center of gravity as they appear in various postures can be noted.

5. Secondary coordination planes and secondary coordinate systems can be

established for the different joints of the extremities by laying planes parallel to the cardinal planes of the body through the center of the joint which then becomes the origin of the secondary coordinate system.

6. The effect of the rotatory component of a muscle (FR = F sin α where α is the angle of application) can also be determined in respect to the coordination planes both relative to the cardinal coordinate system which has the common center of gravity as origin and to the secondary coordinate system with the center of the joint as origin.

7. For example the gluteus medius with an angle of application at the greater trochanter of, say 30°, has a total rotatory component of F sin 30. This rotatory component can be resolved in three components in respect to the coordination planes. If the middle fibers of the muscle lie exactly in a frontal plane, then the component for this plane is F sin 30 cos β_1 where β_1 is the angle between midfibers and frontal plane. Since this angle is 0, cos β_1 is 1 which means that all of the rotatory component of the midfibers of the muscle acts in abductory capacity. On the other hand if the anterior fibers form with the frontal plane an angle of 90° (a theoretical assumption), the cosine 90 being 0, F sin 30 cos β_1 is 0, which means the anteriormost fiber has no abductory effect whatsoever.

8. The amplitude of motion of a joint can be measured easily by the goniometer. For movements in two or three planes perimeters or global joint perigraphs are used.

9. The action of individual muscles on joint movement changes considerably in different positions of the joint. Example: the adductor longus is a flexor of the hip joint up to 70° flexion, and then becomes an extensor.

10. The difficulty of changing angular values of rotation into linear measurements is easily solved if the radius of motion is known; each degree of rotation represents $\dfrac{1}{57th}$ of the length of the radius. ω degrees $\dfrac{\omega r}{57}$ etc.; thus the

Pathways s = $\dfrac{\omega r}{57}$, the velocity v = $\dfrac{s}{t}$ = $\dfrac{\omega r}{57t}$; the acceleration a = $\dfrac{s}{t^2}$ equals $\dfrac{\omega r}{57t^2}$ and the force producing the acceleration F = ma equals $\dfrac{m\omega r}{57t^2}$.

BIBLIOGRAPHY

1. BRAUNE, W., and FISCHER, O.: Über den Schwerpunkt des Menschlichen Körpers, Vol. XV. *Abh. Math. Phys. Klasse Kgl. Sächs. Ges. d. Wiss. 20 VII.* Leipzig, S. Hirzel, 1889.
2. DANN, W.: Perimetrie und Perigraphie der Gelenke. *Ztschr. Orthop. Chir., 39:*148, 1920.
3. STRASSER, H., and GASSMANN, A.: Hilfsmittel und Normen zur Bestimmung und Veranschaulichung von Stellungen, Bewegungen und Kraftwirkungen am Kniegelenke. *Meckle: Bonnet: Anat. Hefte, 2:6/7,* 1893.

Part II

THE TRUNK

Lecture IX

THE MECHANICS OF THE SPINAL COLUMN

I. THE FUNCTION OF THE SPINAL COLUMN

THE spinal column is designed for many purposes. It is, in the first place, a sustaining rod which is implanted into the pelvic ring and maintains the body in upright position; it carries the thoracic cage and sustains the balance between it and the abdominal cavity; it serves as a post of anchorage for many powerful muscles which maintain the balance of the spine and perform all spinal movement, and it furnishes the site of origin for many muscles of the shoulder girdle and the pelvic girdle; the spinal cord is encased in the vertebral column and protected against mechanical injuries.

By virtue of its many articulations and the interposition of the intervertebral disc the spine acts furthermore as a buffer spring which receives and distributes in rapid sequence innumerable jars and jolts associated with the dynamic function of the body.

As an organ of great flexibility it is able to produce and to accumulate moments of force as well as to concentrate and transmit those received from other parts of the body.

The philogenetic development of the upright position and the independent functions of the upper extremity have greatly increased the dynamic demands made on the spinal column. In complying with all intricate requirements, the human spine has developed into a complicated and delicate mechanical unit, the construction of which has placed high demands upon nature's ingenuity.

II. CONSTRUCTION AND MORPHOLOGY

A. THE GROSS CONSTRUCTION

1. Osteology

The spine is made up of a number of links movable against each other and held together by ligamentous and muscular structures.

From the first cervical to the last lumbar vertebra the bodies increase consistently in mass. From the first sacral to the last coccygeal their masses gradually decrease. In the frontal view the spinal column has, therefore, the form of two isosceles triangles, one long and drawn out containing the presacral vertebrae, and one short consisting of the sacrum and the coccyx. Their apexes point in opposite directions and they are joined together at their bases (Fig. 1).

Strasser[7] found that the greatest transverse diameter is that of the fifth lumbar and the greatest antero-posterior that of the second and third lumbar vertebrae. In the cervical, lumbar and sacral spine the greatest diameter is in the frontal plane while in the dorsal spine the sagittal diameter is greater. The cervical vertebrae up to the third carry, at the upper contour, a socket-shaped projection, the marginal or unco-vertebral process (Fig. 2) which may assume pathological significance by encroaching upon the intervertebral foramen.

The column of the vertebral body is joined by the pedicles or their equivalents to the column of the arches. The latter carry three apophyseal processes, the articular, the transverse, and the spinous.

The modal number of the presacral segments of the vertebrae is 24: seven cervical, 12 dorsal, and five lumbar. The caudal portion, the sacral part of which is immovably implanted into the pelvis and forms part of the pelvic

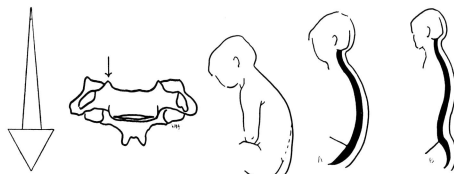

FIG.1. General form of the spinal column in the frontal view.

FIG.2. The marginal process.

FIG. 3. Single posterior curve of the newborn.

FIG. 4. Changes in the spinal curves. *A*. Lordosis of cervical spine as child holds head up. *B*. Lumbar lordosis added as child sits and stands.

ring, consists of 5 sacral and 3 or 4 coccygeal segments so that the total number of the vertebrae amounts to 32 or 33.

Willis[8] found great consistency in the modal number of the presacral vertebrae (96%); *absolute* numerical variations occur in two types: in one there is an increase of the presacral vertebrae from 24 to 25; this is an atavistic tendency and occurs comparatively less frequently; in the other, the modal number is decreased to 23 which would denote a futuristic trait commensurate with the philogenetic upward movement of the pelvis. On the other hand, *relative* numerical variations in which one segment of the spine is increased at the expense of another are much more frequent; for instance an increase of the sacral vertebrae to six at the expense of the lumbar (sacralization) or the increase of the dorsal vertebrae to 13 by inclusion of the last cervical (dorsalization, cervical rib).

B. SPINAL CURVES

1. Development

Seen in the frontal plane the spine appears perfectly straight and symmetrical with the exception of a very slight right dorsal (physiological) curve, probably due to the position of the aorta; seen in the sagittal plane, however, the spine shows four definite curves: a cervical lordosis (convex in front); a dorsal kyphosis (convex behind); a lumbar lordosis (convex in front); and a sacral kyphosis (convex behind). These curves are the result of developmental changes from early embryonic periods to maturity.

In the latter half of the embryonic life and at birth the spine shows only one single curve with posterior convexity from the occiput to the sacrum (Fig. 3). From this common convex curve, counter curves evolve as the static functions of the body develop and the spine adjusts itself to the upright position. The first of these counter curves is the lordosis of the cervical spine which appears when the child begins to hold up its head to enlarge its horizon of vision. The second curve is the lordosis of the lumbar spine which forms as the child begins to sit up and later when he begins to stand. It is this curve which produces the increase in the inclination of the pelvis (Fig. 4). Between the cervical and lumbar lordosis the dorsal spine remains kyphotic, although the curve becomes consistently shorter and flatter as the cervical and lumbar curves expand.

2. Landmarks

For orientation of the spinal column in relation to palpable superficial skeletal points the landmarks given by the late Dr. H. J. Prentiss[6] (Fig. 5) are useful:

A bare finger's width below and in front of the mastoid process is felt the tip of the transverse process of the first cervical vertebra; this process is opposite the angle of the jaw (Fig. 5, 1).

The cricoid cartilage lies opposite the 6th cervical vertebra and its transverse process. It marks the lower limit of the larynx and pharynx and the beginning of the trachea and esophagus (Fig. 5, 2).

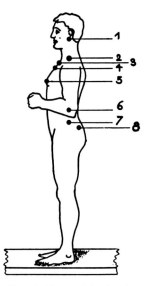

FIG. 5. External landmarks of Prentiss.

The episternal notch lies opposite the second dorsal vertebra (Fig. 5, 3).

The junction of the first and second portion of the sternum, the manubrium and body, or the angle of Louis, is opposite the junction of the 4th and 5th dorsal vertebrae (Fig. 5, 4). The junction of the 2nd and 3rd portion of the sternum (body and xyphoid) is opposite the 9th dorsal vertebra (Fig. 5, 5).

The lowest part of the 10th rib is opposite the junction of the 2nd and 3rd lumbar vertebrae (Fig. 5, 6). A horizontal plane passing through the crests of the os ilei strikes the 4th lumbar vertebra (Fig. 5, 7). The posterior inferior spine marks the termination of the horizontal portion of the sacroiliac joint (Fig. 5, 8).

3. The relation of the curves to the line of gravity

There is a great physiological latitude in relative length and degree of the three presacral curves. Yet each curve is compensatory to its neighbor with the result that the line of gravity as it passes upward from the supporting base intersects with all four curves of the spine at certain levels (Fig. 6).

Owing to the compensatory nature of the curves the spine as a whole approaches the line of gravity as much as the antero-posterior deflection of the curves will permit. In consequence, the body weight is more or less evenly distributed in front and back of the line of gravity (Fig. 7).

The prime function of the spinal column is to hold the body erect and to maintain posture. More shall be said on the subject of the normal and abnormal posture in a following lecture. At this point we merely anticipate

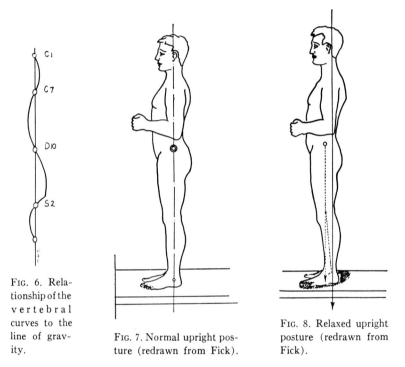

Fig. 6. Relationship of the vertebral curves to the line of gravity.

Fig. 7. Normal upright posture (redrawn from Fick).

Fig. 8. Relaxed upright posture (redrawn from Fick).

the statement that normal posture is based upon the above-mentioned relation of the spine to the line of gravity.

From the *morphological* point of view such a posture described by Goldthwait[3] and others is one in which the head is held erect, the eyes look straight forward, the shoulders are square, the chest is well rounded and is carried forward of the abdomen; there is full extension of the hips and of the knees, the feet are at right angles to the legs and the toes point straight forward.

From the *mechanical* standpoint the definition of normal posture must be based upon the already mentioned relationship of the spine to the line of gravity. Looking at the spine in the frontal plane, the gravitational line rises midway between the heels, strikes the gluteal fold, coincides with all the spinous processes of the vertebrae and intersects with the head at the middle of the occipital bone. It divides the body in two symmetrical halves. When the spine is viewed sideways in the sagittal plane, its relation to the line of gravity is as follows: it arises from the supporting surface between ball and heel in front of the ankle joint; it runs slightly in front of the knee joint axis; in relaxed posture (Fig. 8), through or directly behind the center of the hip joint; then it ascends to cut the upper end of the sacroiliac junction. It then runs upward behind the center of the bodies of the lumbar spine and intersects with the spine at the lumbodorsal junction; it continues in front of the dorsal spine and intersects with the spine again at the cervico-dorsal junction. It then runs slightly behind the cervical spine and finally reaches the head behind the ear at the mastoid process.

There is some latitude in what may be considered normal posture. The inclination of the pelvis shows individual differences within normal ranges and they all affect the angular values of the cervical, dorsal and lumbar spinal curves. Where the dorsal spine shows more convexity a compensatory increase of the cervical and lumbar lordosis accompanies it. Whether these variations are still within normal limits depends upon the relation which the line of gravity bears to the curves. We believe that as long as the intersection points remain at approximately the normal level, one can classify the posture as physiological because the mutual compensatory effect of the spinal curves is still preserved.

4. The posterior column and its adnexa

The posterior column joined to the vertebral bodies by the pedicles consists of the neural arches which carry the articular, the transverse, and the spinous apophyseal processes. The arches enclose the spinal canal and are connected with each other by ligamentous structures. They furnish the site for the origin and insertion of the powerful musculature of the back.

The pedicles are in front of the intervertebral articulations and skirt the intervertebral foramina. They are the pillars of the system of the arches. The intra-articular portion of the arches carries the superior and inferior articular and the transverse processes and forms the keystones of the arch. The laminae of the arches join each other in the midline and form the bases for the spinous processes.

Considering that all axes of motion between vertebrae go through the column of the bodies (the intervertebral disc being the fulcrum), one can understand why the spinal nerves use for their exit the intervertebral foramen in front of the articular processes, where there is less separation of the arches during the spinal movement because the foramina are so much closer to the center of motion. The contact between the arches is confined to the intervertebral articulation which lies immediately behind the exit of the nerves. In relation to the pedicle the posterior portion of the arch appears downward deflected in the dorsal and the lumbar portion of the spine.

The transverse processes arise from the arches between the upper and lower articular processes closer to the lower intervertebral joint. According to Strasser[7] there seems to be some evidence that at least a portion of the body

weight is born by the posterior column of the lumbar spine. This is especially evident in spondylolisthesis where the lower articular process of the 5th lumbar has become separated from the upper and is now held back by the upper articular processes of the first sacral (Fig. 9).

5. The spinal canal

In the mid-dorsal region the canal is round. From there upward it expands laterally; it becomes transversely oval at the 7th cervical. From here upward

FIG. 9. Spondylolisthesis.

again it is triangular and again becomes rounded and quadrangular at the level of the odontoid.

From the mid-dorsal level downward the lumen of the canal increases to the sacrum. It becomes again triangular at the lumbar spine and is widened transversely at the 1st sacral. In the sacral region the spinal canal is flattened anteroposteriorly, the front and back walls becoming almost parallel; from then on the lumen becomes rapidly smaller toward the coccyx.

C. THE ARTICULATIONS

1. Intervertebral

The segments of the posterior column move against each other in the intervertebral articulations. These movements, however, occur about centers of

FIG. 10. Cervical vertebra—intervertebral articulations.

FIG. 11. Dorsal vertebra—intervertebral articulations.

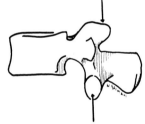

FIG. 12. Lumbar vertebra—intervertebral articulations.

motion which lie outside of the intervertebral articulations of the arches and the posterior column simply participates in motions about a center lying close to or in the intervertebral discs. Therefore the movement in the intervertebral articulations must be a gliding one. Their facets which are lining the articular processes, an upper and a lower facet to each side, serve only as guides or pilots for the glide.

Because of the difference in form and in position of the articulations the direction of motion differs in the various sections of the spine.

In the *cervical* spine the articular facets occupy an oblique plane of gentle slope from above and forward to downward and backward (Fig. 10). Resembling saddle joints they permit three degrees of freedom of motion, namely forward and backward in the sagittal, sideways in the frontal and axial or rotatory in the transverse plane.

In the *dorsal* spine the articular surfaces are also rather flat, but their slope is sharper so that they approach more the frontal plane, and they converge strongly posteriorly. Consequently the articular facets do not move in a flat plane but they describe the surface of a cone. The upper articular processes overhang the lower shingle fashion. Motion is free in flexion and side bending and axial rotation, but backward bending is limited (Fig. 11).

In the *lumbar* section the articulations of the vertebrae are aligned more in the sagittal plane; their facets have cylindrical curves, the upper concave and the lower convex. In gliding over each other they describe parts of the surface of a cylinder. This arrangement provides more for axial rotation and flexion-extension and comparatively little for side motion.

The latter is then taken up by the *sacrolumbar* articulations which are usually arranged in the frontal plane and therefore give more freedom for side bending (Fig. 12).

The *atlanto-occipital* and the *atlanto-axoid* articulations have their own special construction.

The *atlanto-occipital* joint is formed by the condyles of the occiput and the articular surfaces of the lateral masses of the atlas. These articulations converge forward. They are oval in shape and their axes run from backward, upward, and outward, to forward, downward and inward. The atlas carries the concave socket, the facet of the occiput is convex. The principal motion of this joint is flexion and extension, but this is associated by forward and backward gliding of the occipital condyles; they glide forward on the atlas when the head is extended and they glide backward when it is flexed. The symmetrical antero-posterior motion in the atlanto-occipital joint is checked by capsule and adjacent ligaments; extension is checked more than flexion. The range of motion as given by Fick[2] is 20° forward and 30° backward, while Krause[5] gives it a total of only 45° and Strasser[7] only 35° total range.

The *atlanto-axoid* articulation consists of four joints; the two articulations between the lower surface of the atlas and the upper of the axis, the intervertebral articulations proper, and the two articulations of the odontoid process with the anterior ring of the atlas. The process is fastened springily into the anterior arc of the atlas and articulates with its anterior surface. Its posterior surface articulates with the transverse ligament (Fig. 13). The principal motion is rotation about the longitudinal axis of the odontoid process. In motion the lower articular surface of the atlas glides on the upper articular surface of the odontoid.

2. The sacroiliac articulation

Only the first three sacral vertebrae are connected with the os ilei by the sacroiliac articulation. Caudally the connection continues as a syndesmosis. It is a kidney-shaped articulation between sacrum and os ilei with an upper vertical and a lower horizontal portion, the latter approaching the posterior surface of the sacrum directly medially to the posterior inferior spine. It is a true joint with articular facets, synovial lining and capsule, and it is subject to all pathological changes which may occur in any joint; tuberculous, pyogenic, non-specific arthritis; its surfaces, however, are so irregular with their numerous interlocking elevations and indentations that under normal conditions there is practically no motion whatsoever in this joint (Fig. 14).

3. The costovertebral and costotransverse articulations (Fig. 15)

These provide the articular connections of the ribs with the spinal column. The *costovertebral* articulations from the 2nd to the 9th rib are double or

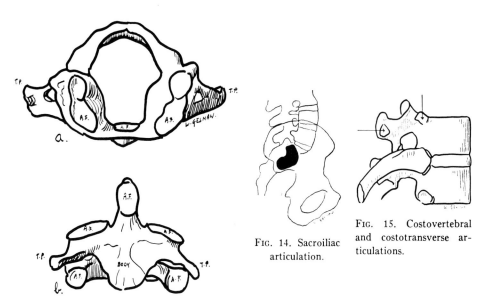

Fig. 14. Sacroiliac articulation.

Fig. 15. Costovertebral and costotransverse articulations.

Fig. 13. Atlanto-axoid articulation.
A.F. Articular facet   T.P. Transverse process

bipartite joints. The heads of the ribs articulate each with the upper and lower border of their adjacent vertebrae by two distinct facets which are separated by a ledge (Crista Capituli) and from which a fibrous strand runs to the intervertebral disc. The first rib articulates with one vertebral body only; so do the 10th, the 11th and the 12th.

The *costotransverse* articulation is situated between the tubercles of the 1st to the 10th rib and the transverse processes of the respective vertebrae. The tubercle of the ribs carries an oval convex facet fitting into a concave facet of the transverse process. The 11th and 12th rib have no costotransverse articulation.

Both articulations have a common axis going through the convexities of their articular facets. Consequently motion in the combined costovertebral and costotransverse articulation is possible only in one plane.

4. The costosternal junction

Of the 12 ribs, only seven enter into a direct relation with the sternum by articulations. Of the remaining ribs, three have their front ends joined together by means of cartilaginous junctions; in two (the 11th and 12th) their anterior ends remain free (floating ribs).

Although the different portions of the sternum fuse together early, the articulations of the true ribs from the 2nd to the 7th are bipartite which is most marked at the junction of the 2nd rib corresponding to the line between the manubrium and the body of the sternum (Fig. 16).

III. THE NON-CONTRACTILE SOFT STRUCTURES OF THE SPINAL COLUMN

A. THE INTERVERTEBRAL DISC

Of all the soft structures of the spine the intervertebral disc carries the greatest responsibility for the preservation of the function of the spinal column. Since all movements between the segments produce a gliding in the intervertebral articulations, the latter are outside of the center of motion. It is in the disc where the center of motion lies and all motion between vertebrae are associated with adaptational changes within the disc itself. This is made possible only because its elastic nature permits the disc to change in shape and to allow a varying degree of tilting and twisting.

Rotatory motion between the vertebrae occurs in three planes and about three perpendicular axes; axial rotation about a longitudinal axis, sidebending about a sagittal axis, and flexion and extension about a frontal axis. The center of the disc, that is the nucleus pulposus thus assumes the character of an amphiarthrosis with 3 degrees of freedom of motion.

The disc consists of a fibrocartilaginous ring, the annulus fibrosus, which is united to the vertebral body by means of a limiting layer of hyaline cartilage. It has a lamelar structure, the fibers running in all directions and containing between their masses small clusters of hyaline cartilage (Beadle[1]). The diversity of the direction of the fibers indicates that the ring has to sustain stresses in all directions.

The center of the disc is occupied by the nucleus pulposus. It is a remnant of the notocord, a gelatinous or hyaline mass, elastic and capable of changing its form as well as its position within the ring. It offers a considerable resistance to compression under which it becomes flattened and broader. The compression by the superincumbent load develops in the disc a considerable springy resistance which is one of the elastic forces contributing to what is called the intrinsic equilibrium of the spine.

Tilting movements of the vertebral bodies against each other occur in flexion, extension or side motion. They produce an asymmetrical unilateral compression of the disc. To this the disc responds by change of its form, becoming flattened and compressed at the site of the concavity. In this maneuver the nucleus pulposus is always forced toward the convexity of the curve (Fig. 17). Hence, in forward bending the disc is compressed anteriorly and the nucleus pulposus is forced posteriorly; in side bending the nucleus pulposus is forced toward the convex side and the annulus is compressed at the concavity. In pure axial rotatory movements between the vertebrae the nucleus does not change its position within the fibrous ring but both are under torsion stresses which are resisted by the circular fibers of the annulus fibrosus.

In all these movements the axis of motion passes into, or at least close enough to, the nucleus to make it for practical purposes the center of motion.

There is a fourth degree of motion possible between the vertebrae, namely, in translatory direction, that is, compression and extension. That the disc is under constant compresion in upright position is shown by the fact that the

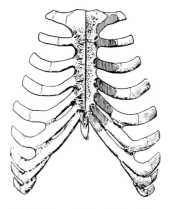

FIG. 16. The costosternal junctions.
(Strasser)

FIG. 17. Tilting of vertebral bodies, forcing nucleus pulposus toward convexity of curve. (After Strasser)

body is actually longer in recumbency, after a night's rest than it is in orthograde position.

The mechanical requirements placed on the disc are so great that the structure fails early and undergoes degenerative changes. These play an essential part in a number of static deficiencies of the spine.

B. THE LIGAMENTS

The many and powerful ligaments uniting the vertebrae can be divided into 3 systems.

1. A longitudinal tract system which binds the vertebrae together into a mechanical unit (intersegmental ligaments).

2. A longitudinal system which secures segment to segment (intrasegmental ligaments).

3. The articular and capsular reinforcements.

1. The longitudinal intersegmental systems

a) Uniting the bodies.

1) *The anterior longitudinal ligament* arises from the tubercle of the occiput as a narrow firm strand. It is fastened to the anterior tubercle of the atlas

and then runs over the anterior surface of the axis and over all the vertebrae down until it is lost in the periosteum of the anterior surface of the sacrum. It increases in width from the axis downward and on crossing the intervertebral discs it becomes firmly woven into their anterior borders. This ligament becomes taut on backward bending and relaxes on forward bending (Fig. 18).

2) The *posterior longitudinal ligament* arises from the basillary portion of the occiput and runs over the posterior surface of the vertebral bodies into the spinal canal its full length downward to the coccyx. It is narrow as it passes

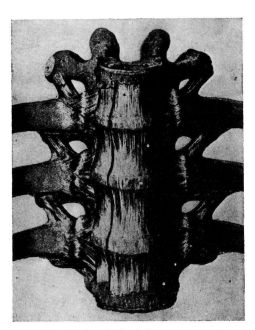

Fig. 18. Anterior longitudinal ligament. (Spalteholz)   Fig. 19. Posterior longitudinal ligament. (Spalteholz)

the vertebral bodies but widens across the intervertebral discs with which it is closely interwoven. It is thicker in the dorsal than it is in the cervical and lumbar region. In the latter it often becomes inadequate to sustain the pressure of a herniating intervertebral disc, especially if degeneration of the latter allows the nucleus or the degenerated portion of the annulus to strain against the ligament. It becomes tight on forward flexion and it is relaxed on extension of the lumbar spine (Fig. 19).

b) *Uniting the posterior column.* The only long intersegmental ligament is the supraspinous. It starts as ligamentum nuchae from the external occipital protuberance in a slightly concave line to the 6th or 7th cervical and from there it continues along the tips of the spinous processes as a round slender strand down to the sacrum. Through its entire length it tightens on forward flexion and relaxes on extension of the spine.

2. The longitudinal intrasegmental systems

a) The ligamenta interspinalia are arranged along the entire spine as firm membranes between the spinous processes. They separate the deep muscle layers of both sides. Between the upper spines they are thin and slender but they become very strong and powerful between the spinous processes of the lumbar vertebrae (Fig. 20).

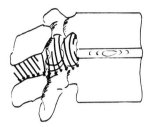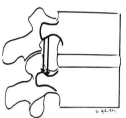

FIG. 20. The ligamentum inter-
spinosum and ligamentum flavum (after Braus).

FIG. 21. Intertransverse
ligament (after Braus).

These ligaments also become tight on forward flexion after a tightening of the supraspinous ligament has occurred.

b) The ligamenta intertransversaria unite the transverse processes. They are poorly developed in the cervical spine and are strongest in the lumbar. In the dorsal section they serve for the origin of the multifidus. They become

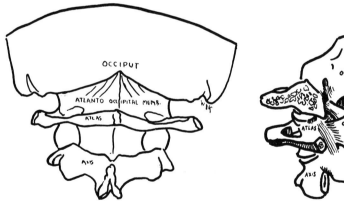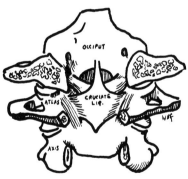

FIG. 22. Atlanto-occipital membrane.

FIG. 23. Cruciate ligament

tight on side bending on the convex side while the concave side ligaments re-lax (Fig. 21).

c) The ligamenta flava are broad flat structures uniting the arches (inter-laminar ligaments). They also tighten in forward flexion after the supra and in-terspinous ligaments. In unstable spines one will find them greatly thickened as they also take care of the anteroposterior stability. They also tighten up on axial rotation (Fig. 20).

d) The reinforcing ligaments of the atlanto-occipital and atlanto-axial joint are modifications of the general ligamentous tracts. The posterior atlanto-occipital membrane (Fig. 22) unites the arches of the atlas to the occiput. It is a broad flat ligament. The upper shank of the cruciate ligament is fastened between odontoid process and occiput (Fig. 23). The ligamenta alaria of the

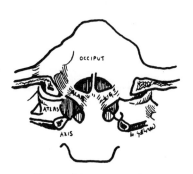

FIG. 24. Ligamentum alare

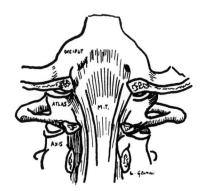

FIG. 25. Membrana tectoria.

odontoid spread sideways from the odontoid process to the anterior ring of the atlas (Fig. 24) and both are covered by a membrana tectoria (Fig. 25).

3. The articular and capsular reinforcements

a) Between atlas and occiput the anterior and posterior accessory ligaments run from the occiput to the lateral portion of the atlas.

b) Between occiput, atlas and odontoid, the odontoid is secured to the occiput by the ligamenta alaria mentioned above which run to the lateral portion of the occipital foramen; also by the ligamentum cruciatum which run from the body of the atlas to the occiput.

The relation between atlas and axis is further controlled by the anterior and posterior atlanto-axial membrane.

The effect of all these ligamentous reinforcements is that the symmetrical antero-posterior movements of the atlanto-occipital joints are checked by the capsule and the adjacent ligaments, while in the lower (atlanto-axoid) joint where the principal motion is rotation, the check is provided mainly by the ligamenta alaria.

c) The capsules of the intervertebral articulations are reinforced anteriorly and medially by the ligamenta intercruralia (ligamenta flava) which stream into these articulations. Secondarily the ligamenta interspinalia and intertransversaria reinforce the intervertebral joint.

d) For the reinforcement of the costovertebral and costotransverse articulations there are several segmental ligaments available; first the radiate ligaments between the heads of the ribs and the vertebrae. They tighten in their respective areas on elevation, depression or rotation of the rib. Then, from the tip of the transverse process, ligaments run to the next lower rib; the anterior and posterior costo transverse ligaments; and finally, also from the tip of the transverse process reinforcing the costotransverse articulation run the ligamenta tuberculi costae to the costal tuberosities (Fig. 26).

e) In the sacrolumbar region there are special provisions for stabilizing this joint.

From the fourth lumbar transverse process strong ligaments run to the posteromedial border of the os ilei, the iliolumbar ligaments (Fig. 27). These hold the 4th and 5th lumbar back from sliding forward and it is particularly important where there is a sacrolumbar instability, a sharp sacrolumbar angle

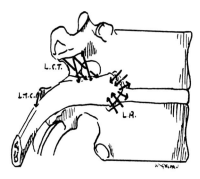

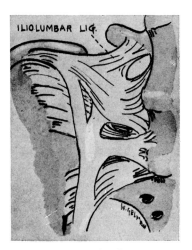

FIG. 26. Reinforcement of costovertebral and costotransverse articulations. (After Braus)

    L.R.   Ligamentum Radiatum
    L.T.C. Ligamentum tuberculi costae
    L.C.T. Ligamentum costotransversa-
          ria, anterior and posterior

FIG. 27. Iliolumbar ligament.

for instance, or a spondylolysis. In these cases one also finds the ligamenta flava considerably thickened.

f) The reinforcing ligamentous apparatus of the sacroiliac junction is represented by the sacroiliac ligamentous mass. It has an anterior and a posterior portion. The anterior portion consists of powerful oblique and transverse fibers covering the entire anterior surface of the junction. The posterior portion which occupies the space between the projecting posterior rim of the os ilei and the sacrum is really a non-descript mass of tough fibers crossing in all directions, though some anatomists distinguish a deep and superficial portion. This ligamentous apparatus is under considerable stress in upright standing position where it has to resist the forward rotating tendency of the sacrum under the gravitational load. It also comes under strain on length rotatory movement of the trunk as the rotatory moment is transmitted from the trunk and upper extremities to the pelvis.

IV. THE STATICS OF THE NORMAL SPINE

The study of the statics of the *resting spine* has two aspects:

First the investigation of the form and architecture of the skeleton insofar as they reflect the stresses imposed upon it.

Secondly, the analysis of the conditions of equilibrium between the different parts in upright standing position. So far as the latter is based on non-contractile tissues such as ligaments and is not due to muscular effort it may be called a passive equilibrium. On the other hand, so far as the equilibrium is maintained by active muscular forces it is a problem of the dynamics of the spine which we reserve for the following lecture. Under normal conditions both passive tension and active muscular contraction are necessary to hold the spine in equilibrium.

A. THE ARCHITECTURE OF THE SPINE

The manifestation of the law of functional adaptation.

According to the law of Wolff we should find in the architecture of the vertebrae trabecular arrangements which correspond to the stresses to which the spine is exposed. A study of these trajectories was made by Japoit and

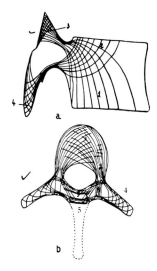

Fig. 28. Architecture of vertebra:
a. Sagittal:
    1. Superior oblique trabecular system
    2. Inferior oblique trabecular system
    3. & 4. Supporting systems for the processes
b. Transverse:
    1. Transverse body system
    2. & 3. Intertransverse system
    4. Oblique system to articular process
    5. Oblique system to spinous process.
(Callois and Japoit)

Gallois[1] who were able to establish the principal trabecular systems and to interpret them in terms of static stresses.

There are recognizable: a principal vertical one and several secondary oblique and horizontal systems. Except for the interruption by the intervertebral disc, the vertical system runs throughout the entire row of vertebral bodies from the odontoid process which represents the first cervical vertebral body to the sacrum. The oblique accessory systems run in four tracts: on each side there is a superior oblique system from the upper articular process of one side downward to the opposite side of the vertebrae; and an inferior articular system which runs from the inferior articular process of one side upward to the opposite side of the vertebrae.

There is also a horizontal accessory system with a right and left tract. It starts in the transverse process and streams into the body of the vertebra where it crosses the midline intersecting with the horizontal trabecular system of the opposite side.

The interpretation of these systems in mechanical terms is as follows: the vertical systems sustain the body weight and all jars and shocks which reach the spine in perpendicular direction. It is the principal system and one that resists atrophy more than the others. In osteoporotic spines the secondary trabeculae undergo atrophy first and their disappearance makes the lines of the vertical systems stand out more sharply. The spirally wound oblique systems resist torsion; they also share with the vertical system in the resistance to bending and shear.

The horizontal system is principally tension resistant and the minor accessory systems of the transverse and spinous processes also resist muscular pull (Fig. 28).

B. THE DISTRIBUTION OF STRESSES THROUGHOUT THE SPINAL COLUMN

If the spinal column were a straight solid rod firmly implanted into the pelvis it would be equilibrated against gravitational stresses comparatively easily, the only condition being that the line of gravity fall in the supporting surface.

If the spinal column were a straight but multiarticulated rod, a complete passive equilibrium could still be maintained. The condition would be that the line of gravity fall in the supporting surfaces and into all partial centers of gravity of the component links.

However, the human spine is neither a straight rigid nor an articulated straight rod; the partial centers of gravity of the respective links do not lie in the line of gravity and neither do the centers of articulations (with the exception of a few) fall into this line. The result is that gravity produces, from vertebra to vertebra, rotational stresses which must be neutralized by opposing forces lest the spine lose its equilibrium and collapse.

These rotatory stresses produced by gravity tend to accentuate the already existing curves. In the dorsal spine where the line of gravity passes in front of the centers of motion, the tendency of the spine is to collapse in forward flexion under the superincumbent weight. Some resistance to the forward collapse is offered by the posterior longitudinal ligament and by the ligaments of the posterior column such as the interspinous ligament and the ligamenta flava. In the lumbar spine the line of gravity runs behind the centers of motion. The tendency is toward increasing the lumbar lordosis under the body weight. Resistance is here furnished by the anterior longitudinal ligament.

The translatory stresses produced by gravity are better provided for. The shock-absorbing action and the buffer effect of the intervertebral discs are able to distribute and to deflect the innumerable jolts and stresses which the spine receives during its everyday activity. So perfect is, under this mechanism, the normal distribution of these translatory stresses that any disagreeable sensation following an ordinary jolt indicates that something is amiss with the mechanism of the discs. Under pathological conditions this mechanism may be gravely impaired and severe strains or even fractures may occur. The normal spine, however, is an elastic system kept in perfect shape of dynamic equilibrium by an extremely finely graded system of distribution and absorption of gravitational and other translatory stresses.

C. THE INTRINSIC EQUILIBRIUM

Incorporated in these soft structures there are certain intrinsic forces which combine to make the human spine a closely knit mechanical unit. The fact that the spine has antero-posterior curves makes it less susceptible to lateral bending than a straight rod would be. Even a small increase in the antero-posterior curve will, in some manner, protect the spine against a lateral curvature and the latter occurs less often in the kyphotic than in the straight spine.

Still this does not explain why the spine, if entirely deprived of musculature does not collapse as easily as one should expect. It still represents a rather rigid column and not only in erect position but it also shows normal limitations of flexibility. The ease with which the spine can be held in erect position and kept from collapsing with a comparatively small amount of muscle force is quite in contrast with the amount of muscle power necessary to hold the lower limbs in equilibrium against gravity. After a long and protracted illness a patient may be able to sit up in bed long before his legs will hold him up because so much less is the muscle power necessary for the upright position of the trunk than is that for the stabilization of the lower extremities.

All this indicates a hidden intrinsic mechanism within the spinal column itself which makes this pluriarticular structure a comparatively stable and rigid mechanical unit. These hidden forces are vested in the elastic properties of the soft non-contractile structures of the spine. They are responsible for what is called the intrinsic equilibrium.

If one separates the entire column of the arches from the column of the body, the column of the arches at once will shrink so that it loses 14% of its original length (Fick[2]). The reason is that as long as the arches are united with the bodies they are spread apart and the ligaments between arches and their apophyses, supra and interspinous flava and so forth are under tension. This tension is released as the posterior column is separated from the anterior. It means that the tension stresses of the ligaments press the vertebrae firmly against each other and thereby contribute materially to the solidity of the spine as a whole. All the intersegmental longitudinal and all the segmental ligaments of the spinal column are under tension in the normal spine.

Another factor of intrinsic stress is vested in the intervertebral discs. These are also fitted tightly into the system, not under tension as the ligaments are but under pressure stress. The weight of the body flattens the intervertebral disc, and just as the ligament resists tension so does the disc resist compression. It becomes an elastic buffer placed between the bodies of the vertebra to absorb pressure stresses of all kinds. Even when it changes its shape as it does on motion of the spine it occurs under elastic pressure resistance (Fig. 29).

If a spine is deprived of its ligaments as well as of its muscles it will automatically collapse. But as long as the ligaments and the discs are intact it preserves a considerable resistance against deformation due to the elastic resistance of the ligamentous structures and the discs. Thus the intrinsic equilibrium of the spine is the resultant of the elastic tension resistance of the ligaments and the elastic pressure resistance of the disc.

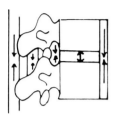

Fig. 29. Intrinsic equilibrium of the spine. (Steindler)

A similar arrangement exists in the thoracic cage. Here also the ribs are forced into the cage under considerable stress much as a rim would be fitted onto a wheel. In addition the elastic cartilaginous portions constantly change their shape with the respiratory movements. The result is that the vertebral bodies find themselves clamped between the vertebral ends of the ribs as in a vise. This arrangement greatly increases the elastic resistance of the thorax and a considerable amount of energy is stored up in it. Some of it is used in the respiratory movement but the greater part is absorbed in the pressure exerted by the ribs against the vertebral column. The elasticity stored in the thoracic cage is therefore another factor contributing to the intrinsic equilibrium of the spinal column.

V. SUMMARY

1. The spinal column performs many functions both static and dynamic.

2. In construction it can be grossly likened to two isosceles triangles joined at base, the upper one long and drawn out representing the movable, and the lower short one the immovable portion of the spine.

3. The modal number of the presacral segments is very constant (96%); most variations such as cervical ribs, sacralization are relative, i.e., one segment of the spine is numerically increased at the expense of its neighboring segment.

4. Of the four antero-posterior spinal curves the lordoses (cervical and lumbar) develop with the assumption of static functions (raising of head, sitting, standing).

5. For the orientation of the spinal column in relation to palpable skeletal points we find Prentiss' landmarks very useful.

6. The relation of the curves to the line of gravity indicates in the normal spine a fair distribution of weight stresses in antero-posterior direction.

7. The movements in the intervertebral articulations are all gliding, about a center located at or near the disc. According to the slope of the articular facets, they describe in rotation the surface of a cone of varying height.

8. Movement in the atlanto-occipital joint is more complex as it is a combination of gliding and rocking.

9. The atlanto-axoid movement provides principally for length rotation by

four articular facets, two between the bodies of atlas and axis and two between anterior ring of atlas and odontoid process.

10. The sacroiliac junction consists of the kidney-shaped sacroiliac joint and the powerful sacroiliac ligamentous masses. There is normally no measurable motion in this junction but it is subject to considerable static stress.

11. The connection between ribs and vertebrae consists of the costovertebral articulations which are bipartite with exception of the 1st, 10th, 11th and 12th, and of the costo-transverse articulations of the 1st to the 10th rib. The 11th and 12th have no costo-transverse articulation. This double junction restricts the costal movement to one plane about the combined axes of the two articulations.

12. Only seven ribs enter in direct relation with the sternum by sternocostal articulations. The 8th, 9th, and 10th have their front ends joined together by means of cartilaginous junctions. The anterior ends of the 11th and 12th are free (floating ribs).

13. The intervertebral disc acts as amphiarthrosis for all intervertebral movements due to its ability to change form.

14. The ligamentous apparatus of the spinal column can be divided in the following classes:

a) A longitudinal tract system binding the vertebrae together into a mechanical unit; the intersegmental ligaments (anterior longitudinal, posterior longitudinal and supraspinous ligaments).

b) An *intrasegmental* longitudinal system between single vertebrae (lig. interspinalia, lig. flava, lig. intertransversaria and the reinforcing ligaments of the atlanto-occipital and the atlanto-axoid junctions).

c) Articular and capsular reinforcements between atlas, odontoid and occiput, the reinforcements of the capsule of the intervertebral articulations and the reinforcing ligaments of the costovertebral and costotransverse articulations, as well as the special ligamentous tracts securing the sacrolumbar and sacroiliac regions.

15. The inner architecture of the vertebral body reflects the different static and dynamic stresses the spine is exposed to. There is a principal vertical pressure sustaining system and several oblique and horizontal trajectories resisting mainly pull and torsion.

16. The rotatory gravitational stresses tend to accentuate the already existing antero-posterior curves. The translatory gravitational stresses are, in the normal spine, adequately absorbed by the buffer action of the disc.

17, What is called the intrinsic equilibrium of the spine is its ability to maintain its shape without collapsing, by means of the discs and ligaments alone. This ability is due to the elastic resistance with which these structures are fitted into the spinal column as a unit: the ligaments under tension and the discs under pressure. This makes them act the same as a system of guy wires would on an upright pole: the tension of the wires acting like the ligaments and the pressure they produce on the pole acting like the pressure resistance of the discs.

BIBLIOGRAPHY

1. BEADLE, O. A.: *The Intervertebral Disc. Med. Research Council,* London, 1931.
2. FICK, R.: *Handbuch der Anatomie und Mechanik der Gelenke.* Jena. G. Fischer, 1910.
3. GOLDTHWAIT, J. E.: Relation of Posture to Human Efficiency. *Boston Med. & Surg. J.,* Dec. 9, 1909.
4. JAPIOT and GALLOIS: Architecture Intèrieure des Vertebres. *Rev. de Chir., Paris, 63:*688, 1925.
5. KRAUSE, C. F.: *Handbuch der Menschlichen Anatomie,* Vol. II. W. Krause, Hannover, 1876.
6. PRENTISS, H. J.: Personal Communication and Anatomical Lectures (not edited).
7. STRASSER, H.: *Lehrbuch d. Muskel und Gelenkmechanik, Vol. II.* Berlin, J. Springer, 1913.
8. WILLIS, T. A.: Backache from Vertebral Anomalies. *S.G.B., 38:*658, 1924.

Lecture X

THE DYNAMICS OF THE NORMAL SPINE

I. ANALYSIS OF MOVEMENTS IN THE SPINAL ARTICULATIONS

A. THE INTERVERTEBRAL JOINTS

ACCORDING to the spinal segment involved, movement in the intervertebral joint varies in type, in direction and in range of motion, and all of these variations are contingent upon the shape and position of the articular facets.

Each of the vertebrae carries two pairs of articular processes, an upper pair and a lower pair. The vertebrae are stacked upon each other in such a fashion that the lower articular process of the upper superincumbent vertebra covers

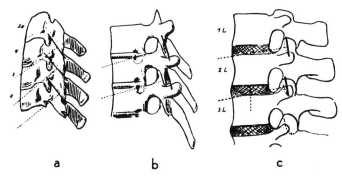

FIG. 1. The change in the slope of the articulations of the spine.
a. Cervical. *b.* Dorsal. *c.* Lumbar. (Strasser)

the upper articular process of the lower. They therefore resemble a number of cups stacked upon each other, their openings directed downward. As the slope of these articulations varies from section to section, we find that these imaginary cups are shallow and rather wide-mouthed in some sections and deep and narrow-mouthed in others.

In all these articulations the facets glide upon each other during motion. Therefore one must look for the center of motion outside of the facets.

This center is the intervertebral disc. It represents a universal joint which permits rotatory motion in three planes: flexion and extension in the sagittal, side bending in the frontal, and axial rotation in the transverse plane. In gliding the joint facets describe circular pathways. The radius of the circle is the perpendicular from the center of the facet to the center of motion in or about the disc. From the difference of the obliquity of these radii one can draw conclusions regarding the change in slope these articulations assume in the various sections of the spinal column (Fig. 1).

1. The general regional arrangement

In the *cervical* spine the articulations are in a gently downward and backward sloping plane which approaches the transverse plane. The articulation is a saddle joint and allows motion in three directions: forward and backward in the sagittal, sideways in the frontal, and axial in the transverse plane (Fig. 1a). The most favored motion is side bending in the frontal plane.

In the *dorsal* spine the articulations form a part of the surface of a cone. They are directed more perpendicular and they approach more the frontal plane. This arrangement provides well for forward flexion and axial rotation, but because of the shingle-like imbrication of the articular processes the dorsal extension of the spine is limited (Fig. 1b).

In the *lumbar* spine the articulations are arranged in a sagittal plane. The lower articular process of the upper vertebrae forms the upper half of the joint and is concave. The upper articular process of the lower vertebrae forms the lower half of the joint and is convex. This arrangement provides more for antero-posterior than for lateral motion and allows only a limited amount of rotation (Fig. 1c) (Braus[1]).

The first and second spinal nerves emerge behind the articulation since the intervertebral foramen lies posterior to these. The lower articular surface of the axis and the articular surfaces of all vertebrae below lie behind the intervertebral foramen so that the spinal nerves emerge in front of them.

The primary division of the spinal nerves into a ventral and a dorsal ramus occurs immediately after they emerge from the intervertebral foramen, the dorsal ones running along the transverse mass of the lateral process and the volar along its costal elements. In the sacral spine, however, the foramen is the shape of a inverted Y (λ); the anterior leg of the λ is the foramen which opens between the fused costal elements (Strasser[10]), and the posterior leg is the opening between the transverse elements of the sacral vertebrae.

2. Special regional differentiations

a) The atlanto-occipital joint (Knese[4]). The angle between the occipital condyles is 55-60° opening dorsally, and the length axis points from ventro-medial to dorsolateral so that the distance between the anterior points of the condyles is only 21-25 mm. while that of the posterior is 30-34 mm. The transverse axis of the condyles at right angles to the length axis runs from medio-dorsal to antero-lateral. The condylar curve is irregular; the ventral portion has a radius of 13 mm., the dorsal portion 21 mm., and frequently there is also some asymmetry between right and left.

The principal motion is flexion-extension in the sagittal plane. Since there is no common frontal axis for both sides, this motion cannot be a pure rocking one but necessitates a gliding of the atlas upon the condyles of the occiput. Also, because of the change of the radius, the axis changes. In flexion, the movement is about the smaller anterior, in extension about the larger posterior radii of the condyles. The different instantaneous axes, therefore, form an evolute, a curved line similar as is the case in the knee joint and the joint tightens in dorsi-flexion and becomes loose in volar flexion.

There is some variance among the authors regarding the amplitude of the antero-posterior motion. According to Knese[4] it is 21.7° while Virchow[11,12] gives a 20-28° in vivo, and Strasser[10] gives it 35°, Mollier[6] 20-35°, with the greater ratio for dorsi-flexion. The amount of for and backward gliding which accompanies flexion and extension movement is 10 mm.

In the frontal plane there is also no common axis and consequently side bending becomes associated with length rotation; bending to the left is associated with rotation to the right and vice versa. Lateral bending thus becomes a circling movement associated with convex side rotation (Strasser[1]). It is maximal in slight volar flexion; there is no lateral bending in extreme extension, neither is there any in extreme volar flexion.

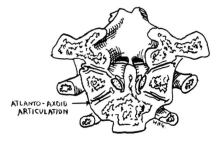

ATLANTO-AXOID ARTICULATION

FIG. 2.

Fick[2] gives the excursion range of the atlanto-occipital joints for lateral bending, both sides combined, as 30-40°.

b) The atlanto-axoid joint. This joint forms with the atlanto-occipital joint, a functional unit with three degrees of freedom of motion. Axial rotation has by far the greatest range amounting to 82.4° for both sides while side bending excursion is only about 10° on both sides and the antero-posterior range, as established by Knese[4] is 25.4° although Virchow[11,12] gives it only between 16° and 21°.

In the frontal plane the facets slope from medio-cranial to latero-caudal (Fig. 2) with a slight concavity pointing cranially. Motion in the sagittal plane, i.e., about the transverse axis, involves here also a gliding in which the arch of the atlas moves upward and forward on the odontoid in dorsiflexion and vice versa in volar flexion. In contrasts to the atlanto-occipital joint, the dorsal excursion range is greater than the volar. In this joint also rotation is associated with side bending.

Taking the atlanto-occipital and the atlanto-axoid joints as one functional unit, one finds that in forward flexion atlas and axis separate posteriorly. They are checked in front by the ligamentous fixation of the odontoid and the ligamenta alaria become tight. In lateral motion the opposite ligamentum alare becomes tight first but because the lateral bending is associated with length rotation the other one also tightens up. In fact it is difficult in axial rotation of the head to suppress side bending.

In upright position the head is in labile equilibrium. The foramen magnum occupies almost a horizontal plane (only 7-21° slope with the horizontal), and the atlanto-occipital joint is situated precisely below the center of gravity of the head so that the latter is balanced upon it. Commensurate with this labile equilibrium of the head in normal upright position the dorsal portion of the occipital condyles is flat for better balance and allows little axial rotation or lateral gliding in contrast to the ventral portion which is more strongly curved and which provides for axial rotation. Both joints combine lateral movement with rotation, but most of the movement in the frontal plane occurs in the lower cervical spine (Ficke,[2] Strasser[10]).

  c) *The temporo-mandibular joint.* The condylar process of the mandible articulates with the temporal bone in a hinge joint of peculiar construction. The capsule is attached to the condyles but it lacks reinforcing ligaments except laterally. The menisci cover the entire cartilage and are attached to a lateral epicondyle by the lower lateral meniscus band and to the base of the skull anteriorly and posteriorly by the anterior and posterior meniscus bands (Fig. 3).

  Both menisci move as one unit. On opening the jaw one sees the condyles

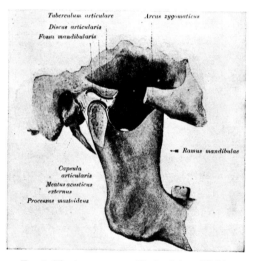

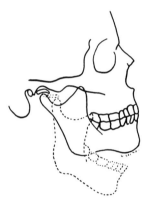

FIG. 3. The temporo-mandibular joint. (Fick)

FIG. 4. Forward movement of the condyle of the mandible.

gaping anteriorly. It is not a pure hinge movement. An antero-posterior shift occurs in combination with it, the pterygoid muscles moving the jaw forward (Fig. 4). A pure antero-posterior movement can be accomplished by the external pterygoid alone. The lateral (chewing) movement is not as highly developed in man as it is in certain herbivorous animals (ruminants). In this movement one condyle rotates axially in its socket while the other describes an arc.

  It has been debated whether the meniscus moves forward with the condyles or whether it is squeezed forward by the condylar pressure. The latter is not likely because the external pterygoid is inserted in the front end of the meniscus and, on contraction, forces the meniscus forward. The closure of the jaw is accomplished with considerable force by the internal pterygoid, the masseter and the temporal muscle. Their combined strength is estimated by E. Weber at 400 kg. In forward and backward movement the external pterygoid, the genio and mylohyoid and the biventer are active. This is a movement particularly well developed in the rodents.

d) The sacrolumbar articulation. There is a departure in this articulation from the general pattern of the intervertebral joints. Because of the slope of the sacrum the lumbosacral angle sometimes approaches 90°. This angle not only causes the intervertebral disc to assume a wedge shape but it also produces a considerable shearing stress upon the articulation itself. Some of it is absorbed by the sacrolumbar articulations being arranged in the frontal plane in contrast to the lumbar joints which are all in the sagittal plane. Such an arrangement gives more latitude to antero-posterior and lateral motion.

II. THE DYNAMICS OF THE SPINE

THE DISTRIBUTION OF MOTION

A. MOBILITY BETWEEN BODIES

Motion between the vertebral bodies occurs in four directions. Three of these are rotatory and one is translatory.

In translatory direction in the long axis of the spine distension and compression are possible (Fig. 5). Distension is resisted by the elasticity of the longitudinal ligaments as well as by the weight of the body. Compression is checked by the elastic buffer action of the intervertebral discs. In this translatory direction the entire mobility amounts to hardly more than 5-10 mm. If one applies traction the effect is much more a flattening of the antero-posterior curves than a true separation of the vertebral bodies. Under the weight the total height of the spine decreases and it becomes larger in recumbency but these fluctuations are only to a lesser degree due to an actual change in the length of the spinal column.

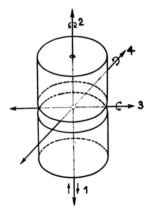

FIG 5. Diagram of movement between vertebral bodies.
1 and *2* Distension and compression axis (translatory), vertical rotation axis.
3 and *4* Rotatory axes, frontal and sagittal (flexion - extension and lateral movement).

The second type of motion between vertebrae is the axial rotation about a perpendicular axis. Here it is the elasticity of the circular fibers of the annulus fibrosis which determines the range (Fig. 5).

The third type is the forward and backward bending, a rotation about the frontal axis. The intervertebral disc becomes compressed in front in forward and behind in backward bending (Fig. 5).

The fourth type is motion about a sagittal axis or lateral bending. The shape of the disc changes here also, becoming flattened on the side of the concavity and widened on the convex side (Fig. 5).

B. MOTION IN THE INTERVERTEBRAL ARTICULATIONS (FIG. 6)

The dorsal spine favors forward flexion and rotation; the lumbar spine backward extension and lateral bending, and the cervical spine including the occipito-atlanto-axoid articulation is suited for all three types: flexion-extension, lateral bending and axial rotation.

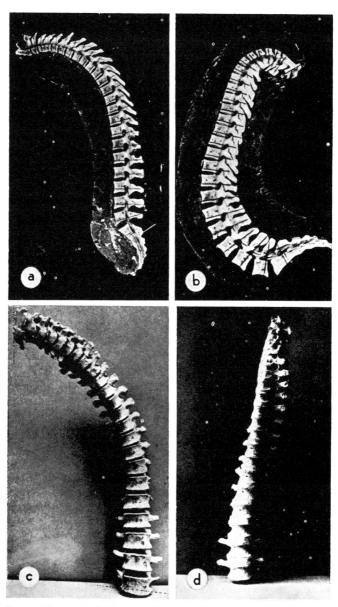

FIG. 6. Motion in the intervertebral articulations. *a.* Forward flexion. *b.* Backward extension. *c.* Side bending. *d.* Axial rotation. (Fick)

Specifically motion is distributed in the living as follows:

In the sagittal plane forward flexion is enacted mainly in the cervical and upper dorsal spine (Fig. 6a), while the cervical and the lumbar spine are the principal backward extensors (Fig. 6b). In the frontal plane the cervical spine again shows the greatest mobility while the lateral mobility of the dorsal spine is considerably less (Fig. 6c). In the transverse plane rotation about the longitudinal axis rests almost completely with the cervical and the dorsal sections. The ability to rotate axially decreases constantly from the occiput downward. In the lumbar spine it is comparatively small (Fig. 6d).

Deprived of all its muscular and ligamentous checks the spine collapses automatically. Yet the shape which the spine assumes in collapse is not a haphazard one but it is determined by the intervertebral articulations. In the dorsal spine, for instance, the slump occurs in forward flexion and in rotation, since these are the two types of motion favored by the arrangement of the dorsal articulation. In the lumber spine which allows only scant lateral motion, the rotational element prevails (Steindler[9]).

While movement between vertebrae originates in the intervertebral disc as center of motion, the intervertebral articulations, because of their anatomical arrangement, have a directing and a restraining effect which makes the motion "trackbound" (Zwangsläufig).

III. THE MUSCLE MECHANICS OF THE SPINE

For the purpose of simplification the complex musculature can be divided into four main groups according to their principal mechanical action (Fig. 7).

A) The system of extensors of the spine.

B) The system of flexors of the spine.

C) The system of lateral benders of the spine.

D) The system of rotators of the spine.

A. THE EXTENSORS OF THE SPINE

1. The deepest layer of this group is formed by the short muscles arranged from segment to segment. In their medial portion they are pure extensors: the rectus capitis posterior major and minor and their continuation, the inter-spinosi (Fig. 8).

The lateral portions are the obliquus capitis posterior and inferior; they continue throughout the spine as inter-transversarii. The rotatores and levatores costarum extend from the spinous to the lateral processes or from the transverse processes to the tubercles of the next lower rib. Only the straight medial portion of these muscles act as pure extensors; the more obliquely placed have a rotatory component which becomes neutralized when they act symmetrically (Fig. 9).

2. Superimposed upon this deep layer are the oblique muscles, the multifidus and semispinalis. Originating from the spinous processes and inserting into the transverse processes of the next or second lower vertebra, the

multifidus is an extensor as well as a rotator and side bender (Fig. 10); a similar arrangement is that of the semispinalis except that it is longer reaching from the spinous processes to the fourth to sixth lower vertebrae (Fig. 11). Therefore its extensory components greatly exceed the rotatory.

3. The most superficial layer is that of the powerful sacrospinalis mass comprising the long muscles of the back. Arising from the sacrum and the lumbodorsal fascia as well as from the spinous processes of the lower lumbar

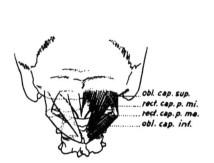

Fig. 8. The deep layer extensory muscles of the head. (Poirier)

Fig. 7. The muscle systems of the spine. (H. Meyer)
a. Extensory
b. Flexory
c. and d. Lateral and rotatory
e. Transverse

Fig. 9. The deep rotatory muscles of the spine (short and long rotators). (Strasser)

vertebrae and from the crest of the os ilii, this mighty muscle mass extends upward to the occiput covering the entire length of the back.

It is divided into two portions, a medial and a lateral one. The medial portion is the longissimus. It fills the groove between the spinous processes and the angles of the ribs. This muscle arises from the transverse processes of the lumbar spine and goes to the transverse processes of all dorsal vertebrae as well as to the angles of the twelfth to the second rib (Fig. 12). The longissimus cervicis and capitis are equivalents of this muscle. Because of its almost strictly longitudinal course its action on the back is mostly extensory.

The lateral portion of this mass is the iliocostalis. It consists of the iliocostalis lumborum which comes from the common sacrospinalis mass and inserts into the angles of the 12th to the 4th rib; the iliocostalis dorsi which arises from the angles of the 12th to the 7th rib and inserts into the transverse processes from the 7th dorsal to the 7th cervical; and, finally the iliocostalis

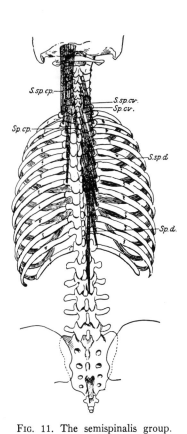

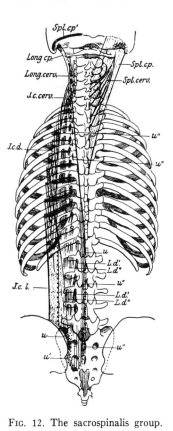

FIG. 12. The sacrospinalis group. (Strasser)
  I.c.l. Iliocostalis lumborum
  I.c.d. Iliocostalis dorsi
  I.c.cerv. Iliocostalis cervicis
  L.d.' Longissimus dorsi
  u-u'. Sacral origin
  Long.cp. Longissimus capitis
  Long.cerv. Longissimus cervicis
  Spl.cp. Splenius capitis
  Spl.cerv. Splenius cervicis

FIG. 11. The semispinalis group. (Strasser)
  S.sp.d. Semispinalis dorsi
  S.sp.cv. Semispinalis cervicis
  Sp.d. Spinalis dorsi
  Sp.cv. Spinalis cervicis
  S.sp.cp. Semispinalis capitis
  Sp.cp. Spinalis capitis

FIG. 10. The multifidus group. (Strasser)

cervicis which originates from the angles of the 6th to the 3rd ribs and inserts into the transverse processes from the 6th to the 4th cervical vertebrae. The splenius capitis and cervicis (Fig. 12) are continuations of this system. The splenius capitis comes from the ligamentum nuchae and, diverging cranially, inserts into the mastoid process and into the upper occipital ridge. The splenius cervicis is the distal continuation of the former, inserting into the posterior tubercles of the transverse processes of the 1st to the 2nd dorsal vertebra. Acting unilaterally the ilio-costalis system develops a rotatory component as well as an extensory.

B. THE FLEXORS OF THE TRUNK

They are also arranged in several distinct systems.

1. The first system is the spino-thoracic group. This group flexes the head and neck against the thoracic cage. The muscles come from the spinal column or from its cranial equivalents and insert into the thoracic cage. The sternocleido mastoid runs from the mastoid process in a forward, downward and inward direction to the sternum and the medial half of the clavicle. Because of its obliquity, it has three components. It flexes the head forward, it inclines it laterally to its side, and it rotates it to the opposite side. It is a pure flexor only if its inclinatory and rotatory components are neutralized by the opposing muscle of the other side.

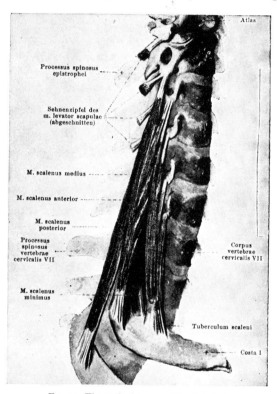

FIG. 13. The scaleni group. (Spalteholz)

The three scaleni, the anterior, the medius, and posterior arise from the anterior and posterior tubercles of the cervical spine respectively to insert into the 1st and 2nd ribs. Acting singly they flex the head forward against the chest and incline it homolaterally; acting conjointy they are pure forward flexors (Fig. 13).

2. A ventral group of flexor muscles is arranged between the mandible, the hyoid and the thyroid cartilages and the sternum. Not taking origin at the spinal column or at the occiput they are indirect forward flexors. The muscles are the mylohyoid, the sternohyoid and the thyrohyoid. They are pure forward flexors of the head against the sternum, or if the mandible is fixed they elevate the hyoid and thyroid bones toward it and thereby elevate the sternum. In this action they become auxiliary inspirators.

3. The third system is the spinal group proper, arising from the front of the spinal column. The longus colli runs upward from the anterior tubercles of the transverse process of the cervical spine to the anterior surfaces of the

cervical bodies higher up. The longus capitis also comes from the transverse process and goes into the basillary portion of the occiput. Both muscles flex the head straight forward (Spalteholz[8]).

The medial and lateral rectus capitis anteriores are continuations of the longus capitis. They come from the transverse processes and lateral masses of the atlas to insert into the occiput (Fig. 14).

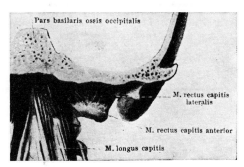

FIG. 14. (Spalteholz) anterior flexor of head.

4. A fourth group are the muscles which flex the trunk against the pelvis and the thighs by way of the thorax. In a manner this group is continuous with the one which flexes the head against the thorax, the sternocleido mastoid and the scaleni. This group is interrupted by the sternum, only the triangularis sterni forming the connecting link. The muscles of this fourth group arise from the lower costal margin and are divided into longitudinal and obliques.

The longitudinal muscle is the rectus abdominis which inserts into the symphysis. Its right and left halves are separated in the midline (Fig. 15). This muscle is a straight flexor.

The oblique muscles are the obliquus abdominis externus and internus. Acting singly they are rotators and side benders; when combined, their lateral and rotatory components are suppressed and then they assist the rectus abdominis in flexing the thoracic cage against the pelvis (Fig. 15).

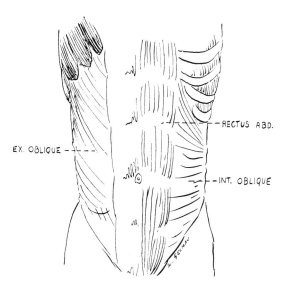

FIG. 15. Anterior and lateral muscles of the trunk.

5. Finally there is the system of the spino-femoral muscles which flex the whole trunk against the thigh. These muscles are the iliacus arising from the inner surface of the os ilii and the greater and lesser psoas which come from the lateral walls of the last dorsal and the upper lumbar vertebral bodies. Both muscles combine into a common tendon inserted into the lesser trochanter of the femur. Although the primary action of these muscles is to flex the thigh against the pelvis, when the former is stabilized they become powerful forward flexors of the trunk (Fig. 16).

C. THE SIDE BENDERS OF THE TRUNK

Because side bending and rotation are always combined, pure side bending exacts the suppression of the rotatory component by the antagonist. The external oblique arises from the 5th to the 12th rib and, taking a course downward, forward and inward, ends in a broad and firm aponeurosis which covers the anterior surface of the rectus abdominus. The lower portion blends with the ilio-inguinal ligament (Fig. 15).

The internal oblique also comes from the lower thoracic aperture but its fibers diverge in their course. The upper portion is directed upward; the

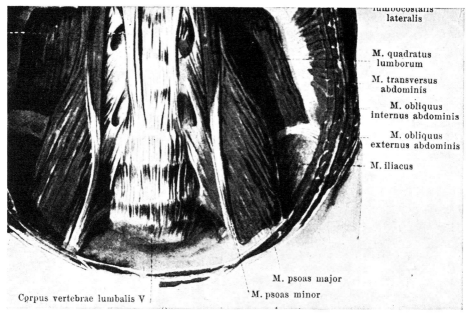

FIG. 16. Spino-femoral muscles. (Spalteholz)

middle portion goes into the aponeurosis of the rectus, splitting into an anterior and posterior lamella and enclosing the rectus as they reach the muscle; the lower third takes a course forward and downward and, ending in an aponeurosis, blends with the anterior rectus sheath only.

Underneath this muscle is the transversus abdominis, a flat quadrangular muscle which comes from the 7th to the 12th costal cartilage and from the lumbodorsal fascia. It runs more horizontally forward with a tendon which blends with the posterior sheath of the rectus.

Neither of these muscles is a pure side bender. The external oblique acting symmetrically is a powerful forward flexor, synergistic with the rectus abdominis. From the direction of its fibers one can infer that its action is very much like that of the sternocleidomastoid; it pulls the rib cage forward, rotates the lower aperture toward the opposite side and bends it laterally to the same side. Consequently pure side bending requires the suppression of both the forward flexory and the rotatory component.

The action of the internal oblique is also forward flexion of the trunk and in that respect it is synergistic with the external; it is also synergistic with the latter muscles in side bending but it rotates the thoracic cage to the same side opposite to the action of the external oblique.

The transversus abdominis has a peculiar action. Operating bilaterally it pulls the lower thoracic aperture down and flattens the abdomen thereby increasing the intra-abdominal pressure. Singly it rotates the trunk toward the opposite side and it becomes synergistic with the external oblique.

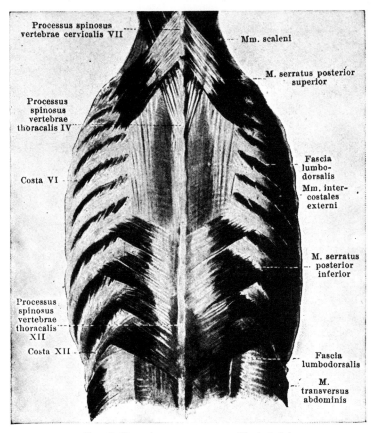

FIG. 17. Posterior serratus group. (Spalteholz)

The quadratus lumborum is a flat muscle which occupies a rectangular space between the lower ribs and the posterior portion of the iliac crest. Symmetrically it acts as a powerful extensor. Its unilateral effect is straight side bending.

D. THE ROTATORS OF THE TRUNK

The axial rotation of the trunk is produced by unilateral contraction or obliquely directed trunk muscles which, because of their obliquity, develop a length rotatory component. Many of the extensory and flexory muscles have such a rotatory component and it may be the dominant one; however their secondary components must always be suppressed if a pure rotatory effect is to be achieved.

The principal rotators are the external and internal obliques, the obliquus capitis superior and inferior and the ilio-costalis. To this should be added the posterior serratus group consisting of the superior and inferior. The superior is a flat quadrangular muscle which arises from the ligamentum nuchae and the spinous processes of the cervical and upper dorsal spine and runs obliquely downward to the angles of the 3rd to the 5th rib. Its symmetrical effect is to elevate the ribs and it is therefore an important respiratory aid. Singly it rotates the ribs backward (Fig. 17).

The inferior is also a flat and square muscle which arises from the lumbo-dorsal fascia at the level of the 12th dorsal to the 2nd lumbar. Its parallel fibers run upward and outward hugging with fat fleshy digitations the 9th to the 12th rib. In symmetrical action the muscle pulls the ribs down and like the superior it is also an auxiliary inspirator. In unilateral action it pulls the ribs down and rotates them backward, at the same time inclining the trunk slightly to the side (Fig. 17).

IV. THE EXCURSION FIELDS OF THE SPINE

A. THE SAGITTAL PLANE (FIG. 18)

Isolated measurements of the ligamentous specimens on cadavers made by Virchow[11] revealed that the flexion-extension range from the 2nd to the 7th cervical vertebrae amounts to 117°, from the 7th cervical rib to the 12th dorsal it is 107.5° and from the 12th dorsal to the 1st sacral 84°.

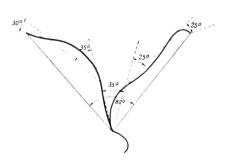

This is somewhat at variance with the later figures of Novogrodski[7] of 90° from the 2nd cervical to the 1st dorsal; 62° from the 1st dorsal to the 1st lumbar, and 52° from the 1st lumbar to the 1st sacral.

FIG. 18. Excursion fields of the spine in the sagittal plane. (Strasser)

While in the isolated specimen the total excursion in this plane is about 360°, we find it much restricted in the living. The greatest individual differences are found in the backward extension range. Lohr's[5] investigations (quoted from Fick, III, page 106) on the excursion field of the spine in the living yielded the following values of the sagittal plane: fore and backward bending of the cervical spine, 161°. The total excursion of the entire spine in the sagittal plane is 245°, which is somewhat different from Weber's figures of 334° for the fore and backward movement of the entire spine.

B. THE FRONTAL PLANE

In lateral motion Novogrodski[7] found the average total range from the 2nd cervical to the 1st dorsal is 62.5°, from the 1st dorsal to the 1st lumbar is

76°, and from the 1st lumbar to the sacral, 49° on both sides. The maximum lateral bending of the head is obtained if combined with length rotation whereby the concave side rotation of the odontoid is compensated by a convex rotation of the atlas in the atlanto-axoid articulation.

Compared with the mobility of the cervical spine the lateral range of the dorsal section is small: a total of 30-40° for both sides. Lateral motion of the dorsal portion of the spine is always associated with convex side rotation. This is due to the peculiar anchoring of the spine in the intervertebral articulations. The facets of the dorsal articulations are not strictly in the frontal plane but they slant backward and toward each other so that in motion they describe the surface of a cone. A straight lateral motion would be conceivable only if the articulations were arranged in a flat plane. If they are part of the curved surface of a cone, side bending must necessarily be combined with length rotation, which usually occurs toward the convexity of the lateral curve. That this occurs more easily if lateral bending is combined with slight forward flexion is plausible because in this position the pressure of the shingle-like contact of the facets of the dorsal spine is released and gliding movement can occur with greater freedom. However the fact that one can bend the spine laterally and rotate it voluntarily to the side of the concavity is a matter of common observation. But this is not the automatic tendency of the spine and it requires a special antagonistic muscle effort to rotate the spine in the opposite direction. What is called a physiological concave side rotation involves a definite muscle effort to neutralize the inclination of the spine as it bends to the side to slide automatically into rotation on the convex side. One may assume, therefore, that the existence of a so-called contralateral or concave side rotation indicates that the spine is adequately mastered by its musculature and that it is in full control of its normal muscular equilibrium.

The maximal number of lateral curves which can be produced by muscle action in the normal spine is four: a lower lumbar, and upper lumbar and lower dorsal, an upper dorsal and a cervical. In the normal spine they are completely under muscular control and are reversible. They correspond to the scoliotic curves where they are fixed rather than being voluntarily reversible.

C. THE TRANSVERSE PLANE

For the axial rotation the figures given by Hughes[3] are as follows: from the 2nd cervical to the 1st dorsal 143.7°; from the 1st dorsal to the 1st lumbar 66.3°; and from the 1st lumbar to the 1st sacral 18.3° in both directions. All figures of the above apply to ligamentous specimens.

The ranges in the living are as follows: head and neck 65° on each side (totaling 130°); dorsal spine 40° to each side (totaling 80°). The range of the lumbar spine is almost negligible except for the movement between the 5th lumbar and the sacrum. The total range from the head to the lumbar spine inclusive is 210°. To this should be added, however, the rotation which the trunk can carry out in the hip joint and which is 90° when standing on both feet or 140° in standing on one foot. A further addition is the deflection of the eyeballs of 45° on each side. This would give a total rotatory excursion for the body as a whole when standing with one foot on the floor and including hip and eye motion of over 400°. With this winding up of the body goes hand in hand a certain shortening of the spine due to the accentuation of its curves. It is reported that one of Napoleon's police agents succeeded in winding himself up or down a difference of 11-13 cm. in height (Fick[2]).

The number of separate axial rotations of the spine is the same as that of the lateral curves, namely four from the occiput to the sacrum. Each of these torsions is brought about by a particular muscle group. The lowest rotational curve involves the lumbar spine to the lower aperture of the thoracic cage. It is produced by the external oblique of one and the internal of the other side, both rotating the trunk in the same direction. A second system of rotation is represented by the muscles of the shoulder girdle. With the shoulder girdle immobilized, rotation of the thorax is carried out by the serratus anterior and the pectoralis muscles. The third system of rotation is represented by the back muscles which are attached to the thoracic cage and to the occiput and lateral processes of the cervical spine. These are the superior and inferior posterior serratus, the ilio-costalis and the semi-spinalis muscles. They are able to produce rotation of the entire thoracic cage up to the upper cervical level. The fourth system is the rotation of the head upon the cervical spine brought about by a set of muscles which use the mastoid process and the occipital ridge for insertion. They are the sternocleido mastoid which rotates the head to the opposite side and inclines it to its side and the splenius capitis which rotates and inclines the head to the same side.

The four rotations are then: lumbar spine and lower thorax, upper thorax in shoulder region, entire dorsal and lower cervical spine, and the torsion of the head in the atlanto-axoid joint. In scoliosis these torsions are represented as fixed and irreversible deformities.

V. SUMMARY

1. The center of motion of the intervertebral joints is the disc. The movement of the articular facets is a gliding one in which they describe part of the surface of a cone, the height of which increases from the cervical to the lumbar section.

2. Special regional modifications exist for the atlanto-occipital and the atlanto-axoid articulations. In upright position the head is in labile equilibrium, its center of gravity balanced upon the atlanto-occipital joint.

3. In the temporo-mandibular articulation both menisci move as one unit. Opening and closing of the jaw is associated with fore and backward gliding of the mandible.

4. The change of the sacrolumbar facets from the sagittal to the frontal plane serves for greater freedom in lateral and backward bending.

5. Between the bodies of the vertebrae motion occurs in all 3 planes, with

the discs as the center, and there is also a slight degree of longitudinal extensibility of the spine.

6. The motion in the intervertebral articulations is distributed as follows: in the sagittal plane flexion is carried out mainly in the cervical and upper dorsal spine; the principal extensory regions are the cervical and lumbar spine. In the frontal plane the cervical spine shows the greatest lateral mobility; rotation in the transverse plane rests almost completely with the cervical and dorsal segment.

7. The complicated musculature of the spine can be divided in four main groups.

a) The system of extensors consists of:
   1) A deep layer (rectus capitis, interspinosi, obliquus capitis, intertransversarii, rotatores and levatores costarum).
   2) The intermediate layer (splenius, multifidus and semi-spinalis).
   3) And the superficial layer of the sacrospinalis (longissimus and iliocostalis).

b) The system of flexors of the trunk consists of:
   1) The spino-thoracic group (sternocleidomastoid, scaleni).
   2) The ventral flexor group (mylohyoid, sternohyoid and thyrohyoid.
   3) The spinal group proper arising and inserting in the spinal column (longus capitis, longus colli; medial and lateral rectus capitis anteriores).
   4) A thoraco-pelvic group flexes the lower aperture of the thorax against the pelvis (rectus abdominus, external and internal obliques).
   5) The system of spino-femoral muscles flexes the trunk against the thigh when the latter is fixed (iliacus, psoas).

c) The side benders of the trunk are the external and internal obliques and the quadratus lumborum, while the transversus abdominis pulls the lower thoracic aperture down and flattens the abdomen.

d) The rotators of the trunk are the external and internal obliques and the posterior and inferior serratus and the iliocostalis, acting singly, while the obliquus capitis superior and inferior rotate in single action head and cervical spine.

8. The excursion field of the entire spine in the living is in the sagittal plane 245° for forward and backward bending according to Lohr; in the frontal plane it aggregates to 182° for both sides and in the transverse plane 210°.

9. The maximal number of lateral curves which can be produced by muscle action is four: a lower lumbar, an upper lumbar and lower dorsal, an upper dorsal and a cervical. They correspond to the lateral curves in scoliosis.

10. The maximal number of axial rotations of the spine is also four: a lumbar, an upper dorsal, a total thoracic and a cervical. They correspond to the rotatory curves in scoliosis.

BIBLIOGRAPHY

1. BRAUS, H.: *Anatomie des Menschen, Vol. I.* Berlin, J. Springer, 1921.
2. FICK, R.: *Handbuch d. Anatomie und Mechanik d. Gelenke.*
3. HUGHES, R. W.: Die Drehbewegung der Menschlichen Wirbelsäule. *Arch. f. Anat. und Entwicklungsgesch.*, 1892.
4. KNESE, KARL H.: *Ztschr. Anat. und Entwicklungsgeschichte. 114:67,* 1948-1950.
5. LOHR: *Muench. Med. Wchnschr.*, 1890.
6. MOLLIER, S.: *Über die Statik und Mechanik des Menschlichen Schultergürtels.* Jena, Festsch. f. C. O. Kupffer, 1899.
7. NOVOGRODSKI, M.: *Die Bewegungsmöglichkeiten der Menschlichen Wirbelsäule.*
8. SPALTEHOLZ, W.: *Handatlas der Anatomie des Menschen, Vol. II.* Leipzig, S. Hirzel, 1921.
9. STEINDLER, A.: *Mechanics of Normal and Pathological Locomotion in Man.* Springfield, Illinois, Charles C Thomas, Publisher, 1935.
10. STRASSER, H.: *Lehrbuch der Muskel und Gelenkmechanik, Vol. II,* Berlin, J. Springer, 1913.
11. VIRCHOW, H.: Die Eigenformen der Menschlichen Wirbelsäule. *Verh. d. Anat. Ges. Giessen.* 1909.
12. ————: Über die Sagittal Flexorische Bewegung im Atlas-Epistropheusgelenk des Menschen. *Arch. f. Anat. u. Entwicklungsgesch.,* 1909.

THE PATHOMECHANICS OF THE LUMBOSACRAL JUNCTION

INTRODUCTION

IN THE course of philogenetic development the upright position and the orthograde gait of man is an attainment of comparatively recent date. It has produced a radical change in the relation of the pelvis to the spinal column. The sacrolumbar junction which represents the transition from the movable to the immovable portion of the vertebral column has become the least stable part of it. Many static disorders centered at the lower portion of the spine have followed in this train of development. Some of these disorders are based upon developmental aberrations; others develop in an apparently normal spine on the basis of constitutional causes or of occupational strain; others again are purely traumatic. At all events this junction is more exposed to static stresses and is less equipped to meet them adequately than any other location in the spinal column.

I. ANATOMICAL VARIATIONS

Anatomical variations of this junction may contribute to static stresses because they restrict or check mobility; in this case the burden of motion is transmitted to other levels of the column; or else the variation affects the stability and, due to this flaw in construction, motion is not checked soon enough. In this case the strain falls primarily upon the soft structures of the sacrolumbar and sacroiliac regions.

A. ANATOMICAL VARIATIONS AFFECTING THE MOBILITY OF THE SPINE

The anatomical relationship of the 5th lumbar vertebra entails a number of variations which have a restraining effect upon mobility.

1. Abnormally long transverse process

It is more commonly found in painful backs than in a normal adult spine. According to O'Reilly[13] it occurs in 25% of the cases of back strain, and Goldthwait[5] finds it more frequent in men than in women. The reason for this is that the lumbar spine in the male pelvis is set so much lower between the hip bones than it is in the female which makes it easier for the transverse process to become impinged against the os ilii.

In many instances a stereo x-ray picture shows that the transverse process does not really impinge upon the iliac bones but it clears them by a safe margin. When it does impinge it is apt to lock motion and cause strain; in

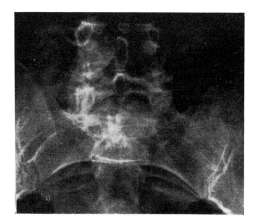

Fig. 1. Enlarged but nonarticulating transverse process.

side bending the opposite ilio-lumbar ligament comes under tension (Fig. 1). Together with these tension stresses on the opposite side pressure stresses are produced on the same side due to impingement.

2. Sacralization

The sacralization of the 5th lumbar represents a higher degree of this enlargement of the transverse process (Fig. 2). Motion of the spine in any direction is now transferred to the junction between the 4th and 5th lumbar; but the articular processes of this junction are directed in the sagittal plane and therefore lack the freedom of rotation which the normal sacrolumbar joint enjoys.

In most instances the superincumbent articulations adjust themselves to the added strain. Still, this sagittal orientation of the 4th lumbar articulation is a frequent cause of rotatory sacrolumbar stresses.

If one side only is sacralized, the other free, no strain occurs in the sacrolumbar junction but is transmitted to the upper vertebrae. On the other hand if the sacralization is incomplete, i.e., if a joint is established between the transverse process and the sacrum (transverso-sacral articulation), then side bending to the opposite side causes tension stresses on the articulation; bending to the same side causes tension stresses of the opposite transverso-iliac ligaments and compression stress of the homolateral transverso-iliac joint. Stresses in the sagittal plane by fore and backward bending are all referred to the junction between the 4th and 5th lumbar. Rotatory stresses are likewise so transmitted but rotation is restricted by the sagittal orientation of the intervertebral articulations of the superincumbent vertebrae.

In years past the pathological significance of the sacralized vertebra in relation to sciatic radiation and

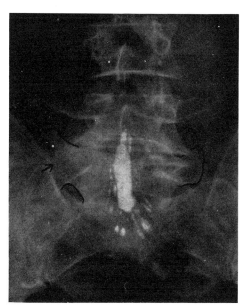

Fig. 2. Unilateral sacralization of 5th lumbar vertebra (retouched).

low back pain have been greatly exaggerated. Bertolotti,[2] Leri,[10] Benassi,[1] and others have tried to explain the sciatic symptoms associated with low back pain entirely on the basis of sacralization. However, since sacralization is found in many cases incidentally and without causing any symptoms whatever, this explanation was soon found to lack foundation.

On the other hand, Casolo[3] found that 58% of the individuals who were x-rayed because of neuralgic symptoms showed some degree of enlarged transverse process or sacralization and that many of these had secondary pathological changes due to strain and stress such as osteoarthritis or calcification of the ligaments.

There are other congenital anomalies which produce strains by restriction

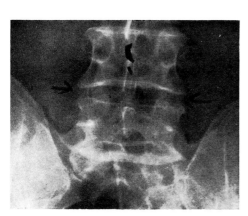

Fig. 3. Tropism (retouched).

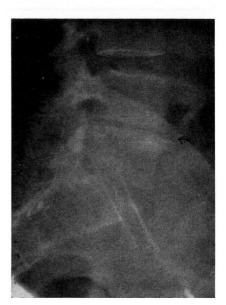

Fig. 4. Osteoarthritic changes in a case of unilateral sacralization.

of motion at this level. There is the increased inclination of the pelvis, for instance, which may or may not accompany sacralization. It is a potent factor in producing strain. The increased shearing stress placed on the lumbosacral junction contributes to early degeneration and an earlier herniation of the 5th lumbar disc with ensuing symptoms of sciatic radiation. It is generally accepted that while the impingement of the 5th lumbar disc is of no particular diagnostic importance, that of the 4th is and indicates degeneration and possible herniation of this structure.

3. Tropism

The name implies that the articular facets of the sacrolumbar junction are asymmetrically arranged. Normally they should be directed in the frontal plane to allow for greater mobility in backward and side bending. Occasionally they are placed in different planes. On one side the articulation is directed as usual in the frontal plane; on the other side it is directed sagitally, thereby assimilating itself to the higher lumbar articulations. The result of this asymmetry is that these articulations mutually lock each other, the one oriented sagittally restricting both lateral and rotatory motion (Fig. 3).

The effect of all the mobility restricting variations is that a greater stress falls upon the more proximal lumbar articulations. In the young this added burden is taken up easily enough by the soft tissues; but in later life when their elasticity fails, the articulations react to the added strains by osteo-arthritic changes (Fig. 4).

4. The impinging spinous process

If the 5th lumbar spinous process is large it may impinge upon the sacrum as the patient bends backward. The interspinous ligament becomes caught

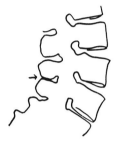

Fig. 5. Impinging spinous processes (tracing from x-ray. Case M. S.).

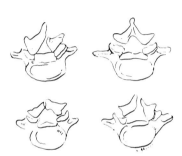

Fig. 6. Clefts and pseudarthrosis. (Redrawn from Steindler)

between the squared-off tips of the process. The patient assumes forward flexion as the position of relief (Fig. 5).

B. ANATOMIC VARIATIONS WHICH AFFECT THE STABILITY OF THE SPINE

1. Clefts and pseudoarthroses (Fig. 6)

The stability of the lumbar spine depends in the first line upon its ligamentous equipment. Any defect in the skeletal construction of the arches will encroach upon the ligamentous apparatus since the latter depends for its attachment upon the completeness of the posterior column. This is illustrated by the incomplete closure of the arches of the 5th lumbar vertebrae and of the first sacral (Fig. 6). In some cases a cleft exists between the lamina and the pedicle. The latter carries the upper articular process while the lower remains with the lamina. In the so-called spondylolysis these articular processes are separated from each other by a cleft which may be a true pseudoarthrosis but more often is a syndesmosis.

This condition facilitates the sliding forward of the upper vertebra upon the lower. A more advanced stage is the spondylolisthesis in which the forward gliding of the superincumbent vertebra has already taken place (Fig. 7). Such an event is only possible because the 5th lumbar has relinquished its anchorage to the sacrum, and because of this defect the stability of the spine at the sacrolumbar junction now depends entirely upon ligamentous structures, particularly upon the ilio-lumbar ligament which holds the 4th and 5th lumbar tethered to the wing of the os ilii. It is, however, too weak and too

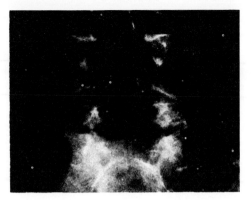

Fig. 6a. Incomplete closure of 5th lumbar vertebral arch.

resilient to provide sufficient stabilizaiton and hence it cannot prevent the articular processes from separating at the isthmus and the 5th lumbar from moving forward upon the sacrum. The 5th lumbar may, in fact, proceed so far forward and caudally that in severe cases it comes to lie with its posterior wall against the anterior surface of the sacrum pushing the latter backward into a more perpendicular position. One actually finds, in extreme cases, the 5th lumbar in front of the sacrum. The latter reacts to it by forming a bony shelf which partially sustains the 5th lumbar and its superincumbent weight, an attempt of nature to put a stop to further gliding.

2. The spina bifida occulta

Here the cleft is in the midline. The central defect of the arch interferes considerably with the development and attachment of the ligamentous apparatus uniting the 5th lumbar vertebrae with the sacrum. Not only is the defect in the neural arches responsible for the ligamentous inadequacy and the subsequent strain, but it leads automatically to early osteoarthritic changes changes which must be construed as nature's measure to increase the stability and to restrict the range of motion to safe limits.

3. The horizontal sacrum

An abnormal horizontality of the upper portion of the sacrum is often found associated with a wedge-shaped 5th lumbar vertebra and a posteriorly compressed 5th lumbar disc. In contrast to spondylolisthesis the 5th lumbar does not abandon its relation to the upper end of the sacrum. The angle between the length axes of the 5th lumbar and sacrum is

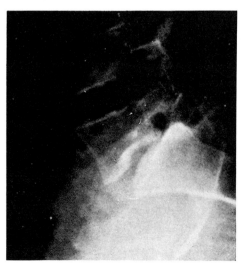

Fig. 7. Spondylolisthesis.

considerably decreased and as the sacrum approaches a horizontal position there is a commensurate increase in the pelvic inclination.

In the antero-posterior x-ray view this appears as a marked foreshortening of the 5th lumbar. In the lateral view the wedge shape of the 5th vertebra and the gaping at the anterior edge of the 5th disc is clearly visible (Lewald[11]) (Fig. 8).

The abnormal forward shearing stress at the lumbosacral junction created

FIG. 8. Horizontal sacrum.

by this condition is commensurate with the sine of the angle between the plane of the junction and the horizontal; at 45° it equals the pressure and at 90° all gravitational stress acts as shear, thus putting a great deal of tension upon the restraining ligamentous structures. Besides, as mentioned above, the 5th lumbar spinous process is likely to impinge upon the sacrum and to put considerable pressure upon the interspinous ligaments. The spine is unstable and particularly susceptible to jars which, in its almost vertical position the intervertebral disc is unable to intercept (Van Lackum[9]).

Some observers believe that the horizontal sacrum is the forerunner of spondylolisthesis, but there is no displacement of the 5th lumbar over the sacrum nor is there any trace of an abnormal cleft between the neural arches and the pedicles. Nevertheless both in the spondylolisthesis and the horizontal sacrum (sacrum acutum or hollow back) there is a great deal of weakness of the ligamentous apparatus at the lumbosacral junction and the iliolumbar ligament is under a great deal of stress. So is also the sacroiliac ligamentous apparatus which resists the forward rotating tendency of the upper end of the sacrum under the superincumbent body weight.

4. The retro-spondylolisthesis

As the name implies, in this condition the 5th lumbar vertebra is displaced backward upon the sacrum. According to Kimberly[8] retroposition of the 5th lumbar was first observed by Ferguson in 1924. Occasionally such a backward displacement is noted between more proximal vertebrae. Junghanns (1930) observed the same condition at the 2nd lumbar in six cases and the 1st lumbar in one case. T. A. Willis[16] who examined over 800 spines anatomically and by x-ray thought retroposition was merely an illusion. The question whether such a type of displacement really does occur and what its mechanical effect is has been raised by many investigators. From our own observation by measurements (Swerdloff[15]) we are inclined to believe that it does occur although the x-ray may be deceptive. It may be simulated by errors in focusing on the sacrolumbar level and torsion or lateral flexion of the spine may also create the impression of such a retroposition. DeForest Smith considered retroposition of the 5th lumbar as a definite clinical entity producing low back pain and sciatic radiation.

The projection of the 5th lumber backward might be due to its greater antero-posterior diameter in comparison with the upper surface of the sacrum. Garland and Thomas[4] found a difference of from 1-4 mm. in width in 73 out of 170 cases.

The point in question is whether such a disalignment has any clinical significance and whether it produces spinal instability. Melamed and Deerfield[12] indeed observed that forward bending causes a backward displacement and similar observations were made by Strasser.[14] Recently Hagelstam[7] in an excellent monograph on the subject was able to prove by an accurate method of measurements the existence of a true retro-spondylolisthesis (Fig. 9). He examined 119 roentgenograms and found a retroposition

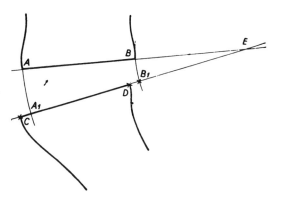

FIG. 9. Hagelstam's method of measuring a retro-spondylolisthesis. (Hagelstam)

of the dorsal margin of the 5th lumbar in 55 cases, and of 101 cases examined, 25 had a visible backward displacement of the 4th lumbar on the 5th. Even in a healthy individual a slight instability of the lumbar vertebrae is observed during flexion and extension and the 5th lumbar can be noticed to move backward and forward; only 1/6 of Hagelstam's cases were absolutely stable. It was also found that the majority of patients with low back pain with or without sciatic radiation showed such instability; in these patients the displacement was more extensive though it seldom exceeded 3 mm. Although there is nothing to indicate that pain is due to the retroposition as such, yet one must assume that some degeneration of the disc precedes the retroposition. An incipient degeneration of the disc does not give radiological signs, and it is therefore impossible to prove it by x-ray evidence. The fact that it is frequent in cases of low back pain with lumbosacral instability would indicate that degeneration of the disc is a primary factor.

C. THE GENERAL CONDITIONS AFFECTING MOBILITY AND STABILITY OF THE LUMBOSACRAL AREA

The greater the restriction imposed upon the spine due to its peculiar anatomical make-up, the greater is also the possibility that the range of motion be transgressed in the ordinary locomotor functions and that strains occur.

1. Anatomic build

Some spines are of an anatomic build which facilitates static stresses. These are particularly the narrow and long backs more often seen in women than in men. The lumber spine sits high upon the sacrum which appears to be longer than normal. There is flattening of the lumber spine which is endowed with rather poor musculature and is apt to develop postural anomalies.

2. The osteoarthritis of the spine

The strain at the junction is caused by the fact that the upper lumbar spine is immobilized by arthritic proliferations and spurs. When the sacrolumbar junction becomes included in this arthrosis all forward and backward movement must be carried out in the hip joint and very little pain is complained of. However in earlier stages before the sacrolumbar junction becomes involved the burden of movement falls upon this one articulation and strain develops easily.

Such conditions as spina bifida and spondylolisthesis likewise lead to early reactive proliferative arthritic changes as a natural protection against excessive motion and in the later stages of this so-called osteoarthrosis, the entire trunk moves in a block against the hip joint. This relieves the back muscles to some extent but it puts additional strain upon the extensors and rotators of the hip.

3. The atrophic type of arthritis

Due to the fact that this condition frequently starts with involvement and obliteration of the sacro-iliac joints, additional difficulties arise to axial rotation of the trunk. In the Strumpell-Marie type, especially, the spine becomes intolerant to rotatory motion and it is only slowly and at considerable cost to his comfort that the patient learns to handle his body without the help of the intervertebral articulations (Fig. 10).

4. A series of conditions in which sacrolumbar strain is a remote effect

Destructive lesions situated at higher levels result in deformation which often indirectly affects the sacrolumbar junction. A tuberculous gibbus of the dorsal spine is compensated by an increased lordosis of the lumbar region which again places a strain on the sacrolumbar junction. A similar situation is the kyphosis of the dorsal spine which follows fracture. There are destructive lesions of the lumbar spine itself which obliterate the normal lordosis and produce stress reactions.

Remote effects upon the sacrolumbar junction may result also from static disturbances of the lower extremity. Congenital dislocation or coxa vara cause sacrolumbar strain by exaggerated lumbar lordosis (Fig. 11). The static flat foot, when associated with flexion of the knee, produces by reason of the forward shifting of the line of gravity a strain of the ligamentous and muscular structures of the sacrolumbar junction. Here the low back strain is condi-

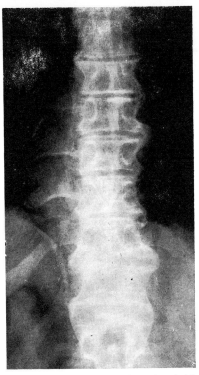

FIG. 10. Shrümpell-Marie type of arthritis of the spine.

FIG. 11. Congenital dislocation of the hips with exaggerated lumbar lordosis.

FIG. 12. Dorsal epiphysitis with increased dorsal kyphosis and lumbar lordosis.

tioned, (1) by the outward rotated hip which produces stress of the iliofemoral ligament, and (2) by the increased inclination of the pelvis.

5. There are certain developmental retardations resulting in malposture which also have an indirect mechanical effect on the sacrolumbar junction and which are a frequent cause of low back complaints. In the kyphosis adolescentium or Scheuermanns-Delahaye's disease the lack of mobility of the spinal column concentrates stresses in the lumbosacral junction in a similar manner as in the ankylotic arthritic spine (Fig. 12).

II. THE PATHOMECHANICS OF LUMBOSACRALGIA

The mechanism by which these above-mentioned morphological conditions cause strain of the soft tissues occupying the sacrolumbar junction as well as the clinical response of these tissues to strain requires further analysis.

Specifically the identity of the several structures affected should be ascertained and the nature and type of strain correlated with the clinical manifestations.

A. THE STRUCTURES UNDER STRESS

Theoretically, since the strain comes about by excessive motion, the structures farther from the center of motion must become strained first since the strain producing force operates on a longer lever arm.

In forward flexion the sequence should therefore be: the supraspinous, the interspinous ligaments, the ligamenta flava, the capsular ligaments, the posterior longitudinal ligament and finally the disc.

In backward extension the situation is different because impingement and compression forces now come into consideration for all structures lying behind the centers of motion as well as distension stresses for structures lying in front of it. The anterior longitudinal ligament would then be under tension stress. The sequence for compression stresses is reversed: namely, the disc, the articulations (impingements) and the spinous processes.

In side bending the tension stresses which are created on the convexity should involve in sequence the quadratus lumborum and lateral musculature, the sacrospinalis, the intertransverse muscles and ligaments.

In rotation the sequence also would follow in concentric direction from the long to the short rotator muscles, then to the interspinous muscles and ligaments, the ligamenta flava, the articular capsular ligament and the disc.

Of all these structures under stress, some of them can be identified by novocainization, especially those which lie superficially and hence are more remote from the center of motion.

If the clinical picture does not quite follow these sequences of manifestation postulated on mechanical and anatomic considerations, it is because of the difference with which these tissues respond to mechanical stretching, i.e., the difference in their sensitivity to strain.

Haboush[6] finds that excessive forward flexion first strains the ilio-lumbar ligament, then comes the intraspinous ligament from the 5th lumbar interspace up, then the strain is felt in the lumbo-dorsal fascia and in the sacral triangle. This is not quite in accord with our theoretical premises insofar as the more peripheral ligamentous structures should come first and hence strain of the lumbo-dorsal fascia and of the sacro-spinalis insertion at the sacral triangle should precede and not follow the deeper ligamentous strain.

In rotation the strain sequence follows closer the theoretical pattern: first the ilio-lumbar ligament, then the quadratus lumborum then the inter-transverse ligament.

In side bending the sequence is: quadratus, ligamenta flava, interspinous ligaments.

In backward bending the impingement signs prevail and the sequence is reversed: the articulations, the interspinous ligament. Because of the anatomic situation the ligamenta flava escape impingement while the interspinous ligaments do not.

It could not be expected that the response of the strained tension follow strictly a purely mechanical pattern. The muscles which are the first line of defense in guarding the articulations are much poorer in protopathic sensory fibers than the ligamentous or aponeurotic tissues.

On the whole the best means to identify the structure involved is to establish the so-called trigger points which indicate the site of the injured structures and whenever feasible to prove it by novocainization.

B. THE EFFECT OF THE LUMBOSACRAL STRAIN ON POSTURE

The so-called antalgic position is either symmetrical forward bending or asymmetrical forward bending associated with lateral listing. Any true antalgic position is protective and is maintained by muscle spasm. One should hesitate to diagnose an organic lesion in cases of exaggerated side list without forward bending such as is sometimes seen in the so-called hysterical scoliosis. In investigating what determines the antalgic position, one should keep in mind the following points:

1. The effect of posture on relieving strained ligaments and muscles.
2. The effect of posture on weight-bearing tolerance.
3. The effect of posture on the patency of the intervertebral articulations.
4. The effect of posture on the spinal roots.

ad. 1. So far as the protective effect of posture on strained muscles and ligaments is concerned symmetrical posture should be expected if the strain is in the midline or symetrical (Fig. 13).

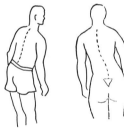

ad. 2. If the strain involves the sacro-spinalis and the gluteals on one side, then the antalgic position will be a list to the opposite side, in order to release weight bearing on the side of affection (Fig. 14).

ad. 3. If there is unilateral encroachment of the intervertebral foramen as it is in osteoarthritis, the position of relief should be a list to the opposite side.

FIG. 13. Symmetrical antalgic position.

FIG. 14. Asymmetrical antalgic position.

ad. 4. The effect of the position in strain of the lumbosacral spinal roots is not clear. It would seem that if it is due to a disc pressing on spinal nerves the list should be homolateral because this relaxes the nerve roots. On the other hand if the impingement is in the foramena the list should be contralateral because the patency of the foramena is thereby increased.

C. THE EFFECT OF LUMBOSACRAL STRAIN ON MOBILITY

During the painful stage the protective antalgic position is rigorously maintained and any efforts to disturb it are resisted.

Most of the so-called leg signs can be explained on these grounds.

1. The straight leg raising or Lasegue's sign causes a retrotorsion of the

pelvis. It is resisted by the spasticity of the extensors. Force increases the low back pain and also the sciatic symptoms (Fig. 15a).

2. The head and trunk raising sign (Ely, Brudzinski) falls into the class of the leg raising sign inasmuch as it also tries to break the protective sacrospinalis spasm (Fig. 15b).

3. Gaenslen's sign causes the os ilii to be rotated forward against the sacrum forcing the sacrum into retrotorsion. This is resisted by the protective sacrospinalis muscle. The sign is particularly competent in showing strain in the sacroiliac junction (Fig. 15c).

Any attempt to correct the protective list forcibly reactivates and augments the signs of strain. The body should never be subjected to force. One of Putti's maxims was to respect the antalgic position and to put the patient in a cast "as is," so as to let the asymmetry disappear spontaneously.

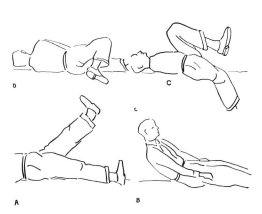

Fig. 15. A. Lasegue sign. B. Ely-Brudzinski sign. C. Gaenslen's sign. D. Ober sign.

5. The Ober sign is a test to decide whether or not the strain of the back muscle is transmitted to the iliotibial band. The contraction of the tensor fascia holds the pelvis in abduction as a protective measure which falls in line with the spasm of the sacrospinalis or the quadratus lumborum. This sign, if positive, denotes transmission of strain through the lumbodorsal fascia to the tensor (Fig. 15d).

Many other signs can be explained on kinetic grounds. Although there are some other features pertaining to posture and restriction of mobility, they have so far still eluded mechanical analysis. We believe, however, that the majority of the signs can be explained on kinetic grounds and that the kinetic interpretation of the clinical signs will go a long way toward early and specific diagnosis and toward establishing a rational method of managing sacrolumbar strains.

III. SUMMARY

1. Anatomical variations predispose the sacrolumbar junction to static strain.

2. There are variations affecting the mobility such as: an abnormally long transverse process, sacralization, hemi- and bilateral, tropism and spinous process impingement.

3. Other variations affect the stability of the junction: clefts and pseudarthroses, including spina bifida occulta, separate neural arch, spondylolisthesis, retrospondylolisthesis and the horizontal sacrum.

4. There are general conditions of the spine which affect both mobility and

stability of the sacrolumbar junction: the flat back, the osteoarthritis as well as the atrophic arthritis of the spine.

5. In another series of conditions the sacrolumbar strain represents a remote static effect: a tuberculous gibbus compensated by increased lumbar lordosis; a fracture deformity at a higher level; static disturbances of the lower extremity such as congenital dislocation of the hip, coxa vara, static pes valgus.

6. Certain developmental retardations resulting in malposture may produce sacrolumbar strain: kyphosis adolescentium (vertebral epiphysitis).

7. The analysis of the pathomechanics of lumbosacralgia requires in the first place the identification of the specific structures which are under strain.

8. From the purely mechanical viewpoint a certain sequence of the structures involved can be anticipated according to their greater or lesser distance from the center of motion which has produced the tension strain and, inversely, in sequence, according to the lesser and greater distance from the center of motion which produces the pressure and impingement strain.

9. The clinical picture does not quite follow this theoretical sequence because of the difference with which the various tissues react to strain.

10. The best way to identify the injured structures, provided they are easily accessible, is by establishing trigger points and suppressing localized tenderness by the novocain test.

11. The best way to appraise the overall picture of the sacrolumbar strain is the analysis of the antalgic position. It is a posture assumed to relieve strain to ligaments and muscles, to open the intervertebral foramina and to alleviate pressure on the spinal roots, all of which conditions can be differentiated by the type of antalgic position.

12. The effect of lumbosacral strain on mobility is also expressed in the so-called leg and trunk raising signs, because all of these mean interference with the antalgic protective position, with the difference only that here tests are made in recumbency.

BIBLIOGRAPHY

1. BENASSI: Lombalizzazione del primo metamero Sacrale. *Chir. Org. Mov. 8:*357, 1924.
   ———: Sacralizzazione e pseudosacralizzazione de la quinta vertebra lombare. *Chir. Org. Mov., 7:*1, 1923.
2. BERTOLOTTI: Contributo alla conoscienza degli vicii di differenziazione regionale dei rhachide con speziale regardo della assimilazione de la 5ª lombare. *Radiol. Med.,* May-June, 1917.
3. CASOLO, G.: Clinical and Roentgenological Study of the 5th Lumbar. *Radiol. Med. Milano., 11:*357, June, 1924.
4. GARLAND, L. H., and THOMAS, S. F.: *Am. J. Roentgen., 55:*275, 1946.
5. GOLDTHWAIT, J. E.: Lumbosacral Articulation. *Boston Med. & Surg. J.,* 1911.
   ———: Variations in Anatomical Structure of Lumbar Spine. *Am. J. Orth. Surg., 2:*416, 1920.
6. HABOUSH, J. S.: *J. Bone & Joint Surg., 24:*123, 1942.
7. HAGELSTAM, LARS: Retroposition of Vertebræ. *Acta chir. Scandinav. Suppl.* 143, Helsingfors, 1949.
8. KIMBERLEY: Radiology. *22:*5, 8, 1934.

9. LACKUM, VAN L.: The Lumbo-sacral Region. An Anatomical Study and some Clinical Observations. *J.A.M.A., 82:*110, 9, 1924.

10. LERI: La 5me Vertèbre Lombaire et ses Variations. *Presse Méd., 15,* 1922.
———: La Sacralization d'apres l'étude radiographique et Clinique de 100 Regions Sacrolombaires. *Bull. et Mém. Hôp., Paris,* July 29, 1921.

11. LEWALD: Lateral Roentgenography of the Lumbo-sacral Region. *Am. J. Roentgenol., 3:*168, October, 1924.

12. MELAMED, A., and ANSFIELD, D. J.: *Am. J. Roentgenol., 58:*307, 1947.

13. O'Reilly, A.: Malformations of Lower Spine, *J. Bone & Joint Surg., 23:*997, 1925.
———: Lumbo-sacral Variations. *J. Bone & Joint Surg., 19:*171, 1921.
———: Backache and Anatomical Variations of Lumbo-sacral Region. *Am. J. Orth. Surg., 3:*171, 1921.

14. STRASSER, H.: Lehrbuch d. Muskel und Gelenkmechanik. Berlin, J. Springer, 1913.

15. SWERDLOFF, H.: Personal Communication.

16. WILLIS, T. A.: Backache from Vertebral Anomalies. *Surg., Gynec. & Obst., 38:*658. 1924.

THE NORMAL AND PATHOLOGICAL MECHANICS
OF THE PELVIS

I. THE NORMAL MECHANICS OF THE PELVIS

IN THE philogenetic struggle of the human skeleton to adjust itself to the upright position, two regions have remained in arrears. One is the sacrolumbar junction we have already dealt with; the other is the pelvis.

From the vertical position in the quadruped it strove toward establishing itself horizontally so that it could serve as a stable base for the upright human spine. This the pelvis has achieved only halfways; it is still considerably oblique, midways between the vertical position of the four-legged animal and the ultimate trend toward horizontality. This made it necessary to install special provisions to secure it against the powerful downward and forward thrust which is transmitted to it by the superincumbent weight of the trunk. It is not surprising, after all, to find that such safeguards frequently fail and that formidable deformations develop whenever there is a let up in the normal stress resistance.

A. ANATOMICAL CONSIDERATION

Three bones on each side constitute the pelvis: the os ilii, the os pubis, and the os ischii. They form a ring into which the sacrum is implanted by means of the sacro-iliac junction. The latter contains in its proximal portion a kidney-shaped articulation while caudally it continues as a syndesmosis.

The facets of the innominate bone, as the half of the pelvis is called, articulate with the three upper lateral masses of the sacrum. The os pubis and os ischii are joined to the iliac bone by a Y-shaped cartilage which later obliterates. At this junction the three bones form a deep cavity, the acetabulum, surrounded on all sides by a high crest. Only at its lower pole is there a cut-out in this crest, the incisura acetabuli. Both the os ischii and the os pubis, in running forward, curve strongly toward the midline so that they are able to join in front with the bones of the other side forming the symphysis (Fig. 1).

FIG. 1. Pelvis from in front.

To effect a juncture between the os pubis and the os ischii the two bones converge toward each other, the os pubis sending a descending ramus downward to meet the ascending ramus of the os ischii; in doing so they enclose between them an oval opening, the thyroid foramen or foramen obturatum situated in front and below the acetabulum and covered by a tough membrane, the membrana obturatoria.

Behind the acetabulum the iliac bone and the adjoining os ischii describe a

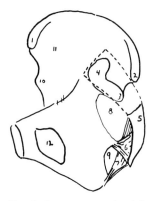

FIG. 2. Inner aspect of pelvis with anatomical relationship to sacrum.

1. Anterior superior spine.
2. Posterior superior spine.
3. Posterior inferior spine.
4. Auricular articular surface of pelvis.
5. Sacrum.
6. Sacrospinous ligament.
7. Sacrotuberous ligament.
8. Greater sciatic foramen.
9. Lesser sciatic foramen.
10. Anterior inferior spine.
11. Iliac fossa.
12. Obturator foramen.

large curve which recesses the posterior border into an arcuar cut-out, the greater sciatic foramen (Fig. 2).

Two ligaments both coming from the sacrum divide this cut-out into two separate compartments. The ligamentum sacro-spinosum runs from the sacrum to the spine of the os ischii; the sacro-tuberosum runs from the sacrum to the tuber of the os ischii. The upper and larger compartment is the greater sciatic foramen through which the sciatic nerve leaves the interior of the pelvis. The space below, between the sacro-spinosum and sacro-tuberosum ligaments, is called the lesser sciatic notch (Fig. 2).

The sacro-iliac articulation has the shape of a kidney with an upper perpendicular and a lower horizontal portion; the latter runs backward toward the posterior inferior spine of the os ilii (Fig. 2). This joint is covered by hyalin cartilage and has a regular synovial lining. All the remaining space between sacrum and os ilii of the sacro-iliac junction is occupied by powerful ligaments, the interosseous sacro-iliac ligamentous apparatus. It is further enforced by strong ligamentous strands which cover the anterior and posterior surfaces of the junction (Fig. 3). The posterior sacro-iliac ligaments are the more powerful ones. The fibers of this mass show a great diversity of direction crossing at all possible angles (Fig. 4).

When one looks at the pelvis from behind one notices that the posterior crest of the os ilii adjacent to the sacrum projects strongly backwards, forming with a sacrum a deep recess. This space is filled with the massive posterior ligamentous apparatus and overlying this mass is the sacro-spinalis muscle.

The manner in which the sacrum is fitted between the os ilii is significant from the mechanical point of view. In vertical direction the sacrum forms a keystone because its cranial base is wider than its tip. Thus it is secured against downward sliding. In the antero-posterior direction, however, the anterior surface is broader than the posterior. Consequently there is no osseous check against the sacrum penetrating forward into the pelvis and it devolves upon the ligamentous apparatus to prevent its forward displacement.

Furthermore, the superincumbent weight tends to produce a rotation of the sacrum in the sagittal plane about a frontal axis trying to force the upper end downward and the lower end backward, separating the latter from the tuber ossis ischii.

Safeguards against this event are the two ligaments which secure the lower sacrum against tilting backward, the sacro-tuberous and the sacro-spinous ligament (Fig. 3, 4) (Fick[3]).

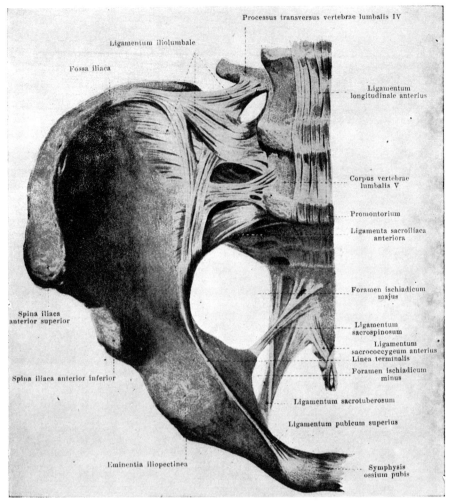

Processus transversus vertebrae lumbalis IV

Ligamentum iliolumbale

Fossa iliaca

Ligamentum longitudinale anterius

Corpus vertebrae lumbalis V

Promontorium

Ligamenta sacroiliaca anteriora

Foramen ischiadicum majus

Spina iliaca anterior superior

Ligamentum sacrospinosum

Ligamentum sacrococcygeum anterius
Linea terminalis

Foramen ischiadicum minus

Spina iliaca anterior inferior

Ligamentum sacrotuberosum

Ligamentum pubicum superius

Eminentia iliopectinea

Symphysis ossium pubis

FIG. 3. Sacro-iliac ligamentous masses, anterior view. (Spalteholz)

The tendency of the upper end of the sacrum to rotate forward about this axis under the body weight is neutralized by the ilio-lumbar ligaments which extend from the transverse processes of the 4th and 5th lumbar to the posterior rim of the os ilii (Fig. 3). They are particularly important in cases of horizontal sacrum in which the forward shearing stress of the 5th lumbar is increased.

The bones are fitted into the pelvic ring under considerable elastic resistance which results in storing up of a great amount of latent energy. An intrinsic equilibrium is thereby created between the tendency of elastic expansion of the bones and their being held in check by the opposing resistance of the articulations and of the syndesmoses of this ring. It is similar to that which we have encountered in the spinal column or in the thoracic cage. If this intrinsic equilibrium is destroyed by severing the pelvic ring at any point, for instance, by a symphysiotomy or episiotomy, the elastic energy stored up is released and the ring springs apart, the same as does the thoracic cage if its continuity is interrupted.

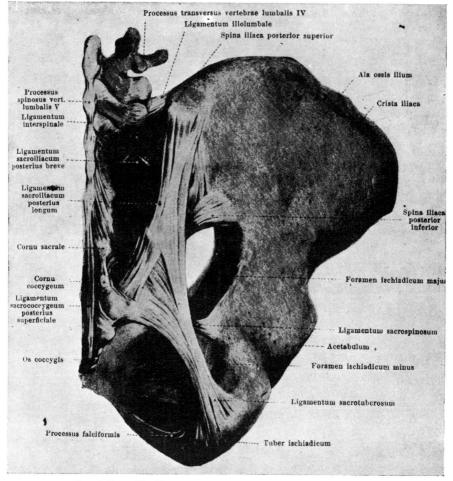

FIG. 4. Sacro-iliac ligamentous masses, posterior view. (Spalteholz)

B. THE ORIENTATION PLANES OF THE PELVIS

To establish the normal relationship of the pelvis to the rest of the body certain landmarks are used as points of reference in relation to the generally accepted body planes. Through these points planes are laid which determine the degree of obliquity of the pelvis against the trunk as well as against the lower extremities: the *special pelvic orientation* planes.

1. Relation of the pelvis to the frontal plane

If the individual stands upright a plane laid through both anteror superior spines and through the upper edge of the symphysis is a frontal plane parallel to, and in front of, the cardinal frontal plane of the body which contains the center of gravity (Fig. 5).

2. The relation of the pelvis to the transverse plane

The individual stands upright and a plane is laid through the two anterior

superior spines in front and the two
posterior inferior spines behind. This
plane is practically transverse (Fig. 5),
parallel and slightly below the cardinal
transverse plane which also contains the
center of gravity.

3. Relation of pelvis to frontal and transverse planes

To establish the relation of the pelvis
to these two planes we set up special
pelvic orientation planes as follows:

In the upright position a plane is laid
through the upper edge of the symphysis and the two posterior inferior spines.
This plane represents the inlet of the
pelvis. It is inclined 30-40° to the frontal and 60-50° to the horizontal plane
of the body. The latter value is used to
indicate the angle of inclination of the pelvis (Fig. 5).

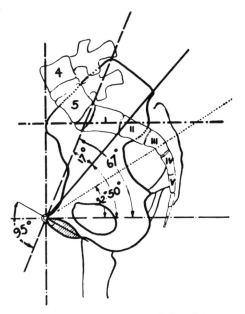

FIG. 5. Orientation planes of the pelvis.

4. Upright position

In upright position a second plane is laid through both anterior superior
spines and both tubera ossis ischii. This is the axial plane of the pelvis which
contains the axis of the pelvic canal. It is at right angles with the above named
pelvic inclination plane, being inclined 50° to the frontal and 40° to the
horizontal plane of the body.

5. The plane of the sacrolumbar junction

This is determined radiographically. The longitudinal axis of the 5th lumbar
vertebra forms with the axis of the upper sacral segment an angle of 120-
135° which opens posteriorly. This angle may decrease to a right angle so that
the upper portion of the sacrum appears almost horizontal. The smaller the
angle, the greater the weight stress component which passes through the junction as shearing stress (Fig. 5). The greater shearing stress then increases
the tendency of the 5th lumbar to slide forward.

C. THE ANALYSIS OF THE STATIC CONDITION OF THE PELVIS AS A WHOLE. THE PELVIS AT REST. STANDING ON BOTH LEGS

1. In the frontal plane

For the purpose of analysis we consider the pelvis viewed in the frontal
plane as a horizontal beam. The ends of this beam are resting on the femoral

heads while the middle of the beam carries the superincumbent load. The pressure reaction of the femoral head is a continuation of the reaction from the floor. As long as the beam is symmetrically and concentrically loaded, these reactions are equal and opposite in relation to the superincumbent weight. The condition for these reactions being equal is merely that the weight born by the beam applies exactly in the midline between the two supporting posts. However the gravitational reactions from the floor are transmitted through the femoral neck following its oblique course. This force can be considered the resultant of a horizontal and a vertical component. The vertical pushes upward while the horizontal presses against the socket toward the midline. In straight upright standing both horizontal components are equal

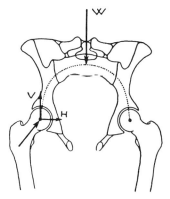

FIG. 6. The horizontal and vertical components of the gravitational reaction of the body upon the hip joint.

FIG. 7. Horizontal components directed toward midline greatly increased by strong abduction.

FIG. 8. Horizontal components directed away from midline due to strong adduction.

and opposite; they neutralize each other and the equilibrium in this respect is automatic (Fig. 6).

Let us now suppose that the two extremities are not perpendicular and parallel. If both are abducted to an equal degree, the gravitational reactions to these obliquely placed extremities also produce their horizontal and perpendicular components. The former also creates a horizontal thrust against the midline; the latter is directed upward (Fig. 7). The difference is only in the size of the components. They represent the summation of the two obliquities; namely that of the neck of the femur which is oblique against the shaft and that of the shaft which is oblique against the line of gravity. The pressure against the acetabulum is greater the greater the degree of abduction, while the upward thrust becomes correspondingly smaller.

Let us further asume that the two lower limbs are placed obliquely in the same degree, but in adduction. Here again there is the obliquity of the lower extremities which produces a horizontal as well as a vertical component of the gravitational reaction. The horizontal components are equal and opposite on both sides, but they are now directed away from each other and away from the midline (Fig. 8).

Consequently in the first instance the horizontal component exerts a pressure which is transmitted as such through the hip joint to the sacro-iliac junction. In the second instance the horizontal component acts as tension transmitted as such to the sacro-iliac articulation. Under normal conditions the resistance of the pelvic structures will absorb these stresses.

The next situation is one in which both legs are oblique but they are in different directions. One of the legs is in abduction and the other one is in adduction. This represents a frequently assumed position of rest (Fig. 9). The line of gravity now shifts closer to the adducted hip which consequently bears more weight while the other, the abducted hip, is resting. With the greater degree of gravitational stress falling on the adducted hip, its horizontal component in direction away from the midline becomes commensurately greater.

The other hip is in abduction and it bears the lesser amount of weight being farther from the line of gravity which passes closer to the adducted hip. Here also a horizontal component is operative directed toward the midline. Both horizontal components, therefore, act in the same direction and no longer neutralize each other. Equilibrium is no longer automatic. It requires active cooperation of the abductors of the adducted leg which bears the greater

Fig. 9. Horizontal component directed away from midline in adducted hip, and toward midline in abducted hip, both in same direction.

weight to produce a horizontal component to counteract the horizontal component of the other, abducted, leg. In this effort the abductors of the adducted hip are assisted by the adductors of the other, abducted, hip. The result is that a person with paralysis of the abductors of one side and adductors of the other cannot stand or rest himself on one leg, while a person with paralysis of either the adductors or the abductors can do so. The closer the line of gravity is to the adducted leg and the farther it is from the abducted, the smaller is the horizontal shifting component of the latter.

2. The sagittal plane

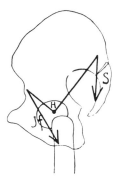

We next proceed to analyze the static condition of the pelvis in the sagittal plane while the individual stands upright on both legs. In the normal comfortable stance the line of gravity runs from the floor upward to 4 cm. in front of the ankle joint axis, then slightly in front of the knee joint axis and through or slightly in front of the hip joint. One can see that the conditions of equilibrium are here entirely different from those in the frontal plane. Here we had a concentrically loaded beam supported on both ends. In the sagittal plane, however, the situation is that of the superincumbent trunk precariously balanced upon the hip joint. Automatic equilibrium obtains for this joint only for the moment when the line of gravity falls through it. Because this can occur only momentarily

Fig. 10. Rotatory gravitational forces in the sagittal plane about the hip joint.

gravity practically always produces moments of rotation which have to be neutralized by muscle action (Fig. 10).

These rotatory moments produced by gravity depend upon the distance of the line of gravity from the center of motion. The more one bends forward the farther the line of gravity moves away from the hip joint and the greater, therefore, is the muscular effort required of the extensors of the hip joint to keep the line of gravity within the bounds of the supporting surface. In upright standing position this line of gravity constantly fluctuates back and forth even when one stands as quietly as possible. These fluctuations are under constant regulation by the extensory and flexory apparatus of the hip joint, and a constant change in the pelvic inclination accompanies these adjustments. As the angle of inclination diminishes and the pelvic ring becomes more horizontal, the lumbar spine flattens and the hip joint goes more into extension. An increase of the pelvic inclination on the other hand causes a flexory position of the hip joint and an increase in the lumbar curve.

There exists, therefore, a close inter-relationship between the position of the line of gravity, the inclination of the pelvis, and the position of the hip joint The obliquity of the acetabular plane which looks downward, outward, and forward causes the thrust of the reaction to develop a backward as well as an upward component, even if the limbs are held perfectly perpendicular. The pelvic inclination is increased as the gravitational components act on the upper and posterior rim of the acetabulum. The backward rotatory component is now commensurate with the sine of the angle of inclination. It would be 0 if the pelvis is completely horizontal and it is maximum if the angle of inclination is 90°, which position is possible only under certain pathological conditions. Between these extremes the backward thrust increases or decreases with the pelvic inclination.

3. The horizontal plane

A rotation of the pelvis in the horizontal plane can be encompassed only if the applying force is not at right angle to this plane. In upright standing the long axis of the extremities are perpendicular to the horizontal plane of the pelvis. The obliquity of the head and neck caused by the inclination to the shaft develops no rotatory moment in respect to the transverse plane, as the force transmitted goes to the center of rotation and therefore the lever arm, which is the distance of the force from the center of rotation, is zero.

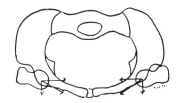

Fig. 11. Rotatory forces in horizontal plane.

It is somewhat different with the declination angle of the head and neck, i.e., the angulation back and forward. If we again construct the two components of the back or forward rotated neck we arrive at one directed upward which moves the pelvis in the frontal plane and the other which is directed forward and which rotates the pelvis in the horizontal plane (Fig. 11).

So long as the stance is symmetrical, the two horizontal components of the gravitational reactions transmitted through the forward or backward bent neck will neutralize each other and no muscular force is required to maintain the equilibrium. But as soon as the situation becomes asymmetrical, either because of a movement of the trunk and pelvis against the stationary extremities or a movement of the extremities against a stationary pelvis causes a relinquishment of the symmetry, then the components of the gravitational stresses no longer will neutralize each other automatically and it takes muscular effort to maintain the position in equilibrium.

In regard to the horizontal plane, let the first assumption be that trunk and pelvis rotate together so that their mutual relationship is undisturbed. All rotation then occurs in the hip joint. If now an asymmetrical position is adopted by rotating trunk and pelvis to one side, then, the feet being firmly on the ground, the hip joint on the side of the backward rotated pelvis will be inward rotated and on the side of the forward rotated pelvis it will be outward rotated. Both components try to rotate the pelvis in the same direction and therefore they have to be neutralized by muscle action; but because in this situation these components are relatively small in relation to the upward thrust, only a minimal amount of muscle force is required.

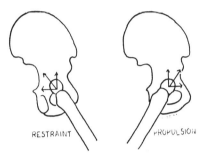

Fig. 12. Forces developed by restraint and propulsion. Restraint causes pelvis to rotate backward, and propulsion causes pelvis to rotate forward.

However, the situation becomes different when the limbs are no longer parallel in the sagittal plane but form a divergent angle as is, for instance, the case in the gait where in the period of double support one leg is planted diagonally forward and the other diagonally backward. The horizontal components now become greatly increased. The backward leg develops a forward and the forward leg a backward rotating thrust. These forces are essential in producing the propulsion and restraint of the human gait. The propulsion of the right leg rotates the pelvis in the same sense as restraint rotates it on the left. But restraint and propulsion of the same leg rotate the pelvis in opposite direction (Fig. 12). One can see that the obliquity of the leg has a similar effect on increasing the horizontal component of the gravitational force which it has in the frontal plane when the leg goes into ab or adduction.

Next let us consider the situation where hip joint and pelvis are so rigidly held together that they form a single mechanical unit, and that rotation occurs now between pelvis and trunk. Any rotatory movement carried out by the trunk against the pelvis or vice versa is now intercepted by the lumbosacral articulation which divides the movable from the immovable portion of the spine. We have referred to this event in one of the foregoing chapters as one of the factors producing low back strain. However it should be pointed out that if this lumbosacral transition plane is not strictly horizontal but is oblique as is the case when the 5th lumbar vertebra is wedge shape, then the rotating force develops a forward directed component in the sagittal as well as a length rotatory in the horizontal plane.

D. THE SPECIFIC ANALYSIS OF STATIC STRESSES OF THE PELVIC JOINTS WITH THE BODY AT REST

1. The sacro-iliac joint

In relaxed posture the line of gravity runs slightly behind the hip joint and then in front of a frontal plane which is laid through the anterior edges of the sacro-iliac joints. It then intersects with the spine somewhere at the lumbo-dorsal and cervico-dorsal junction. When the body is in symmetrical position gravity produces a rotatory stress upon the sacro-iliac joint both in the frontal and in the sagittal plane. As the line of gravity sways backward and forward even when a person stands quietly, the rotatory stresses vary with the distance of the line of gravity from the center of the joint. Since there is no visible motion in the sacro-iliac joint, the tension falls mainly upon the ligamentous structures which secure the joint, The same condition arises in the frontal plane as the line of gravity sways sideways.

In the transverse plane no rotatory stress occurs so long as the position is symmetrical. Any asymmetry, however, necessarily produces a rotatory effect upon the sacro-iliac articulation. This strain also has to be resisted by the ligamentous structures.

The principal ligaments which are resisting stresses are as follows:

In the sagittal plane the forward tilting of the upper end of the sacrum forward is checked by the sacro-iliac ligamentous apparatus as well as by the iliotransverse ligaments; the backward tilting of the lower end by the ligamenta sacro-spinosum and sacrotuberosum (Fig. 3,4).

In the frontal plane the gravitational stress on the sacro-iliac joints is, under normal conditions, amply resisted by the triangular-shaped sacrum implanted into the pelvis and by the resistance of the sacro-iliac ligamentous apparatus.

2. The sacrolumbar articulation, body at rest

In the sagittal plane the line of gravity falls somewhat in front of this junction. Because of the latter's obliquity, gravity produces a forward shearing stress varying with the degree of pelvic inclination. The line of gravity may be displaced considerably forward as in certain arthritic conditions or backward as in some types of malposture.

The iliolumbar ligaments act as immediate checks against the forward sliding of the 5th lumbar; but it requires the integrity of the entire ligamentous apparatus including the intervertebral disc to prevent the forward gliding and there are many anomalies which interfere with the security of the lumbosacral junction.

In the frontal plane complete automatic equilibrium obtains and no rotatory stress develops so long as the spine remains at rest in symmetrical position.

3. The symphysis

This junction is secured by the superior and inferior arcuate ligaments; under normal conditions there is no motion in this joint.

E. SPECIFIC ANALYSIS OF DYNAMIC STRESSES OF THE PELVIC JOINTS. THE BODY IN MOTION

1. Sagittal plane

A) THE SACROLUMBAR ARTICULATION

Skeletal checks of forward and backward movement are achieved by the sacrolumbar articular processes. Normally they are arranged in the frontal plane which gives freedom to forward flexion. Backward extension, however, is under a greater restraint to which the impingement of the 5th lumbar spinous process against the sacrum often adds a further impediment.

The muscular inhibition of forward flexion is taken up by the sacro-spinalis and the quadratus lumborum. Ligamentous restraints are in succession the supraspinous, the intraspinous ligaments and the ligamenta flava. Muscular checks to extension are the abdominal wall and the ilio-psoas muscle.

B) THE SACRO-ILIAC JOINT

In the sagittal plane the restraints are provided by the sacro-iliac ligamentous masses, the iliolumbar ligament and the sacrospinous and sacrotuberous ligaments. In addition, the mighty masses of the sacrospinalis as well as the iliopsoas muscles protect the joint before it really comes to a ligamentous strain. With the back in motion, particularly in the rhythmic back and forward swing incidental to locomotion, the protective role of these muscles becomes more important. The momentum which is imparted to this junction by the movements of the body must be held in check and it is the function of these muscles in the first place to do so. A second factor is the longitudinal thrust upon the junction in jumping or in violent concussion by falls. These jolts are first taken up by the sacrolumbar junction, but any degree of shock not absorbed by the latter immediately transmits itself to the sacro-iliac junction forcing the upper part of the sacrum to rotate forward.

2. The dynamic stresses in the frontal plane

A) THE SACROLUMBAR JOINT

The principal side benders are the oblique abdominals and the quadratus lumborum. Side bending is restrained by many structures. There are first the lumbosacral articulations themselves with their reinforcing capsular ligaments. Then the ligamenta flava and the ligamenta intertransversalia and the tension of the oblique abdominal, the quadratus lumborum and the sacrospinalis muscles.

B) The Sacro-iliac Joint

The same applies to this articulation which receives and sustains all side movements so far as they are not absorbed by the sacrolumbar articulation. How much of the momentum is taken up by the sacrolumbar articulation and absorbed in visible motion, depends to a great extent upon the position of its articular facet. In case of sagittal orientation of one or both of the sacrolumbar articulations (tropism) which restrain the range of lateral motion, a much greater amount of momentum reaches the sacro-iliac junction and their protective ligaments and muscular apparatus.

3. Dynamic stresses in the transverse plane

There is good reason to give this situation closer consideration. When the body rotates about its longitudinal axis the stresses are transmitted to the sacrolumbar joints and from there to both sacro-iliac junctions. As explained above, length rotation produces a particularly powerful momentum because the inertia which resists rotation ($I = m \rho^2$) is comparatively small, since the radius of gyration (ρ) is computed to be only $\frac{1}{3}$ of the diameter of the body taken as a cylinder. Hence the acceleration (a) developed by the rotatory momentum (M) is proportionally greater ($a = \dfrac{M}{I}$). In contrast the movements of the trunk in the frontal and sagittal planes are performed about transverse axes. Here the radius of gyration (ρ) which determines the size of the inertia resistance is about $\frac{1}{3}$ of the length axis of the body. In addition, the rotation axis in these frontal and sagittal planes does not go through the center of gravity of the trunk but occurs about the end of the spine. This increases the inertia, by a further amount. This is equal to the product of the square of the distance (e) of the center of gravity from the axis of motion times the mass. So the total inertia in this case equals $I = m \rho^2 + me^2$. In other words the inertia is a great deal larger than it is when rotatory movement occurs about the length axis of the body.

The result is then that length rotatory movements, because they are encountering so much less resistance are much easier to carry out and the acceleration which is developed by the same force is so much greater in these length rotatory movements. That much greater is also the stress on the lumbosacral or the sacro-iliac joints which any rotatory force can generate in movement about the length axis provided that such motion is not spending itself in visible motion before it reaches and is intercepted by these articulations.

A) The Sacrolumbar Joint

The lumbar spine has only a small rotatory range and the force is promptly transmitted to the sacrolumbar articulations. Whether it exhausts itself here in visible motion again depends upon the orientation of the articular facets as it does in lateral motion. If they are directed in the sagittal plane instead as usuallly in the frontal, then rotation becomes locked sooner and the remaining force is then transmitted to the sacro-iliac junction.

B) The Sacro-iliac Junction

This junction has no visible motion. The rotating force is intercepted from above and produces stresses of the ligamentous apparatus of the sacro-iliac junction. It then depends upon the stability of the pelvis whether or not this stress must be absorbed by the ligaments or whether the force is spent in visible motion. In the latter case, what remains of the force is now transmitted through the pelvis to the hip joints and from there to the extremities. An example is the following: a man swings a golf club with his feet planted on the ground. If he uses good technique the swing is initiated by rotating the body backward on the right side and is then followed by the swing of the club and the striking of the ball.

Then the remaining energy is spent by rotating the body to the opposite side; here the hip rotates inward and finally he pivots on the ball of the left foot. The experienced golfer holds his body supple and pliant so that most of the remaining motor energy is transformed in visible motion and the ball is hit hard.

Let us assume, on the other hand, that his form is poor, that his legs are glued firmly to the ground; the ball of the foot does not pivot nor does the hip rotate. Then the momentum that comes from the swing of the body is absorbed by rigid restraint. It is not transformed into visible motion of rotation of the hip and pivoting on the ball of the foot, but it ultimately spends itself in the resistance which the ground develops against the firmly implanted feet. It is a case of a torsion effect by two couples of forces, one originating from the upper end and the other acting in opposite direction from the lower portion of the body, the critical point being the sacro-iliac and the sacro-lumbar articulations. No wonder that this often results in a severe strain of one or the other of these articulations. The same situation may arise in other motor performances such as in hammer throwing or shot putting or javelin throwing which depend on acceleration produced by powerful length rotatory movements of the body.

The crux of the situation is always whether or not a rotatory impulse can be transformed in visible motion or whether such motion is being blocked and is absorbed by skeletal or ligamentous resistance.

That this principle has a very wide application both in physical education and in orthopedic surgery is evident. The athletic trainer must distinguish between a short, heavy and rigid spine and pelvis and a slender and pliable one when he has to decide what type of sport his subject is fitted for. Above all things it must be realized that the "spinning" or "winding up" part of the performance, as for instance in baseball pitching places a high demand on the pliability and suppleness of the body and that clumsy performances not only fail in delivering the acceleration desired for the athletic performance but they also expose the body structures to considerable stresses, ligamentous as well as muscular.

By the same token we know from clinical experience that it is particularly the length rotatory movements of the trunk which, when blocked in their purpose of carrying through visible motion by an intrinsic rigidity of the body or by external resistance, are most apt to lead to structural damage.

II. THE PATHOMECHANICS OF THE PELVIS

Developmental aberrations, injuries, inflammations and other causes leave their impact upon the shape of the pelvis and cause mechanical dysfunction of the pelvic organs on the basis of pelvic deformities. The stress resistance of the pelvic bones may be so lowered that deformation occurs without external influences; or the pelvis changes shape secondary to a disalignment of the spine or of the hips, or, the deformation is entirely due to external factors which destroy the normal relationship of the pelvic bones.

A clincal grouping of the vastly heterogenous and dissimilar types would be as follows:

A. Developmental Factors
B. Degenerative Conditions
C. Static Changes
D. Traumatic Deformities

A. DEVELOPMENTAL FACTORS

1. Abberations by developmental defects

A) THE ANTERIOR CLEFT OF THE PELVIS

A most uncommon deformity is the diastasis of the anterior pelvic ring. The separation of the symphysis may reach 10 cm. or more in the adult and it

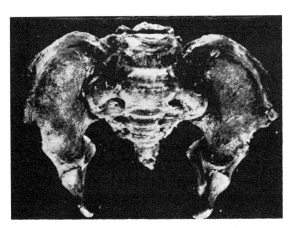

FIG. 13. Anterior cleft of the pelvis. (Putschar)

is often associated with ectopy of the bladder (Fig. 13). As a result of the static stresses, the sciatic notch becomes narrower and, to make up for the lack of support by the anterior portion of the pelvic ring a compensatory hypertrophy of the posterior pelvic wall develops (Braus and Kolisko[1]), and the ligamentous reinforcements of the sacro-iliac joint appear markedly over-developed (Putschar[10]).

B) THE POSTERIOR CLEFT OF THE PELVIS

Smaller degrees of central clefts of the arches of the 5th lumbar and 1st sacral vertebrae are not uncommon. They do not change the configuration of the pelvis as a whole but they are a factor in lumbosacral instability. However in cases of complete cleft of the sacrum one finds a particularly strongly developed anterior portion of the pelvic ring seen even in the non-viable fetus (Fig. 14). Gruber[4] describes a posterior pelvic cleft with dorsal ecstrophy of the intestines in which there was a strong increase of the transverse diameter and a flattening of the pelvis antero-posteriorly.

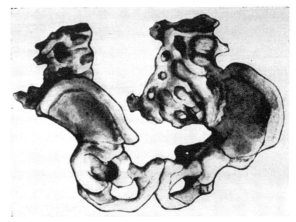

FIG. 14. Posterior cleft of the pelvis. (Putschar)

One wonders what effect these anterior and posterior clefts have on the statics of the pelvis. No doubt in the anterior cleft the bulk of the body weight is placed upon the posterior portion of the pelvic ring and this can be accomplished only by making the sacral region more perpendicular, that is, by decreasing the pelvic inclination. This again places the hip in a more extensory position, with most of the weight passing through the postero-superior portion of the femoral head. In the most unusual cases of a wide posterior cleft it would seem that the opposite should take place, namely that the pelvic inclination is increased and that the hip joint assumes a position of flexion and that weight is borne by the antero-superior portion of the head.

C) THE SPONDYLOLISTHETIC PELVIS

In simple spondylolysis or in cases of spondylolisthesis with only slight forward displacement of the 5th lumbar there are no pelvic changes. In higher degrees of this deformity, however, one finds the pelvic inlet considerably narrowed by the forward sliding of the 5th lumbar vertebra. Nature tries to meet the increased forward shearing stress by forming a buttress and supporting exostoses against further displacement (Fig.

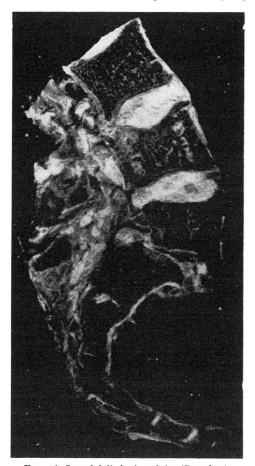

FIG. 15. Spondylolisthetic pelvis. (Putschar)

15). One will find the spondylolisthetic pelvis rather large, the bones massive, the sacrum particularly strong, and the sacrolumbar intervertebral disc thinned out and considerably degenerated.

The static effect of the displaced 5th lumbar is that it forces the upper end of the sacrum backward while the sacrospinous and the sacrotuberous ligaments hold the lower end of the sacrum forward thereby increasing its concavity. As an effect of the static changes it is also noted that the os ilii becomes more massive conforming with its more perpendicular position. The anterior superior spines are deflected inward. So is also the horizontal ramus of the os pubis until, at the symphysis it forms with the opposite ramus a sort of duck bill very much as one sees in higher degrees of osteomalacia.

2. The statics of the pelvis in errors of differentiation

The so-called assimilation pelvis is one in which the 5th lumbar is a transitional vertebra. It may show a complete or incomplete sacral assimilation so that the sacrum really has 6 vertebrae (Fig. 16). The sacrum appears long, the promontory is high, and the pelvis is flattened in antero-posterior direction (Braus and Kolisko[1]).

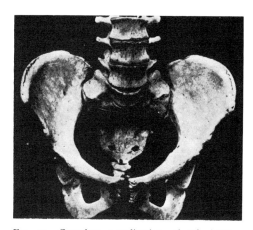

FIG. 16. Complete sacralization of 5th lumbar vertebra. (Putschar)

The pelvic deformity is more marked in unilateral sacralization where the assimilated vertebra often is the keystone of a lumbosacral curve. There is really a double promontory, a higher one on the assimilated side and a lower one on the other side. The asymmetry of the pelvis corresponds to the lumbosacral scoliosis which is often compensatory to a higher lumbar curve. That means that the pelvis is wider on the side of the sacral convexity. Weight-bearing is unevenly distributed in contrast to the symmetrical bilateral sacralization. In either case there is a change in the dynamic situation since motion is now transmitted to the junction between the 4th and 5th lumbar vertebrae which have a more restricted range. It is a condition definitely predisposing to strain.

B. DEGENERATIVE CONDITIONS

1. Congenital growth anomalies

THE CHONDRODYSTROPHIC PELVIS

Braus and Kolisko[1] distinguish two types. In one the dimensions of the pelvis are reduced uniformly; it is smaller than normal but shows no definite static deformation; the other is the flat chondrodystrophic pelvis which is similar to the rachitic and which shows definite static changes (Putschar[10]).

The flattened pelvis is the usual type. An important feature of it is the reduced conjugata vera. The normal ratio of the conjugata to the transverse diameter of the pelvis is as 100:123. In the cases mentioned the ratio is often 100:164 and may even reach 100:280 in extreme cases.

In addition the acetabulum is flat and its floor is very thick. The most striking

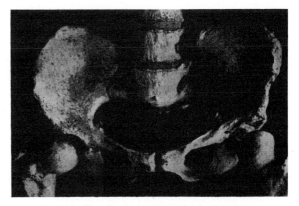

FIG. 17. Chondrodystrophic pelvis. (Putschar)

changes are in the sacrum. The promontory is high and projects strongly; the sacrum stands almost horizontal; the pelvic inclination is greatly increased. Associated with this is a marked narrowing of the sacral canal. The increased lordosis produced by the horizontal sacrum puts a strong shearing stress on the lumbosacral junction (Fig. 17).

2. Degenerative malacic condition

THE RACHITIC PELVIS

In form and dimensions it must be called a flat pelvis. Braus and Kolisko[1] subdivided it into a typical flat pelvis and a generally reduced flat pelvis. Unfavorable static conditions develop because the line of gravity is shifted forward of the common axis of the hip joint and the strong tilt of the sacrum diminishes its stability. The trunk is thrown backward by the lumbar lordosis. The pelvic inclination is not really increased but the sacral inclination is such that the lordosis is due to the sacral tilt rather than to the dropping of the whole pelvic inlet.

That static factors have a decided influence upon the configuration of the rachitic pelvis is shown by the fact that in the first year only general growth disturbances are seen; then in the second year the typical deformations of the rachitic pelvis develop. Much of the deformation is smoothed out by natural growth process, but much of it remains. Some ascribe all deformation to static causes, especially the pressure which reaches the pelvis through the acetabulum (Mayer[8] and Litzmann[6]). However, considering the greatly increased pliability of the skeleton, muscular forces as well as gravitational must be considered.

THE OSTEOMALACIC PELVIS (PUTSCHAR[10])

The striking feature here is the duck bill deformity. This is due to the lateral component of the pressure on the hip joint which forces the acetabula against each other (Fig. 18). It results in a narrowing of the anterior portion of the pelvic inlet, and in extreme cases this inlet assumes a regular clover leaf shape

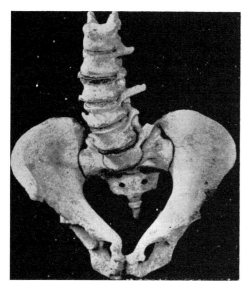

FIG. 18. Osteomalacic pelvis. (Putschar)

(Fig. 19). In addition the sacrum together with the superincumbent spine sinks downward and forward between the iliac bones, rotating about a transverse axis. The tilt may be so strong as to force the sacrum into a horizontal position. The projecting promontory further encroaches upon the pelvic inlet. The conjugata is not really shorter because of the projecting duck bill shape of the symphysis but the pelvic inlet is so narrowed that it constitutes a serious obstetrical problem.

The mechanical stresses which produce this deformity can be analyzed as follows:

Under the weight of the trunk the upper sacral vertebrae are forced downward and forward. At the hip sockets the gravitational reaction forces the acetabula upward and inward. The so-called Otto[9] pelvis (Braus and Kolisko[1]) or the intrapelvic protrusion of the acetabulum is produced in osteomalacia by inward thrust of the femoral heads against the acetabular floor. Such a pelvis is also found in senile involution or in cases where some abnormal static demands are made upon the hip joint.

C. STATIC CHANGES

Ordinary static changes which are associated with malposture do not produce pelvic deformations; their only effect is the change in the inclination of the pelvis. More severe spinal deformities, on the other hand, are apt to have a remote effect on the configuration of the pelvis.

1. The kyphotic pelvis

The closer the gibbus is situated to the pelvis the greater is the pelvic deformation. A kyphos of the upper or mid dorsal spine does not produce any morphological changes of the pelvis. The only change is the position of the sacrum if a compensatory lumbar lordosis is present. If the gibbus is situated in the lumbar spine the sacrum becomes more perpendicular as the normal lumbar lordosis is obliterated. There are also some changes in the sacro-iliac junction. The sacrum stands more dorsally be-

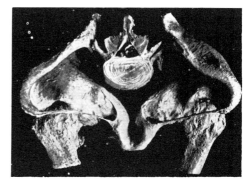

FIG. 19. Extreme osteomalacic pelvis. (Putschar)

tween the iliac wings and the articular joint surfaces are displaced in the same direction. In addition to this retroposition the sacrum actually rotates backward with its upper end about a frontal axis. This again causes an increase in the length of the conjugata and the pelvis seems to be more narrowed sideways which corresponds to the lateral narrowing and the antero-posterior deepening of the lower thoracic cage.

While the wings of the os ilii are farther separated between the sacrum, the os ischii moves closer to it and the pelvic outlet now appears narrowed. It is the abnormal distribution of the body weight due to the kyphosis which causes the change in the position of the pelvic bones and of the sacro-iliac articulation. The latter loses some of its contact area resulting in the loss of its cartilaginous covering and even subsequent fusion. When the 5th lumbar is destroyed by an inflammatory process (Putschar[10]), a lumbosacral kyphosis develops (Fig. 20). There is no real promontory; the pelvis becomes funnel-shaped and has a much narrowed outlet.

According to Braus and Kolisko[1] two mechanical factors are operative in

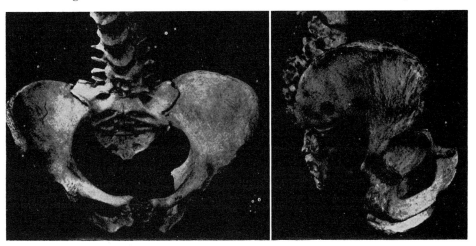

Fig. 20. Pelvis in lumbo-sacral kyphosis. (Putschar)

this pelvic deformation. First there is the angular gibbus which diverts the gravitational force of the superincumbent weight into an oblique course. This force then becomes the resultant of two components: a horizontal one which forces the sacrum back, and a perpendicular one which forces it downward.

2. The scoliotic pelvis

On the whole the pelvic changes are not as drastic as they are in the kyphotic pelvis. According to Braus and Kolisko[1] they depend on the relation of the sacrum to the scoliotic curve in the first place. If the sacrum forms part of the lumbar curve the 5th lumbar is the keystone of this curve. If the sacrum forms a counter curve below the lumbar, then the 5th lumbar is a transitional vertebra. In both instances the pelvis is affected. Forming part of either a lumbar curve or a sacral counter curve, the sacrum shows the

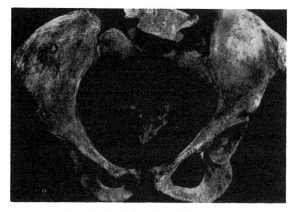

FIG. 21. Scoliotic pelvis. (Putschar)

characteristic scoliotic deformities. On the convex side it is lengthened and thinned out and on the concave side it is shortened and more massive (Fig. 21). On the other hand, if the sacrum is not included in any curve and stands symmetrical there is no effect upon the pelvis.

We find the pelvic deformity more pronounced in the rachitic kyphoscoliosis where the scoliotic deformity is superimposed upon the rachitic (Fig. 22).

3. The pelvis in dislocation of the hip

The normal gravitational stresses are no longer transmitted from the pelvis to the head of the femur but they pass in front of it. As a result the pelvis is suspended between the femoral heads in a hanging position in which it must find a new gravitational equilibrium. In this situation, some of the pelvifemoral muscles, especially the psoas and the internal obturator act as hanging straps. The result is a much increased pelvic inclination and an exaggerated lumbar lordosis. The pathological features of the pelvis in dislocations of the hip are not particularly marked. It is wid-

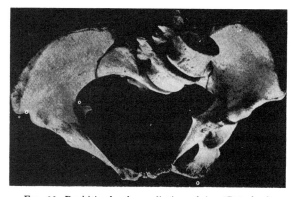

FIG. 22. Rachitic kypho-scoliotic pelvis. (Putschar)

ened, especially in its inlet, since the femoral heads no longer exert their pressure against the acetabulum. If a dislocation is unilateral the pelvis will be wider and the innominate bone more arcuar on the dislocated side and the horizontal ramus of the os pubis will be more slender and more drawn out. The obturator foramen also appears larger (Fig. 23). The static strain on the lumbo-sacral junction produces secondary arthritic reactions in later years.

4. The coxalgic pelvis

The abolition of motion in the hip joint has a profound effect upon the formation of the pelvis, especially when the pathological event occurs in the years of growth. As a rule the os ilii is smaller on the affected side. The inlet is egg shape with the broad side pointing to the coxalgic side. The sacro-iliac

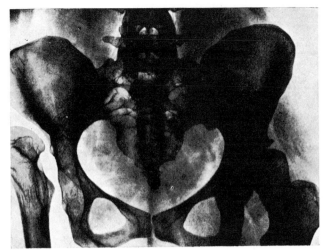

FIG. 23. Pelvis in unilateral congenital dislocation of the hip. (Putschar)

articulation may show signs of reactive exostosis. In flexion ankylosis of the hip the lordosis increases the forward shearing stress.

Pelvic asymmetry is absent in double ankylosis (Fig. 24). While the pelvic deformation is less obvious than in unilateral cases, the widening of the pelvic inlet is unmistakable.

5. The paralytic pelvis

In paralysis the pelvic changes are similar to what one finds in coxalgia being due to the absence of weight pressure which reaches the

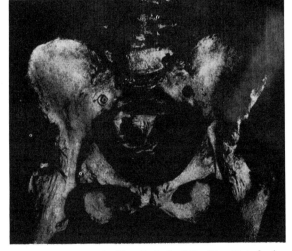

FIG. 24. Pelvis in double ankylosis of the hip. (Putschar)

pelvis through the femoral heads. But it requires a rather extensive paralysis, one in which weight bearing is eliminated and which occurs in the growing years to produce any noticeable degree of asymmetry (Fig. 25).

On the whole the affected side shows the usual aspect of a non weight bearing pelvis, namely the absence of the acetabular projection,

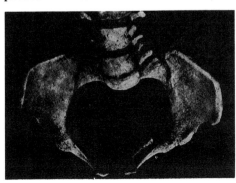

FIG. 25. Poliomyelitic pelvis. (Putschar)

giving the innominate line of the paralyzed side a more rounded form and the anterior pelvic ring is atrophic on the affected side. In bilateral paralysis Braus and Kolisko[1] describe the pelvic inlet as flattened with a large transverse diameter.

D. TRAUMATIC DEFORMITIES

On the whole one must concede that the discontinuities of the pelvic bones caused by trauma are as unpredictable as the mechanism of the trauma itself. Yet there are certain critical areas of the pelvis which withstand the traumatic forces less than the rest of the structure.

Three distinct groups can be recognized.
1. The apophyseal separations.
2. The injuries which separate the junctions.
3. True fractures.

1. Apophyseal separations

These are almost exclusively minor lesions most often occurring in sport injuries. For instance running or sprinting may produce avulsions of the anterior superior spines. One sees such injuries in football players where the avulsion occurs by the sudden tension of the sartorius or the tensor fasciae latae. It is interesting to note that muscle strain results in bony separation without tearing the muscle itself only in young and vigorous individuals. Another instance is the apophyseal separation of the tuber ossis ischii which is produced by the hamstring.

2. Injuries to the pelvic junctions

From the mechanical viewpoint they are of interest because under normal conditions it requires a very large force and a suitable mechanical arrangement to produce such a result. This applies both to the sacro-iliac junctions and to the symphysis. Fractures are more frequent than the true separations with the possible exception of the sacro-coccygeal junction where separation may occur by a fall on the buttocks.

THE SEPARATION OF THE SYMPHYSIS

Normally this is a very strong junction. Even in childbirth where it is relaxed the passage of the child's head alone cannot produce a separation. But symphyseal tears may occur when the legs are abducted during parturition as a combination of tension and shearing stresses.

More important and formidable are the separations resulting from crushing injuries where the diastasis may reach as much as 9 cm. One must distinguish between actual separation of the symphyseal joint and one in which the hyalin cartilage layers become detached from the bone. This latter is really a para-symphyseal fracture. Such a fracture does not necessarily have any effect upon the sacro-iliac junction in the back.

In a true symphyseal separation which occurs between the two layers of cartilage of the symphyseal joint the halves of the pelvic ring spring apart and act as levers upon the sacro-iliac joints which are often found strained in combinations with the symphyseal injury.

The Sacro-Iliac Separations

Considering the strong reinforcements which this junction receives from the massive sacro-iliac ligamentous apparatus, one understands that pelvic fractures are seen more often than actual sacro-iliac separations. A separation may occur by a fall from a great height in which case an entire half of the pelvis may be pushed upward on the sacrum with simultaneous separation and displacement of the symphysis. Complete separation of both junctions is most unusual.

When the pelvic ring has been separated either at the symphysis or at the sacro-iliac junction, nature subsequently tries to secure stability by the hypertrophy of the ligamentous apparatus and the formation of bridges and exostoses.

3. The mechanics of pelvic fracture

The different types of pelvic fractures, marginal fractures, fractures of the sacrum, fractures through the anterior or posterior pelvic ring, may all have

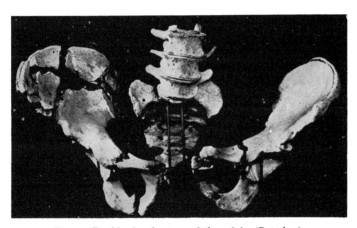

Fig. 26. Double ring fracture of the pelvis. (Putschar)

their specific mechanical premises which cannot be established accurately and one has to depend largely on cadaver experiments. Much information can be gained from an old monograph of Messerer[7] which deals with the elasticity and resistance of human bone. A few details may be mentioned here:

Pressure exerted upon the anterior pelvic ring in antero-posterior direction will at the strength of 250 kg. produce a symmetrical fracture of the ascending ramus of the os pubis. On the other hand if a force of 180 kg. is applied to the os ilii in the frontal plane, that is crosswise, a separation of the sacro-iliac joints occurs. Furthermore transverse pressure at the level of the acetabula produces first an elastic deformation which is almost completely reversible up to an amount of 27 mm. deformity. Only at the pressure of 290 kg. could a fracture of the pelvis be produced in the form of a double ring fracture; the anterior ring fractured around the foramen obturatum, the posterior ring in the sacro-iliac region. A second experiment showed that the double ring fracture occurred at 250 kg. pressure (Fig. 26).

A mechanical factor of importance is the position of the individual. In standing the superincumbent weight rests on the so-called femoral-sacral arch which transmits the weight from the sacrum to the femur. In sitting, however, the weight is transferred from the sacrum to the tuberosity of the os ischii so that weight stresses and reactions from below meet at the lower and posterior portion of the pelvic ring, namely the ischiosacral arch. The resistance of either portion of the ring is considerable.

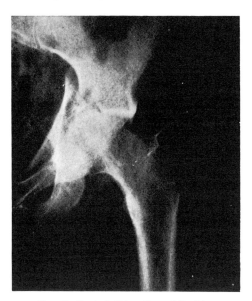

FIG. 27. Central dislocation of the hip.

Lesshaft[5] estimated the weight resistance of the two flanks of the pelvic ring at 1254 kg. and the bursting of the ring under this load occurred in the roof of the acetabulum. The resistance of the sacro-iliac joint against lateral distension was examined by Fessler[2] and estimated at 213 kg. and the resistance of the symphysis against lateral distension was estimated by the same author at 197 kg.

In longitudinal direction the so-called impaling fracture where the sacrum is jammed into the pelvis should be mentioned. Of particular interest are the torsion fractures. They occur if the pelvic ring is fractured in front or if the symphysis is separated so that the os ilii rotates forward twisting the lateral masses of the sacrum.

Bending fractures (Messerer[7]) occur by direct compression of the pelvic ring as do the compression fractures. Probably the best known mechanism is that of the lateral compression of the pelvis. Such a force first produces a compression fracture of the os pubis, then a twisting of the iliac bone and then a fracture of the posterior ring of the pelvis.

One condition deserves special consideration, namely the fracture of the acetabulum. This may occur by a fall on the legs, or by a blow against the knee with the hips flexed, the so-called dashboard fracture. It may also follow a direct violent blow against the greater trochanter. While the dashboard fracture involves the articular rim only, the fractures from direct violence or a fall on the abducted hip will produce what is called a central dislocation, that is, a penetration of the head into the pelvis through the fractured acetabulum (Fig. 27). It depends upon the degree of force and on the individual bone resistance how far the head will penetrate into the pelvis. The fracture is more likely to occur when the force applies in the direction of the femoral neck; otherwise a fracture of the neck itself would be the more probable event.

III. SUMMARY

A. THE NORMAL PELVIS

1. The construction of the pelvis is on the order of a skeletal ring, supported at the hip joints, and formed by the confluence of the three separate bones which constitute the pelvic half or the innominate bone.

2. Three articulations interrupt the continuity of the ring: the two sacro-iliac joints between which the sacrum is implanted into the ring and the symphysis where the anterior halves of the ring join each other in the midline.

3. The manner in which the sacrum is fitted in between the two os ilii secures it against downward sliding but in antero-posterior direction its security against forward penetration into the pelvis is entrusted to the reinforcing ligaments of the sacro-iliac junction.

4. The bones are fitted into the pelvic ring under considerable elastic tension; if this intrinsic equilibrium is destroyed by severing the ring, it springs apart, similarly as the thoracic cage.

5. Certain accessible landmarks can be used to ascertain the relation of the pelvis to the cardinal planes of the body as well as to determine the planes of the inlet and the axial plane of the pelvis.

6. The reactions from the floor transmitted through the oblique femoral neck to the pelvis develop both horizontal and vertical components; the former exerting a side thrust, the latter an upward thrust.

7. Standing upright with legs in symmetrical position the horizontal components of both sides neutralize each other and automatic equilibrium obtains.

8. If legs are abducted equally, the horizontal component is directed toward the midline, as it is when legs are parallel, but this component increases with the degree of abduction.

9. If legs are adducted equally, the horizontal component of the gravitational reaction is directed away from the midline.

10. If one leg is abducted, the other adducted (position of rest) the line of gravity is shifted toward the adducted hip. Equilibrium is no longer automatic: the abductors of the adducted, and the adductors of the abducted hip cooperate in shifting the pelvis toward the abducted side; they are neutralized by the adductors of the adducted and the abductors of the abducted hip which cooperate in shifting the pelvis toward the adducted side.

11. In the sagittal plane the equilibrium is labile, being automatic only if the line of gravity passes through the center of the hip joints. Otherwise active tension of the flexors or extensors is necessary to hold the equilibrium against gravitational rotatory stresses.

12. In the horizontal plane the for or backward angulated neck produces a rotatory component; so long as the stance is symmetrical, the two sides neutralize each other; if asymmetrical, muscular effort is required to maintain equilibrium.

13. If one leg is placed forward, the other backward, the backward leg develops a forward, the forward a backward rotatory thrust. This is an essential feature of the restraint and propulsion in the human gait.

14. When thigh and pelvis move as one unit in case of an ankylosed hip joint and the rotatory impulse of the body cannot resolve itself in visible motion of the hip, the stress becomes concentrated at the sacrolumbar and sacro-iliac junctions.

15. Any degree of impulse not absorbed by the sacrolumbar junction falls upon the sacro-iliac syndesmosis.

16. In the sagittal plane the skeletal checks of for or backward movement come from the sacrolumbar articular processes. The muscular inhibition is taken up by the sacrospinalis and quadratus lumborum. The sacro-iliac joint is restrained by the sacro-iliac ligamentous masses, the ilio-lumbar ligament and the sacrospinosum and sacrotuberosum.

17. In the frontal plane the skeletal checks of the sacrolumbar joint are likewise the articular facets while the capsular ligaments, the flava and intertransversaria as well as the oblique abdominals, the quadratus and sacrospinalis restrict the range. A similar ligamentous and muscular restraint applies to the sacro-iliac articulation.

18. In the transverse plane the stress situation is of particular importance because of the powerful momentum developed by length rotation of the trunk. This is due to the inertia being much smaller for length rotatory motion and, consequently, the acceleration being so much greater for length rotation.

19. The sacrolumbar joint having only a small length rotatory range the momentum is readily transmitted to the sacro-iliac joint. The latter has no visible motion and therefore transmits the rotatory momentum to the hip joint, and the lower extremity to be eventually resolved in visible motion by pivoting about the ball of the foot. Any circumstance preventing this release in visible motion is liable to result in strain of the sacro-iliac and sacrolumbar joints.

B. THE PATHOMECHANICS OF THE PELVIS

1. The situations can be grouped clinically in congenital anomalies, static disturbances, traumatic and inflammatory or degenerative conditions.

2. Aberrations by developmental defects are: diastases of the anterior ring

(symphyseal separations) with compensatory hypertrophy of the posterior pelvic wall; posterior clefts ranging from small central clefts to wide and total separations, the latter usually concerning non-viable fetuses; and lateral separation of the neural arches, either simple or combined with forward gliding of the 5th (occasionally the 4th or 3rd) lumbar vertebrae; spondylolisthesis. In the latter case there are considerable changes of the pelvic inlet.

3. Error of differentiation is represented by the so-called assimilation pelvis, hemi- and bilateral sacralization. Pelvic deformation is seen only if the sacrum is involved in a scoliotic curve.

4. Among degenerative conditions there is the chondrodystrophic pelvis, usually the flattened type with a much shortened conjugata vera; the rachitic pelvis of the typical flattened or generally reduced type and the osteomalacic pelvis with the characteristic duck bill deformation and the intrapelvic protrusion of the acetabula (Otto pelvis).

5. In kyphosis of the lumbar spine following destructive lesions, especially tuberculosis, the so-called kyphotic pelvis shows a decrease of the pelvic inclination, a retroposition and retrotorsion of the sacrum and a laterally narrowed inlet with antero-posterior deepening and increase of the conjugata.

6. The scoliotic pelvis shows changes corresponding to the scoliotic vertebra: lengthening and thinning out on the convex, and shortening and massiveness on the concave side. If the sacrum is not included in the curve and stands symmetrical, there is no pelvic deformation.

7. In congenital dislocation the inclination of the pelvis is increased and on the dislocated side the inlet is widened.

8. The changes are similar in the coxalgic pelvis, especially during the years of growth. The inlet is egg shaped with the broad side pointing to the coxalgic side. In flexion ankylosis the inclination of the pelvis is greatly increased.

9. If weight bearing is eliminated on one side due to paralysis, the paralytic pelvis shows the widening of the innominate line on this side and absence of the acetabular projection into the pelvis.

10. The traumatic deformities of the pelvis can be divided into: apophyseal separations, separations of the junctions and fractures.

11. The most common apophyseal separations are those of the anterior superior spine and of the tuber ossis ischii.

12. Symphyseal tears may occur during parturition but more formidable are those caused by crushing injuries.

13. True sacro-iliac separations require an excessive force such as fall from a great height, in which case the entire half of the pelvis may be pushed upward on the sacrum.

14. Pelvic fractures are marginal, fractures of the sacrum, fractures through the anterior or posterior ring, all having their specific mechanical premises. According to Messerer's experiments, pressure in antero-posterior direction produces fractures of the ascending ramus of the os pubis; a double ring

fracture is produced around the obturator foramen and in the sacro-iliac region by transverse pressure at the level of the acetabula. Fracture of the acetabular rim occurs by a blow on the knee with the hip flexed, the so-called dashboard fracture. Direct violence to the trochanter or fall on the abducted hip produces what is called a central dislocation, i.e., a penetration of the femoral head into the pelvis through the fractured acetabulum.

BIBLIOGRAPHY

1. BRAUS and KOLISKO: *Die Pathologischen Beckenformen.* Leipzig, F. Deuticke, 1904.
2. FESSLER, J.: Die Festigkeit der Menschlichen Gelenke mit Besonderer Berücksichtigung des Bandapparates. *Munich,* 1893.
3. FICK, R.: *Handbuch d. Anatomie und Mechanik der Gelenke, Vol. I.* Jena, G. Fischer, 1904.
4. GRUBER, G. B.: Über das Becken eines Hochgewachsenen Eunuchoiden Akromegalen. *Wien. Klin. Wchnschr. 38:*114, 1925.
5. ———: Vorweisungen zur Frage der Totalen Wirbelsäulenspaltung mit und ohne Darmekstrophie. *Verhandl. deutsch path. Ges. 19:*339, 1936.
   LESSHAFT, P.: *Die Architektur des Beckens.* Wiesbaden, 1893.
6. LITZMANN, H.: Ein durch Mangelhafte Entwickelung des Kreuzbeines quer verengtes Becken. *Arch. f. Gynaek, Berlin, XXV:*31, 1884-1885.
7. MESSERER, O.: Über die gerichtlich Medizinische Bedeutung verschiedener Knochenbruchformen. *Friedreich's Bl. f. gerichtl. Med. Nürnberg, XXXVI:*81, 1885.
8. MAYER, H. V.: *Missbildungen des Beckens unter dem Einfluss abnormer Belastungsrichtung.* Jena, 1886.
9. OTTO, A. W.: Neue seltene Beobachtungen zur Anatomie, Physiologie und Pathologie gehörig. Berlin, A. Rücker, 1824.
10. PUTSCHAR, W.: Spezielle Pathologie des Beckens, Henke Lubarsch, *Handb. d. Speziellen Pathol. Anatomie. 914:*430. Berlin, J. Springer, 1939.

Lecture XIII

THE MECHANICS OF RESPIRATION

INTRODUCTION AND DEFINITION

RESPIRATION provides for the oxygen intake into and the release of CO_2 from the circulatory system and is a vital function of the body, the cessation of which is incompatible with life and any impairment is detrimental to health.

The mechanical premise behind respiration is the rhythmic increase and decrease of the volume capacity of the thoracic cage. To accomplish this the cage must be able to change its form.

The subject of this analysis is to explain the mechanical events which produce the changes of the thoracic cage essential for the respiratory exchange and to scrutinize the effect of some pathological conditions on the respiratory function.

I. THE MECHANICS OF THE THORAX AT REST

In the change of the volume capacity of the thorax, the most important but not the only factor is the movement of the ribs. Anteriorly they move against the sternum and posteriorly against the spinal column. While the vertebral column remains practically stationary, the movement of the ribs is transmitted to the sternum causing it to take part in the respiratory excursion.

A. THE MOVEMENT BETWEEN RIBS AND VERTEBRAE

1. The anchorage

All ribs, with the exception of the last two, are joined to the vertebrae by a double articulation; one between the head of the rib and the vertebral body, and the other between the transverse process of the vertebra and the tubercle of the rib. While each articulation taken by itself would permit 3° of freedom of motion, this double anchorage restricts motion to 1° about a combined axis. This axis runs through the center of the capitulum of the rib and through its tubercle, these being the convex bodies of the articulations (Fick[4]).

The costovertebral articulation between the heads and the bodies of the vertebrae is bipartite (Fig. 1). The capitulum is inserted between the lower border of its proper vertebral body and the upper border of the next lower vertebra, the two being separated by an intervertebral disc. This point is marked at the capitulum by a transverse crest.

The axes of the costovertebral articulations are related to the cardinal planes of the body as follows: they are oblique to all three planes, but their obliquity

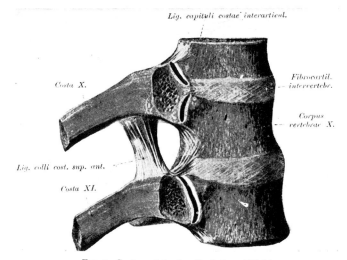

Fɪɢ. 1. Costovertebral articulation. (Fick)

changes from rib to rib. The combined axis of the upper ribs approaches more the frontal, of the lower ribs more the sagittal plane. Consequently, the upper ribs, as they move about their frontal axis, swing up and down. The lower ribs, in moving about their axis which approaches more the sagittal plane, flop outward and inward.

The planes of the costal ring itself are oblique to the cardinal planes of the body. They slant forward and downward in relation to the spinal column. The lesser this slant, the more the rib ring approaches the horizontal plane, the higher and deeper is the thoracic cage and the more it is adapted for respiratory expansion. The inclinations of the ribs diminish from the first rib downward. Volkmann[13] found the inclination angle with the horizontal 50° for the first rib, 43° for the fourth, and 42° for the seventh. However there are wide individual variations which make it difficult to set up a normal standard for the inclination of the ribs. Under certain pathological conditions the inclination angles may be excessively large, producing a flat

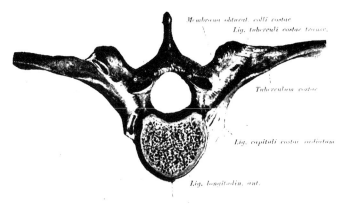

Fɪɢ. 2. Costovertebral articulation. (Fick)

thorax such as is seen in tuberculosis; or they may be unusually small as one sees in the barrel shaped thorax of the emphysematic.

2. Ligamentous reinforcements and restraints

Considering the uninterrupted movement which goes on in these articulations one must expect them to be amply reinforced by an extensive ligamentous apparatus.

a) The costovertebral articulations are secured by the radiate ligaments which support the capsular apparatus (Fig. 2).

b) The ligamenta tuberculi costae reinforce the articulation between the transverse process and the tubercle of the ribs (Fig. 2).

c) The ligamenta colli costae anterior and posterior portions run from the upper edge of the neck of the rib to the vertebra above (Fig. 3).

d) A costotransverse membrane fills the space between the transverse process and the neck of the rib (Fig. 2).

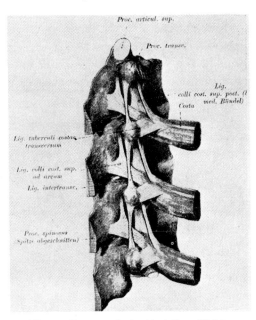

FIG. 3. Costovertebral articulation. (Fick)

All these ligaments are needed to safeguard the two articulations between the ribs and vertebrae. Their degeneration and calcification in senescence or in arthritis considerably encroaches upon the respiratory capacity of the thorax.

B. THE JUNCTION BETWEEN RIBS AND STERNUM

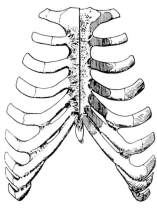

FIG. 4. Sterno-costal junction. (Strasser)

1. The anchorage

At the front end the ribs are joined to the sternum as follows: there is no articulation between the 1st rib and the manubrium. In many cases from the 2nd to the 7th rib, but always from the 2nd to the 5th rib, there is a sternocostal junction in form of a true articulation. Up to puberty the sternum consists of 5-6 segments corresponding to the attachment of the 2nd to the 5th rib. It is not surprising, therefore, to find that from the 2nd to the 4th and sometimes to the 5th rib the articulations are bipartite similarly as are the articulations between the heads of the ribs and the vertebral bodies. The fibrous plate which divides these articulations is continuous with the fibrous junctions uniting the segments of the sternum (Fig. 4).

2. The ligamentous reinforcements and restraints

These articulations also have their elaborate ligamentous apparatus. The most important are the radiate ligaments, anterior and posterior reinforcing the articulations.

In the intercostal space and in close relation to the intercostal muscles are spanned the internal and external intercostal ligaments. the fibers of which parallel their muscular namesakes (Fig. 5).

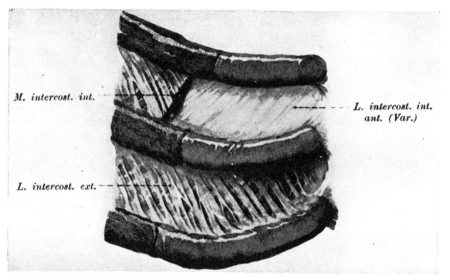

FIG. 5. Internal and external intercostal ligaments. (Fick)

C. THE AUTOMATIC EQUILIBRIUM OF THE NON-CONTRACTILE STRUCTURES OF THE THORAX

This is the equilibrium maintained by elastic forces stored up in the thoracic cage. One could call it also the intrinsic equilibrium. Due to the tension under which spine, ribs and sternum are joined together, a good deal of intrinsic energy is hoarded up by the elastic resistance of the parts. The situation compares with the intrinsic equilibrium existing in the spinal column. Part of the hidden force comes from the elasticity of the ribs themselves. They are fitted into the thorax under elastic resistance. This increases or decreases as in inspiration or expiration the ribs straighten or buckle. If the thoracic walls were cut longitudinally on both sides the thorax would spring wide open. A second factor in the intrinsic equilibrium of the thoracic cage are the ligaments which reinforce the junctions between the ribs and vertebrae behind and the ribs and sternum in front. These ligaments are also fitted into the thorax under elastic resistance. For example, the anterior and posterior radiate ligaments of the costovertebral articulations are both under tension which decreases and increases alternatingly with the movements of the ribs.

It is not surprising under the circumstances that the thorax can stand a great deal of pressure before the ribs become cracked. This resistance compares with that of an elastic switch in contrast to a dried-out branch of the same thickness. A well known athletic stunt is to have a board weighted down with a number of persons placed on the chest of the performer, but one will not fail to note that in these exhibitions the performer is very careful to keep the thorax in deep inspiration in which position the elastic resistance of the ribs is at its best. Even children may sustain a weight of 200 pounds or more upon the chest without crushing of the ribs.

To establish the neutral level of this intrinsic equilibrium, one must find the position in which all non-contractile structures are under equal elastic tension. Yet for each structure this neutral position is different. This makes it difficult to speak of a neutral position common to all structures involved; one has to be content with an average position of equilibrium. For the thorax as a whole this is somewhat between full inspiration and full expiration.

Experiments show that if two thoracic segments are severed horizontally in the upper thoracic wall, then the upper one automatically moves forward and upward. If the same is done in the lower half of the thorax the upper ring wanders outward which is in keeping with the different direction in which the axes of the costovertebral articulations are oriented.

This means that active muscular effort is necessary both for full inspiration and for full expiration and that the latter is by no means an automatic event. As a matter of fact full expiration requires a downward pulling of the rib cage by gravity and by the tension of the abdominal muscles to overcome the intrinsic elasticity of the structures, ligaments, cartilages and muscles.

For the whole thorax the neutral point lies somewhat on the side of inspiration which means that in the state of rest the thorax is adjusted slightly in inspiratory position. Each of the upper 5 rib rings, after being separated from the lower, assumes in the costovertebral junction a mid position between extreme elevation and extreme depression.

Helmholz[7] believed that the neutral position is expiratory which led to the erroneous opinion that the elastic forces of the thorax alone are capable to taking care of expiration, at least in quiet breathing. It is quite true that in the cadaver the position of the thorax is definitely expiratory. But Fick[4] and Strasser[11] believe that in the living even for quiet expiration a certain muscular effort is necessary. Strasser[11] in fact distinguishes two phases of the expiratory act. First comes the slow gradual exhalation which is then followed by a more explosive expulsion of air. It is this second period which represents the actual muscular effort of expiration. How important this is in certain pathological conditions in which the mobility of the thoracic cage is impaired needs no further emphasis.

The exact level of the neutral position varies individually and with age. In the newborn it lies more in the inspiratory field than it does in the adult, since the ribs of the newborn occupy a more horizontal position. In certain individuals who assume a military posture and who have a greater vital capacity, it is also more on the inspiratory side. There are, besides, some pathological conditions where the neutral level is more inspiratory, for instance in pneumonia, in hydrothorax, in pneumothorax, in emphysema, or in tumors of the thoracic cavity. In pregnancy the contents of the abdominal cavity are pushed upward and the diaphragm stands high which likewise forces the thorax to take up a more inspiratory position at rest. On the other hand, a more expiratory position associated with a greater slant of the costal rings is seen in persons with a dorsal kyphosis, in malposture, in old individuals, and in people with phthisical habitus.

II. THE THORAX IN MOTION

A. THE MOVEMENT OF THE RIBS

It was stated that the essential dynamic factor of respiration is the rhythmic movement of the ribs. During this motion the ribs change their shape, being longer in inspiration and shorter in expiration (Fig. 6).

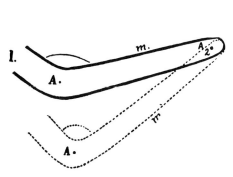
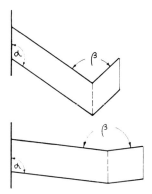

FIG. 6. Adaptive changes of the costal cartilage in respiration. Costal angle at A in inspiration (full line); in expiration (dotted line); A_2 articulation of cartilage m with sternum. (Fick)

FIG. 7. Changes of the cartilaginous rib in respiration. $\alpha\alpha$ sternocartilaginous angle, $\beta\beta$ costo-cartilaginous angle.

Investigating these changes which the ribs themselves undergo during the respiratory act at the chondro-costal junction, we find that the costo-chondral angle increases in inspiration and decreases in expiration: the rib stretches out in inspiration and buckles in expiration. The expiratory smaller angle is closer to the natural angle of the rib while it takes force to stretch the rib during inspiration. But in full expiration again the angle is smaller than it is in the resting position. Consequently in full expiration the elasticity opposes further kinking, thus as it opposes further straightening of the rib in inspiration; therefore full expiration is not automatic but requires a certain muscular force.

The more the thorax goes into inspiration the higher the upper ribs rise upward and the more they straighten out (Fig. 7). Similarly, in the lower half of the thorax, the deeper the inspiration is, the greater is the outward flare of the ribs and the more they straighten at the costo-chondral angle (Fig. 8).

Not only must the up and down movement of the upper ribs be imparted to the sternum but the periodic lengthening and shortening of the rib must also be adjusted at the sterno-costal junction. Where there are no joints between cartilage and sternum, the changes in length are taken care of by the pliability of the chondro-costal junction. In any case the sternum receives a motor

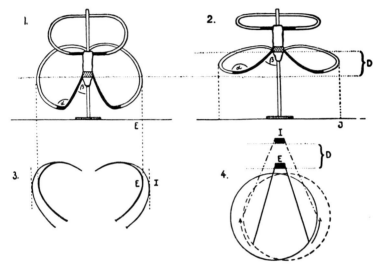

FIG. 8. Fick's model of the excursion field of the thoracic rings in respiration (Braus). 1. Expiration, 2. Inspiration, 3. Mid-rib excursion, lower ring, 4. Sternal excursion, upper ring.

impulse either through the sterno-costal junctions or through the sterno-costal articulation.

B. THE MOVEMENT OF THE STERNUM

Taking in consideration the movement of the ribs as such irrespective of how they change form during respiration, the question arises what effect this movement has on the sternum. So far as the outward flair is concerned as long as movement is symmetrical the sternum maintains a stationary position. The up and down movement of the upper ribs, however, must necessarily take the sternum with it. Since the ribs change their length during respiration, both the upper during their up and down movement and the lower as they swing outward and inward, it implies that the sternum gets an impact from the ribs to which it must accommodate itself.

The movement imparted to the sternum is threefold: (1) a translatory perpendicular motion up and down; (2) a rotatory motion in the sagittal plane about a frontal axis which causes the upper end of the sternum to move forward and the lower to move backward; and (3) another translatory movement in the sagittal plane which carries the whole sternum backward and forward. The forward and upward movement is procured by the elevation of the upper ribs in inspiration, and the downward and backward movement occurs with the depression of the lower ribs in expiration (Fig. 9).

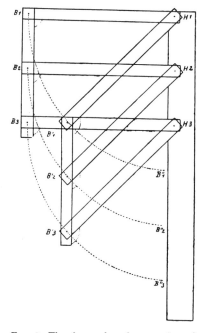

Strasser[11] gives the ranges of sternal movements as follows:

At the upper border the elevation is 14-17 mm., the forward movement is 9-10 mm.; at the lower border the elevation is 11-12 mm., and the forward movement is 21-24 mm. When the respiration is of the lower rib cage type, there is more lateral flare of the ribs and the elevation of the upper end of the sternum is diminished while the excursion of the lower end of the sternum is increased and there is more a straight forward movement than elevation.

The point of practical importance is that the integrity of the costal cartilages is essential for the normal breathing process. If the normal elasticity of the rib is lost due to calcification or ossification, or if there is an ankylosis of the costosternal articulations, the

FIG. 9. The forward and upward, and downward and backward movement of the sternum as the result of rib excursions. (Fick)

ribs can no longer change their length according to the respiratory position nor can they adjust themselves to the sternum. If such changes are extensive, thoracic breathing is seriously impaired. As early as 1859 Freund[5] correlated pulmonary disease with primary anomalies of the costo-cartilaginous junction and according to him pulmonary tuberculosis of the apex was often caused by defective development or abnormal stiffness of the cartilage of the 1st rib.

III. THE DYNAMICS OF RESPIRATION

A. THE INSPIRATORS

1. The diaphragm

This is the most important of all inspiratory muscles. It is a flat muscle which has its tendon located at its center surrounded by a fringe of muscle fibers. These are attached in front to the xyphoid process and to the inner surface of the anterior half of the 7th to the 12th ribs. In the back the fibers are attached to the anterior surface of the lumbar vertebrae from the 1st to the 3rd. The muscle forms a large dome, the concavity of which points downward and forward. Consequently in contraction the impulse imparted to the contents of the abdominal cavity is directed downward against the pelvis and forward against the abdominal wall. Being inserted into the lower thoracic aperture, the muscle also tries to pull the lower end of the sternum and the costal arch inward. The latter effect is neutralized by the anterior and lateral abdominal muscles which stabilize the costal arch against being pulled into the thoracic cage by the contracting diaphragm (Braus[2]). The ancients knew of no other respiratory muscle than the diaphragm. Galen added to it the intercostals, but it was left to Borelli[1] to recognize the respiratory descent of the diaphragm. Duchenne[3] observed that contraction of the diaphragm pulls the ribs inward and closes up the thoracic cage if the abdominal pressure is lacking, and he concluded that abdominal pressure is essential for diaphragmatic function. Also due to Duchenne we know now that respiration can be sustained by the diaphragm alone, in contrast to Borelli who insisted that respiration could not take place without the help of the intercostal muscles.

That the fixation of the costal arch by the abdominal muscles is essential for proper diaphragmatic action can be seen in rickets where contraction produces a sucking-in of the arch, the Harrison's groove. The excursion of diaphragmatic movement amounts to the length of one thoracic vertebra at the base of the heart (Strasser[11]). The diaphragm occupies the highest position when the muscle is completely relaxed.

2. The intercostals

The controversy over the action of these muscles goes back to Galen in the second century A.D. He was the first to mention the intercostals as respiratory muscles. That they are important in respiration was shown by Duchenne[3] who cites cases in which adequate respiration was possible in spite of total paralysis of the diaphragm and all the auxiliary inspiratory muscles with only the intercostals remaining. Cases are also reported in which expiration was possible with total paralysis of the abdominal and of the auxiliary expiratory muscles by using the intercostals alone. This illustrates the importance these muscles have, both as inspirators and expirators.

The old concept was that the internal intercostals are inspirators and the external expirators. This has long been disproven. A model of Bayles (quoted by Fick[4] (Fig. 10) shows that when the ribs are elevated the in-

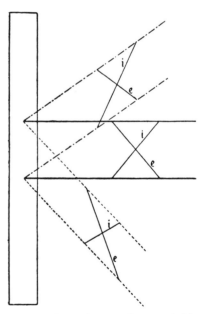

Fig. 10. Internal (*i*) and external (*e*) intercostals. When ribs are elevated the external intercostals are contracted. When ribs are depressed the internal intercostals are contracted. (Fick)

insertion of the external intercostals come close together. Today the general opinion is that the external intercostals contract in inspiration and that the internal intercostals are inspiratory muscles only in their intercartilaginous portions.

The work which these intercostals can perform has been determined by Weber who found that the weight of the external intercostals on each side is equal to that of the biceps and that their physiological cross section area is 100 cm.^{2} which is as much as the combined cross section area of the gluteus medius and minimus. When one calculates the shortening of 2 mm. and the absolute muscle power set at 3.6 kg. per cm.^{2} (Recklinghausen), then the total working capacity of the external intercostals amounts to 0.72 kgm.

3. The auxiliary inspirators

a) THE SCALENI

These muscles arise from the transverse processes of the cervical vertebrae and they are attached to the 1st and 2nd ribs. In forced inspiration they raise the ribs.

b) THE STERNOCLEIDOMASTOID

This muscle acts as an auxiliary inspirator by elevating the sternum. To accomplish this the head must be stabilized in extension by the posterior neck muscles.

c) THE STERNOHYOID, STERNOTHYROID AND MYLOHYOID MUSCLES

In a remoter sense the sternohyoid and the sternothyroid muscles as well as the mylohyoid are also auxiliary inspirators because they elevate the sternum. There is also the serratus posterior superior which elevates the upper ribs and pulls them backward.

d) THE THORACOHUMERAL AND SCAPULOTHORACIC MUSCLES

In addition there is the group of thoracohumeral and of scapulothoracic muscles which are called upon eventually to act as auxiliary inspirators. The muscles of the first group are the pectoralis major and minor. They elevate the ribs and widen the thoracic cage. Those of the second group are the rhomboids, the levator scapulae and the trapezius. These muscles pull the scapula backward in order to increase the tension of the auxiliary thoracohumeral inspirators.

In analyzing the efficiency of these auxiliary inspirators, one finds that the spino-thoracic muscles such as the sternocleido and scaleni come into play only in situations of severe dyspnoe. Their contractile strength is not overly great and if they are to act upon the thoracic cage, head and neck must be firmly stabilized which involves a considerable output of muscular energy.

On the other hand the powerful humero-thoracic and scapulo-thoracic muscles are more effective. They do require for their action upon the thoracic cage the stabilization of humerus and scapula, respectively. But this is easily accomplished by resting the extremity against an external resistance. We see therefore that after strenuous exercise the individual will lean his arms against a table thereby immobilizing the upper extremities and enabling the thoraco-humeral muscles to move the thorax against the humerus by elevating the ribs instead of moving the humerus against the thorax as is the usual function of these muscles.

One makes use of this principle also in the Swedish gymnastic exercises. Here arms and shoulders are moved backward and upward, and fixed in this position the pull of the thoraco-humeral muscles on the ribs widens the thorax and deepens inspiration. In the same sense the energetic swinging and circling exercises of the arm by means of Indian clubs or dumbbells causes the humero-thoracic muscle to exert a vigorous pull upon the rib cage. For this very reason such exercises are employed in malposture or in such congenital chest deformities as the funnel chest where it is expected that the pull of the muscle on the thoracic wall will widen its inspiratory excursion and increase the vital capacity.

B. THE EXPIRATORS

One of the principal expiratory muscles is the interosseous portion of the internal intercostals. All intercostal fibers which descend forward are inspirators which includes the entire external intercostals; and all fibers which ascend forward are expirators which means the interosseous portions of the internal intercostals, while all fibers of the inter-cartilaginous interspaces of the ribs whether descending or ascending forward are inspirators. This means the intercartilaginous portions of the internal intercostals since the external intercostals have no intercartilaginous portion (Fig. 11).

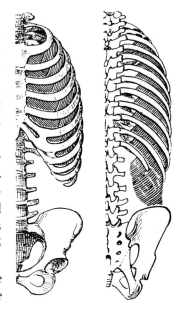

The interosseous portion of the internal intercostals represents a muscle mass of 75 g. weight. It has a cross section of 50 cm.² and has a shortening distance of 3 mm. On the above mentioned basis for the unit strength of muscle power this would give the muscle a working capacity of 0.5 kgm.

The most powerful expiratory muscles are those of the abdominal wall: the rectus abdominus, the internal and external obliques and the transversus abdominis. All these muscles pull the lower ribs down and depress and narrow the costal arch.

Fig. 11. External and internal intercostal muscles. (Strasser)

Between the thoracic cage, the diaphragm and the abdominal wall there exists a sort of three-cornered equilibrium. Duchene[3] observed that on descent of the diaphragm the widening of the lower aperture of the thorax which should accompany this descent does not take place when the abdominal pressure produced by the muscles of the abdominal wall is lacking, as is the case, for instance, in paralysis, where the costal arch is drawn inward. Again, in expiration, when the muscles contract and the abdominal pressure is increased, the diaphragm is forced upward into the thoracic cavity. This is what is called the expiratory "backstroke." The center of this three-cornered equilibrium is the costal arch. It must be well stabilized if the muscles, either diaphragm or the abdominals, are to perfrom to the best advantage. Abdominal and thoracic muscles combine to stabilize this arch against the pull of the diaphragm. Sometimes this cannot be accomplished because the ribs are too soft and pliable, as in rickets. Then the costal arch is pulled in producing the above-mentioned Harrison's groove. It is therefore the abdominal wall in the first line which produces the expiratory "backstroke" when the diaphragm is relaxed; and it is also a stabilizer of the costal arch and secures it against being pulled inward by the contracting diaphragm.

The diaphragm and the abdominal muscles are the forces which control abdominal respiration. Pure abdominal breathing occurs where there are no other active forces left to enlarge the thoracic cage. The abdominal protrusion during inspiration is caused by the descent of the contracting diaphragm and the relaxation of the abdominal wall. The intra-abdominal pressure being distributed evenly, the bulging of the wide anterior wall must be greater than that of the narrow flank.

Auxiliary expiratory muscles are the triangularis sterni which always pulls the lower ribs down and thereby decreases the costal angle; the quadratus lumborum which flattens the bulging lateral and posterior abdominal wall; and the serratus posterior inferior which forces the lower ribs downward and inward (Strasser[11]).

IV. RESPIRATORY EXCHANGE AND VITAL CAPACITY

In quiet breathing the amount of air exchanged is about 500 cc. For emergencies nature establishes a very considerable reserve. Forced inspiration can add another 1600 cc. and it is possible after ordinary expiration to expel another 1600 cc. from the lungs. This, with the ordinary exchange of 500 cc. will add up to a vital capacity of 3700 cc. It has been known that under particularly strenuous efforts this capacity can reach 7000 cc.

This forceful expulsion of residual air naturally requires the cooperation of all expiratory muscles including the auxiliaries. The first half of the expiration corresponds to the release of the elastic tension of the thoracic walls, but the last and more forceful half is a definite muscular effort which engages the cooperation of all expiratory muscles. Fick[4] states that even the normal expiration is a more or less active procedure and that it requires particularly the oblique and the transverse abdominals to cooperate with the interosseous portions of the internal intercostals. It is possible that in very quiet breathing the neutral position of the thorax is reached in expiration without any particular expiratory effort, but the usual expiration is an active process.

On the other hand, if the neutral line is on the inspiratory side, it may be assumed that it saves inspiratory effort. But this is also true only in a restricted sense; because as a rule the inspiratory requirement is much higher and it is questionable whether even in most quiet breathing the inspiratory position represented by the intrinsic equilibrium is sufficient. Consequently inspiration is virtually always an active procedure.

V. RELATION OF RESPIRATION TO THE SPINAL COLUMN

A. BACKWARD BENDING

The sternum is elevated; the abdominal wall is under increased tension and approaches the spinal column, and the spread of the ribs is increased. It is preeminently the first rib which elevates the sternum partly because the rib is short and has a greater excursion range than the others (Haines[6]). When the costal cartilages calcify the movement of the ribs becomes restricted. Since they can no longer adjust themselves to proper length through this cartilaginous portion, rib and sternum must now move more or less as a single unit.

In backward bending the lower ribs are also held down by the tension of the abdominal wall. This increases the lung space in longitudinal direction but the sagittal diameter of the chest is decreased. In this position the excursion of the ribs is considerably diminished, partly by the position of the spine, partly by the action of the abdominal wall upon the lower ribs.

B. FORWARD BENDING

Here the spinal column recedes from the abdominal wall, but the ribs are crowed in front and they converge; the sternum recedes backward especially at its lower end. The lungs are crowded and their excursion is greatly limited as it is in hyperextension.

C. LATERAL BENDING

The ribs spread on the convex side and the thoracic space is increased; on the concave side, however, they are crowded and the lung space is diminished. This offsets the increase on the convex side and the respiratory excursion on the whole is decreased.

D. LONGITUDINAL ROTATION

Longitudinal rotation also influences respiration, though less so than the other positions; but since it is often combined with lateral bending it shares the disadvantages of the latter in regard to the respiratory function.

E. THE NORMAL UPRIGHT POSITION

There is no doubt that this position is the most serviceable for the function of the respiratory system. The most favorable upright position is the military or attention position, although from the standpoint of muscular effort in general it is by no means an economic one.

VI. THE TYPES OF RESPIRATION. THORACIC AND ABDOMINAL BREATHING

The ordinary respiratory mechanism is a combination of thoracic and abdominal breathing but their relative ratio varies with the individual.

The earmark of *thoracic* breathing is the elevation of the ribs in inspiration, the upward movement of the sternum and the flattening of the abdominal wall. Sometimes in forced thoracic breathing the wall is pulled in so that a hollow appears below the costal arch.

The characteristic feature of *abdominal* breathing is that in inspiration the anterior abdominal wall bulges out following the contraction of the diaphragm which increases the intra-abdominal pressure. This bulge disappears in expiration, particularly in strong expiration where the abdominal muscles are pulled in by the tightening of the transversus abdominis. The costal arch and the ribs themselves remain stationary. Inspiration in abdominal breathing is affected exclusively by the contraction and a flattening of the diaphragm. Expiration, on the other hand, is entirely due to the contraction of the abdominal wall which furnishes the expiratory backstroke against the relaxed diaphragm.

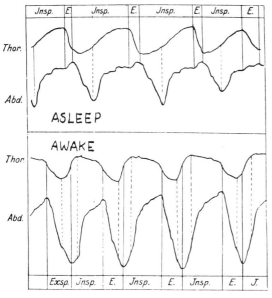

FIG. 12. Respiration, awake and asleep. (Strasser)

In the male, respiration is more abdominal; in the female, it is more thoracic. Unconsciously one often uses more the abdominal or more the thoracic type according to whether more attention is paid to one or the other. Hultkrantz[8] estimates the thoracic respiratory exchange at 490 cc. against 170 cc. of the abdominal.

There is also an interesting difference in breathing between being awake and being asleep. In sleep the respiratory excursions are greatly diminished, often only to 1/10 of the waking excursions (Mosso[9]). Both sexes show a relative increase in thoracic and a decrease in the abdominal breathing during sleep (Strasser[11]) (Fig. 12).

Costal breathing is still possible even if the movement of the sternum has become more difficult because of lack of mobility of the first rib, but then it is effected mostly by the lateral excursions of the middle and lower ribs.

Two types of costal breathing are distinguished by Strasser.[11]

1. The *normal costal* breathing: all ribs move parallel, the upper ribs move

up and down, the lower flair out laterally, the sternum moves upward and forward in inspiration.

2. The *lateral costal* breathing: a purely lateral flapping movement of the lower ribs. The reasons for it are twofold: in some cases it may be an abnormal flexed position of the head, for instance, in Strumpell-Marie's disease which impedes the elevation of the first rib and of the sternum even in upright position. In others the costal cartilage of the first rib is ossified or some other change in the costovertebral and sternocostal articulation of the rib interferes with its excursion.

There is also a difference in costal respiration between lying in prone and in side position. In the latter the lateral rib movement becomes more prominent.

The most regular and uniform excursions of the sternum take place in the normal upright position. It is here where thoracic breathing prevails most and diaphragmatic breathing becomes subordinate. It is the most favorable position for a healthy and efficient respiratory mechanism.

VII. THE PATHOMECHANICS OF RESPIRATION

Mechanical difficulties in respiration arise under the following pathological conditions:

a. From obstruction of the upper air passages,

b. From insufficient ventilation due to congenital malformation of the thoracic cage,

c. From deformation caused by abnormal pliability of the thorax,

d. From restrictions of the motion of the thoracic cage based on ligamentous or articular degenerative lesions,

e. From loss of respiratory muscle power,

f. From diseases or neoplasm of the organs of the thoracic cavity.

A. THE OBSTRUCTION OF THE AIR PASSAGES

Such obstructions are mostly the fault of enlarged tonsils and adenoids; less frequently they are due to an enlarged thyroid which compresses or deflects the trachea. The respiratory rhythm is not greatly disturbed; mouth breathing supplements the insufficient air intake through the nose, but some embarrassment arises on exertion which is usually avoided. Attention should be paid to the effect on posture which arises from the obstruction of the upper air passages by tonsils and adenoids. It is one of the frequent causes of malposture. The difficulties coming from compression of the trachea are more serious and may call for surgical relief.

B. CONGENITAL THORACIC DEFECTS

A congenital thoracic deformity is the funnel chest often associated with anatomical anomalies of the diaphragm. The retraction of the lower end of the sternum due to the contracting diaphragm is facilitated by the softness and pliability of the ribs (Fig. 13).

The mechanical effect is much the same as in rickets since the abdominal muscles are flabby and lack tone. This deformity also causes micrognathia but the salient point is the depression of the sternum which forms a more or less deep groove and the thorax appears flattened. The deepest point of the hollow is at the junction between body and ensiform process. The respiratory excursion is diminished and the whole thoracic cage assumes more of an expiratory position as indicated by the slant of the ribs. The posture is decidedly asthenic. Abdominal respiration prevails. The muscles of the anterior abdominal wall are relaxed. The thorax is held backward, the abdomen protrudes, there is malposture with a round and hollow back (Steindler[10]) (Fig. 14).

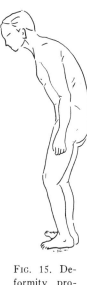

FIG. 13. Funnel chest. (Steindler)

FIG. 14. Malposture.

FIG. 15. Deformity produced by ankylosing arthritis of the spine.

Early synostosis of the sternum may also impair the movement of the thorax. Normally a cartilaginous junction is found where the second rib joins the sternum. If it is absent then a similar cartilaginous junction of the sternal segment is found at the junction of the third rib with the sternum. Since the mechanism of the thorax requires some give of the upper portion of the sternum against the lower, it follows that early synostosis of the sternal segment has some effect on respiration. The sternum is shorter which is unfavorable because then the lower cartilaginous ribs must be longer and move movable in order to mediate the costo-sternal movement.

Congenital absence of ribs is almost always associated with some vertebral anomaly but it does not seem to affect the respiration unless there is a sizable thoracic defect. The vital capacity is not impaired because the organism very early accommodates itself to the changed mode of breathing. In large defects one sees on inspiration a retraction of the thoracic wall. Congenital defects of the thoracic muscles are very rare. Absence of the pectoralis major is seen on rare occasions and has been reported in conjunction with congenital deformities of the upper extremity and congenital scoliosis (Tedeschi[12]).

C. RICKETS

Rickets has a marked effect upon the configuration of the thorax. Owing to its pliability, the equilibrium between it, the diaphragm and the abdominal walls cannot be maintained. The thorax gives in to the pull of the diaphragm and the so-called Harrison groove develops. The breathing becomes more abdominal.

D. STATIC DEFORMITIES AND MALPOSTURE

Static deformities and particularly malposture have definite effects upon the thoracic breathing mechanism and abdominal breathing prevails.

E. DEGENERATIVE CHANGES

Changes which impair the mobility of the thoracic articulation such as senile calcification of ligaments or arthritic changes profoundly affect the respiratory mechanism. In chronic arthritis of the spine, especially in the ankylosing type, the rib movement may be entirely abolished. Even the simple calcification or ossification of the cartilaginous ribs or the ankylosis of the costosternal articulations has a detrimental influence upon breathing. It may even reach the point where thoracic breathing becomes entirely eliminated. That there is a decrease in respiratory capacity early in the course of the chronic ankylosing arthritis of the spine is obvious and in many cases the decrease of the respiration is the first objective sign of the arthritic process (Fig. 15).

F. TUBERCULOSIS OF THE SPINE

This condition may lead to thoracic changes which are especially detrimental to respiration. They are more marked if the spinal disease is situated in the upper half of the dorsal spine. In this event the deformed thorax is flattened in antero-posterior direction and the ribs take a sharp slant downward. The sternum protrudes more at its lower end.

On the other hand, if the gibbus is located in the lower dorsal or lumbar spine then the slant of the ribs is not changed materially. In fact the lower ribs become more horizontal and the thoracic cage appears lengthened in antero-posterior direction and flattened from side to side (Steindler[10]) (Fig. 16).

G. EMPHYSEMA, HYDRO- OR PNEUMOTHORAX, OR TUMORS

In emphysema and in hydro- or pneumothorax or in tumors within the thoracic cavity the thorax approaches the shape of the newborn, being maintained in a more inspiratory position.

Fig. 16. Tuberculosis of the spine and its effects on the thoracic cage. *A*. Dorsal tuberculosis. *B*. Lumbar tuberculosis. (Putschar)

H. SENILE KYPHOSIS

A more expiratory position is found in the senile kyphosis of the dorsal spine. Here the slant of the ribs is more precipitous and the thorax approaches in form that of the upper dorsal tuberculosis. It is particularly the rigidity of the cage which affects respiration because the calcification of cartilages and ligaments interferes with the elevation and depression of the ribs. The vital capacity is low.

I. PARALYSIS

The most formidable respiratory difficulties are those found in paralytic conditions, especially in poliomyelitis, and in traumatic and ascending myelitis.

1) Paralysis of the diaphragm

Since the diaphragm cannot descend in inspiration the thoracic space cannot increase in downward direction. Breathing becomes entirely costal. In unilateral paralysis the asymmetrical descent of the diaphragm can be noticed. As long as the intercostal muscles are acting, paralysis of the diaphragm is still compatible with life though the respiratory reserve is reduced to a point where any exertion is out of question. It is only when the intercostals are paralyzed also that a fatal outcome precipitated by bronchitis or bronchial pneumonia may be expected. Duchenne[3] observed paralysis of the diaphragm in hysterical patients causing the same respiratory difficulties.

2) Paralysis of the intercostals

In paralysis of the intercostals the thoracic respiration is abolished and abdominal and diaphragmatic respiration take over. In anterior poliomyelitis the inhibition of thoracic breathing may be the result of spasm of the intercostals and not due to true paralysis.

3) Paralysis of the muscles of the abdominal wall

Paralysis of the abdominal muscles interfering with expiration is very common in anterior poliomyelitis. As the tone of the muscles is lost, they become distended by the pressure of the viscera which are crowding downward and forward during the inspiratory movement. The abdomen protrudes and a condition develops similar to that in visceroptosis, which, in fact, it is. Abdominal inspiration is still possible by the action of the diaphragm but the great respiratory difficulty lies in the loss of the expiratory "backstroke." In unilateral paralysis where abdominal expiration is still operative on the sound side, the paralyzed side of the abdomen becomes distended during inspiration as the contents of the abdominal wall are crowded toward this side.

4) Paralysis of back muscles

The paralysis of the back muscles, as well as asymmetrical paralysis of the muscles of the abdominal wall, frequently results in spinal asymmetry, a paralytic scoliosis. Such asymmetries of the spine carry their effect to the thorax which shares in the deformation of the spinal column. The result is that on the concave side of the spinal deformity the capacity of the thorax is greatly reduced and costal breathing is achieved mostly on the convex side. In severe cases convex side breathing is never sufficient to make up for the deficiency on the concave side, and the respiratory reserve remains much lower.

VIII. SUMMARY

1. Movement between ribs and spinal column is restricted to one plane by the double anchorage to vertebral bodies and transverse processes.

2. The movement of the upper ribs is up and down, that of the lower sideways.

3. Slant of the ribs decreases from the upper to the lower.

4. A number of ligaments both reinforce the articulations and restrain their movement (radiate lig., lig. tuberculi costae, lig. colli costae, costotransverse lig.).

5. From the 2nd to the 7th rib the anchorage to the sternum is by articulation.

6. Reinforcing ligaments are the radiate and the internal and external intercostal ligaments.

7. An automatic intrinsic equilibrium of the thoracic cage is provided for by the elasticity of the ribs, especially of the cartilaginous portions and of the reinforcing ligaments.

8. The neutral position of this intrinsic equilibrium lies somewhat in the inspiratory field. Expiration is therefore not entirely automatic.

9. The level of this neutral position varies with age; it lies more in the inspiratory field in the newborn; also in certain postures (military) and in certain pathological conditions (emphysema, hydrothorax).

10. In inspiration the chondro-costal angle is increased, in expiration it is decreased in relation to the neutral position.

11. The movement of the sternum is threefold: a translatory perpendicular motion up and down, a rotatory motion in the sagittal plane, and a translatory motion forward and backward.

12. The integrity of the costal cartilages is essential for normal breathing.

13. The most important inspiratory muscle is the diaphragm. In contracting it imparts to the contents of the abdominal cavity an impulse directed forward and downward against the abdominal wall. The pulling in of the lower end of the sternum and of the costal arch is neutralized by the anterior and lateral abdominal muscles. Where this fixation fails, as in rickets, contraction of the diaphragm causes a drawing in of the arch, the so-called Harrison's groove.

14. Next important inspirators are the external intercostals and the intercartilaginous portion of the internal intercostals.

15. Auxiliary inspirators are the scaleni, the sternocleido-mastoid and, in a remoter sense, the sternohyoid and mylohyoid as well as the thoraco-humeral and scapulo-thoracic muscles.

16. The principal expirators are the interosseous portions of the internal interosseous and the muscles of the abdominal wall.

17. Between thoracic cage, diaphragm and abdominal wall there exists a three-cornered equilibrium. When the abdominal muscles contract, they increase the intra-abdominal pressure, forcing the diaphragm up into the thoracic cavity by the expiratory "backstroke." At the same time they secure the costal arch against being pulled inward by the contracting diaphragm during inspiration.

18. Auxiliary expiratory muscles are the triangularis sterni, the quadratus lumborum and the posterior inferior serratus.

19. The respiratory exchange at rest is about 500 cc., but in emergency can rise to 3700 cc. by adding 1600 cc. to inforced inspiration and 1600 cc. to inforced expiration.

20. *Backward* bending of the trunk increases the tension of the abdominal wall and the spread of the ribs which are held down by the tense abdominal muscles. The rib excursions are considerably diminished and the sagittal diameter of the chest is decreased. Respiration becomes more difficult.

In *forward* bending the spinal column recedes from the abdominal wall but the ribs are crowded and converge. Excursion of the lungs is greatly limited as it is in hyperextension.

In *lateral* bending the ribs spread on the convex side and are crowded concavely, more than offsetting the convex side increase of the thoracic capacity.

Longitudinal rotation interferes least with respiration unless combined with lateral bending.

21. The upright normal position is most favorable for the respiratory function.

22. The ordinary respiratory mechanism is a combination of thoracic and abdominal breathing. In the male respiration is more abdominal, in the female more thoracic. In sleep the respiratory excursion is greatly diminished, often to 1/10 of the waking, active hours. Strasser distinguishes a normal costal and a lateral costal breathing type; the latter is a purely lateral flapping of the lower ribs and becomes more noticeable when lying on the side.

23. Mechanical difficulties to respiration arise:

a) From obstruction of the upper air passages, enlarged tonsils and adenoids, less frequently from an enlarged thyroid pressing on the trachea.

b) From congenital thoracic deformities such as a funnel chest. Here the mechanical effect is similar as in rickets. Abdominal respiration prevails.

c) From early synostosis of the sternum which also impairs thoracic movement because of the lack of "give" of the upper portion of the sternum against the lower.

d) Congenital absence of rib does not affect respiration unless there is a sizable thoracic defect.

e) In rickets breathing becomes more abdominal, because the equilibrium between diaphragm, abdominal wall and thorax cannot be maintained due to the pliability of the latter. A Harrison's groove develops.

f) Malposture has a definite effect on respiration; abdominal breathing prevails.

g) Senile calcification of ligaments or arthritic changes of the spine profoundly affect respiration. Thoracic breathing may become entirely eliminated.

h) A tuberculous gibbus of the dorsal spine flattens the thorax anteroposteriorly and greatly interferes with respiration; the lumbar or lower dorsal gibbus does not.

i) The most formidable respiratory difficulties are found in paralytic conditions.

1) In paralysis of the diaphragm the breathing becomes entirely costal. It is still compatible with life but the reserve is low and any exertion must be avoided.

2) In paralysis of the intercostals, abdominal breathing takes over. In poliomyelitis the intercostal deficiency may be due to spasm.

3) Paralysis of the abdominal wall abolishes the expiratory backstroke. The abdominal muscles become distended and visceroptosis follows.

4) Paralysis of the back muscles frequently results in paralytic scoliosis with greatly reduced thoracic capacity on the concave side. The convex side takes over thoracic breathing but in severe cases it is not sufficient and the respiratory reserve remains low.

BIBLIOGRAPHY

1. BORELLI: *De Motu Animalium*. Lugduni Batavorum, 1679.
2. BRAUS, H.: *Human Anatomy, Vol. I*. Berlin, J. Springer, 1921.

3. Duchenne, G. B.: *Physiologie des Mouvement*. Paris, Baillières, 1867, translated B. Kaplan, Phila. Lippincott, 1949.

4. Fick, R.: *Handbuch der Anatomie und Mechanik der Gelenke*. Jena, G. Fischer, 1911.

5. Freund, W. A.: *Der Zusammenhang gewisser Lungenkrankheiten und Primären Rippenanomalien*. Erlangen, F. Enke, 1859.

6. Haines, R. W.: Movement of the First Rib. *J. Anat., 80-81:*94, 1946-47.

7. Helmholtz, H.: Über die Bewegungen des Brustkastens, Verh. d. Nat. Hist. *Vereins d. Preuss. Rheinlande, 13,* 1856.

8. Hultkrantz, J. W.: Über die respiratorischen Bewegungen des Menschlichen Zwerchfells. *Scandinav. Arch. f. Phys.,* 1890.

9. Mosso, A.: Über die gegenseitigen Beziehungen der Brust und Bauchatmung; duBois Reymonds. *Arch. f. Phys.,* 1878.

10. Steindler, A.: *Diseases and Deformities of Spine and Thorax*. St. Louis, C. V. Mosby, 1929.

11. Strasser, H.: *Lehrbuch der Muskel und Gelenkmechanik*. Berlin, J. Springer, 1913.

12. Tedeschi, E.: Über angeborene Brustmuskeldefekte. *Arch. f. Orth. u. Unfallschir., 13:*276, 1914.

13. Volkmann, A. W.: Zur Theorie der Intercostalmuskeln. *Ztschr. f. Anat.,* 2, 1877.

Lecture XIV

THE MECHANICS OF POSTURE

I. THE NORMAL POSTURE

A. DEFINITION

AS A PEOPLE we have become posture conscious. Profession and laity alike have for decades been waging a campaign for good posture in the interest of health. It is argued that good posture is a sign of vigor and of perfect and harmonious control of the body.

What is good posture? It is described as one in which the head is held erect, the chest is forward, the shoulders are drawn back, and the abdomen is re-tracted (Goldthwait[3]). This definition covers much more than the purely aesthetic aspect because it implies a relationship of the parts of the body to each other which is the most favorable one for the proper function of the respiratory, circulatory and digestive systems.

Nevertheless the definition of posture on the basis of the relative position of the body parts to each other does not lend itself to a mechanical analysis. The latter requires the setting up of points of reference which are independent of the human body. Gravity is such an extraneous factor and the center and line of gravity can be made to serve as a reference. The objective is, in other words, to define posture in terms of the relationship of the parts of the body to the line and center of gravity.

The two most essential relations are first, that which the spinal column bears to the line of gravity, and secondly that of the pelvis to the cardinal transverse plane of the body. Under normal conditions the levels at which the line of gravity intersects with the spinal column are fairly constant. We call these the conventional levels and they are approximately the cervico-dorsal, the dorsolumbar and the lumbosacral junction. Furthermore, in upright position the pelvic inlet forms with the horizontal plane an angle of about 50°. Both of these relationships are interdependent; deviations from the normal posture show changes both in the intersection levels of the line of gravity and in the angle of inclination of the pelvis.

The development of upright posture in man has a long philogenetic history. The chief factor was the development of the extensory muscles of the back, hip and knee. It is not surprising that human posture, being still in the evolutionary stage, shows physiological variations both in type and in degree. Such normal fluctuations make it difficult to establish a strictly standard type unless one is willing to concede a certain latitude to normal posture. Classification on posture based on very rigid standards regularly turn out so that only a small minority of individuals are able to satisfy what is more or less arbitrarily set up as normal.

For this reason it seems better from the phsiological as well as from the physical point of view to judge posture in more accurate geometrical terms. To put it more succinctly, definition of posture, normal or pathological, should consider not only the morphological relationships of the parts to each other, but also what the relationships mean in terms of gravitational stresses.

B. LINE OF GRAVITY AND NORMAL POSTURE

For the normal posture we postulate that the line of gravity intersects with the spinal column at certain levels (Fig. 1) and that the inclination of the pelvis is within normal limits, with the understanding that these two requirements have a mutual relationship.

If this standard for posture is accepted whereby the line of gravity intersects with the spinal column at the cervicodorsal and lumbodorsal junction and at the same time falls in front of the sacroiliac articulation, a certain latitude can be given as to the degree of the antero-posterior curves which is still considered as normal. For instance, one may accept as normal a higher degree of lordosis of the lumbar spine so long as it is compensated by a commensurate kyphosis of the dorsal spine, and so long as the line of gravity continues to intersect the spinal column at the conventional levels and so long as it still falls between hip and sacroiliac articulations.

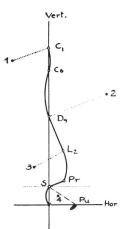

Fig. 1. Relation of line of gravity to the antero-posterior curves (Redrawn from Fick). 1, 2, 3, 4. centers of the cervical, dorsal, lumbar and sacral curves. Intersections at C_1, C_6, D_9, S.
Pr. promontory
Pu. os pubis (symphysis)

The increase of the lumbar lordosis also accentuates the pelvic inclination. On the other hand, the change in pelvic inclination causes the hip joints to assume a flexion position which produces a protrusion of the abdominal wall and a relaxation of its muscles.

Conversely, if the pelvic inclination is decreased by a flattening of the lumbar curve and the line of gravity is shifted forward, the hips go into a position of extension. Such a forward shift of the line of gravity occurs, for instance, in arthritic malposture.

Theoretically one may thus distinguish between one posture group in which the intersection points are at the normal levels but the inclination of the pelvis is changed. Some types considered as round hollow backs may fall in this class, but it is a question whether these postures should be considered as pathological; the other is the truly pathological group in which the line of gravity is definitely deflected and the pelvic inclination is correspondingly increased or decreased.

1. The types of physiological posture

Hermann V. Meyer[6] distinguished a relaxed or normal and a military position, while Braune and Fischer[1] differentiate between a normal, a relaxed and a military position. All these postures are physiological.

In the ordinary normal posture (Fig. 2) the center of gravity of the body portions above the femur stands vertically over the common axis of the hip joint. The line of gravity runs from in front of the ankle joint to slightly in front of the kneejoint, through the hip joint or slightly behind. Obviously any definite posture can be maintained only momentarily because of the backward and forward swaying of the line of gravity, but on the whole the posture varies only in narrow limits.

In the relaxed posture (Fig. 3) the body is carried slightly backward which means that the line of gravity falls behind the hip joint; it runs in front

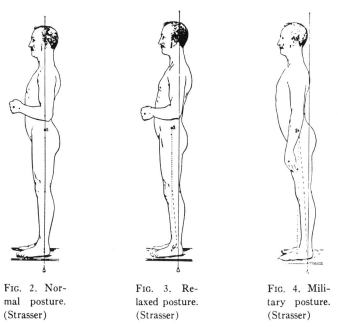

FIG. 2. Normal posture. (Strasser)

FIG. 3. Relaxed posture. (Strasser)

FIG. 4. Military posture. (Strasser)

of the ankle joint causing some tension of the calf muscles to resist the rotatory gravitation stresses at this joint.

In the military posture (Fig. 4) the thorax is carried forward, the abdomen is drawn in, the pelvic inclination is increased, the knees are held rigidly straight, the head is stretched forward. This position throws the center of gravity forward. The line lies in front of the hip joint, knee and ankle joint. Consequently the tension of the extensors of the hip joint and of the plantar flexors of the ankle joint is considerably increased, as these muscle groups neutralize the rotatory effect of the gravitational stresses. Only in the knee joint the extensors are relaxed because the line of gravity falls in the front of the knee joint axis.

It is not a comfortable position because it requires so much muscular effort to maintain it. It is a position of attention out of which, due to the tension of the gluteals and of the calf muscles, the body is held ready for a sudden forward spring. This is in contrast to the relaxed posture where the chest is carried backward and the abdomen is slightly protruding. Here, because of the line of gravity falling behind the hip joint, the gluteal extensors are relaxed while, in front the ilio-femoral ligament comes under tension. Also, because the line falls behind the knee joint there is tension of the extensors of the knee. The abdominal muscles likewise assist in preventing the body from falling backward and consequently the extensors of the back are relaxed. The calf muscles have to maintain a certain tension because the line of gravity still falls in front of the ankle joint axis.

The substance of the matter is that all three positions blend into each other without exceeding the limits of psysiological posture. The closest to a pathological state is the relaxed posture. This is fully in accord with Lovett's[5] statement that although it is comparatively easy to recognize symmetry in the frontal plane, it is difficult to decide whether a body seen from the side maintains a normal attitude or a pathological one.

This should make it more than ever convincing that any reliable method of analysis of posture must take into account the relation of the body not only to its support but also to the line of gravity, if one wishes to establish an earmark of the normal body balance in upright standing position.

II. THE PATHOMECHANICS OF MALPOSTURE

A. KINETIC CONSIDERATION

Two groups of posture were mentioned as deviating from what is called the normal type. In the first group the relation of the line of gravity to the spine is not changed but the pelvic inclination is increased. The intersection points were still approximately at normal levels; only the angular values of the antero-posterior spinal curves had increased.

In the second group the line of gravity is deflected forward or backward so that in lateral view the pelvis is no longer aligned straight under the shoulders but it is displaced forward or backward and the pelvic inclination is accordingly either increased or decreased.

We have considered the first group as still physiological; possibly it is a transition type. The second group is definitely pathological. We can advance a purely geometrical concept as the basis for this distinction; it is the principle of compensation.

In the first type, one will observe that as the angular value of the lumbar curve increases, that of the dorsal counter curve also becomes larger so that the deflection of the line of gravity caused by one is offset in the same measure by the other curve. It is a sort of compensation which is accomplished within the territory of the spine itself. The shoulders still remain aligned with the pelvis, the latter following the increase of the lumbar curve and assuming a higher degree of inclination, but the intersection points remain at normal levels (Fig. 5).

The second type bears a different relationship to the line of gravity. In the lateral view the shoulders no longer appear aligned over the pelvis and the pelvis is displaced backward or forward. The shoulders and pelvis are adjusted to the feet only so far as it is necessary to hold the center of gravity over the supporting surface and to let the line of gravity fall into it. Because the spine is deflected backward or forward this line of gravity can no longer intersect the spinal column at the conventional levels.

Yet, these deflections of the trunk must be compensated in some manner if the upright position is to be maintained. The difference is, however, that this time compensation is encompassed not within the spine itself but the entire body has a part in it. The backward throwing of the trunk produces a forward thrust of the pelvis with hyperextension of the hip joint. There are indeed two curves mutually compensatory but only the upper one has its apex in the dorsal spine while the lower has its apex in the ilio-femoral joint. It is not, therefore, an intraspinal compensation, but rather one which takes in the entire length of the body. This represents the essential physical difference between a normal or a transitional posture on one hand and a decidedly pathological one on the other (Fig. 6).

A similar principle of distinction can be introduced in scoliosis which represents a spinal asymmetry in the frontal and horizontal plane. The difference is only that any deviation from symmetry in the frontal plane is pathological, whereas in the sagittal plane the spine normally shows four alternating curves.

B. THE PROPENSITIES TO MALPOSTURE ACCORDING TO BUILD

Goldthwait[3] related the different types of malposture to variations in body build. He distinguishes the normal type, the carnivorous narrow back type, and the broad back or herbivorous type (Fig. 7).

1. In the normal type with a moderately rounded thorax, a high diaphragm, the shoulders are drawn backward, the thorax is in inspiratory position and a favorable balance is maintained between abdomen and thorax. This is most suitable for the respiratory function (Fig. 7).

2. The narrow back type is tall, slender and delicately built. The person has a large head, a high palate with often hypertrophied tonsillar or adenoid tissue. The chest is narrow and small; the thorax is long and slender, the ribs are slanting. The spine of this type shows great flexibility in the lumbar portion (Fig. 7).

3. The broad back or stocky type is built on heavy lines. The skeleton is large in structure and there is an excess of fat about the body. The chest is large, the costal angle is wide; the diaphragm stands high, the lumbar region is short, the abdominal cavity is broad and deep. The vertebrae are more flat and show stronger articular processes. This type has a somewhat more restricted mobility of the lumbar spine. One will observe that these types fit into the anthropological classifications of a mesomorph (normal), an ectomorph (slender), and an endomorph (stocky) type.

Fig. 5. Compensated round hollow back.

Fig. 6. Malposture. Dorsum rotundum (epiphysitis). Line of gravity deflected backward.

Fig. 8. The flat back.

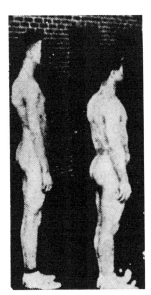

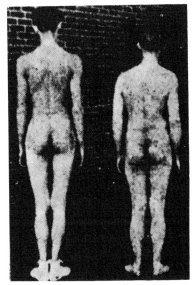

Fig. 7. Slender and normal types of anatomic build.

Each type is endowed with a propensity to certain kinds of malposture. The round upper and the round hollow backs develop from the normal build; the slender anatomic type develops with preference two malposture types: the flat back (Fig. 8) and also the round back; from the heavy anatomic type arises also the round hollow back, but in contrast to the round hollow back coming from the normal type, there is no compensatory kink at the lumbo-sacral junction, but rather the whole of the lumbar spine is engaged in a compensatory lumbar lordosis.

Everything being equal, a spine built on slender lines will give more easily under a certain load than one stockily built. But not all slenderly built persons acquire malposture and, on the other hand, many of the stocky type do. Consequently the type of build alone cannot furnish a satisfactory explanation of the deformity. One must look deeper for underlying causes.

C. PATHOLOGICAL FACTORS

1. Disturbance of the epiphyseal growth of the spinal column

The round back of vertebral epiphysitis is a classical example. At about the eleventh or twelfth year of life the epiphyses of the vertebrae appear which finally change the contour of the vertebral bodies from the biconvex shape of

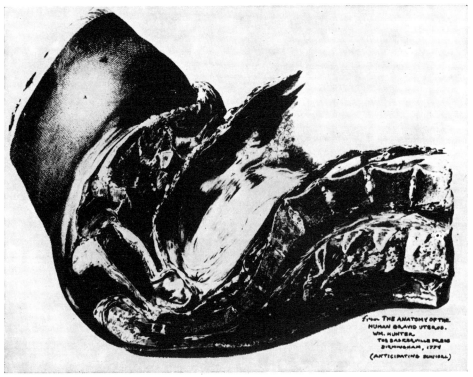

Fig. 9. An early demonstration of Schmorl's node. Wm. Hunter, 1774.
(courtesy Dr. Robert Moes, Los Angeles)

infancy to the biconcave mature form. When the epiphyseal growth is retarded or inhibited as it occurs foremost in the dorsal spine, the vertebrae appear wedge-shape in the sagittal plane, its posterior wall being higher than the anterior.

While normally some difference in height must exist between the anterior and the posterior wall of the dorsal vertebrae in order to produce the physiological kyphosis of the dorsal spine, these growth inhibitions greatly accentuate the curve and the result is a dorsum rotundum, the so-called kyphosis adolescentium (Fig. 6). The intervertebral disc plays a considerable role in this deformity. It becomes inadequate to sustain the pressure stresses of the superincumbent weight. From the investigations of Schmorl[7] we learn that the vertebral epiphysitis of Delahaye and Scheuermann is, at least in part, due to nucleus pulposus material penetrating through the cartilaginous end plates of the disc into the spongious substance of the vertebrae. These appear in the vertebral substance as the well known Schmorl nodes. Incidentally, one can see a remarkable reproduction of such nodes in W. Hunter[4] (Anatomy of the Gravid Uterus 1774) (Fig. 9).

That the degeneration of the disc has its part in many deformities of the back is common knowledge. It is particularly seen in the senile form of kyphosis which develops past middle age.

It may be justly argued that the kyphosis adolescentium is more than a posture anomaly, that it is a definite spinal disease. The head is stretched forward, the chest is flattened. The long dorsal curve is only partially compensated by the lumbar lordosis which is not extensive enough to prevent the backward shift of the line of gravity. The hips are consequently in hyperextension. There is a slight protrusion of the abdomen.

2. The flat back and visceroptosis

This type usually develops in the long-backed slender build. The dorsal spine is almost straight and the body is thrown backward by hyperextension of the lumbar spine. This gives the lumbosacral junction a sharp kink and increases the inclination of the pelvis. We refer to this type again because it is one which leads most often to abdominal imbalance and visceroptosis.

The abdominal cavity is enclosed by tissues different from those forming the thoracic cage. Its walls are formed by rigid as well as by elastic structures. The spine, the pelvis, and the lower thoracic opening represent the rigid skeletal part of it. The lateral and anterior abdominal wall is formed by elastic musculature; the diaphragm and the pelvic floor also represent an elastic muscular wall. The position of the diaphragm which under normal conditions fluctuates with the respiratory phase attains its lowest level and always occupies a lower rest position in visceroptosis. Between the extreme position of inspiration where it is the lowest and of expiration where it is the highest, the difference amounts to as much as 4 cm. The lower aperture of the abdominal cavity is formed by the pelvic diaphragm. This is made up by: (1) muscles coming from the anterior surface of the sacrum and from the inner surface of the innominate bone and going to the femur: the piriformis, the obturators and the gemelli; and (2) by the perineal muscles proper which may be described as a muscle sling running from the sacrum and innominate bone forward to the anterior pelvic ring and giving passage to the intestinal and urogenital tracts. The principal muscle of this latter group is the levator ani, a quadrangular thin and flat muscle which has a pubo-coccygeal and an ilio-coccygeal portion. In contracting this muscle group produces a pressure directed upward and forward. The anterior portion of the perineum is closed by the ischio-cavernosus and bulbo-covernosus muscles which surround the urogenital outlet.

The weight of the intra-abdominal organs is partially borne by the anterior and lateral enclosure of the abdominal cavity and by the oblique flare of the innominate bone. A greater portion of the weight stress, however, descends into the pelvic cavity proper where it is resisted by the tension of the muscle mass which forms the pelvic diaphragm. When this diaphragm relaxes the weight of the abdominal contents produces a downward stress and leads to a displacement of the organs, especially of the genito-urinary tract, a viscero-ptosis.

So far as the abdominal wall is concerned it must display a proper amount of tension to hold back the weight of the abdominal contents. Normally these muscles furnish the expiratory backstroke as the diaphragm relaxes. In this process the transverse and the internal oblique in contracting flatten the protruding wall of the hypogastrium. When the rectus abdominis contracts it does not follow the straight line from the costal angle to the os pubis but it is drawn in against the vertebral column at the level of the umbilicus. This is facilitated by the transversus abdominis and the internal oblique muscles as they act upon the tendinous inscriptions of the rectus. The latter acts as a hypomochlion against which the lateral muscles of the abdominal wall produce their flattening effect.

In this manner an intra-abdominal pressure is created. The pelvic diaphragm is transmitting pressure upward, the anterior and lateral abdominal muscles send it inward and the rectus backward, and the thoracic diaphragm in contracting sends it downward. Consequently an equilibrium must be established among these four forces. Under normal conditions this equilibrium is provided by the tension of the muscles of the abdominal wall and the two diaphragms.

The intra-abdominal pressure which increases with the weight of the organs from above downward causes the abdomen to be convex in all three planes, frontal, sagittal and transverse. It assumes the shape of a bell. The function of the abdominal muscles is to maintain a rhythmic rise and fall of the intra-abdominal pressure during respiration, never permitting undue relaxation or distension of the muscular wall.

In forward bending the anterior abdominal wall is concave held by the rectus and external oblique. The transversus and internal obliques assist in holding the wall back. The resultant of all these muscles is directed backward holding the equilibrium to the intra-abdominal pressure. In backward bending the abdominal wall becomes convex and its muscles develop components which are directed upward in the upper and downward in the lower portion. Since the latter are stronger the rectus becomes more curved caudally.

In the type of malposture in which the abdominal wall is overextended by the backward thrown trunk and the hyperextended hips, the flat back malposture, this abdominal equilibrium is gravely disturbed. The relaxed abdominal wall is no longer able to hold back the weight of the intra-abdominal organs.

The sum of all vertical components acting upon the abdominal wall equals the weight of the contents of the abdominal cavity. The resultant goes through the center of gravity of the abdominal contents and is directed downward. The horizontal component of the total weight pressure and the general intra-abdominal pressure neutralize each other (Strasser[8]). The intra-abdominal pressure increases gradually from above downward, but at a given level it is equal in all directions and hence the horizontal curves encircling the abdominal cavity should have the same angular values.

In cases in which the pelvic inclination is decreased, that is where the pelvic inlet is more horizontal, there develops, in addition, an increased pressure over the pelvic diaphragm. Under normal conditions a portion of the intra-abdominal weight is taken up by the psoas shelf. This is the projection produced by the promontory and the overlying ilio-psoas muscle. However the portion of the weight borne by this shelf is comparatively small. The greater the forward flexion of the pelvis, the more weight of the abdominal contents rests on the anterior wall. The lesser the pelvic inclination due to the hyperextended hip joint the greater is the strain upon the pelvic diaphragm. A downward displacement of the organs follows.

3. Malposture in celiac disease

A good example of relaxation of the abdominal wall is seen in celiac disease. Here the muscular relaxation is of metabolic nature due to the lack of fat absorption. The children develop a round hollow back entirely due to muscular deficiency. On proper dietary regime and vitamin application we have seen this malposture straighten out in an incredibly short time (Fig. 10).

4. The arthritic malposture (Fig. 11)

Ankylosing arthritis of the spine develops a type of malposture which is characterized by the forward flexion of the trunk and the increase of the pelvic inclination. The head is thrust forward, the chest is flattened, the trunk is carried forward, and the lumbar lordosis is flattened. In spite of this the pelvic inclination is not decreased as would be expected but it is relatively increased because the hips as well as the knees are held in flexion.

The weight bearing center and the weight line have moved forward in relation to the spine. The adaptation of the line of gravity to the supporting surface is accomplished by a compensatory flexion in hips and knees. The shape of the back is similar to that seen in the juvenile kyphosis but the body is carried forward and there is no compensatory lumbar lordosis.

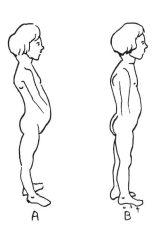

Fig. 10. Malposture from coeliac disease. Case T. M.
A. 10-19-28, before treatment
B. 11-24-28, after treatment.

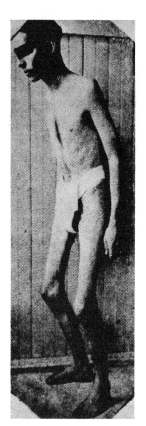

Fig. 11. Malposture in osteoarthritis. Line of gravity deflected forward.

D. THE BREATHING MECHANISM IN MALPOSTURE AND VISCEROPTOSIS

1. The respiratory back stroke

Probably the most striking feature of the relaxation of the abdominal wall in malposture is the effect it has on forceful expiration. It is the combined strength of the muscles of the abdominal wall which forces the diaphragm upward and produces forced expiration provided the airways are open. If the airways are closed or obstructed then the effect of the contraction will be increased air pressure in the thoracic cavity and then the concerted action of the abdominal wall and of the diaphragm produces a considerable rise of pressure in the abdominal cavity. This pressure then causes a downward stress against the pelvic diaphragm. It is the same pressure which is used for the evacuation of bowels and bladder.

In malposture this mechanism is greatly impaired. The protrusion of the abdominal wall already present in a neutral state of respiration becomes more marked with the inspiratory descent of the diaphragm and the weakened muscles fail to produce the expiratory backstroke that is necessary for the complete ventilation of the lungs.

2. The abdomino-thoracic balance in malposture

A most distressing factor is that the relaxed and protruding abdominal wall no longer affords the necessary stabilization to the costal arch. The inward pull of the descending diaphragm in inspiration is no longer adequately resisted by the abdominal muscles. Duchenne's[2] experiments show that when the diaphragm is made to contract by stimulating the phrenic nerve it descends and widens the lower thoracic aperture only so long as the abdominal muscles are intact. If the animal is eventrated contraction of the diaphragm causes the drawing in of the ribs and the narrowing of the aperture.

Something similar occurs in malposture where the abdominal muscles are greatly weakened and even more so in conditions of abdominal muscle paralysis. The tension of the muscles of the wall no longer suffices to stabilize the lower costal margin against the contracting diaphragm and, in consequence, the costal aperture is drawn inward. In addition, the arrangement of the muscle fibers of the diaphragm at the posterior abdominal wall is apt to accentuate the distending effect of malposture upon the anterior wall muscles. The fibers are so arranged that those of the posterior portion which come from the lumbar spine are the longest. Therefore the diaphragm descends lower in its posterior than in its anterior portion. The result is that the thrust of the diaphragm is not directed downward but downward and forward in such a manner that with each contraction a powerful thrust is exerted against the lower anterior wall. If this wall is already distended by the malposture position the result will be a considerable weakening of the respiratory efficiency.

III. SUMMARY

1. Posture, normal or pathological, usually described from the morphological angle, is best determined in terms of gravitational stresses, the criterion being the relationship of spine and pelvis to the planes of the body, the intersection levels of the spine with the line of gravity and the pelvic inclination.

2. Normal posture is one in which these intersection points remain at the conventional cervico-dorsal and lumbodorsal level and the pelvic inclination is within the normal limits.

3. This leaves a certain leeway for the angular values of the dorsal and lumbar curves, so long as they are mutually compensatory.

4. The types of physiological posture are: the normal, the relaxed and the military. In the normal the center of gravity stands vertically over the common hip joint axis; in the relaxed, the line of gravity falls behind the hip joint; in the military it is in front of it; all three positions blend into each other without exceeding the limits of physiological posture.

5. The purely geometrical concept of a definitely pathological posture is that the spine no longer intersects with the line of gravity at conventional levels. This implies: a) that shoulders are no longer aligned over the pelvis, the latter being displaced forward or backward; b) that the deflection of the trunk is not compensated within the spine itself as it is in the normal posture or in the borderline cases, but that the compensation encompasses the entire body in a manner in which the apex of the dorsally deflected spine is compensated by a lower anterior convex curve which has its apex in the protruding pelvis.

6. Goldthwait's distinction between a normal, a slender and a stocky build implies propensities to different clinical types of malposture.

7. The round upper and the round hollow backs develop with preference from the normal build; the slender build develops the flat back and the round back; and the round-hollow back comes from the heavy anatomic type.

8. Disturbance of the epiphyseal growth in vertebral epiphysitis develops a dorsum rotundum, the kyphosis adolescentium.

9. The flat back developing from the long-backed slender type with a sharp sacro-lumbar kink predisposes to visceroptosis. The diaphragm occupies a lower level; the abdominal wall is relaxed and protruding; the pelvic diaphragm is exposed to undue downward pressure and relaxes, permitting the abdominal contents to bear downward. The backward thrust of the trunk overextends the abdominal wall, the hips are hyperextended and the abdominal equilibrium is gravely disturbed. The relaxed abdominal wall is no longer able to hold back the weight of the intra-abdominal organs.

10. If the pelvic inclination is decreased (hyperextended hips) there develops, in addition, an increased pressure upon the pelvic diaphragm.

11. In celiac disease, due to lack of fat absorption, the abdominal wall is relaxed and the abdomen protrudes.

12. Spondylarthritic malposture is characterized by a forward flexed trunk. Because of the flexed hips the pelvic inclination is increased.

13. The most striking feature of the relaxed abdominal wall is the deficiency of the expiratory backstroke.

14. Another is that because of this relaxation the costal arch is no longer sufficiently stabilized and the tendency of the contracting diaphragm to pull the arch inward is no longer adequately resisted.

BIBLIOGRAPHY

1. Braune, W., and Fischer, O.: Über den Schwerpunkt des Menschlichen Körpers. *Abh. d. Math-Phys. Klasse d. Kgl., Sächs ges. d. Wiss.* Leipzig, S. Hirzel.
2. Duchenne, G. B.: *Physiologie des Mouvements.* Paris, Ballière, 1867.

3. GOLDTHWAIT, J. E.: The Opportunity for the Orthopedist in Preventative Medicine through Education Work on Posture. *J. Orth. Surg., 14:*443, 1916.
———: Relation of Posture to Human Efficiency. *Am. J. Orth. Surg., VII:*371, 1909.
4. HUNTER, W.: *Anatomy of the Gravid Uterus.* Birmingham, Baskerville Press, 1774 (Courtesy Dr. Moes, Los Angeles, Calif.).
5. LOVETT, W. R.: *Lateral Curvature of the Spine and Round Shoulders.* Philadelphia, Blakiston, Beard & Co., 1907.
6. MEYER, H. VON: *Static und Mechanik des Menschlichen Knochengerüstes.* Leipzig, 1873.
———: Das Aufrechte Stehen und Gehen. *Arch. f. Anat. und. Phys., Wiss. Med.,* 1853.
7. SCHMORL, G.: Über bisher nur wenig beobachtete Eigentümlichkeiten ausgewachsener und kindlicher Wirbel. *Arch. f. Klin. Chir., 150:*420, 1928.
8. STRASSER, H.: *Lehrbuch d. Muskel und Gelenkmechanik, Vol. II.* Berlin, J. Springer, 1913.

Lecture XV

THE PATHOMECHANICS OF SCOLIOSIS

INTRODUCTION

I T IS indeed unfortunate that our knowledge of the causes of scoliosis has not gone much beyond the observational stage: we know little more about it than what clinical and x-ray examination reveals. This is true at least for the habitual scoliosis except for some highly conjectural theories which have been developed in relation to the causes of this deformity.

In the case of the paralytic scoliosis we can recognize the muscular imbalance and we know that muscular deficiency is one of the causes. In the case of congenital scoliosis we have recognized a hereditary tendency to prenatal deformities as a link in the chain of causal connections. But in what is called idiopathic or habitual scoliosis our insight into the etiology has not advanced beyond the morphological level unless it be that more recent investigations of hormonic or metabolic nature should be able to deepen our knowledge.

What we do acknowledge, however, is that in all cases mechanical factors are at work which have their share in the formation of the deformity. We realize that these mechanical factors are universal but that only in certain spines, for reasons largely unknown, a predisposition exists to become deformed under the influence of these mechanical stresses. We may speak of a lessened resistance or a spinal insufficiency, but if such a term means anything at all, it is at best only a road sign into the unknown.

Be that as it may, we know that mechanical, principally gravitational forces are necessary to bring forth a habitual scoliosis in spines so predisposed. The object of this lecture is to investigate the working of these forces and their effect upon the spinal column.

I. GENERAL MECHANICAL CONSIDERATION

A. RELATION TO THE COLUMN THEORY. THE EFFECT OF THE LOAD

To approach the problem of spinal curvature from a mechanical point of view we start with the simple concept that the spine represents a long and comparatively thin column. The theory of Euler postulates that under pressure such a column will become deformed as would any other column which is excentrically loaded and that the type of the curves of deflection depends upon the manner in which the column is implanted.

If both ends of the column are *free* to turn about a parallel horizontal axis, as is the case for instance in the femur viewed in the sagittal plane, then the line subtended to the deformation curve (1) equals the length of the entire bone (L), or l (length of the curve) = L (length of the column). The tibia as seen in the frontal plane is a column which is *fixed* at both ends; in this case the curve is only half the length of the column itself $(l = \dfrac{L}{2})$.

The spine may be represented as a column which is *fixed* at the lower end into the pelvis. At the upper end, however, it is *free* to move in all directions, forward, backward or sideways. Under the column theory the deformation should be a curve which is about ⅔ of the length of the entire spine (Boyd and Folk[1] (Fig. 1).

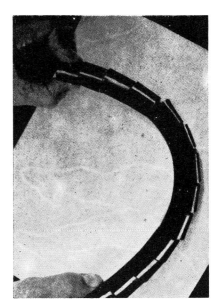

FIG. 1. Representation of spine as a column fixed at lower end, free at upper end. (Boyd & Folk)

FIG. 2. Risser's segmented rubber tube showing longitudinal rotation with side bending. (Risser)

It would hardly do to carry the analogy with a loaded column and the application of Euler's theory of columns any farther, beyond stating that under a superincumbent load a curvature of the spine must be expected unless counteracted and controlled by muscle forces. For one thing the spine is neither a homogenous column as the theory would suppose nor is the relation in length to breadth of the column a constant factor since the vertebrae increase in mass from above downward. Secondly, while the spine is implanted firmly into the pelvis, the latter itself is not immovable but is in constant motion during gait and even standing. Thirdly, the spine is not a straight rod but it has its natural antero-posterior curves.

B. THE EFFECT OF GRAVITATIONAL REACTIONS

A point to be taken in consideration in the mechanical analysis is the forces which reach the spinal column through the gravitational reactions from the floor. These reactions are of static as well as of dynamic nature. So long as the superincumbent load is placed concentrically upon the pelvis which can be considered as a horizontal beam, the system is equilibrated. But the implantation of the spine into the pelvis is not always strictly horizontal; it is often oblique, the 5th lumbar being a wedge vertebra. In this event no automatic equilibrium maintains but the obliquity of the pelvis will cause the reactions from the floor to develop both a horizontal and a vertical component. The vertical component is absorbed by the resistance of the structure; the horizontal component produces lateral bending of the lumbar spine.

Furthermore, dynamic effects develop in the gait. The pelvis oscillates with alternating obliquities both in the frontal and in the horizontal plane. Since normally these oscillations are symmetrical, the lumbar spine alternates in its lateral deviation and no permanent lateral deformity is permitted to develop, provided again that the spine is placed squarely upon the sacrum.

C. THE EFFECT OF SIDE BENDING ON ROTATION

Now the question presents itself, what is the effect of side bending in a rod which is not straight but which, as the human spine, has antero-posterior curves. For strictly mechanical reasons pure side bending can occur only when the bending force is applied in a frontal plane.

When a curved rod with antero-posterior curves such as the human spine is bent sideways, that is perpendicular to the plane of its natural curves a longitudinal torsion appears. We know already from the experiments of Schulthess[12] that the very excentricity of the superincumbent load makes pure side bending without rotation improbable. Risser[11] uses a segmented rubber tube to show that when bent laterally a longitudinal rotation occurs (Fig. 2).

D. HOW THE NORMAL SPINE MANAGES THE PROBLEM OF COMBINING SIDE BENDING WITH ROTATION

The earmark of a normal spine is its ability to revert from all asymmetrical positions to the normal upright position with promptness and precision. It can bend laterally without rotation or reverse rotation from the convex side as it occurs in scoliosis to the concave side. This, however, is not automatic but requires special muscular effort. Lovett[9] found that in normal individuals side bending in flexion is associated with convex side, and side bending in extension with concave side rotation. Why then does the scoliotic spine always rotate convexly? According to Lovett[9] when the spine is flexed laterally it must rotate about its vertical axis and lateral flexion does not exist as a pure movement but is part of a combined lateral and rotatory movement.

In position of hyperextension the upper and mid dorsal spines are locked and then the whole thoracic spine twists around the lumbo-dorsal junction. In flexed position, however, the dorsal spine is unlocked and then it rotates so far as the limitations of intervertebral mobility permit.

Consequently rotation occurs at a higher level in flexion and at a lower level in extension. In the normal spine the question whether in side bending rotation will be convexly or concavely depends upon the direction of the sagittal axis about which the side bending occurs. Strasser[13] states that if this axis has a downward slope rotation will be on the concave side, and if it has an upward slope it will be convexly. On this point, however, the literature shows many discrepancies (Fick[4]).

II. SPECIAL MECHANICAL CONSIDERATIONS REGARDING THE SCOLIOTIC SPINE

1. Without prejudging any of the more remote causes this much can be stated that the scoliotic spine is one in which the intrinsic equilibrium of the spinal column has been lost, since the preservation of such equilibrium would preclude development of the deformity.

2. The principal elements of the deformity are the lateral deviation in the frontal and the longitudinal rotation in the transverse plane.

3. A third element is the loss of the normal relation of the dorsal spine to the thoracic cage and of the lumbar spine to the abdominal cavity and the pelvis. The reason for this is that neither thoracic cage nor pelvis follow the spine in its rotation but they stay behind.

4. There is also a disalignment between thorax and pelvis in the frontal plane. The former is no longer placed squarely over the pelvis, but thorax and shoulder girdle hang over to one side while the pelvis protrudes to the opposite. This disalignment which develops in the course of the deformity is called decompensation.

A. THE MECHANOGENETIC ELEMENTS OF SCOLIOSIS: INCLINATORY AND COLLAPSE

There are two distinct elements of deformity so far as the mechanogenesis is concerned. One shows that a section of the spine as a whole is bent against a neighboring section. Each of these sections comprises a number of vertebrae and the resultant curve is therefore a long arc. This type of curve is seen, for instance in earlier stages of paralytic scoliosis (Fig. 3). It is called the inclinatory type.

In the other element the arc is short, the deviation between single vertebrae is greater; they seem to be displaced sideways against each other and there is also a greater amount of rotation. This is the collapse type. The first type can, to a large degree, be produced voluntarily by inclination of the body; the second type cannot be so produced, and any purely lateral shift between vertebrae is unphysiological (Fig. 4).

From the mechanical point of view it is interesting to know how this second or collapse type is brought about. Obviously such a collapse cannot occur without the loss of the intrinsic equilibrium, i.e., without impairment of function of the intervertebral disc and of the ligamentous structures which normally prevent displacement of the vertebral bodies among each other.

The distinction between an inclinatory and collapse element applies only to the more initial stages of the deformity. As the deformity increases, signs of collapse with sideways displacement of the vertebrae appear also in a

FIG. 3. Inclinatory type of curve.

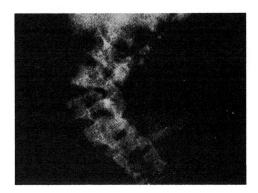

FIG. 4. Collapse type of curve.

scoliosis which has started as an inclinatory deformity, as for instance the paralytic scoliosis (Farkas[2]).

From the therapeutic point of view it is important to know that only the inclinatory, or as we may call it, the contractural element, is amenable to correction, while the collapse element in which there is already a translatory shift of the vertebrae against each other is not.

B. THE MECHANICAL EFFECT OF GROWTH PLATES AND OF THE DISC

1. The lateral curve

As the lateral curvature increases we notice the following morphological changes of the vertebrae: some become wedge shape, the concave side being lower than the convex. Others show in addition a distortion in the frontal view, the normal rectangular contours being changed into a rhomboid under the effect of the side shift; and all vertebrae show rotatory distortions.

The point is to establish the connection between these morphological changes and the gravitational stresses under which they develop. Whatever the deeper causes may be which make the particular spine incapable to resist weight and other stresses, the stresses themselves are always a necessary factor for the production of the deformity.

The wedge shape deformity of the vertebra is the product of the asymmetry in growth. Whether this asymmetry precedes the development of the deformity and thus becomes the etiological background has no direct bearing on the mechanics of scoliosis. What does concern us here is the influence the curved spine has upon the growth of the vertebrae. What is the effect of the increased pressure on the concave side upon the activity of the epiphyseal growth plates of the vertebral body?

Experiments carried out by Haas[7] and later ones by Nachlas have shown that one can produce a lateral curvature of the spine by suppressing one-half of the growth plates while the other half continues to grow. The question is now can the increased pressure on the concavity of the curve produce the same inhibition of epiphyseal growth? So far as the experiments are based on partial suppression of the growth producing epiphyseal plates, they can be taken as evidence that the growth factor plays an important part in the production of the deformity. The force of growth varies with age. It is intensified in certain age periods and experience shows that the idiopathic scoliosis increases rapidly during the years of rapid growth particularly in the prepuberty years from 10-15. According to Risser[11] the changes in the epiphyseal plate which appear in form of fragmentation are most noticeable in the apical vertebrae. This could well account for the later wedge formation which we find in the lateral deformity where pressure reaction on the side of the concavity inhibits epiphyseal growth. This effect of pressure has already been recognized since the work of Maas[10] in contrast to the idea of functional apposition of bone under the influence of pressure as one might assume under Wolff's law. We know that epiphyseal plates in general are sensitive to constant pressure and respond to it by growth inhibition (Scaglietti). This explains scoliosis as a deformity of the growing age. Not only does pressure inhibit growth but the forces producing the deformity have a direct effect upon the bone structure itself even though the epiphyseal disc plays the primary role in the deformity of the vertebra.

2. The rotatory deformity

It would be difficult to explain on epiphyseal grounds alone the architectural changes of the vertebral bodies as well as of the arches which are caused by the rotatory curve. There is more at work than the epiphyseal growth disturbance can account for. There must be, above all, an overthrow of the intrinsic equilibrium of the spine which under normal conditions results from the reciprocal action of discs and spinal ligaments. In scoliosis the ability of the disc to restrain motion between the vertebral bodies is impaired and pathological displacements occur. Already Schulthess[12] observed early changes of the discs in scoliosis. He found them flattened and protruding on the concave side and higher and somewhat retracted convexly. They were also drawn out and distorted by the translatory or side shift of the vertebrae. As a matter of fact, as early as 1896 Ghillini[6] called attention to the importance of the disc in the pathogenesis of scoliosis and recently Farkas[2] in a study on paralytic scoliosis pointed out that the early degeneration of the disc is the prime factor in the production of the scoliotic collapse.

From the clinical viewpoint the distinction between the contractural or inclinatory and the translatory or collapse element of scoliosis is most important. The former is based on muscular imbalance while the latter involves degenerative causes such as growth inhibition and disintegration of the disc. Both factors are intrinsically associated in the fully-developed structural scoliosis but they may vary in ratio and in sequence. For instance the paralytic scoliosis starts with purely inclinatory changes; in the idiopathic scoliosis collapse also follows the inclinatory element. The inclination is represented by the long scoliotic arc, the collapse by the short arc with sharp translatory shift. As early as 1907 Feiss[3] called attention to the fact that there exists no voluntary trunk movement which can produce such a short arc curve. All voluntary motion produces long arc or inclinatory curves. All short arc curves are pathological.

3. The relation of the rotatory to the lateral curve

As stated above length rotation necessarily accompanies side bending because of the natural antero-posterior curves which make straight side bending without rotation mechanically impossible.

For the understanding of the connection between lateral bending and rotation the experiments of Lovett[9] on the cadaver are especially important. He showed:

(1) That lateral bending in extension of the spine is associated with concave side rotation and lateral bending in flextion with convex side rotation.

(2) Since lateral movement is invariably associated with rotation, no rotation can occur without lateral deviation.

So long as this rotation occurs within physiological limits, no changes in the function of the intervertebral disc nor in the form of the intervertebral articulation which restrict and regulate rotatory motion is to be expected. But with the progression of the deformity we see that rotation becomes excessive, that is ultraphysiological. Then one must look for definite morphological changes and structural transformations of the skeleton which parallel those produced by the lateral curve.

The mechanical effect of the lateral curve on the structure of the vertebrae is a condensation of trabeculation of the concave and a thinning out of the trabeculae on the convex side in response to the changed pressure distribution. The suppression of epiphyseal growth as we have stated above is a pressure effect leading to the ultimate formation of the wedge vertebrae.

The rotatory curve also develops under the gravitational influence because side bending cannot occur without rotation. However if side bending alone would be the dominant factor which produces the rotatory deformity, one should expect a constant ratio between lateral and rotatory angulation. Such a relationship does not exist. One sees cases in which rotation develops early and progresses out of proportion with the degree of lateral bending.

The longitudinal rotation of the spine begins in the intervertebral articulations with the discs as centers. As the limit of rotatory excursion is reached, the rotatory force is transmitted to the vertebral bodies and causes structural changes of torsion. The sagittal axis of the vertebra shifts its anterior end to the convex side; the cross section of the vertebral foramen is egg-

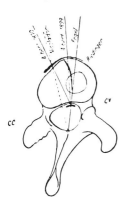

FIG. 5. Torsion effect on vertebral body.

shaped with the larger pole pointing convexly. The intervertebral articulations have changed their direction in conformity with both the rhomboid obliquity and the wedge shape of the vertebral bodies. Everything is condensed and stubby on the concave side and is thinned out or drawn out convexly. There are two features which particularly reflect the torsion effect. One is the deflection of the arch in relation to the body. It does not follow entirely the convex side rotation of the body but it is deflected against it toward the concave side (Fig. 5). The other, more impressive, is the fact that due to the rotation the perpendicular trajectories of the vertebral bodies assume now a spiral course. On the surface this manifests itself by a twist between the upper and lower plane of the vertebral body giving its surface a sort of fluting spiral aspect.

While lateral and rotatory curves do not develop upon identical mechanical premises and do not always keep pace in degree of development, they still are interdependent in the sense that both rotation and lateral deviation have their maximum at the apex of the curve.

4. The muscular imbalance

What is the role of the muscular imbalance in the mechanogenesis of scoliosis? In the paralytic scoliosis it is no doubt the immediate mechanical factor; but in the course of development changes occur which can no longer be explained on the basis of muscular imbalance alone. The latter represents only the contractural or inclinatory elements of the scoliotic deformity. The muscles on the convex side may show atrophy and degeneration while on the concave side they are contracted. This applies especially to the asymmetrical weakness of abdominal muscles in the case of the lumbar and of the sacro-spinalis in the case of a dorsal scoliosis. On the other hand, the large flat muscles of the thorax, the trapezius, the latissimus, the rhomboids, and the serratus do not take part in the contractural deformity and they show little change (Virchow[14]). However, since no muscle contracture can produce a translatory shift nor the excessive rotation which one sees in advanced scoliosis, it must be assumed that the collapse element based upon structural changes enters sooner or later into the picture of all types of scoliosis including the paralytic.

C. THE RELATION OF SCOLIOTIC DEFORMITY OF THE SPINE TO THE THORACIC CAGE AND TO THE PELVIS

Under normal conditions the spinal column and the thoracic cage move together as one mechanical unit both in side bending and in rotation. Thus in any position the body may assume, the spinal column divides thorax and abdominal cavity in two approximately symmetrical halves and occupies the midline between them.

In the scoliotic spine the condition is essentially different.

1. So long as the curve is strictly of the inclinatory type this symmetry will obtain as it does in the normal spine where lateral bending and rotation can be produced voluntarily.

2. As soon as the element of collapse enters, however, the scoliotic spine penetrates into the convex half of the thorax or the convex half of the abdominal cavity. Therefore convex and concave sides are no longer symmetrical.

3. In rotation, likewise, the thorax of the normal spine goes with the spinal column so that the symmetry of the right and left half of the thorax or of the abdominal cavity is preserved.

4. In contrast, in the scoliotic spine, neither the thorax nor the pelvis rotate commensurately with the spinal column. They stay behind on the concave side and thereby abandon their normal relationship with the spine. So far as the rib cage is concerned, this discrepancy affects the normal attachments of the ribs to the spinal column where they are anchored by the costo-vertebral and the costo-transverse articulations. The pelvis is fastened to the spine by the sacro-iliac syndesmosis. This junction, supported further by the lateral and anterior abdominal muscle acts as support of the spinal column while the latter again maintains the thoracic cage in position.

The relaxation of the anchorage of the ribs is not the only, nor the essential, feature which prevents the rib cage from following the convex side rotation of the spinal column and which causes it to stay behind on the side of the concavity of the curve. The more essential reason is that during the process of spinal rotation the ribs, still attached to the column, undergo deformations of their own. They are rolled up posteriorly on the convex side with the result that the costal angle becomes markedly sharpened and a costal prominence develops. On the concave side, posteriorly, the ribs are flattened out; conversely, in front this drawing out of the costal rings causes a flatness of the convex and a prominence of the concave side. The rolling up of the posterior end of the convex side of the ribs was first described by Albert and by Lorenz,[8] the latter attributing it to a disturbance of the epiphysis at the posterior end. It has been argued whether the spine pulls the ribs into distorsion or whether the ribs force the spine into rotation. From the mechanical point of view this is an idle question because it is clear that there is a mutual mechanical effect in operation. The ribs are both actively deforming and passively deformed.

The relation of the pelvis to the scoliotic lumbar spine is similar. If the sacrum is included in the scoliotic curve, then a deformation of the pelvis develops similar to that which the thorax undergoes in scoliosis of the dorsal spine. It is the shape of the 5th lumbar vertebra which determines the changes of the pelvis. If the lower contour of this vertebra is horizontal, then the sacrum is not included in the curve, and the curve ends above the pelvis.

The latter stands horizontal and there is no pelvic deformation. On the other hand if the lower contour of the 5th lumbar is obliquely planted into the pelvic ring, then it serves as the keystone of a curve which includes the sacrum as well as the lumbar spine. Then the pelvis is tilted downward on the side of the convexity of the lumbar curve. A third possibility which occurs rather rarely is that the sacrum forms a separate counter curve to a lumbar curve above. In this case the 5th lumbar acts as a transitional vertebra. The

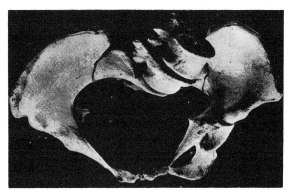

pelvis naturally follows the sacral curve and is then tilted downward on the side of the sacral and not of the lumbar convexity. Since these are on opposite sides it appears as though the pelvis was dropped on the side of the lumbar concavity. This is called a paradox scoliosis and whenever the sacrum is downward tilted on the concave side of the lumbar curve one

FIG. 6. Deformation of the scoliotic pelvis. (Putschar)

may be sure that a sacral counter curve exists.

The deformation of the sacrum corresponds to that of the vertebrae. It is stubby and shortened on the concave side and it is drawn out and thinned out convexly. The pelvis also shows deformation corresponding to that of the thoracic cage in dorsal scoliosis, only to a lesser extent. When it becomes distorted the greater diameter runs obliquely from convex backward to concave forward (Fig. 6).

III. THE PROGRESSION OF THE SCOLIOTIC DEFORMITY UNDER MECHANICAL INFLUENCE

To maintain that scoliosis progresses without restraint until well into the adult life (Lorenz[8]) would be underestimating nature's ability to bring the deformity to an early halt. The essence of nature's efforts is to develop compensatory curves in order to create a balance with the primary curve both in lateral and rotatory sense and to regain a proper realignment of the trunk over the pelvis. In this effort nature succeeds only in a minority of cases of idiopathic (about 20%) and only in exceptional cases of paralytic scoliosis. Mechanically such a compensation would mean that a compensatory curve is developed within the spine itself. To be fully compensatory, such a curve must have the same length and the same angular value as the primary curve.

Good realignment means that the shoulders are poised squarely over the pelvis and no overhang of the body exists. From then on it is the function of the muscular apparatus to maintain this state of compensation. Whether such a natural compensation can develop or not depends upon a number of factors: the site and length of the primary curve, the rigidity of the structural curve, the intensity of growth, the pliability of the remaining portions of the spine. If the primary curve is dorsal it must be short enough to allow the establishment of a secondary lumbar curve of adequate length and adequate angular value.

In the majority of cases spontaneous compensation does not occur at all or if it does it becomes lost and decompensation with overhang of the body and protrusion of the pelvis on the opposite side reappears. We must assume that in cases which are seen in later stages a period of natural compensation had existed but that compensation had been lost and that attention had been

FIG. 7. Scoliosis, compensated.

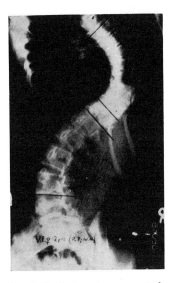

FIG. 8. Scoliosis, not compensated.

first called to the deformity when overhang and protruding hip made their appearance.

It has been stated that compensation means a compensatory curve established entirely within the length of the spinal column and that this curve must have adequate length and adequate angular value (Fig. 7). Decompensation, by the same token, is a state in which the remaining portion of the spine is unable to produce an adequate countercurve. The latter is not long enough or it does not have sufficient angular value to neutralize the effect of the primary curve on the balance of the body (Fig. 8). Yet some compensation must occur if the total balance of the body over the supporting surface is to be preserved. This time, however, the compensatory curve is not confined to the spine itself but it takes in the pelvis and the lower extremities. The protruding pelvis is now actually the keystone of a curve which includes the entire lower limbs. As the pelvis moves toward the side of the concavity of the primary curve, the concave side hip joint goes into adduction while the convex side is in abduction (Fig. 9). The same principle applies to the rotatory curve. In compensation the convex side rotation of the dorsal spine is compensated by a contra-lateral convex side rotation of the secondary curve.

Here we have a perfect analogy to what appears in malposture in relation to the antero-posterior curves of the body. In the round hollow back the dorsal and lumbar curve if of proper length and proper angular value compensate each other so that in the sagittal view shoulder and pelvis are again superimposed and the line of gravity retains its normal relationship to the spine. For this reason we have classified this type as a borderline type of malposture.

FIG. 9. Effect of scoliosis on hip joints: Concave side, hip into adduction; convex side, abduction.

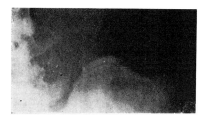

FIG. 10. Completion of iliac apophysis.

In true malposture on the other hand, the trunk is thrown backwards, the pelvis and abdomen are forward analogous to the overhang and the pelvic protrusion in the lateral curve. Compensation in this situation also exists in a form of a counter curve to the primary dorsal; but this compensatory curve likewise includes the pelvis and the lower extremities, the hyperextended hip joints being the keystone of this curve.

The first indication of decompensation is that the primary curve increases in length and in angular value or in both while the secondary curve flattens out its ends being drawn into the lengthened primary curve. Consequently both the angular values and the length of the secondary curve decrease. This discrepancy between the primary and secondary curve becomes greater as decompensation progresses. The straightening of the lumbar without the corresponding straightening of the dorsal curve does not indicate an improvement but rather a deterioration of the scoliotic deformity. Any attempt to straighten the secondary curve without assurance that the primary curve can also be straightened adequately will necessarily end in failure.

In its natural course the scoliosis which is not spontaneously compensated or in which compensation is lost progresses to adult life. Risser[11] maintains that it stops at the time of full maturity and he accepts as the sign of the latter the completion of the iliac epiphysis as it reaches the sacro-iliac angle (Fig. 10). This is true for the great majority of idiopathic scoliosis, but not for the paralytic which can progress far beyond the age limit of 16-18 years. At any rate when the progress stops at the time of maturity, the spine is already in a state of gross decompensation except in the small minority of automatically and permanently compensated cases.

IV. THERAPEUTIC CONCLUSIONS BASED UPON THE PATHOMECHANICS OF SCOLIOSIS

The essential point to keep in mind is that, of the two elements which constitute the scoliotic deformity, namely the contracture and the collapse, only the contracture is amenable to correction or improvement by mechanical means. The scoliosis which starts as inclinatory, or contractural, deformity can, to a certain degree, be improved or even straightened. In later stages when the collapse element is added to it only the inclinatory part of the deformity will yield to such treatment. Actual deformation of the vertebral bodies and particularly their translatory shift are not amenable to correction.

For example a primary dorsal curve still in the incipient stage of pure inclinatory contracture can be straightened; but it must be remembered that only a complete reversal of the curve can be taken as evidence of full correction. Therefore if alignment cannot be accomplished by efforts to correct the contracture by whatever means may be employed, traction, hammock, cast and so forth, one must attempt to develop adequate compensatory curves. At this stage all attempts to correct the primary curve fail. What is accomplished is merely the establishment of a compensatory counter curve. Not all portions of the primary curve are equally resistant to mechanical correction. The apex shows the most and the ends the least resistance. Therefore the corrective attempts draw the ends of the primary curve into the counter curve while the middle remains in its original state.

So far as the thoracic deformity is concerned one must keep in mind that the thorax has remained at the concavity of the curve. Consequently any attempt to correct rotatory deformity of the thorax by pushing forward the posterior prominence is illogical and can only serve to accentuate the discrepancy between the thoracic cage and the spine (McKenzie Forbes[5]).

All efforts of the so-called compensation treatment are directed toward establishing an adequate compensatory lumbar curve. This requires a lumbar section of sufficient length and sufficient pliability. For this reason compensation treatment fails in long and rigid dorsal curves. This type of scoliosis, therefore, is not suitable for the method; on the other hand, a short dorsal curve or an S curve showing already some compensatory secondary curve is suitable.

The compensatory method has no direct effect upon the maintenance of the compensation when once it is established. This is entirely a problem for the systematic development of the musculature. Hence most paralytic curves will not be served by compensation alone. One may or may not succeed in the idiopathic scoliosis to maintain compensation by muscle reeducation. If one does not and the primary curve increases and compensation becomes lost soon after it has become established, then the case must be committed, even though reluctantly, to operative fusion. Fortunately this applies only to a minority of idiopathic scoliosis cases. The decision depends entirely upon whether or not it can be shown that in spite of compensation having been satisfactorily established and in spite of all measures to develop the musculature of the back, decompensation threatens. This requires trial and observation and it seems incomprehensible that operative indications should be made on the spot as is done so often.

In this connection it is also well to remember that an increase in angular values does not in itself constitute failure of the compensation treatment. It is the *discrepancy between the angular values and the length of the primary and secondary curves* and not the angular value itself which constitutes failure. Thus an evenly increased angular value with equality of length of the curves may well be a success. On the other hand a decrease in the angular value and a shortening of the secondary curve without corresponding change of the primary curve undoubtedly is a failure.

V. SUMMARY

1. Whatever the underlying cause, gravitational stress is indispensable as eliciting factor in the production of the deformity.

2. So far as the column theory can be applied to the scoliotic spine the nearest situation is that of a column firmly implanted below and free above with a curve occupying $2/3$ of the length of the column. This represents a long dorsal curve.

3. Because of the free movements of the pelvis, the gravitational reactions from the floor reaching the pelvis through the hip joints likewise influence the spine. Implantation of the spine into the pelvis is not always strictly horizontal. The obliquity of the pelvis may be caused by a wedge shaped 5th lumbar acting as the keystone of a lumbosacral curve.

4. Because of the normal antero-posterior curves side bending is necessarily associated with longitudinal rotation of the spine.

5. The earmark of a normal spine is its ability to revert from any asymmetrical position to complete symmetry with promptness and ease.

6. Automatically the normal spine goes into convexside rotation in forward flexion and in concave side rotation in extension (Lovett). But this can be voluntarily reversed by muscle action.

7. The development of a structural curve presupposes loss of the intrinsic equilibrium of the spine.

8. The 3 elements of the scoliotic spine are the lateral curve, the rotatory

curve and the loss of the normal relationship of the spine to the thoracic cage and the abdominal cavity and pelvis.

9. In addition there is a disalignment between thorax and pelvis which are no longer superimposed upon each other, the thorax overhanging on the side of the convexity and the pelvis protruding on the side of the concavity of the curve. This so-called decompensation is cosmetically the most disturbing feature of the deformity.

10. In the scoliotic curve two elements participate: the inclinatory and the collapse element. The inclinatory is the bend of one section of the spine against the other; in its beginning it is within physiological limits; the collapse element is a disalignment with lateral shift between single vertebrae; any degree of lateral shift is pathological.

11. In the lateral deformity the wedge shaped vertebra is the product of asymmetrical growth. Lateral deformity can be produced experimentally by suppressing one side of the growth plate.

12. The rotatory deformity cannot be explained on grounds of asymmetrical epiphyseal growth. It presupposes a complete overthrow of the normal intrinsic equilibrium of the spine, especially degenerative changes of the disc.

13. The distinction between the inclinatory and the collapse element of the scoliotic curve is of clinical importance because the inclinatory element lends itself to improvement and even correction by mechanical means, while the collapse element resists all such efforts.

14. The inclinatory curve is long, the collapse curve short; both types melt into each other in the course of development of the deformity.

15. There is no strict relationship between the lateral and rotatory deformity in scoliosis. The length rotation begins in the intervertebral articulations with the discs as centers and, as the limit of excursion is reached and the force is transmitted to the vertebrae, structural changes appear; convex shift of the sagittal axis of the vertebra, egg shaped cross section of the spinal canal with the larger pole on the convex side; condensation of bone on the concave and thinning out on the convex side, spiral course of the perpendicular trajectories of the bodies. Some interdependence of the lateral and rotatory deformity is shown by the fact that both rotation and lateral deviation have their maximum at the apex of the curve.

16. The muscles on the convex side show atrophy; those on the concave side are contracted. Asymmetrical weakness of the abdominal muscles is seen in lumbar, and of the sacro-spinalis in dorsal scoliosis. The large flat muscles of the thorax show little change.

17. Under normal conditions spinal column and thorax move together as one unit and so long as the scoliotic curve is strictly in the initial inclinatory stage, the symmetry whereby the spinal column divides the thorax in two symmetrical halves still maintains.

18. With the entering of the collapse element this symmetry is destroyed as the scoliotic spine penetrates into the convex half of the thorax or the

abdominal cavity. The same applies to the rotatory deformity.

19. Neither the thorax nor the pelvis rotate commensurately with the scoliotic spine but they stay behind on the side of the concavity.

20. Aside from the loosening of the anchorage of the ribs, the more essential reason for the thoracic deformity is the deformation of the ribs which are rolled up posteriorly on the convex side and flattened on the other.

21. The relation of the pelvis to the scoliotic lumbar spine is similar, the pelvis showing deformations analogous to those of the thorax, provided that the sacrum is included in the lumbar scoliosis.

22. In a minority of cases the idiopathic scoliosis (about 20%) nature succeeds in establishing and maintaining an adequate compensatory curve before the stage of permanent decompensation with overhang of the trunk is reached.

23. The first sign of loss of compensation is increase of the primary curve in length or in angular value or both.

24. Compensation means that the primary curve is offset by a (lumbar) counter curve of equal length and angular value so that the readjustment of the shoulders squarely over the pelvis is accomplished within the length of the spine itself.

25. Decompensation means that such adjustment does not occur within the confines of the spine but that a lower counter curve involves the pelvis as well as the lower extremities in a manner whereby the protruding pelvis is the keystone of the counter curve and there is a disalignment between shoulder and pelvis. It will be noted that this distinction of compensation and decompensation can be applied on the same basis to normal posture and malposture.

26. A number of therapeutic conclusions can be drawn from the pathomechanical analysis of scoliosis.

a) Only the early inclinatory element yields to mechanical treatment so far as actual improvement or correction of the primary curve is concerned; the opportunities of treating scoliosis in this early stage are far too few.

b) Once a structural curve is established corrective attempts do not accomplish abolition of the primary curve but lead to formation of a compensatory curve. This means that in suitable cases the primary curve can be shortened as its ends are drawn into the secondary curve and that the realignment of the trunk and the disappearance of the overhang and the pelvic protrusion can be achieved.

c) Attempts to correct the rotatory deformity of the thorax by pushing the costal gibbus forward are illogical because this merely accentuates the already existing concave side position of the thorax.

d) The object of the compensation treatment is to establish below the primary curve a secondary curve of equal length and equal angular value.

e) The maintenance of the achieved compensation is a problem of the spinal musculature which may or may not be developed to the point of holding the spine definitely in position of compensation.

f) So long as the conformity of length and angular value of the two curves persists, compensation is maintained.

g) By the same token straightening or shortening one curve without the other is doomed to failure.

BIBLIOGRAPHY

1. BOYD, J. E., and FOLK, S. B.: *Strength of Materials,* 5th Ed. New York, McGraw-Hill Book Co., 1950.
2. FARKAS, A.: *Bedingungen und auslösende Momente der Skoliosenentstehung.* Stuttgart, F. Enke, 1925.
3. FEISS, O.: Mechanism of Lateral Curvature. *Am. J. Orth. Surg., 5:*152, 1907.
4. FICK, R.: *Spezielle Gelenk und Muskelmechanik.* Jena, G. Fischer, 1913.
5. FORBES, MCKENZIE: *Am. J. Orth. Surg., 2:*509, 1920.
   ———: *New York State J. Med.,* 1912.
   ———: *Am. J. Orth. Surg., 11:*75, 1913.
6. GHILLINI, C.: Experimentelle Knochendeformitäten. *Arch. f. Klin. Chir., 56:*850, 1896.
7. HAAS, S.: *J. Bone & Joint Surg., 21:*963, 1939.
8. LORENZ, A.: *Rückgratsverkrümmungen.* Vienna, Urban & Schwartzenberg, 1889.
9. LOVETT, R. W.: The Mechanics of Lateral Curvature of the Spine. *Boston Med. & Surg. J.,* Boston, Danwell & Upham, 1900.
   ———: *Lateral Curvature of the Spine and Round Shoulders,* 5th Ed. Philadelphia, P. Blakiston & Co., 1931.
10. MAAS: Über die Einwirkung abnormer Zug und Druckspannungen auf das Knochenwachstum. *Virchows Arch., 163:*2.
11. RISSER, J.: Instructional Course. *Am. Acad. Orth. Surgeons, V:*250, 1948.
12. SCHULTHESS, W.: Über den Zusammenhang der Physiologischen Torsion der Wirbelsäule. *Zeitsch. F. Orth. Chir., X-XVI:*455, 1902.
13. STRASSER, H.: *Lehrbuch der Muskel und Gelenkmechanik.* Berlin, J. Springer, 1913.
14. VIRCHOW, H.: Der Zustand der Rückenmuskulatur bei Skoliose und Kyphoskoliose. *Ztschr. f. Orth. Chir., 34:1:* 1914.

Part III

THE EXTREMITIES

MECHANICS OF THE HIP JOINT

I. ORIENTATION

THE hip joint is an amphyarthrosis; it has 3 degrees of freedom of motion because it permits movement in three planes perpendicular to each other. The three perpendicular planes are a frontal, a sagittal and a transverse plane; if the center of the hip joint is chosen as the intersection point of these three planes it becomes the origin of a three coordinate system. The position of any point can then be expressed numerically by the coordinates of this system.

A. THE POSITION OF THE ACETABULUM

If a plane is laid through the circumference of the acetabular cavity in upright standing position, this plane intersects with the sagittal plane of the body at a backward opening angle of about 40°. Consequently one may say that the acetabular cavity looks obliquely forward (Fig. 1). The same plane intersects with the transverse plane of the body at an angle of 60° opening laterally. Therefore this cavity also looks obliquely outward. Being inclined both against the sagittal and the transverse plane the acetabulum is also inclined against the frontal plane with which it forms an angle opening downward. Therefore the cavity looks obliquely downward.

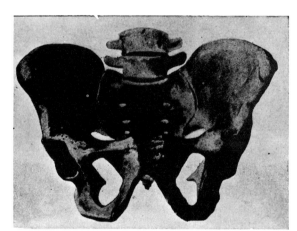

FIG. 1. Pelvis from in front showing position of acetabulum. (Spalteholz)

The continuation of the acetabular cavity is provided by the cartilaginous limbus or labrum glenoidale which circles this cavity except for a small area at its lower border, the so-called incisura acetabuli. Here the ring is completed by the transverse acetabular ligament (Strasser[10]) (Fig. 2).

B. THE POSITION OF THE FEMORAL HEAD

We consider the mid position of the head in the acetabulum one from which excursions can be carried out to equal degrees in all directions; this means a position in which the center of the surface of the head is in contact with the center of the acetabulum.

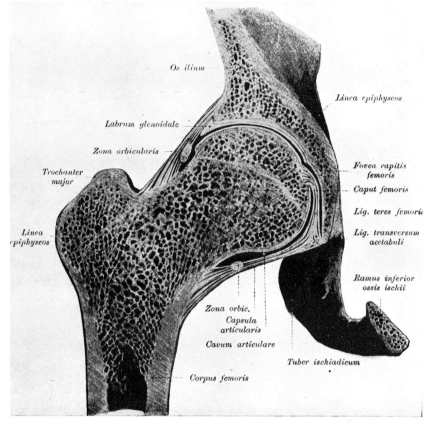

FIG. 2. Frontal section through hip joint showing continuation of acetabular cavity
by labrum glenoidale and transverse acetabular ligament. (Fick)

The head is almost a perfect globe. If in this concentric position a plane
is laid through the rim of the cartilage of the head, this plane is also inclined
40° to its sagittal and 60° to the horizontal plane, the same as the plane laid
through the margin of the acetabulum. Both joint surfaces are approximately
congruent, having an angular value of about 180°.

C. THE POSITION OF THE FEMUR

When the head is thus centralized in the acetabulum, the position of the
femur is one of abduction, flexion and outward rotation. This presents the
neutral or mid position of the joint.

However it is not the normal position that the femur assumes in upright
standing. For practical purposes it is much more suitable to relate all positions
of the femur to the one it occupies in upright position rather than in mid
position and to consider the normal position as the starting point from which
all excursions of the joints are measured.

Because this position deviates so much from the one in which head and
socket are concentrically adapted it must be expected that the anterior portion

of the cartilage of the head is no longer harbored in the acetabular cavity. It has slipped out of it because in assuming the upright position the head of the femur rotates externally. This outward rotation is the result of the developmental changes which occur from the earlier embryonal position of the outward rotation and flexion. Gradually during embryonic development and even after birth which the assumption of static function, the femoral shaft goes into inward rotation and adduction, as the result of which head and neck become angulated against the shaft in two planes, namely the frontal and the horizontal. In the frontal plane the angle between neck and shaft indicates the adaptation of the femur from this embryonal abduction position into the adduction position which conforms to the parallelism of the legs in upright standing position. This is termed the angle of inclination (Fig. 3). This angle amounts in the newborn to 150° but it decreases in the early years and in the adult it is an average of 125°.

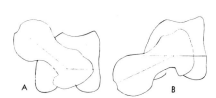

Fig. 3. Angle of inclination of neck and shaft of femur, adult.

Fig. 4. Angle of declination of femoral neck.

Fig. 5. A. Anteversion. B. Retroversion. (Fick)

In the transverse plane there is also an angulation between neck and shaft of the femur which is called the angle of torsion or declination. It is caused by the fact that neck and head rotate outward against the shaft and the shaft remains in inward rotation against the neck.

When the femur is put down so that the femoral condyles lie flush with the table, head and neck are twisted forward away from this plane (Fig. 4). This angle of torsion is likewise a product of embryonic development. It denotes a change of the shaft of the femur from being in outward rotation in mid position to the inward rotation against the neck in normal standing position. This angle is also subject to physiological and pathological variations. Kingsley[9] who examined 630 femora (bone specimens) found the average angle of torsion in the adult is only 8.021°. The angle of antetorsion in infants and young children was much greater; for females the average was 26° and for males, 23°. In the past the angle of antetorsion has been believed to be much greater, the mean average in the literature varying between 11.9°-25°. The values derived from Kingsley's more careful measuring methods should be accepted.

Under certain pathological conditions this angle of antetorsion may be greatly increased, for instance in congenital dislocation of the hip where it may amount to as much as 90° (anteversion or antetorsion). In other conditions such as the coxa vara it may be decreased or even reversed so that head and neck are actually rotated backwards (retroversion or retrotorsion), in which case the humeral shaft would appear outward rotated against the head and neck (Fick[3]) (Fig. 5).

Even normally there are considerable racial and individual variations of this angle. According to Miculicz, antetorsion varies in wide limits and retrotorsion occurred in as much as 10% of his 134 cases examined. The differences

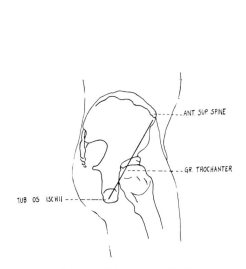

FIG. 6. Roser-Nelaton line. (Fick)

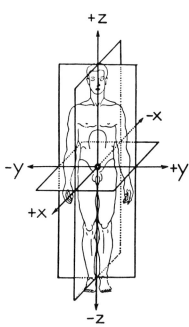

FIG. 7. Orientation planes of the body. Center of gravity. Origin of three coordinate system.

varied from 37° of antetorsion to —25° of retrotorsion. In two-thirds of the adult cases the variations were from +4 to +20 antetorsion and 12° of antetorsion may therefore be accepted as the mean value of the adult torsion angle. Hultkrantz found that the Fire Islanders of South America who have extensive squatting habits have a greater antetorsion of the head and neck than the Caucasians.

D. THE ORIENTATION AXES AND REFERENCE PLANES OF THE HIP JOINT

In upright standing the hip joint lies at about the level of the tip of the greater trochanter. At this level the head of the femur can be palpated in the groin immediately lateral to the femoral artery. If a horizontal line is drawn from the tip of one greater trochanter to the other, it represents approximately the common hip joint axis in the frontal plane. The relation of the trochanter to the hip joint varies little with the movement about the frontal axis, that is in flexion and extension.

It is different with the ab and adductory motions of the joint which are carried out about the sagittal axis. In abduction the trochanter moves upward and in adduction downward in relation to the center of the hip joint. If one therefore measures this relationship by the so-called Roser-Nelaton line drawn from the anterior superior spine to the tuber ossis ischii, the femur should be placed in neutral position. In this position the line normally touches the tip of the greater trochanter (Fig. 6) and it runs above the tip of the trochanter if the limb is adducted and below if it is abducted (Fick[3]).

The next point is to establish the exact position of the hip joint relative to the center of gravity of the whole body (Fig. 7).

In the frontal plane the Y coordinate of the three-coordinate system gives the distance of the hip joint center to the center of gravity. Its value is 8.5 cm on the right and —8.5 cm on the left side.

The X coordinate which denotes the distance of the hip joint from the center of gravity in the sagittal plane is 0 if the line of gravity falls through the common hip joint axis as is the case in the ordinary stance. It means that the hip joints are situated in the line of intersection of a horizontal with a frontal plane.

The Z coordinate denotes the elevation of the hip joint above the common center of gravity as about —3 cm. meaning that it is this distance below the center of gravity.

In the relaxed position where the line of gravity falls behind the hip joint the X coordinate assumes a small positive value. In the forward bent military position, on the other hand, the line of gravity falls in front of the hip joint so that the X coordinate has a small negative value. The appearance of definite values of the X coordinate means that gravity develops in the sagittal plane a rotatory force which has to be neutralized by ligamentous or muscular tension. In the frontal plane the rotatory moments produced by the definite values (+8.5 and —8.5) in upright symmetrical position mutually neutralize each other and no muscular force is therefore necessary to maintain equilibrium.

E. CONTOUR AND SHAPE OF THE HIP JOINT

1. The head of the femur

The head of the femur represents a true hemisphere of 180°; the acetabulum has the same angular value; therefore in mid position the articular surfaces cover each other perfectly. The cartilaginous limbus which arises from the acetabulum cups itself beyond the equator of the head since the free diameter of its opening is somewhat smaller than the diameter of its insertion. This accurate fit of the articular bodies is an important factor in the cohesion of the joint. However, as mentioned before, if the cartilaginous limbus is removed and the joint is entirely skeletonized, that is if all soft tissues are removed, one finds that the joint is still held close together by the atmospheric pressure which amounts to more than the total weight of the limb. The experiments of the Weber brothers[12] of drilling the acetabulum and thereby releasing the atmospheric pressure, showed that this pressure amounts to 11.25 kg., and if the distracting weight acts vertically to the acetabular plane, then the pressure may be as high as 18.25 kg. (Strasser[10]).

Considering the largest diameter of the femoral head covered by the limbus the theoretical atmospheric pressure should be about 22 kg. This is much in contrast to other joints, for instance the shoulder joint where the pressure is only 3.09 kg. and the elbow joint where it is only 3.75 kg. It appears, there-fore, from the Webers' experiments that the atmospheric pressure is amply capable of carrying the weight of the entire limb without any muscular help. Fick[3] believes that there are pressure variations produced by some slight incongruencies of the surfaces. Cruvelhier pointed to the communication of the psoas bursa with the hip joint as a factor of regulating the atmospheric pres-sure. Lately D. Fuiks[4] has shown that on strong traction an air-filled vacuum appears which can be demonstrated in the x-ray.

In vivo other forces assist in maintaining the head firmly in its place. In extension and abduction there is the tension of the anterior and upper capsule which prevents any distal displacement of the head. In the gait, especially, muscle tension is active both in forward and backward swing which assist the atmospheric pressure to maintain the head in the socket.

2. The femoral shaft

The contours of the shaft of the femur are rather irregular. Its upper end carries the greater and the lesser trochanter between which the femoral neck is inserted at the inclination angle of 125°.

The anterior intertrochanteric line and the posterior intertrochanteric crest unite the trochanters, the latter forming a deep groove with the posterior aspect of the neck. All crests and prominences serve for the insertion of powerful muscles.

Only in its upper half is the cross section of the femoral shaft cylindrical. Toward the lower end it becomes more oblong and above the condyles it is triangular. Looking at the bone as a whole it appears curved both in the frontal and in the sagittal plane. In the frontal plane, neck and upper end of the shaft form a sharp curve pointing outward. In the sagittal plane, the entire length presents a mild curve bowing forward. Both of these curves have a static significance. The outward curving of the upper end of the femur in the frontal plane greatly increases the ability of the bone to receive weight stresses from above and transmit them to the knee joint; the forward curving of the femur in the sagittal plane increases its resistance to bending and shearing stresses which it receives in sagittal directions such as in walking, running or jumping. This will be referred to later in the study of the architecture of the bone.

F. AXES AND DETERMINATION PLANES OF THE FEMUR

We distinguish an axis of the femoral neck and an axis of the shaft. The axis of the neck runs from the center of the head in midline of the neck to the implantation of the neck between the trochanters. The axis of the shaft is a line drawn from midway of the trochanteric region to the middle of the knee joint. Both axes enclose, as stated above, in adults an angle of 120-135°. The mechanical axis of the whole femur is a line drawn from the center of the head to the midpoint between the condyles (Fig. 8).

The determination plane of the femur is a plane which contains the longitudinal axis of the femur and the horizontal line which unites both epicondyles. This plane does not contain the axis of the femoral neck because of its ante or retrotorsion. The mechanical axis of the femur forms with the anatomical axis an angle of 5-7° Fig. 8).

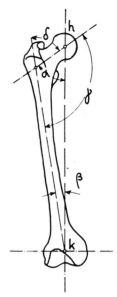

FIG. 8. Mechanical and anatomic axes of the femur.

II. THE STATICS OF THE HIP JOINT

A. THE ARCHITECTURE OF THE FEMUR

The study of the structure of the upper part of the femur is of historical interest. It inspired in the mathematician Colman the idea of comparing it with the construction of a crane and it stimulated J. Wolff's classical investigations of the trabecular systems.

On a former occasion it has been discussed to what extent the texture of this bone is built on strictly mechanical lines. The trabeculae of the proximal femoral end form several tracts which are significant for the interpretation of the static function of this bone. In the frontal plane we distinguish two principal trabecular systems. The medial one, coming from the medial cortex, is the real weight bearing system. It streams in divergent layers into the medial half of the neck and ends in the head at the subchondral layer. The lateral system arises from the lateral cortex and goes to the outer half of the neck, crossing the inner system at right angles at the upper portion of the neck and ending at the lower inner quadrant of the head (Fig. 9).

Secondary trabecular systems are: one radiating from the medial trabecular system and ending in the region of the greater trochanter, and another arising from the lateral trabecular system of the shaft and also going into the trochanter crossing the first at right angles (Fig. 9).

In the sagittal plane one trajectory system runs from the anterior cortex into the head crossing at right angles another reaching the head from the posterior cortex (Fig. 10). It is the medial trabecular system which bears

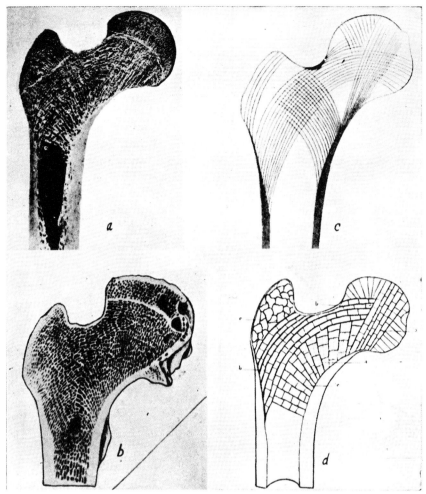

FIG. 9. The trabecular system of the femoral head and neck.
a, *b*, Specimen Section; *c*, *d*, Diagrammatic.

most of the weight, being more directly in line of gravitational pressure than any other. All trabecular systems cross at right angles thus giving the structure the greatest possible pressure and bending resistance.

V. T. Inman[8] finds that the weight bearing trajectories correspond in direction to the angle of reaction. This is the angle which the reaction from the floor as represented by a line from the floor contact to the weight bearing portion of the femoral head forms with the perpendicular. When standing on one leg this angle is about 169° from the vertical, the difference between the angle of the neck and that of the medial trabecular system being 21.1° (Baehr[1]) (Fig. 11). Farkas called attention to the fact that in old people the atrophy of the principal weight bearing system is the reason why fractures of the neck occur under slight provocation.

FIG. 10. The trabecular system of the femoral head and neck in sagittal plane.

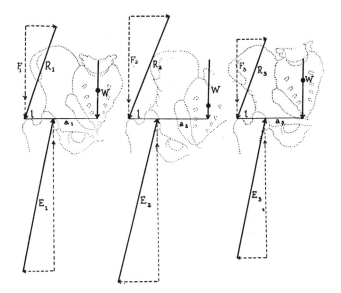

FIG. 11. Weight bearing trajectories correspond in direction to the angle of reaction (angle between perpendicular and weight bearing portion of femoral head). (Inman)

B. ANALYSIS OF THE STATIC FORCES OPERATIVE UPON THE FEMUR

1. The frontal plane

It is now generally believed that the femur in antero-posterior aspect does not represent a crane in the engineering sense but rather an excentrically loaded column (Baehr[1]), the critical point for bending stresses being in the neighborhood of the inter-trochanteric line. If this is the case then Euler's theory of the column should be applicable to it with such restrictions, of course, as are imposed by the irregularities of contour and cross section (Ghillini and Canovazzi[5]).

In the frontal plane the femur represents a column anchored at the lower end and freely movable at the upper end about one or more horizontal axes (Fig. 12). In this situation the expected deformity is a cosine curve (1) occupying two-thirds of the length of the bone (L) ($1 = 2/3L$). It is possible to recognize such a curve in the femur considering the region between the trochanter and the shaft as the apex of the curve. This would refute Wolff's original concept that the femur represents a crane in the frontal plane (Grunewald[6]).

Seen in the sagittal plane the situation is also that of an excentrically loaded column but one which is movable at both ends, namely at the hip joint and the knee joint. Here the expected deformation should be a cosine curve which occupies the full length of the column ($1 = L$) (Fig. 13). Again the aspect of the femur seen from the side suggests such a curve.

But whether one accepts the column theory for the femur or whether one rejects it on the ground that in form and density it deviates too much from a geometrical structure, it is certain that the femur, in receiving the superincumbent load is subject not only to pressure stresses but also to bending and shear. Because of the significance these stresses have in the production of fractures, it might not be amiss to dwell upon the subject with more leisure.

So far as axial pressure is concerned it merely changes with the superincumbent weight. It is the shear and the bend at different levels of the bone which

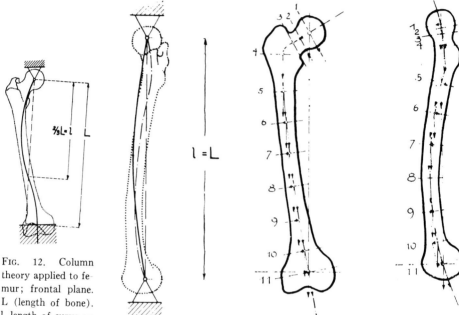

FIG. 12. Column theory applied to femur; frontal plane. L (length of bone). l, length of curve = 2/3L.

FIG. 13. Column theory applied to femur; sagittal plane. l (length of curve) = L (length of femur).

FIG. 14. Axial pressure changes with superincumbent weight only. The shear and bend at different levels of the bone vary greatly.

vary greatly. The bending stress develops a compressing and a distending component on opposite sides of the neutral axis (Fig. 14).

The relation between axial pressure and shear and between bend and shear at different levels of the femur (Fig. 14) have been calculated by Grunewald.[6]

a) THE AXIAL PRESSURE AND SHEAR

If the femur is weighted down by 30 kg. then the transmitted pressure has a considerable shearing component because of the deflection of the femoral neck. At the upper levels (1-3 in Fig. 14) the axial pressure is calculated to be 21 kg. whereas the shearing component is 22.5 kg. Immediately below the greater trochanter (Level 4) down to about the middle of the shaft (Level 8) we find the axial component (pure pressure) has risen to 30 kg. which represents the superincumbent load, and it remains much the same to the end of the bone (Level 10) while the shearing component decreases constantly being only 6.5 kg. at the mid shaft and then increasing only slightly at the end of the bone to 9.5 kg. This indicates that, while the axial pressure component is almost constant, the greatest shearing stress is at the upper end of the femur close to the head and smallest at the middle, and from there it increases again slightly toward the lower end.

b) The Bending and Shear

It is well known that in a loaded beam there exists a mathematical relationship between the bending and shearing stress as produced by a concentrical load. The bending stress has its maximum in the middle of the beam; the shearing stress has its maximum at the end. The shear again is minimum at the middle and the bending stress is minimum at the ends. Inasmuch as the femur is regarded as an excentrically loaded column, this principle may be applied to it. Grunewald[6] thus found the shear greatest at the upper end amounting to 22.5 kg. under a 30 kg. load.

c) Compression and Distension Components of the Bending Stresses

According to the premises that bending would have its minimum at the end, we find at the upper end of the femur (Level 1-3) the bending stress is only 1.4 kg. per cm.^{2} whereas at a somewhat lower level the intertrochanteric region rises to its highest point of 74.6 kg. per cm.^{2} Just above the femoral condyle (Level 10) it decreases to 27.2 kg. per cm.^{2} These figures apply to the compression component of the bending stress. For the distension component the stress distribution is similar. Here the minimum, according to Grunewald is 1 kg. per cm.^{2} at the level of the center of the head and the maximum of 63.4 kg. per cm.^{2} is just below the intertrochanteric region; above the condyles (Level 10) it falls to 4.8 kg. per cm.^{2} (Fig. 14).

Thus we have a maximum bending stress, both of the compression and the distension component, in the subtrochanteric region and the maximum shear in the upper portion of the femoral neck. Bending stresses are at the minimum at both ends of the femur where shears are at the maximum, while axial pressure remains practically the same. From the practical point of view this means that neck and head are more subject to shear, the subtrochanteric regions more subject to bending stresses.

2. The sagittal plane

Here the distribution is similar. Because of the more regular distribution of mass in this plane and the fact that the deformation represents a cosine curve of the length of the femur ($1 = L$), the stress curves should correspond more accurately with theoretical values. At any rate the maximum bending stress should be at the middle of the shaft. That is where the femur usually bends if the breaking force strikes in strictly perpendicular direction from in front or behind (bumper fracture). Also in cases where the resistance to bending is reduced (Paget's disease), we see straight transverse fractures of the mid shaft.

3. The transverse plane

Here the femur is taxed for torsion. Torsion is produced by two couples of forces which are arranged in parallel planes. The two couples rotate the body in opposite direction (pure torsion). The rotating force times the radius (that is the distance from the neutral axis) is called the torque. Between the torque and the shearing stress there is a mathematical relationship.[1] As in the shear the resistance of the bone lamellae to torque depends upon their obliquity, that is the angle they form with the long axis. The torque has its maximum at the upper end of the femur.

All in all this graphic analysis of the internal resistance of the femur to external stress leads one to the conclusion that in the fracture of the neck of the femur the shearing element prevails, while in the fracture of the trochanteric region and below, it is the bending element. While our knowledge of the trajectory resistance is still very incomplete, one should be able in cases of atrophy or destruction of certain lamellar systems, to anticipate from the x-ray evidence what type of fractures would be likely to occur in a given case.

C. THE EFFECT OF MUSCLE FORCE ON THE MOULDING OF THE HUMAN FEMUR (GRUNEWALD[6])

In the sagittal plane the hamstrings mold the anterior convexity of the femur against the quadriceps in spite of the discrepancy of their respective cross sections. Only the biarticular muscles are able to bend the femur in its whole length.

In the frontal plane the tensor and the vastus lateralis being stronger than their oppenents cause the normal valgity of the lower femur and the upper tibia.

D. THE CAPSULAR REINFORCEMENT OF THE HIP JOINT

The synovial attachment is arranged outside of the labrum glenoidale which projects freely into the joint. In this respect the arrangement is similar as in the shoulder joint. At the femur the capsule reaches down to the narrow point of the neck farther distally anteriorly than posteriorly.

The joint is amply protected by capsular as well as ligamentous reinforcements. The inner side of the capsule has a circular reinforcement, the zona orbicularis, which projects into the joint cavity and is formed by crossed fibers. The capsular fibers arising from the os ilii check torsion around the axis of the neck of the femur before tension of the circular fibers of the capsule occurs (Fig. 15).

---

[1] The torque is proportional to the shearing stress and the inertia and inversely proportional to the radius. The shearing force is proportional to the torque and the radius and inversely proportional to the inertia.

$$T = \frac{S\,I}{r}; \qquad S = \frac{T\,r}{I}$$

The capsular apparatus also makes provision against infolding and impingement of the synovial membrane. Medially under the psoas the capsule folds in its anterior superior portion. Anteriorly the capsule folds under the iliofemoral ligament on abduction and flexion while the rectus femoris pulls the fold out of the joint. Posteriorly the membrane folds under the obturator internus. In adduction there is tension of the proximal portion of the capsule while in abduction the lower distal portions are under tension.

E. THE LIGAMENTOUS REINFORCEMENTS OF THE HIP JOINT

The ligamentous apparatus of the hip joint is very powerful as must be expected from the wide range of motion in all planes and from the powerful dy-

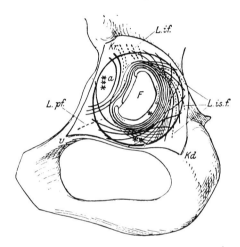

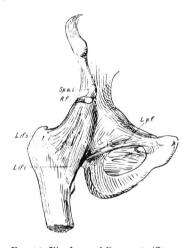

FIG. 15. Zona orbicularis which reinforces joint capsule. (Strasser)

  L.if. ligamentum iliofemorale
  L.pf. ligamentum pubofemorale
  L.is.f. ligamentum ischiofemorale
  Kd caudal triangle
  Kr cranial triangle
  V ventral triangle

FIG. 16. Ilio-femoral ligament. (Strasser)

  L.if.i. ligamentum iliofemorale inferius
  L.if.s. ligamentum iliofemorale superius
  L.p.f. ligamentum pubofemorale

namic effect which some motions have upon the hip joint. The ligaments act principally as checkers of motion.

1. The ilio-femoral ligament (Fig. 16)

This ligament is divided into a superior or lateral portion which has a tensile strength of 250 kg., and an inferior or medial portion with a tensile strength of 100 kg.

The checking effect of this ligament is as follows: It checks adduction especially with its upper lateral portion (Fig. 17). All fibers tighten in extension particularly the lower lateral portion, and its tension increases markedly toward the end of extension at which stage it winds around the neck of the femur so that it approaches more the anterior rim of the acetabulum. The maximum ten-

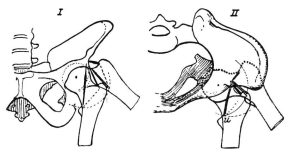

FIG. 17. Checking effect of the ilio-femoral ligament in abduction (*I*) and adduction (*II*). (Strasser)

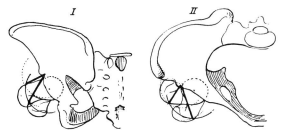

FIG. 18. Checking effect of the ilio-femoral ligament in extension combined with internal rotation (*I*), and in flexion combined with outward rotation (*II*). (Strasser)

sion occurs as the head of the femur, in full extension, presses forward against the ligament and as the pelvis at the same time rotates backward. The iliofemoral ligament becomes tight also in extension combined with inward rotation and in flexion combined with outward rotation (Fig. 18).

2. The ischio-femoral ligament (Fig. 19)

This ligament lies close to the lower border of the ilio-femoral. It tightens in extension, in inward rotation associated with extension, and in abduction.

3. The pubo-femoral ligament (Fig. 16)

When the hip is in extension it checks the outward rotation. It also, together with the ischio-femoral ligament, checks abduction. When the hip is in flexion, the ligament checks abduction and outward rotation.

4. The ligamentum teres

This intra-articular ligament has little mechanical function although it has considerable tensile strength (15-57 kg. according to Braune). It checks abduction of the femur in extension but not before the ilio-femoral ligament becomes tight. The main function of this ligament is to carry blood vessels. Its mechanical role is unimportant.

III. THE MUSCLE MECHANICS IN STATIC CONDITIONS OF THE HIP JOINT. EQUILIBRIUM

Under certain static conditions demands are made by gravity and other forces upon the muscles which maintain the hip joint in static equilibrium. The analysis of the situation centers about two points. What are the respective components of a muscle force acting in the different planes, and how does the equilibrating force change in the individual muscles under diverse static conditions?

In the globus phantom of the hip joint Strasser and Gassman[11] illustrate the rotation moments of muscles which are oblique to all three planes or axes of reference by a number of rods representing the individual muscles. In this

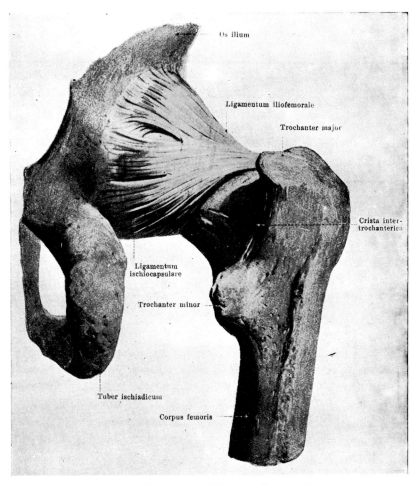

Os ilium

Ligamentum iliofemorale

Trochanter major

Crista inter-
trochanterica

Ligamentum
ischiocapsulare

Trochanter minor

Tuber ischiadicum

Corpus femoris

FIG. 19. Ischio-femoral ligament. (Spalteholz)

device the length of the rod is commensurate to the rotation moment, i.e., the muscle tension times the lever arm for each of the planes (Fig. 20).

The changes of these respective rotation moments wich occur when the joint abandons its neutral position could be determined in terms of coordinates for the components of the muscles in different planes.

A. CONDITIONS OF EQUILIBRIUM IN THE FRONTAL PLANE

In the equilibrium between abductors and adductors the principal muscles of the abductor group are the gluteus medius, minimus and tensor fasciae. Since all three muscle have components in the sagittal plane, that is in forward and backward movement, these have to be neutralized if the muscles are to maintain the frontal plane equilibrium against gravitational stresses.

In the frontal plane the gluteus medius is the principal abductor of the hip joint as almost all its fibers are abductory and it is only in extreme forward flexion that the foremost fibers change to adduction. Besides, the anteriormost fibers have a flexory and the posteriormost fibers an extensory component. This muscle is important as maintainer of the equilibrium, especially in the gait. When it becomes insufficient it fails to assist in holding the rim of the pelvis of the standing side against the trochanter.

However the principal stabilizer of the pelvis in this position is the tensor fasciae. Let us assume that in standing on one leg with the knee extended the mechanical axis of the extremity from the center of the hip joint to the foot coincides with the line of gravity. In this case no rotatory momentum develops either in the frontal or in the sagittal plane, or, for that matter, in the horizontal plane.

Now let the line of gravity move lateral or medial from the hip joint center so that it still lies in the same frontal plane but does not coincide with the extremity axis. In this case no moment develops in the sagittal plane but a moment develops in the frontal plane. These moments cause abductory or adductory rotation.

As remarked above, the tensor fascia is, according to Fick[3] and Strasser[10] and corroborated later by Inman,[8] the principal stabilizer of the pelvis when standing on one leg. The tensor is an abductor in normal position and a flexor of the hip in all positions. It prevents, as Inman has shown, the pelvis from dropping on the unopposed side (Trendelenburg's symptom). It is then only a question of suppressing the flexory component of this muscle. If, as we have mentioned above, the line is in the same frontal plane as is the center of the hip joint and if it does not fall in front of it, no neutralization of the flexory component of the tensor by the extensors of the hip will be necessary.

Fig. 20. Globus phantom of Strasser and Gassmann. (Strasser)

If the line of gravity falls behind the hip joint then the gluteus maximus is relaxed. With the forward bending of the trunk, however, this changes. The line of gravity comes to lie in front of the hip joint and the equilibrium requires neutralization of the flexory effect of gravity by the gluteus maximus.

Returning now again to the abductory and adductory balance in the frontal plane, let us again start from the proposition that the line of gravity passes through the center of the hip joint. As stated above no rotatory component develops in the frontal plane. There is no abductory or adductory force active and so far as this plane is concerned, the equilibrium is automatic and no muscle force is required to maintain it. But this situation is only a momentary one because the body, in standing on one leg, constantly sways sideways so that the line of gravity falls either medially or laterally to the center of the hip joint.

Let us consider what happens if the line of gravity passes medially to the hip joint into the supporting area of the foot. Gravity becomes an adductory force. In order to illustrate the relative strength of the muscles which hold the equilibrium when the line of gravity falls medially to the hip joint center, it may be mentioned that according to Fick[3] the rotatory moment of the gluteus medius and minimus, i.e., the product of their tensile strength times their distance from the center of motion is 18 kgm. against 7.6 of the tensor. This in-

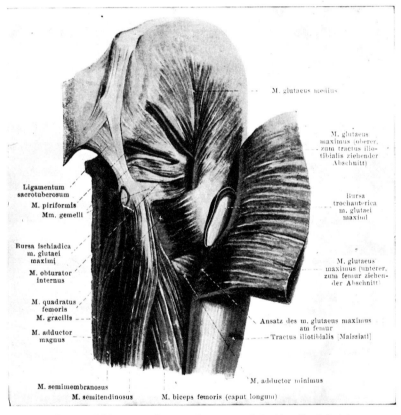

Fig. 21. Adductors and outward rotators of the hip. (Spalteholz)

dicates that the tensor is much more of a stabilizer of the hip joint than the gluteals because the distance of its axis from the hip joint center is much smaller, as the smaller rotation mement shows; therefore its action is axial or stabilizing rather than rotatory in contrast to the gluteals. On the adductory side, the adductors must be relaxed in this condition since gravity acts as an adductory force and it is the job of the abductors to neutralize it. This situation of the line of gravity falling medial to the hip joint applies to the frequently assumed posture of resting on one limb.

Next we assume that the line of gravity falls lateral to the hip joint center so that gravity becomes an abductor. It is now up to the adductors to establish the equilibrium while the abductors relax. The principal adductors are the ad-

ductors longus, brevis, magnus, the gracilis, and the pectineus. The longus is
an adductor in all positions and a flexor up to 70° flexion; then it becomes an
extensor. Its principal action, however, is adduction which it carries out in all
positions. Its working capacity is 5 kgm. The brevis is an adductor in all positions and is a flexor up to 50° and then it becomes an extensor. It is also said
to become a weak outward rotator of the hip in position of extension. The adductor magnus is an adductor in all positions, a flexor up to 50° and then it
becomes an extensor. The gracilis is an adductor in all positions and a flexor
up to 20-40° when it changes to an extensor, and the pectineus is in all positions an adductor and a weak flexor and outward rotator (Flick[3]). The magnus

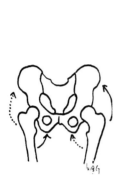

Fig. 22. Synergistic ac-
tion of abductors and
adductors of hip.

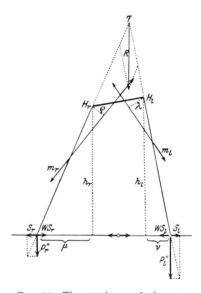

Fig. 23. The resultant of the perpendicular and horizontal components
in their prolongation falls into the
intersection point of the two leg
axes. (Strasser)

is the most powerful of this muscle group. Its rotatory action is slight and consists mainly of inward rotation. The total working capacity of the magnus is 12
kgm. All these muscles take their origin at the ischio-pubic part of the pelvis
from a triangular area of which the horizontal and descending ramus of the os
pubis and the tuber and ascending ramus of the os ischii form the two sides
(Fig. 21).

   *The conditions of the hip joint equilibrium in the frontal plane when stand-
ing on both legs.*

   We start from the symmetrical position where the line of gravity falls into
the midline. The abductor and adductor group of both sides are mutually
antagonistic so far as the position of the hip and pelvis is concerned. The abductors of one side are synergistic with the adductors of the opposite side and
so are the adductors of one side synergistic with the abductors of the opposite

side (Fig. 22). In symmetrical position, therefore, abductors and adductors, respectively, are under equal tension on both sides.

However, differences in muscle tension appear when the line of gravity is no longer in midline but is deflected to one side or the other. The closer the line of gravity moves to one hip joint, the lesser is the rotatory moment of gravity or the bending moment if we consider that the pelvis is a horizontal beam; and the greater becomes the translatory weight stress of gravity until when the line of gravity falls through one hip joint all the superincumbent weight rests upon this joint.

As the line of gravity shifts toward one or the other hip joint, the horizontal components furnished by the muscles must be neutralized if an equilibrium at a certain point is to obtain. For instance a contraction of the abductors on the right and of the adductors on the left will shift the pelvis from the right side to the left side. As a result we see that if either the abductors or the adductors become grossly deficient in both legs, the patient is still able to stand on both legs but the adjusting side movements of the body are inhibited. This explains the lack of stability in cases of abductor and adductor paralysis to which we shall have occasion to refer in subsequent lectures.

There are therefore two principal frontal plane situations to be considered in asymmetrical standing on both legs so far as the relation to the center of gravity is concerned.

The three partial centers of gravity, that is trunk, right and left leg, are in the same frontal plane with the common hip joint axis. Here the gravitational stresses develop horizontal components in the frontal plane at the hip joints (Hl and Hr).

a) Gravity Has no Effect on the Lateral Shift

An automatic equilibrium of these gravitational horizontal components exists only if the resultant of the perpendicular and horizontal components (Hr and Pr and Hl and Pl) of the gravitational force (Pr' and Pl') in their prolongation fall into the intersection point of the two leg axes (Fig. 23).

b) Gravity Does Have an Effect on the Lateral Shift

The equilibrium here requires for abductors and adductors to act as neutralizers, the adductors producing the shift toward the midline, the abductors producing the shift away from the midline (Strasser[10]). These gravitational side stresses come in operation as soon as the position becomes asymmetrical and they have to be neutralized by coordinate action of the ab and adductors.

B. CONDITION OF EQUILIBRIUM IN THE SAGITTAL PLANE

The common center of gravity lies in front or behind the hip joint plane. It now becomes a problem for antero-posterior balance. Here the equilibrium is produced by the antagonistic play between extensors and flexors.

1. The extensors

a) THE GLUTEUS MAXIMUS (FIG. 24)

It is the principal extensor of the thigh in upright position where it acts purely extensory. The uppermost anterior fibers have a slight flexory effect because they are running in front of the transverse axis of the hip joint. It has been explained why in upright position the muscle is not under tension so long as the line of gravity does not fall in front of the common hip joint axis.

The secondary function of the muscle is adduction because the greater part

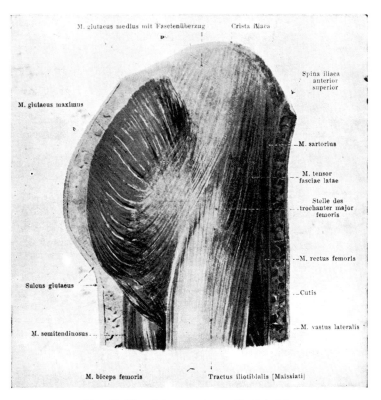

FIG. 24. The gluteus maximus. (Spalteholz)

of the muscle lies medial to the center of the joint. Only the uppermost forward portion is abductory in normal position acting synergistically with the gluteus medius and minimus.

In the transverse plane all portions of the muscles are outward rotators assisting the pelvi-trochanteric muscles. It is only the foremost fibers which can produce an inward rotation and that only in position of strong flexion.

The principal action of the muscle, however, is extension. It is the largest single muscle of the body with a cross section area of 66 cm.^{2} giving it a strength of 238 kg. If we consider the lifting height of this muscle to be 15 cm. the total working capacity amounts to as much as 34 kgm.

b) THE GLUTEUS MEDIUS AND MINIMUS (FIG. 21)

The frontal plane component of these muscles has already been referred to. The anterior fibers lying in front of the transverse axis of the joint are flexors and the posterior lying behind are extensors. The gluteus minimus has a similar function. The medius has a cross section of 40 cm.² and a lifting height of 11 cm., which gives it a working capacity of 16 kgm. The minimus has a cross section area of 15 cm.², a lifting height of 9 cm. with a working capacity of 4.9 kgm.

c) THE HAMSTRINGS

This group of muscles includes the biceps, the semitendinosus and semimembranosus (Fig. 25). All except the short head of the biceps are biarticular muscles and their effect must be considered separately for hip and knee joints. When the knee is fixed, all muscles are extensors of the hip joint in all positions. Their extensory effect also increases with the degree of flexion position of the hip. If hip and knee are free, then they act simultaneously as flexors of the knee and extensors of the hip. When the knee is fixed in extension, the hamstrings by tilting upward the anterior pelvic ring, increase the tension of the rectus femoris and become then synergistic to the extensors of the knee. When the pelvis is fixed the

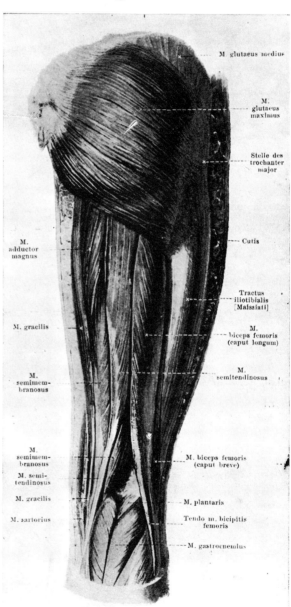

FIG. 25. The hamstrings. (Spalteholz)

hamstrings flex the knee and, by transmitting this motion of the hip joint, bring the thigh forward against the abdomen. Consequently, in this situation they may be said to become synergistic with the flexors of the hip. All hamstrings have an adductory component which increases with increased abduction. Their rotatory action is negligible and the total working capacity of all hamstrings, the medial as well as the lateral, sums up to 22 kgm.

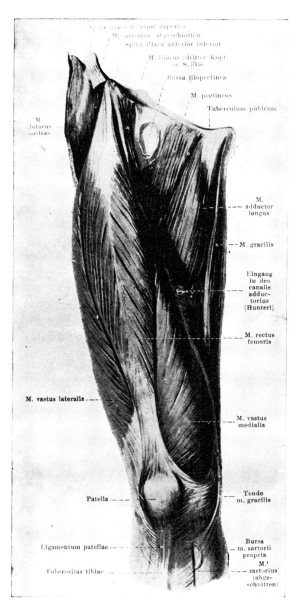

FIG. 26. The flexor group. (Spalteholz)

2. The flexor group (Fig. 26)

a) The tensor fasciae, besides being the most important stabilizer of the hip and a powerful abductor, also has a considerable flexory component.

b) The principal action of the ilio-psoas is to flex the thigh against the trunk. There is some controversy whether this flexion is associated with adduction although from the direction of the muscle it should be assumed to be the case. At any rate, its adductory component is negligible. A further question is whether it is an outward or an inward rotator. Duchenne[2] and others believe that it has a strong outward rotatory action and no doubt this is true in most positions.

c) The rectus femoris is also a powerful flexor of the hip and as such it is a true synergist of the ilio-psoas, the tensor, and also the sartorius. Its side actions are as follows: in extreme extension it is an abductor, in extreme flexion it is an adductor; it has a working capacity of 5 kgm. (Strasser[10]).

d) The sartorius is principally a flexor of the hip; only 1/10 of its strength acts as an abductor. It also is an outward rotator; its working capacity is 1.5 to 2 kgm.

3. The analysis of the conditions of equilibrium in the sagittal plane therefore sums up as follows:

a) Standing on one leg.

1) The line of gravity goes through the hip joint. Gravity produces no rotatory component. The task of maintaining the line of gravity in that position falls upon the concerted action of the extensors and flexors.

2) The situation mentioned above, as is also the corresponding one in the frontal plane, can only be maintained momentarily because of the constant forward and backward swaying of the body. This swaying calls for adjustment between extensors and flexors which consequently are almost always on the alert, always in action. As the line of gravity wanders forward the tension increases in the extensor group of the hip; as it wanders backward the flexors increase their tension. A sort of automatic equilibrium is provided by the ilio-femoral ligament which goes into tension.

b) Standing on both legs.

1) Here also the proposition that the line of gravity is in the same frontal plane as the common hip joint axis has only a theoretical meaning because of the constant backward and forward swaying of the body.

2) Therefore the oscillation of the line of gravity call for a reciprocal play between extensors and flexors, the same as standing on one leg. There is, however, some difference. Because of the much wider supporting surface the adjustments in the frontal plane are much easier. As a result we see that persons with gluteal paralysis on one side have no difficulty in adjusting themselves to backward and forward swaying because it can be fully accomplished by the muscles of the other side, an important point in some cases of infantile paralysis. In unilateral paralysis of all hip muscles such an adjustment can be made easily by the muscles of the sound side. Consequently unilateral paralysis of the hip muscles does not prevent standing on both legs. Furthermore, in bilateral paralysis of the gluteals, as one sees in progressive muscular dystrophy, the patient can still stand on both legs, throwing his body backward so that the ilio-femoral ligament provides the neutralization of the gravitational backward rotation. But it would be impossible for either the poliomyelitic or the dystrophic patient to actually stand on the affected leg although in walking he may still manage a very short period of single support on this leg.

C. THE CONDITIONS OF EQUILIBRIUM IN THE TRANSVERSE PLANE

The conjoint action of the extensors not only extends the hip joint but it also develops a horizontal component forcing the joint forward; the conjoint action of the flexors not only flexes the hip but it also produces a horizontal component which thrusts the hip joint backward (Strasser[10]). Hence a rotatory movement occurs in the horizontal plane about a perpendicular axis. The forward thrust hip goes into outward rotation, the backward thrust hip goes into inward rotation. The equilibrium between out and inward rotation is entrusted to the rotators of the hip joint.

1. Outward rotators (Fig. 27)

Principal outward rotators are the pelvi-trochanteric muscles as follows:

a) The piriformis. It is an outward rotator in all positions since its fibers run behind the longitudinal axis of the femur. In the sagittal plane it is also an extensor in all positions and in the frontal plane it is an abductor, its abductory power being five times greater than its extensory power.

b) The obturator internus. Its principal action is in the transverse plane where it acts as a strong outward rotator; like the piriformis it loses some of its

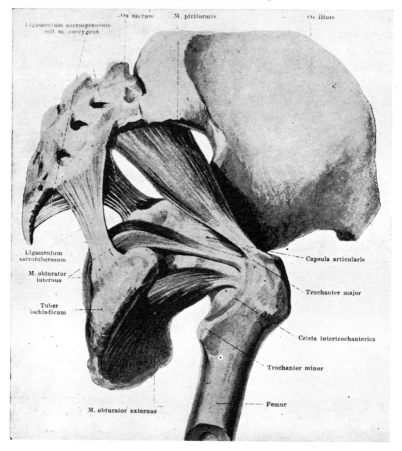

FIG. 27. The outward rotators. (Spalteholz)

external rotatory power with increasing flexion. It is a weak adductor and a weak extensor but its extensory power increases with strong forward flexion of the hip.

c) The quadratus femoris is a small muscle which runs from the tuber ossis ischii to the inter-trochanteric crest. Its principal action is the outward rotation and it is an adductor in almost all positions.

d) The obturator externus is also principally an outward rotator and likewise an adductor in all positions.

e) The action of the gemelli is the same as that of the obturators.

2. The inward rotators (Fig. 26)

In practically all of these muscles the inward rotation is only a secondary function. To this group belong the tensor fascia and the anterior portions of the gluteus medius and minimus. To this one must add the inward rotatory component of the adductor magnus and of the pectineus. At all events the total power of the inward rotators is much less than that of the outward rotators of the hip. According to Fick[3] the relation of the moments between flexors and extensors is 251:290; that between adductors and abductors is 210:347; that between external and internal rotation is 146:54. We observe that the outward rotators are almost three times stronger than the inward rotators.

Outward rotation of the hip joint is an important feature not only of the standing position but also of the gait. As the pelvis swings forward with the swinging leg the hip goes in outward rotation while the opposite standing leg rotates inward. At rest the mass of the thigh is so distributed that it falls into outward rotation automatically. With the toes touching the ground the internal rotators balance the outward rotatory component of the abductors (Duchenne[2]).

IV. THE STATIC PRESSURE AND THE SHEAR EFFECTS PRODUCED BY MUSCLE ACTION

In the computations of the stresses sustained by the head of the femur in standing the weight stresses are only one factor. To this must be added the stresses transmitted by the tension of the hip muscles, especially that of the abductors and the tensor. Inman[8] computed that when standing on both legs the muscular tension exceeded the body weight at a rate of 1.4:1 up to 1.9:1, so that the minimum static pressure on the head of the femur when standing on one leg would be 2.4 to 2.9 times the body weight.

The stabilizing component of the total tension of the abductors, which produces this static pressure on the hip joint, naturally changes with the joint position, the component being commensurate with the cosine of the angle of application. As this changes even in standing position with the shifts of the pelvis, the rotatory and stabilizing components of the abductors which neutralize the adductors change accordingly (Strasser[10]) (Fig. 28).

A. THE PRESSURE AND SHEARING STRESS CAUSED BY MUSCLE ACTION

1. The sagittal plane (Fig. 29)

Just as the force of gravity, so does the muscle tension develop axial and rotatory stresses. The former act as axial pressure, the latter as bending and shearing stresses. In the sagittal plane the muscles involved are the rectus and the hamstrings. The axial pressure is fairly constant. The bending moment has its maximum at the middle, the minimum at the ends; the maximum of the shear corresponds to the minimum of the bend.

2. In the frontal plane (Fig. 30)

In this plane the effect of muscle tension is as follows: the normal (perpendicular) or rotatory component of the adductors develops bending stresses which tend to increase the angle between neck and shaft in the sense of a coxa valga. The tension produced by the abductors, on the other hand, particularly the gluteus medius, tends, with its normal component, to decrease this angle in the direction of the coxa vara.

One can easily understand that under pathological conditions when a deficiency of muscle tensions exists the result is either a decrease or an increase of the angle of inclination of the femur depending upon whether the abductors

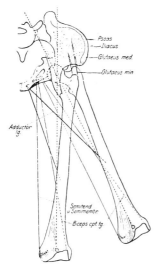

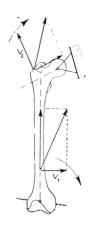

FIG. 28. The adductor group. Changing adductory component of the adductors with angle of application. (Strasser)

FIG. 29. Force analysis of femur. Muscular tension. Sagittal plane.

FIG. 30. Force analysis of femur. Muscle tension. Frontal plane.

or the adductors are involved. In infantile paralysis the lack of the abductors may lead to a coxa valga.

Inman[8] measured by spring scales the rotation moment of the abductors which he found as high as ten times the body weight in males and 9.4 times in females (Fig. 31), coming very close to the theoretical values. The muscles involved are the gluteus medius, minimus and tensor in the ratio of 1:2:4. The tensor remains tight until the pelvis is elevated 10-15° on the non-weight bearing side. With the pelvis horizontal the gluteals and the tensor share equally in resisting the gravitational rotation; with the pelvis sagging the resistance comes from the tensor fasciae; with the pelvis elevated about 10-15° resistance comes from the gluteals.

V. MUSCLE DYNAMICS

A. ORIENTATION

The study of muscle dynamics concerns the ranges and the types of actual visible movements carried out by the muscles. This is in contrast to the above-discussed muscle statics which is concerned primarily with the conditions of equilibrium.

To present ranges of motion in the anatomic specimen graphically, it is practical to use a system of parallel and meridian wire globes as first introduced by

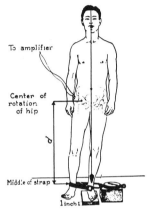

To amplifier

Center of
rotation
of hip

d

Middle of strap

1 inch

Torque = d x scale reading

FIG. 31. Inman's method of measuring rotation moment of the abductors. (Inman)

Albert (Fig. 32). The joint would then be in the center of the globe and the excursions are noted on the periphery of the globe in three planes. Motion in the three orientation planes can then be described in terms of the coordinates of any point in space relative to the center of the joint as origin of the system, and the ranges of excursion can thus be registered.

When we examine the hip joint motion in the living, however, we note very soon that simple motion in one plane only is the exception. Most of the time the movements are carried out in combination: abduction with extension and outward rotation, for instance. Such a combined movement seems to fit better into the scheme of purposeful motor acts which may represent a sequence of continuously changing directions. Take for instance the circling of the thigh in the hip joint. It is a combination of abduction and adduction, inward and outward rotation, flexion and extension in the hip joint in due sequence.

If one follows such movement in the wire globe using the knee as a pointer one finds that it describes a circle, the plane of which does not coincide with any of the orientation planes because its center is not the center of the globe. It is rather a circle around the pole of the globe. This motion is therefore called a *circumpolar movement*. In it the movement crosses meridians and parallels in contrast to the *circumcentral movements* in which movements follow meridians or parallels (Fig. 33).

In following this parallel the thigh has described in proper sequence flexion, abduction, external rotation followed by extension, adduction and internal rotation. Most of the movement we carry out with our extremities are of this circumpolar type.

B. THE EXCURSION RANGES

The combined circumpolar motions are the component product of instantaneous movements in more than one of the orientation planes of the joint. In order to analyze the circumpolar combinations it is necessary to break down the components into their circumcentral elements in the orientation planes, and then to determine the ranges in these respective planes. On this basis the ranges of the combined movements can then be established, which after all must be within the limits of the excursion fields of their separate constituent motions.

In the end we shall arrive at the full appreciation of the interdependence of one motion field upon the other which brings forth information as to the

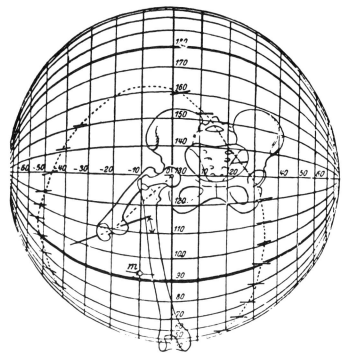

FIG. 32. Orientation of excursion field of hip joint. Parallel-meridian wire globe. (Strasser)

most suitable position for certain motor acts, for instance, kicking a football or skating a figure, etc.

1. The anatomical ranges

In the specimen the flexion extension range totals 140°, the abduction and adduction range 88°. The abduction and adduction and flexion extension ranges are interdependent in so far as the abduction and adductions are maximal when the hip joint is flexed to about 40° below the horizontal and flexion extension are maximal when the hip joint is in 5° abduction. The axial rotation range is given variously from 60-90° and the greatest degree is obtained with the femur in 90° flexion.

With the hip in 30° abduction and 20° inward rotation the flexion extension range is 40° less than the maximum. In 35° abduction and 20° inward rotation the flexion extension range lacks 60° of maximum. All these values apply to the ligamentous specimen where only the tension of the capsule and of the ilio-femoral ligament limits the excursion.

2. The ranges in the living

a) FROM THE NORMAL POSITION

In addition to capsular and ligamentous checks there is the restraining effect of the musculature to consider. This is especially noticeable in the flexion extension range. With the knee extended flexion of the hip is limited by the hamstrings. With the knee flexed the rectus femoris resists extension of the hip. Abduction of the hip is limited by the tension of the adductors and inward rotation is limited by the pelvi-trochanteric muscles. The greatest flexion extension range in the living is 120° flexion and 20° extension from the normal position. The abductory and adductory range in the normal position is 74°. The maximum rotatory motion inward and outward is 90°.

b) THE RANGES IN COMBINED POSITIONS

The greatest possible flexion extension range, namely 140°, can be accomplished when the hip is in 5° abduction and in neutral rotation. The smallest

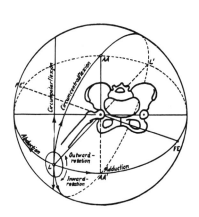

FIG. 33. Meyer's analysis of circumductory motion of the femur, combining the parallel (ab- adduction) and meridian (flexion-extension) component with instantaneous axial rotation (Redrawn from Strasser).

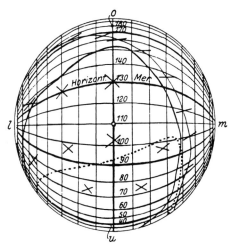

FIG. 34. Strasser's analysis of circumductory motion. The horizontal cross lines through the excursion field outline (thin line) indicate the position of the knee joint axis. (Strasser)

range is in 35° abduction and 20° inward rotation. Abductory and adductory maximum range is 74° in the living and it occurs in slight outward rotation and in slight flexion. The maximum rotatory range is 90° and occurs in flexion, and it is only 30° in normal position (Fig. 34).

These combinations prove the following:

The easiest route for forward flexion is starting from a position between neutral and slight abduction. The easiest path for outward rotation is from slight flexion; the easiest for inward rotation is a flexion position of up to 90°.

c) THE CHANGE OF RANGES WITH POSITION (STRASSER[10])

1) *The normal position (Fig. 35).* All adductors excluding gracilis, obturator

externus, pectineus and quadratus femoris have besides their adductory a slight
flexory action. The upper portion of the quadratus is an extensor. Of the extensors of the hip, the semitendinous, the semimembranous and biceps are also
adductors. Outward rotators in this position are the obturator externus, the
quadratus femoris, most of the adductors except the magnus. The gluteus maximus is an outward rotator and adductor aside from being an extensor. Flexion
is carried out by the ilio-psoas, the rectus femoris, and the tensor, the latter
being also an inward rotator as are also the anterior fibers of gluteus medius
and minimus. The ilio-psoas remains an outward rotator.

  2) Position of extreme abduction (Fig. 36). In this position the extension

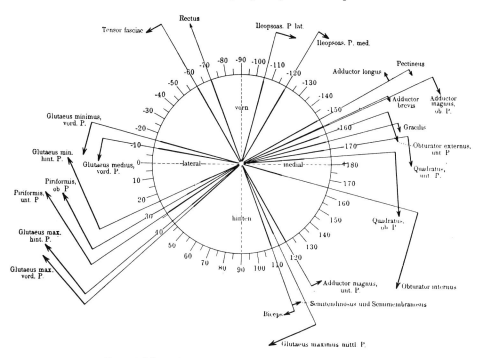

FIG. 35. Muscle arrangement in normal position. (Strasser)

is now concentrated upon hamstrings, quadratus, obturator externus and lower
portion of the adductor magnus. Most of these muscles also have acquired a
strong inward rotatory component. Abduction is now vested in the glutei and
flexion is concentrated on the ilio-psoas, rectus femoris, gluteals (anterior
portion) and tensor. The ilio-psoas has become an adductor and together with
the pectineus has intensified its outward rotatory component.

  3) The position of extreme flexion (Fig. 37). The greater amount of muscle
acts now in adductory capacity. The flexor group which includes the ilio-psoas,
the tensor, and portion of the gluteus medius and minimus, rectus femoris,
pectineus as well as the obturator externus now are all adductors. The ilio-
psoas is a strong adductor and outward rotator; the adductors change from
flexors to extensors as they pass from 50° or 70° flexion. The extensory group

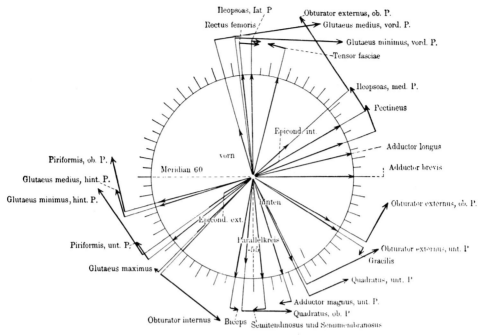

FIG. 36. Muscle arrangement in extreme abduction. (Strasser)

in this position is widely distributed over the gluteus maximus and the hamstrings which at the same time, in contrast to extreme abduction, have now a strong outward rotatory component.

d) *The inversion of function (Strasser[10])*. All this illustrates the changes

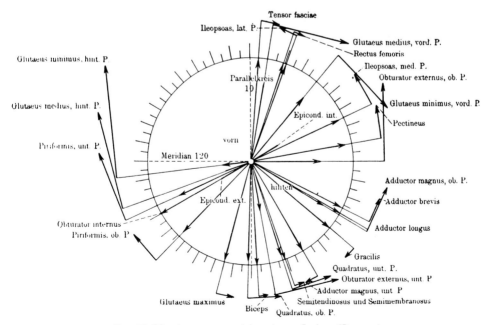

FIG. 37. Muscle arrangement in extreme flexion. (Strasser)

in the secondary muscle functions which occur as the position of the hip joint changes. Even complete reversals are noted. At right angle flexion all adductors, magnus, longus and brevis become extensors of the hip instead of flexors. In normal position we see the anterior portion of the adductory mass is directed backward toward the shaft of the femur, the posterior is directed forward. Consequently the posterior portion becomes extensors while the anterior group of the adductors, namely pectineus and adductor brevis become flexors. Now as the thigh is flexed forward, the axis moves, and more and more of the adductor group falls behind the hip joint center and therefore more of it becomes extensory (Fig. 38).

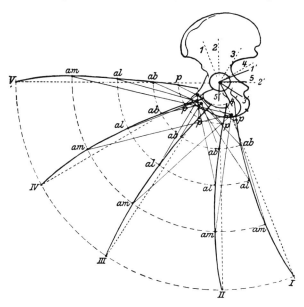

FIG. 38. Inversion of secondary muscle function of the adductors. (Strasser)

Other instances of inversion of function are as follows:

The gluteus maximus is primarily an extensor and secondarily an outward rotator. The secondary function changes to inward rotation when the hip is strongly flexed.

The gluteus minimus in normal position is an abductor but when the hip is strongly flexed it becomes an adductor. The piriformis in normal position is an outward rotator; when the hip is strongly flexed it becomes an inward rotator.

These are just a few examples to illustrate the changes which muscles undergo in their secondary function as soon as terminal positions are approached either in the plane of the secondary function or in any other plane. They explain the kaleidoscopic changes in the muscle grouping which occur with the changing of positions and how muscles constantly gain or lose in their secondary function as the joint is carried toward the extremes of motion. It further proves that on the whole the ranges are greater for the hip joint in or near the normal position, a point which should be well heeded in regard to muscle education as well as in problems involving surgical reconstruction.

VI. SUMMARY

1. The position of the acetabulum is forward, lateral and downward.

2. The mid position of the femoral neck, when the center of the acetabulum coincides with the center of the surface of the head is abduction, flexion and outward rotation. In normal position of the legs the anterior portion of the cartilage of the head slips out of the acetabular cavity.

3. In the course of embryonic development and even after birth the femoral shaft goes into inward rotation and adduction in relation to the neck. The adduction produces the angle of inclination which is 150° in the newborn and decreases to 125° in the adult. The inward rotation of the femoral shaft produces the angle of declination or torsion which is given variously from 8° in the adult to 26° in the infant.

4. Under certain conditions the forward torsion angle is greatly increased, up to 90° antetorsion, for instance in congenital dislocation; in others it is decreased or even reversed to retrotorsion as in coxa vara.

5. In upright standing position with legs parallel the hip joints lie at the level of the tips of the greater trochanters. In abduction the trochanter moves upward, in adduction downward in relation to the hip joint level.

6. In terms of the coordinates of the 3 coordinate system which has as its origin the center of gravity the Y coordinate (frontal plane) of the hip joints is +8.5 and —8.5 cm. in the adult. The X coordinate (sagittal plane) is zero if the line of gravity falls into the common hip joint axis; the Z coordinate (from the transverse plane) is —3 meaning the joint stand 3 cm. below the center of gravity.

7. In *relaxed* posture the X coordinate becomes positive as the line of gravity falls behind the hip joint axis; in military posture the X coordinate becomes negative as the line of gravity falls in front of the hip joint axis.

8. The head of the femur represents a hemisphere of 180° corresponding to the same angular value of the acetabulum.

9. The atmospheric pressure holding the head in the socket amounts to 18.25 kg. according to Strasser, amply capable of carrying the weight of the entire limb without ligamentous or muscular assistance.

10. On strong traction an air filled vacuum is formed in the joint demonstrable in the x-ray.

11. In vivo capsular, ligamentous and muscle tension assist in keeping the head in the socket.

12. The contours of the femoral shaft are irregular; it shows in the frontal plane an outward curve involving head and neck and upper end of the shaft; in the sagittal plane the entire length presents a mild forward curve.

13. The axis of the neck runs from the center of the head to midways between the trochanters; the axis of the shaft runs from midway of the trochanteric line to the middle of the knee joint. Both axes enclose in the adult an angle of 120-135°. The mechanical axis of the femur unites the center of the head with the midpoint between condyles. The determination plane of the femur contains the longitudinal axis and the line uniting the epicondyles.

14. Two principal trabecular systems reflect the static functions of the femur; a medial one from the medial cortex is the real weight bearing system; a lateral one from the outer cortex crosses the former at right angles. Secondary trabecular systems arise from the medial and lateral cortex streaming into the trochanter. The weight bearing trabecular system has, according to Inman, the direction of the so-called angle of reaction, i.e., the angle between the

weight bearing portion of the head and the perpendicular.

15. In the frontal plane the femur represents an excentrically loaded column rather than a crane, anchored distally and the expected curve is a cosine curve occupying 2/3 of the length of the bone. Both axial pressure and bending and shear develop.

16. In the frontal plane, the axial pressure remains fairly constant through all levels. The shear is greatest at the upper end; the compression and distension components of the bending stress have their maximum near the middle, below the trochanteric region.

17. In the sagittal plane the femur represents a cosine curve occupying the whole length of the shaft; here the maximum bending stress is in the middle.

18. In the transverse plane the femur is subject to torsion stress produced by two couples of forces rotating in opposite direction; the upper one is generated by length rotation of the trunk transmitted to the femur, the lower one is represented by the resistance from below.

19. The biarticular muscles, rectus and hamstrings in the sagittal plane, and the tensor and vastus lateralis in the frontal plane have, by virtue of their axial components, a share in molding the femur.

20. The joint is well protected by capsular and ligamentous reinforcements. The ilio-femoral ligaments checks adduction and extension; the ischio-femoral, extension, inward rotation and abduction; the pubo-femoral, outward rotation and abduction.

21. Equilibrium in the frontal plane between ab and adductors: The gluteus medius is the principal abductor; the principal stabilizer of the pelvis is the tensor. If, in standing on one leg, the line of gravity passes through the hip joint center, gravity develops no rotatory component and automatic ab and adductory equilibrium obtains. If the line passes medially to the hip joint center, gravity becomes an adductory force and has to be resisted by the abductors. If it falls laterally to the hip joint center, gravity becomes an abductory force and has to be resisted by the adductors.

22. The principal adductors are the magnus, longus, brevis, and the pectineus. Of these the magnus is the most powerful with a working capacity of 12 kgm.

23. Standing on both legs in symmetrical position ab and adductors are under equal tension on both sides and neutralize each other. The abductors on one side are synergistic with the adductors of the other and vice versa.

24. No such automatic equilibrium obtains if the line of gravity is deflected to one side. The closer it is to one hip joint, the lesser is the rotatory effect of gravity and the greater is the translatory stress. The hip closer to the line of gravity is in adduction, the other in abduction. The pelvis is shifted to the adducted side. The horizontal components of the muscles which shift the pelvis must be neutralized. The abductors of the adducted hip must neutralize the adductors of the abducted hip.

25. Geometrically expressed, automatic equilibrium obtains if the resultant

of the perpendicular and horizontal components of the force of gravity falls in its prolongation into the intersection point of the two leg axes.

26. Gravitational side stresses are not neutralized and tend to shift the pelvis sideways whenever the position becomes asymmetrical.

27. In the sagittal plane the antero-posterior balance requires the antagonistic play between extensors and flexors.

28. The principal extensor is the gluteus maximus; it is under tension whenever the line of gravity falls in front of the hip joint. The muscle is an adductor in the frontal and an outward rotator in the transverse plane. It has a cross section of 66 cm.[2] and a tension strength of 238 kg.

29. Gluteus medius and minimus are secondary extensors with their posterior fibers. The hamstrings are extensors of the hip and flexors of the knee. Their extensory effect increases with the degree of hip flexion. Knee fixed, all hamstrings (except short head of biceps) extend the hip; hip fixed, all flex the knee; knee fixed, hamstrings by tilting pelvis backward become synergistic with extensors of knee; pelvis fixed, hamstrings by flexing the knee bring thigh forward and become synergistic with flexors of the hip.

30. The flexor group consists of the tensor, the ilio-psoas, the rectus femoris, sartorius and the anterior fibers of the gluteus medius and minimus.

31. In standing on one leg the maintenance of the equilibrium in the sagittal plane falls upon the concerted action of the flexors and extensors of the standing hip; the tension of these muscles increases and decreases as the line of gravity sways backward or forward of the center of the hip joint.

32. In standing on both legs balance is much easier because of the greater supporting surface. Furthermore, in case of paralysis of the hip muscles on one side the adjustment in respect to the line of gravity can be made by the muscles of the sound hip and standing on both legs is possible.

33. In the transverse plane the extensors develop a horizontal component which forces the hip joint forward; the flexors produce a horizontal component which gives the joint a backward thrust. The forward hip goes into outward, the backward into inward rotation. The rotators of the joint take care of the equilibrium. Outward rotators are: the piriformis, the obturator internus and externus, the quadratus femoris and the gemelli; inward rotation is practically always a secondary function: the tensor, the anterior portions of the gluteus medius and minimus, the adductor magnus and pectineus. The power ratio between outward and inward rotators is 146:54.

34. The muscular tension, when standing, produces by its stabilizing components a static pressure upon the hip joint which greatly increases the gravitational pressure of the superincumbent weight.

35. Just as gravity develops in the femur axial or pressure and bending stresses and shear, so also do the hip muscles develop axial as well as rotatory stresses by virtue of their stabilizing and rotatory components.

36. In the frontal plane the rotatory component of the adductors develops a bending stress which tends to increase the angle between neck and shaft in

the sense of coxa valga. The tension of the abductors tends to decrease this angle in the sense of coxa vara.

37. Actual everyday movement in the hip joint is seldom carried out in one plane alone. The usual directed movement is composite involving all 3 planes simultaneously. For instance a circling movement of the thigh is a combination of ab and adduction, inward and outward rotation, flexion and extension. In contrast to the simple movement in one plane which follows one meridian through a series of parallels or one parallel through a series of meridians, this composite movement describes on the globe a circle in which both meridians and parallels change consecutively. This movement is called circumpolar, in distinction from the single plane circumcentral movement.

38. The excursion ranges of the hip joint in the living are as follows: the greatest flexion range is 120° flexion and 20° extension in normal position, or in 5° abduction with neutral rotation. The smallest flexion extension range is in 35° abduction and 20° inward rotation. The maximum ab and adductory range is 74° in slight flexion and outward rotation. The maximum rotatory range is 90° in flexion.

39. Inversion of secondary functions of muscles with changing position is exemplified as follows: at right angle flexion all adductors change from flexors to extensors; the gluteus minimus, in normal position an abductor, becomes an adductor in strong flexion; the piriformis in normal position an outward rotator, becomes an inward rotator in strong flexion.

BIBLIOGRAPHY

1. BAEHR, F.: Beobachtungen über die Statischen Beziehungen des Beckens zur unteren Extremität. *Ztschr. f. Orth. Chir., V:*52, 1898.
   ———: Der Oberschenkelknochen als Statisches Problem. *Ztschr. f. Orth. Chir., VII:* 522, 1900.
2. DUCHENNE, G. B.: *Physiology of Motion.* Transl. E. B. Kaplan, Philadelphia, J. B. Lippincott, 1949.
3. FICK, R.: *Handbuch der Anatomie und Mechanik der Gelenke, Vol. III.* Jena, G. Fischer, 1911.
4. FUIKS, D.: Personal Communication.
5. GHILLINI, C., and CANOVAZZI, S.: Über die Statischen Verhältnisse des Oberschenkelknochens. *Ztschr. f. Orth. Chir., X:*III, 14, 1902.
6. GRÜNEWALD, J.: Über Beanspruchung der langen Röhrenknochen des Menschen. *Ztschr. f. Orth. Chir., 39:*27, 1919, and *39:*129, 1919.
7. IDEM: Über den Einfluss der Muskelarbeit auf die Form des Menschlichen Femur. *Ztschr. Orth. Chir., XXX.XXX:*551, 1912.
8. INMAN, V. T.: Functional Aspect of the Abductor Muscles of the Hip. *J. Bone & Joint Surg., XXIX.*3:607, July, 1947.
9. KINGSLEY, D. C.: A Study to Determine the Angle of Anteversion of the Neck of the Femur. *J. Bone & Joint Surg., XIX:* A, 3, 745, July 1948.
10. STRASSER, H.: *Lehrbuch der Muskel und Gelenkmechanik, Vol. III.* Berlin, J. Springer, 1917.
11. STRASSER, H., and GASSMANN, A.: Hilfsmittel und Normen zur Bestimmung und Veranschaulichung von Stellungen, Bewegungen und Kraftwirkungen am Kugelgelenk. *Mechle-Bonnet, Anat. Hefte, 2:*6/7, 1893.
12. WEBER, W., and E.: Über die Mechanik der Menschlichen Gehwerkzeuge nebst der Beschreibung eines Versuches über des Herausfallen des Schenkelkopfes aus der Pfanne im luftverdünnten Raume. *Ann. Phys. u. Chem., 40,* Leipzig, 1887.

Lecture XVII

THE PATHOMECHANICS OF THE STATIC
DISABILITIES OF HIP AND PELVIS

I. THE PHYSIOLOGICAL CURVES OF THE LONG BONES
OF THE LOWER EXTREMITY

T HE line of reference for the deflections of the long bones in the three cardinal planes of the body is the knee joint axis. In the upright standing position this axis is not only strictly horizontal but is also in the frontal plane; in other words the common axis of the knee joint forms an intersection line of a horizontal and a frontal plane with the patella looking straight forward.

The normal femur shows deflection both in the frontal and in the horizontal plane.

1. In the frontal plane the angulation between neck and shaft forms the cervico-femoral or the inclination angle which causes the shaft to become angulated against the sagittal plane and to assume a position of obliquity to it. In order that this deflection preserve the horizontality of the knee joint axis, there must be a compensatory tibio-femoral valgity.

2. The deflection of the femoral neck in the horizontal plane, i.e., back and forward, is called the angle of declination. Here also this rotation of the femoral neck must be so compensated by a converse torsion of the shaft that the position of the common knee joint axis retains its direction in the frontal plane.

Any interference with the position of the knee joint axis, either in the frontal or the horizontal plane, causes static disturbances; examples are: the inward rotation of the knee in the spastics where the knee axis no longer occupies the frontal plane; or the adduction and outward rotation contracture of the hip in the osteoarthritic where likewise the normal orientation of the knee joint axis in the frontal plane is lost.

A. PHYSIOLOGICAL VARIATION

1. The femur

A) THE FRONTAL PLANE, ANGLE OF INCLINATION

The normal angle of inclination of the adult femur is 128° which is a reduction from 150° of the newborn occurring under the effect of static functions. A line drawn through the base of the epiphyseal plate forms with the shaft of the femur an angle of 41.5° (Alsberg[2]) (Fig. 1a).

A decrease of this angle beyond the physiological limits is called a coxa vara (Fig. 1c); an increase is a coxa valga (Fig. 1b).

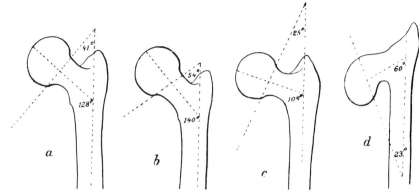

Fig. 1. Alsberg's angle. (Alsberg)

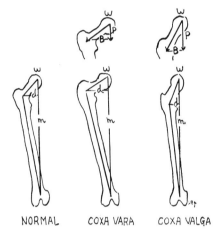

NORMAL   COXA VARA   COXA VALGA

Fig. 2. *d* in coxa vara greater than *d* in coxa valga; bending moment greater in coxa vara than in coxa valga.

m mechanical axis
W weight
B bending component of moment of force
P pressure component of moment of force

If the angle of inclination is reduced to 90° (Alsberg[2]), Alsberg's angle would be zero since the epiphyseal line and the shaft of the femur would then be parallel. The smallest angle of inclination observed in the coxa vara was 60° (Fig. 1d).

An increase of this angle beyond physiological variation means a coxa valga. It may reach a straight angle of 180° in which case the angle between the line through the epiphyseal base and the shaft of the femur would be 90° (Alsberg's angle).

The relation of the femoral shaft to the neck can also be described as an adduction of the neck against the shaft in coxa vara, and an abduction in coxa valga when compared with the normal inclination angle.

The superincumbent weight resting upon the head of the femur produces a bending moment which has its maximum at the junction between neck and shaft. In coxa vara this moment is increased because the distance of the mechanical axis of the femur which is drawn from the head to the mid knee is greater. In coxa valga, on the other hand, this distance is decreased. As the bending component of the weight stress increases, the longitudinal pressure component decreases and vice versa (Fig. 2).

B) THE HORIZONTAL PLANE; THE ANGLE OF DECLINATION

The angle of declination (ante- or retroversion or ante- and retrotorsion in the horizontal plane) fluctuates in the adult femur normally between 14-19° antetorsion (Guintini[9]) and it also varies with the different races. It is smallest in the white race and highest (24°) in the Fire Islanders (Fig. 3).

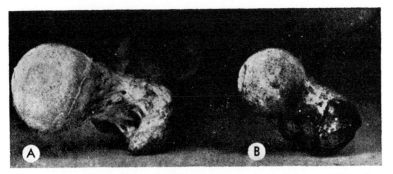

FIG. 3. Angle of declination fluctuates: Slight anteversion of the European (A); large angle of declination in the Fire Islander (B). (Giuntini)

Under certain pathological conditions this angle of declination is increased or diminished. It is increased, for instance, in congenital dislocation of the hip where it may reach 90°; a decrease of this angle or its reversal to a negative value is a pathological retroversion or retrotorsion as seen, for instance, in a congenital coxa vara. Here the neck is inward rotated against the femoral shaft, and the latter outward rotated against the neck.

2. The acetabulum

The socket also shows a double inclination, i.e., both in respect to the horizontal and the sagittal plane. It is inclined at an average angle of 60° to the horizontal and 30° to the sagittal plane (Badgley[3]).

This angle of antetorsion of the acetabulum corresponds to the declination angle of the femur (Giuntini[9]) and it also conforms with the longitudinal axis of the foot (Fig. 4).

While the acetabular position is more stable and does not change

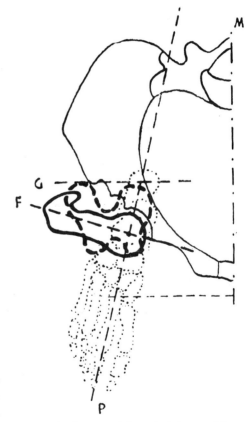

FIG. 4. The femoro-pelvic angle is in complete relationship with the angle between the femoral neck and the longitudinal axis of the foot (right angle). (Giuntini)

much with the moderate decreases or increases of the angles of inclination or declination in coxa vara or coxa valga, definite changes in the relation of the acetabulum to the orientation planes of the body are observed in aplasias such as congenital dislocation of the hip.

II. THE PATHOMECHANICS OF COXA VARA

A. SKELETAL

1. With the decrease of the cervico-femoral angle the weight bearing area of the head of the femur moves upward to the upper quadrant of the head. Whether the coxa vara is of congenital nature or whether it is caused by a slipped epiphysis (Coxa vara adolescentium) or is of rachitic origin, the static effect of the deformity is the same, namely an increase of the bending moment under which the bone yields at the convexity of the bend. With the shift of the weight bearing area to the upper outer quadrant of the femoral head, the latter becomes more deformed.

As the increase of the bending stress is commensurate with the increase of the angle between the axis of the shaft and the mechanical axis of the femur, so is also the increase of the shear except that shear has its maximum at the head while the bending stress is maximal at the junction between neck and shaft.

2. With the decrease of the angle of declination or its reversal to retrotorsion the shaft becomes outward rotated against the neck and the weight bearing pressure is shifted to the anterior aspect of the upper outer quadrant of the head.

3. With the decrease of the cervico-femoral angle, the greater trochanter moves upward so that it appears above Nelaton's line even if the hip is adducted. One must therefore look for a restriction of the abduction because of the impingement of the greater trochanter against the os ilii. Furthermore, because the femoral shaft is relatively outward rotated against the neck, inward rotation of the femur becomes limited.

B. THE LIGAMENTOUS RESTRICTIONS

The three principal reinforcing ligaments of the normal hip joint are the ilio-femoral, the pubo-femoral and the ischio-femoral.

The ilio-femoral ligaments become tight in extension. Retrotorsion of the neck increases the tension so that complete extension of the hip may not be possible. The result is a flexion position of the hip which is compensated by an increased tilt of the pelvis with lumbar lordosis. The erect position of the patient is therefore only apparent. The pubo-femoral ligament normally checks abduction. Both the tension of the pubo-femoral ligament and the impingement of the trochanter limit abduction. The ischio-femoral ligament which in the normal hip restricts outward rotation is always under tension in coxa vara.

According to Manz[16] the tension of the capsule as well as of the pubo-femoral ligament restricts abduction. All ligaments which ordinarily restrict extension are, in addition, stretched by the anterior curving of the retroverted femoral neck.

C. THE MUSCULAR RESTRICTIONS

The muscular inhibitions are largely due to the high position of the trochanter and to the outward rotation of the shaft.

The high position makes for insufficiency of the abductors; gluteus medius and minimus are relaxed and the Trendelenburg sign is positive. The outward rotation of the shaft causes the outward rotators, especially the piriformis, obturator and quadratus femoris to become contracted (Alsberg[2]) (Fig. 5). The decrease of the abductory range becomes more marked as the hip goes into flexion (Fig. 6).

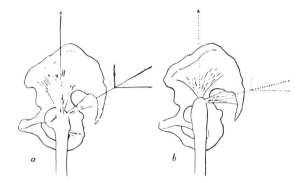

FIG. 5. *a*. Outward rotation of shaft causes outward rotators to become contracted. *b*. High position of trochanter causes insufficiency of abductors. (Alsberg)

While the passive abduction is restricted mainly by skeletal blocking, in active abduction the failure of the normal abductors plays the principal role in coxa vara deformity.

The stress received by the upper end of the femur is a combination of the gravitational load and muscle tension. So far as the latter is concerned it has

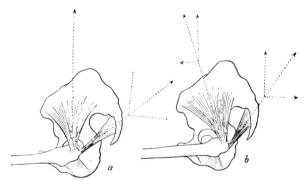

FIG. 6. In flexion the abductory range decreases in coxa vara because the adductors shorten to a greater extent. *a*. normal. *b*. coxa vara. (Alsberg)

been shown by Inman[11] that the amount of muscle tension required to maintain equilibrium many times exceeds the weight stress, while the full gravitational load, when standing on one leg, does not exceed the total weight of the body minus that of the standing leg.

III. THE PATHOMECHANICS OF THE COXA VALGA

When coxa vara is a deformity in which the bending stresses produced by the weight load and by muscle tension exceed the pressure, the coxa vara is a deformity in which, due to the increase of the cervico-femoral angle, the bending stresses are diminished in favor of pressure.

Coxa valga is a collective term as is the term coxa vara. There are many different underlying causes. Albert[1] mentions paralysis, knee ankylosis, rachitis, osteomalacia and dislocation as some of the causes and Lange[13] distinguishes a congenital, a traumatic, and a muscular coxa valga as well as one following elimination of weight bearing. Kumaris[12] recognizes a compensation coxa valga in poliomyelitis, a rachitic coxa valga in which the deformity is primarily

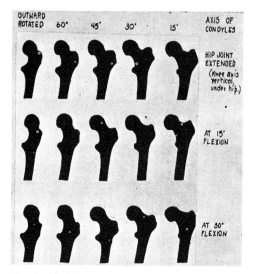

FIG. 7. Variability of the angle of the neck of the femur by rotation in extension and flexion. (Storck)

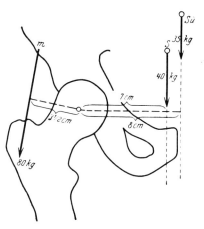

FIG. 8. Coxa valga, showing relationship of femoral head, greater and lesser trochanters, and fovea capitis. (Storck)

diaphyseal, a primary congenital coxa valga, an early embryonal deformity and a coxa valga luxans in congenital subluxation of the hip.

Whatever the cause, from the mechanical point of view the important facts are:

(1) The incresed translatory pressure and the decreased shear and bending stress caused by the straightening of the neck-shaft angle;

(2) The shift of the weight bearing area of the head to the upper inner quadrant.

The true angle is difficult to determine from the x-ray because any rotation of the shaft may simulate a straightening of the angle up to 180°. The smallest angle which can be demonstrated roentgenographically by rotating the femur is the true cervico-femoral angle both in coxa valga and coxa vara (Storck[21]) (Fig. 7).

A. THE SKELETAL FACTORS

Storck[21] made an exhaustive study of the mechanical situation of coxa valga, investigating the relation between upper pole of the femoral head, the greater and lesser trochanter and the fovea capitis (Fig. 8). He found that in coxa valga the distance between the upper pole of the head and the greater trochanter is diminished. So also is the diameter of the head and neck (Fig. 9).

Brandes[5] reports an interesting observation of a patient on whom the greater trochanter had been removed for tuberculosis and who develops coxa valga. Attributing this to the loss of tone of the abductors he then suggested the resection of the greater trochanter and the elimination of the tension of the pelvi-trochanteric muscles for the treatment of coxa vara.

In contrast to coxa vara there are no skeletal checks in coxa valga to restrict

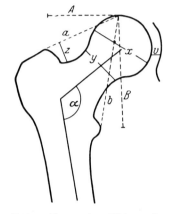

Fig. 10. Coxa valga.
P pressure component
B bending component
ma mechanical axis
aan anatomical axis neck
aas anatomical axis shaft
W weight
Distance between mechanical axis and cervico-femoral junction diminished; pressure component increased and bending component decreased.

Fig. 9. Coxa valga. Distance between trochanter and head shortened. Diameter of head and neck decreased. (Storck)

the ranges of motion. However the relative lengthening of the limb causes a pelvic obliquity which places the hip on the affected side in adduction. The weight bearing contact of the head is now the upper inner quadrant and the joint is subluxated.

The steepness of the neck diminishes the horizontal component of the reaction from the floor in favor of the perpendicular weight transmitting component. In the same sense the bending component of the excentrically loaded superincumbent weight is diminished in favor of the weight pressure component as the distance between the mechanical axis of the femur and the cervico-femoral junction is diminished (Fig. 10).

B. THE MUSCLE MECHANICAL SITUATION IN COXA VALGA

1. The abductors

Irrespective of the fact that the distance between greater trochanter and pelvis is diminished, the situation is unfavorable for the abductors because the straightening of the neck shortens their lever arm. Storck[21] studied the effect of the abductors as well as of the outward rotators on the mechanogenesis of coxa valga. He finds that in coxa valga when standing on one leg the tension of the abductors which holds the equilibrium against the gravitational force is much increased over the normal because of the lessened distance of the axis of the muscles from the center of motion. This upward pressure of the abductors unites with the superincumbent load in the production of a bending stress. One would expect that this tends to decrease the cervico-femoral angle in the sense of a coxa vara, but this stress is absorbed by the resistance of the bone and during growth it is overcome by the turgor of the epiphyseal plate (Schulthess).

What happens now if the pelvi-trochanteric muscles are eliminated as for instance in infantile paralysis? The result is a deflection of growth in the direction of a coxa valga. Storck's[21] x-ray observation on the changes of the femoral epiphysis seem to bear out this point. While the femoral epiphysis has normally its greatest height opposite the center of the joint, in coxa valga this summit is displaced and the head assumes a shape very much like that in congenital subluxation. This change in direction of growth is due to the greater growth

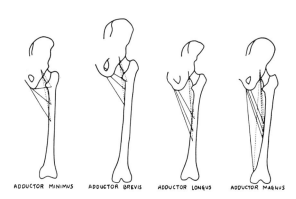

ADDUCTOR MINIMUS   ADDUCTOR BREVIS   ADDUCTOR LONGUS   ADDUCTOR MAGNUS

FIG. 11. Effect of adductor spasm on femoral neck producing coxa valga. (Storck)

activity of the medial portion of the epiphyseal plate. It is in keeping with the concept of Roux that the greater pressure (upper inner quadrant) produces the more active growth which is true, however, only within certain physiological limits.

2. The outward rotators

Storck[21] points out that not only the abductors but also the external rotators of the hip, piriformis, gemelli, obturators and quadratus femoris develop perpendicular components acting upon the neck of the femur. In case of paresis or paralysis of these muscles the neck assumes a more straight position and the lack of the external rotators has a similar effect on the growth as does that of the abductors. The hip goes into inward rotation and with the loss of the abductors the greater pressure falls upon the medial half of the epiphyseal plate. The result is increased growth on the medial side of the plate and a straightening of the angle between neck and shaft.

3. The adductors

In spastic paralysis the spasm of the adductors may develop a coxa valga, due to the fact that the tension of the adductor group is no longer being equalized by an adequate tension of the gluteals (Storck[21]) (Fig. 11).

IV. THE PATHOMECHANICS OF THE DYSPLASIA OF THE HIP JOINT

In this condition both the upper end of the femur and the acetabular cavity show developmental changes. From the point of mechanogenesis the most significant of these is the anteversion of the head and neck.

As early as 1901 Friedlander,[7] in discussing the abduction position of the femur in the early embyro, pointed out that the following inward rotation of the thigh produces the physiological anteversion and the subsequent adduction the inclination of the femoral neck. The normal contact between head and socket exists only so long as this inward rotation of the shaft in respect to the neck is within physiological limits.

According to Badgley[3] the dysplasia of the joint is the result of failure of the normal rotation of the femur. This causes a secondary anteversion (antetorsion) of the head and neck and a flattening of the socket. The anteversion precedes the following dislocation. In the early embryo the femur stands in abduction and outward rotation which levels the head into the developing acetabulum. If the limb remains in outward rotation for some time the head is displaced anteriorly. Then the following belated inward rotation of the femoral shaft against the neck causes an increased outward rotation of the latter. Normally the rotatory change in position between early embryo and birth amounts to 90° which is the difference between the early outward rotation and a definite inward rotation of the femur. In congenital dysplasia the head is already anterior and then as the femur rotates inward, the head, instead of limiting its anteversion to 35° at birth, now becomes excessively anteverted and this abnormal degree is a definite factor in the production of a dysplasic acetabulum. LeDamany[14] considers an anteversion which is more than 60° as definitely pathological and a dislocation may then occur as the hip goes from the flexed to the extended position. Badgely[3] believes that the dysplasia of the hip is entirely produced by this abnormal anteversion of the neck and that the lateral displacement of the capital epiphysis which has been emphasized by Putti[20] as an early sign of congenital dislocation is evidence of anteversion of the neck. At any rate the anteversion constitutes a factor unfavorable for the retention of the head in the cotyloid cavity (LeDamany[14]). It is obvious that an extreme anteversion will strain the head against the anterior and superior portion of the capsule with the posterior inferior border of the acetabulum acting as hypomochlion.

In accord with these deformations of the upper femur as a predisposing factor of the congenital dislocation is the hypoplasia of the acetabulum. The acetabulum in the newborn looks forward, outward and downward, its outlet forming an angle of 60° with the horizontal and 30° with the sagittal plane. In congenital dislocation the plane of outlet of the acetabulum is more sagittally directed (LeDamany[15]). The posterior line of the neck abuts against the posterior rim of the acetabulum. The gap thereby produced anteriorly is filled by the ligamentum teres. The head exerts abnormal pressure against the upper acetabular rim flattening it out. The acetabulum becomes elongated and oval and thus facilitates the subluxation. Murk Jansen[18] insisted particularly upon the shallowness of the acetabulum as being the cause of all congenital dysplasias of the hip. Faber[6] also saw in the dysplasia of the acetabulum the primary cause though many now (Guintini[9]) believe it to be a condition secondary to the malformation of the femoral head. Anteversion was found by Massie and Howorth[17] in 80% of the cases of congenital dislocation of the hip treated by open reduction.

From the mechanical point of view an interesting observation on the result of the shelving operation is one made by Ponseti.[19] The distance of the center of the acetabulum from the midline (Y coordinate) varies according to Braune and Fischer between 5.5 cm. in children and 9 cm. in the adult, with the adult average being 8.5 cm. Ponseti finds that if the Y coordinate of the affected hip exceeds that of the normal for the particular age no more than 1.2 cm., the results of the shelving were good; beyond this difference they were poor and the Trendelenburg sign was positive.

V. THE PATHOMECHANICS OF THE FIXED PELVIC OBLIQUITY

In so far as this obliquity is due to pathological conditions of the spine it has been discussed with the pathomechanics of scoliosis. We are concerned here only with such obliquities as are produced by pathological conditions involving the hip joint or its surrounding musculature.

A. THE PELVIC OBLIQUITY IN OSTEOARTHRITIS OF THE HIP

1. In stance

The effect of pathological changes is the fixation of the hip in ab or adduction, flexion and outward rotation. In the endeavor to attain parallelism of the lower extremities the pelvis is forced to assume the following asymmetrical positions.

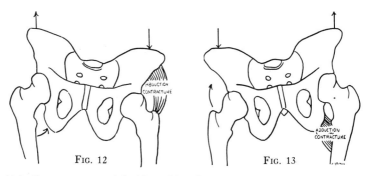

FIG. 12. Abduction contracture of the hip and its effect on the pelvis and opposite hip; lowers pelvis and adducts opposite hip.

FIG. 13. Adduction contracture of the hip and its effect on the pelvis and opposite hip; raises pelvis and abducts opposite hip.

a) The abduction contracture causes a lowering of the pelvis on the affected side which raises the unaffected half and places the contralateral hip in a position of adduction (Fig. 12).

b) The adduction contracture raises the affected side of the pelvis and places the contralateral hip in abduction (Fig. 13).

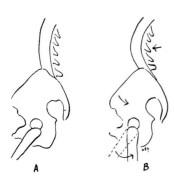

A   B

FIG. 14. A. Hip fixed in flexion. B. Extension of hip fixed in flexion causes increase in lumbar lordosis and pelvic tilt.

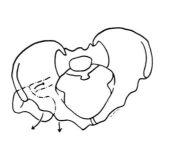

FIG. 15. Forward rotation of pelvis due to external rotation contracture of the hip.

FIG. 16. Positive Trendelenburg. Pelvis drops to unsupported side.

c) The flexion contracture forces the lumbar spine in a position of a compensatory lordosis (Fig. 14).

d) Outward rotation causes the affected side of the pelvis to rotate forward (Fig. 15).

2. In walking

More detailed description will be given in the lecture on pathological gait. Here we mention only the following essential points:

a) The abduction contracture lengthens the leg and forces it to carry out the forward swing by a circumductory movement of the leg about the sound side as a pivot.

b) The adduction contracture raises the side of the pelvis thereby shortening the limb and forces it into equinus position during the supporting period and causes it to pivot about the sound hip during the swing.

c) Flexion contracture produces alternating lordosing of the lumbar spine.

d) Outward rotation contracture forces for and backward rotation of the pelvis in the other hip joint as pivot.

B. THE DYNAMICS OF PELVIC OBLIQUITIES IN COXA VARA AND CONGENITAL DISLOCATION

The difficulties are largely due to insufficiency of the pelvi-trochanteric muscles.

1. Standing

The Trendelenburg sign on the affected side is positive. The body becomes more inclined to the standing side so as to throw the line of gravity lateral to the hip joint center and to make gravity an auxiliary force in abduction (Fig. 16).

2. In the gait

The gait is the characteristic waddle gait, in bilateral cases. In the unilateral case side swaying occurs toward the affected side using gravity as an auxiliary abductor for the standing period.

VI. SUMMARY

1. The knee joint axis is the line of reference for the deflections of the long bones of the lower extremity.

2. In the frontal plane the angulation between femoral neck and shaft caused by the adduction of the latter forms the cervico-femoral angle or angle of inclination. The compensation for the adduction of the femur is the femoro-tibial valgity.

3. In the horizontal plane the angulation between neck and shaft caused by inward rotation of the latter forms the angle of declination.

4. Any interference with the orientation of the knee joint axis in the frontal and horizontal planes causes static disturbances.

5. The angle of inclination in the adult is 128°. The smallest observed in coxa vara was 60. The superincumbent weight resting on the head of the femur produces a bending moment which is increased in coxa vara and decreased in coxa valga.

6. The angle of declination varies normally from 14-19° antetorsion. Under certain pathological conditions (coxa vara) it is diminished or even reversed (retrotorsion).

7. The acetabulum is inclined 60° to the horizontal and 30° to the sagittal plane; its antetorsion position corresponds to the declination angle of the femur.

8. Coxa vara represents in the frontal plane an excentrically loaded column, distally anchored with a pathologically increased lateral angulation which tends to become more accentuated under body weight and muscular tension.

9. In the transverse plane the neck assumes a forward convexity which forces the head into a backward and downward displacement. This backward pressing of the neck is accentuated by the ilio-femoral ligament. In addition the acetabulum changes its position to an oblique sagitto-frontal direction which causes the head to be forced further backward independently of the backward twist of the neck (Grunewald[8]).

10. In coxa valga the bending stresses from the superincumbent load are diminished in favor of the axial pressure. The weight bearing area of the head shifts to the upper inner quadrant.

11. The distance between the upper pole of the head and the greater trochanter is diminished as is also the diameter of head and neck.

12. In contrast to coxa vara there are no skeletal checks to abduction in coxa valga. Because of the steepness of the neck, which approaches the axis of the shaft, the bending component of the gravitational stress is much diminished; when standing on one leg the tension of the abductors must be increased because the lever arm, i.e., the distance of the muscle axis from the center of motion is decreased.

13. If the pelvi-trochanteric muscles are eliminated the result is a deflection of growth in direction of coxa valga.

14. Not only the abductors, but the external rotators also develop bending components which act upon the neck of the femur. In paralysis of these muscles the neck assumes a straighter position; the cervico-femoral angle is increased.

15. Spasm of the adductors may develop coxa valga, the adductor tension prevailing over that of the abductors.

16. Dysplasia of the hip joint is the result of failure of the normal rotation of the femur which causes abnormal antetorsion of the head and flattening of the socket. The normal inward rotation of the femoral shaft is delayed and the neck rotates outward (antetorsion) while the shaft is still in embryonal position of outward rotation. Then the belated inward rotation of the shaft greatly exaggerates the antetorsion of the neck. Anteversion (antetorsion) of more than 60° is definitely pathological. The normal at birth is 26-35°.

17. In shelving operations for dislocation or subluxation of the hip, the Trendelenburg sign remains positive and the operative outlook is poor according to Ponseti, if the distance of the head from the midline (Y coordinate) exceeds the normal by more than 1.2 cm.

18. Fixation of the hip joint by contracture or ankylosis forces the pelvis in asymmetrical position when standing or walking is resumed: abduction contracture causes a lowering of the pelvis on the affected side; adduction raises the pelvis; flexion contracture forces the lumbar spine into compensatory lordosis; outward rotation contracture causes the affected side of the pelvis to rotate forward.

19. In walking the adduction contracture forces the affected limb into equinus position. During the swing it pivots around the sound hip; the flexion contracture produces an alternating lordosing of the lumbar spine; in outward rotation contracture the pelvis is swung forward pivoting about the sound hip joint.

20. In coxa vara and congenital dislocation of the hip the difficulties are largely due to the insufficiency of the pelvi-trochanteric muscles. In standing on one leg the Trendelenburg sign remains positive on the affected side and the body tilts to the affected side more than normally. This tilt to the affected side which engages gravity as an auxiliary abductory force becomes more marked in the gait. In bilateral cases the gait becomes a symmetrical waddle.

BIBLIOGRAPHY

1. ALBERT, E.: *Zur Lehre der sogenannten Coxa vara and Coxa valga.* Wien. Alfred Holder, 1899.

2. ALSBERG, A.: Anatomische und Klinische Betrachtungen über Coxa vara. *Ztschr. f. Orthop. Chir., VI:*106, 1899.

3. BADGLEY, E. C.: Correlation of Clinical and Anatomical Facts Leading to a Conception of the Etiology of Congenital Hip Dysplasia. *J. Bone & Joint Surg., XLI:*503, July 1943.

4. ———: Etiology of Congenital Dislocation of the Hip. *J. Bone & Joint Surg., 31:2,* 1949.

5. BRANDES, M.: Vorschlag zu einer physiologischen Behandlung der Coxa vara. *Arch. Orthop. u. Unfallchir., XXII:*409, 1924.

6. FABER, A.: *Untersuchungen über die Aetiologie und Pathogenese der angeborenen Hüftverrenkung.* Leipzig, E. Thieme, 1938.

7. FRIEDLANDER, F. v.: "Über die Entstehung der angeborenen Hüftverrenkung," Ztschr. f. Orthop. Chir. IX, 515, 1901.

8. GRÜNEWALD, J.: *Ztschr. f. Orthop. Chir., XXXVIII. XIII:*470, 1918.

9. GUINTINI, L.: *Etiopatogenesi della displasia congenita dell'anca.* Bologna, L. Cappelli, 1951.

10. HART, V. L.: Primary genetic dysplasia of the Hip with and without Classical Dislocation. *J. Bone & Joint Surg., XXIV:3,* 1942.

11. INMAN, V. T.: Functional Aspect of the Abductor Muscles of the Hip. *J. Bone & Joint Surg., XXIX:3,* 607.

12. KUMARIS, J.: Ein Beitrag zur Lehre der Coxa Valga. *Arch. Klin. Chir., 87:*625, 1908.

13. LANGE, F.: *Ztschr. f. Orthop. Chir.* Vol. 41.

14. LeDAMANY, P.: *La luxation congenitale de la hanche.* Paris, Flammarion, 1923.

15. ———: Die angeborene Hüftverrenkung; ihr Mechanismus, ihre Ursachen; ihre Anthropologische Bedeutung. *Ztschr. f. Orthop. Chir., XXI:*129, 1908.

16. MANZ: Die Ursachen der Statischen Schenkelhalsverbiegung. *Beitr. Klin. Chir., Vol. 28.*

17. MASSIE, K. W., and HOWORTH, H. B.: Congenital Dislocation of the Hip, *J. Bone & Joint Surg. XXXII:*A.3, 1950 and XXXIII. A.1, 1951.

18. MURK, JANSEN M.: *Platte Hüftpfaune und ihre Folgen.* Stuttgart, F. Enke, 1925.

19. PONSETI, I.: Pathomechanics of the Hip after Shelf Operation. *J. Bone & Joint Surg., XXVIII:2,* 229, April 1946.

20. PUTTI, V.: *Anatomia della lussazione congenita della anca.* Bologna, L. Cappelli, 1935.

21. STORCK, H.: Coxa valga. *Arch. Orthop. u. Unfallschir., XXXII:2,* 133, 1932.

Lecture XVIII

THE PATHOMECHANICS OF THE
PARALYTIC HIP JOINT

INTRODUCTION

IN THE upright position it is the function of the hip joint to stabilize the pelvis and the rest of the body against the stresses of the superincumbent weight. For this purpose the joint must be equipped as an amphiarthrosis with 3 degrees of freedom of motion which means movement in three cardinal planes (sagittal, frontal and horizontal) as well as in all intermediate planes.

Paralysis of the hip muscles produces certain asymmetrical positions; a unilateral paralysis of the abductors causes asymmetry in the frontal plane by adduction contracture; a paralysis of the gluteal associated with flexion contracture of the hip produces a lordosis in the sagittal plane.

All these asymmetrical positions which under pathological conditions are forced upon the hip joint can, under normal conditions, at least to a great extent, be assumed voluntarily and are, in fact, so assumed innumerable times during the activities of everyday life.

Thus it becomes necessary to examine first the conditions of equilibrium in asymmetrical positions assumed voluntarily before the pathological situations are taken into consideration. As a preliminary step we proceed to the analysis of the normal stresses as they project themselves successively into the three orientation planes, the frontal, the sagittal and the horizontal.

I. THE NORMAL HIP JOINT
A. THE FRONTAL PLANE

1. Symmetrical

a) *With the legs parallel* the reaction from the floor develops a horizontal component due to the obliquity of the femoral neck through which the gravitational stresses are transmitted (Fig. 1). This is completely neutralized by the opposite side.

b) *With the legs symmetrically apart* the gravitational reaction also develops a horizontal component from the floor; this produces at the hip joint an increase of the normal medially directed component, or in other words, an increased pressure against the acetabulum (Fig. 2).

c) *With the legs crossed symmetrically,* the horizontal components developed at the floor are in opposite direction from those with the legs abducted. Then they were directed medially while this time the horizontal component is directed laterally, i.e., the pressure against the acetabulum is decreased (Fig. 3).

Theoretically in these three situations of perfectly symmetrical position of

311

the legs, no muscular action should be necessary to maintain the equilibrium because the gravitational reaction from the floor develops on each side components which are perfectly equal and opposite and therefore mutually neutralize each other.

Yet as one tests the adductor and abductor muscles while standing erect in these three positions, legs parallel, apart or crossed, one notices that these muscle groups contract rhythmically. This is because the pelvis constantly sways in the frontal plane to and from the midline, and there is a constant adjustment of the body to the position of symmetry. One also notices that this regulatory function of the muscles is least active when the legs are symmetri-

Fig. 1. Horizontal components directed toward midline when the legs are parallel.

Fig. 2. Horizontal components directed toward midline greatly increased by strong abduction.

Fig. 3. Horizontal components directed away from midline due to strong adduction.

Fig. 4. Horizontal component directed away from midline in adducted hip, and toward midline in abducted hip, both in same direction.

cally abducted; it is somewhat more noticeable in straight upright position and it is most active when the legs are crossed. This shows that in spite of symmetrical position in which the horizontal components of the gravitational reactions are equal and opposite and in which no muscular effort should therefore be needed to maintain equilibrium in the hip joint in the frontal plane, nevertheless the oscillations of the body in this plane are unavoidable even in absolute rigid upright position, and they require the regulatory action of both the adductors and the abductors; furthermore it shows that of the three positions mentioned, that with symmetrical abduction of the limb is the most stable and calls for the least regulatory effort.

2. Frontal plane; asymmetrical position (Strasser[10])

We next consider the situation where the two extremities are not symmetrically poised and where the pelvis is shifted in relation to the trunk and the extremities. The hip joint at the side of the shift is in adduction and the other in abduction (Fig. 4).

The weight being shifted to the adducted leg, the vertical component of the gravitational reaction is greater on this side. At the abducted side the total floor reaction is smaller but its horizontal component is directed toward the adducted side. The pelvis is pushed toward adduction. The abducted side of the pelvis is lowered, and the adducted is elevated. In the abducted hip, the abductors shift the pelvis to the opposite, the adductors shift it to the same side which means that the action of the abductor decreases and the adductor action increases the shift to their respective sides (Fig. 5). In plain words the abductor of one and the adductor of the other side act as synergists in the shifting of the pelvis. To restore the position of symmetry one must therefore strengthen the action of the abductor at the adducted or the action of the adductor at the abducted side.

B. SAGITTAL PLANE

1. Symmetrical

a) The center of gravity falls into the hip line. In this case no rotatory moments are developed and no muscular effort is necessary to obtain equilib-

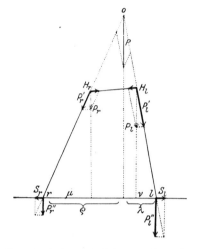

FIG. 5. Action of abductors decreases, action of adductors increases pelvic shift to their respective sides. (Strasser)

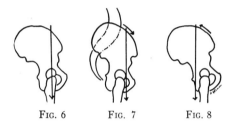

FIG. 6   FIG. 7   FIG. 8

FIG. 6. Line of gravity falls into hip line.

FIG. 7. Line of gravity in front of hip line. Pelvis tilts forward, lumbar spine in lordosis.

FIG. 8. Line of gravity falls behind hip line. Pelvis rotates backward.

rium. However, just as is the case in the frontal plane the line of gravity must be constantly adjusted to this purpose; therefore there is a constant interplay between extensors and flexors according to whether the line sways forward or backward (Fig. 6).

b) In ordinary stance the line of gravity is somewhat in front of the hip line. Consequently the extensors (gluteus maximum and hamstrings) are under tension. The extensory component of their tension must neutralize a rotatory moment equal to the gravitational stress times its distance from the center of the hip joint.

In order to produce the proper tension in the extensor group the pelvis is tilted forward and, holding the line of gravity as closely as possible to the center of the hip joint, the lumbar spine goes into lordosis (Fig. 7).

c) In the case of the line of gravity falling behind the hip joint line as in relaxed posture, the flexors of the hip and especially the ilio-femoral ligament must come under tension. This is facilitated by the backward rotation of the pelvis (Fig. 8).

2. Asymmetrical stance

We assume now that the lower extremities are not vertical under the pelvis but are oblique, one posterior and the other anterior oblique. It is now obvious that the posterior extremity will develop a forward and the anterior a backward horizontal component; the equilibrium requires the respective horizontal components to be equal and opposite. Again theoretically no muscle force should be required if the line of gravity bisects the angle between the two extremities and goes through the hip joint axis (Fig. 9). This is, however, only an imaginary situation and again the adjustment of the line of gravity is accom-

Fig. 9

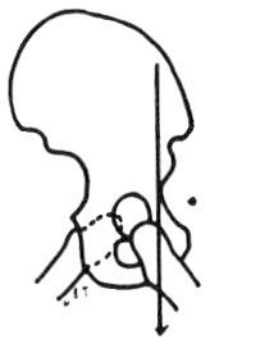

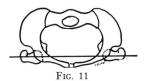

Fig. 11

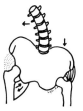

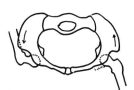

Fig. 10

Fig. 12

Fig. 13. Abductor paralysis. Pelvis drops on unsupported side, trunk inclines to side of paralysis.

Fig. 9. Line of gravity goes through common hip joint axis, bisects angle between legs.

Fig. 10. Line of gravity in front of common hip joint axis.

Fig. 11. No rotary moment with hips in frontal plane and at right angles to horizontal.

Fig. 12. Extremities oblique, developing rotatory moment. Direction the same for the posterior oblique extremity on one side and the anterior oblique on the other.

plished by the opposing muscle groups, the flexors of the posterior limb transmitting a forward and the extensors of the anterior limb, a backward thrust by which the equilibrium between the two hip joints is regulated (Fig 10).

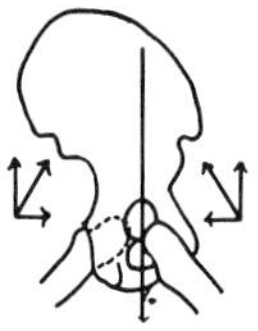

C. THE HORIZONTAL PLANE

1. Symmetrical

As long as the extremities are vertical in the frontal plane and at right angles to the horizontal plane, no rotatory moment develops in this plane. This also applies to the case where the legs are not parallel but are ab or adducted so long as the plane laid through both legs is frontal, i.e., perpendicular in both the horizontal and the sagittal plane (Fig. 11).

2. Asymmetrical

This situation changes when the extremities are oblique to the horizontal plane and to the frontal plane. Being oblique to these two, they are also oblique to the third plane which is the sagittal (Fig. 12).

Again the obliquity itself may be symmetrical, i.e., the angle which the axis of the limb makes with the horizontal plane may be equal on both sides. In this case each oblique limb produces in the horizontal plane a rotatory moment. These moments are equal if the obliquity of the limb and therefore the reaction to the limbs from the floor are equal. As long as one limb is as obliquely backward as the other equally oblique forward the horizontal components are equal; but they are not opposite because they rotate the pelvis in the same direction. Consequently, if the pelvis is to be held at rest, rotation in the horizontal plane must be neutralized by muscle action: by the inward rotators of the forward and the outward rotators of the backward leg.

If for some reason the freedom of motion in one hip joint is impaired, the rotation of the pelvis in a horizontal plane follows the same direction except that the excursions of the pelvic half are not identical, those on the free side being larger. That means that the pelvis does not rotate about an axis lying in the midline but the axis lying close to the affected side (Strasser[10]).

II. THE PELVIC MOVEMENT IN PARALYSIS

We cannot assume that paralysis of any muscle will cause a shift in the position of the pelvis in one plane only; almost all muscles have some obliquity to all three planes and therefore affect the relationship of the femur to the pelvis in all three. But to make this presentation clear we must again analyze the pathological disequilibrium successively in all three planes.

A. THE FRONTAL PLANE

1. Paralysis of the abductors

A) UNILATERAL

When standing on one leg or in walking during the period of single support, the function of the abductors is to stabilize the pelvis and to hold the standing leg in abduction.

Paralysis of the abductors causes the pelvis to drop on the unsupported side. This again necessitates an inclination of the trunk to the side of the paralysis. The pelvis is shifted to the affected side by the unopposed adductors (Fig. 13).

B) BILATERAL

The same pelvic drop on the unsupported side and the same shift to the supported side is observed. The homolateral tilt of the body repeats itself with every single step.

To check the unopposed action of the adductors, the trunk is inclined to the affected side which causes the line of gravity to fall into the center of the standing hip joint so that no rotatory moment is developed in the frontal plane, or to fall lateral to the hip joint center so that gravity also acts as an abductory force. By controlling the center and line of gravity the oscillations of the trunk adjust the abductory requirements for equilibrium. The result is the well known uni or bilateral gluteus medius gait.

2. The paralysis of the adductors

A) UNILATERAL

It is unusual without accompanying paralysis of the abductors. In the latter case the hip is flail, unstable in the frontal plane, and the equilibrium cannot easily be regulated. The adjustment of pelvic shift is up to the contralateral sound ab- and adductors. Unilateral abduction contracture in adductor paralysis is unusual in anterior poliomyelitis.

B) BILATERAL

The same applies to the bilateral adductor paralysis. Since the sound abductors can to an extent direct the line of gravity to fall into the center of the standing hip joint, there is no great difficulty in keeping balance when standing on both legs.

B. THE SAGITTAL PLANE

1. The extensors

With the back and forward oscillations of the trunk, gravity comes to the aid of the paralyzed extensors so readily that there is little difficulty in holding a standing balance. The characteristic gluteus maximus posture develops. The pelvis is rotated forward by the unopposed flexors of the hip and therefore the lumbar lordosis is increased. The line of gravity now passes behind the hip joint line and gravity becomes an extensory force (Fig. 14).

2. The flexors

These are the ilio psoas, the tensor fasciae, the anterior portions of the gluteus medius and minimus and the sartorius. One must not overlook the flexory component of the adductor longus and magnus which persist up to a flexion of 70° and 50° respectively, at which point it changes into an extensor.

Isolated paralysis of the flexors causes a backward rotation of the pelvis and it forces the trunk to lean forward so that the line of gravity passing in front of the hip joint acts in a flexory direction.

C. THE HORIZONTAL PLANE

1. Inward rotators

If the inward rotators (anterior gluteals, tensor and adductor magnus) are paralyzed, the hip is forced in outward rotation. The pelvis is rotated forward on the affected side and backward on the sound side. The inward rotator of the sound side thus becomes synergistic to the outward rotators of the paralytic side (Fig. 15). The result is that the pelvis is kept in oblique position with the anterior spine on the affected side pointing forward. The situation is similar to that in malum coxae senilis where the hip is fixed in outward rotation. When the legs are placed one obliquely forward and the other obliquely backward, the backward thrust which the forward leg should exert upon the pelvis is locked, so to speak, by the contracture of the outward rotators; the forward thrust of the backward leg can materialize only if the paralyzed forward leg pivots on the heel as the well leg deploys for the take-off and then swings forward.

In bilateral outward contracture of the hips due to paralysis of the inward rotators the pelvis can carry out no independent movement in the horizontal

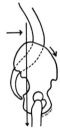

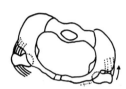

FIG. 14. Paralysis of extensors of trunk. Pelvis rotates forward and lumbar lordosis increases. Line of gravity falls behind hip joint.

FIG. 15. Paralysis of internal rotators of hip with affected side in external rotation and pelvis on same side pointing forward.

plane and the patient is obliged to pivot on the respective heels of the standing leg, similar to what occurs in bilateral malum coxae senilis.

2. The outward rotators

The paralysis of the outward rotators (principally the pelvitrochanteric muscles) should place the hip in strong inward rotation and cause the pelvis to rotate backward on the affected side. As long as the inward rotators are not contracted, which is usually the case, there is little difficulty in the for and backward oscillations of the pelvis. If the inward rotators are contracted (especially the tensor fasciae), then the pelvis on the affected side must be carried backward while the other side swings forward.

Here is the difference between this situation and the one where the internal rotators are paralyzed and the external rotators are contracted. Normally the pelvis changes from *forward to backward* rotation, i.e., the hip joint from outward to inward rotation. However contracture of the external rotators prevents this change which should occur during the standing phase. The change from *back to forward* rotation of the pelvis occurs in the swinging phase and here the hip joint goes from inward to outward rotation. Contracture of the inward rotators prevents it. This change does not occur in the standing, but in the swinging phase; therefore the suppression of this change has no effect on the static equilibrium of the body.

Bilateral contracture of the internal rotators is much more common in the spastic than in the paralytic. As in the bilateral external rotator contracture, the pelvis remains in symmetrical position in standing but in the gait the extremities must pivot alternatingly over the standing heel.

III. THE PARALYTIC IMBALANCE OF THE HIP JOINT ACCORDING TO DIFFERENT MUSCLES

A. GLUTEALS AND TENSOR

1. Gluteus maximus

Paralysis of the gluteus maximus is frequent in the lower extremity paralysis. In our series of 300 cases of the lower extremity (J. Sher[9]) it was found in 66%.

The principal action of this muscle is extension of the hip in all positions. Only the inferior fibers contribute to the adduction (Duchenne[1]). When the muscle is paralyzed the body is thrown back to let the line of gravity pass behind the hip joint. It has been found by Duchenne[1] and confirmed by Inman[2] that in upright position or in standing on one leg the muscle becomes completely relaxed as the line of gravity is thrown backward.

2. The gluteus medius and minimus

These muscles are even more frequently involved in lower extremity paralysis, the percentage in our series being 75%. They are the principal abductors; their flexory and extensory as well as their in and outward rotatory components fluctuate with the position of the joint, and they are comparatively small except for the flexory effect of their anterior portions which is greater.

Paralysis of these muscles causes the well known Trendelenburg sign, the pelvis dropping on the unsupported side. To be able to stand on one leg, the body has to lean to the standing side until the line of gravity passes lateral to the hip joint center. Then gravity becomes an abductory force supplanting the muscle. In the gait this leads to the well-known duck gait of the paralyzed gluteus medius and minimus, especially when the paralysis is bilateral.

3. The tensor fasciae

As a flexor of the thigh the muscle acts synergistic with the ilio-psoas. It is also an inward rotator. Its principal action, however, is as abductor and stabilizer of the pelvis against the thigh when standing on one leg as has been shown conclusively by Inman's[2] investigations. Paralysis of this muscle is usually associated with that of the gluteus medius and minimus.

The principal effect of this paralysis is the instability of the pelvis which fails to be stabilized against the thigh and the Trendelenburg sign becomes positive. The loss of flexion is not so noticeable so long as there is a good ilio-psoas, and so long as the rectus femoris and sartorius are functioning. But there is a definite loss of inward rotation (Duchenne[1]) although the anterior fibers of the gluteus medius and minimus can eventually substitute for it.

B. THE FLEXORS

1. The ilio-psoas and tensor fasciae

The ilio-psoas is a flexor and outward rotator; it is a synergist to the tensor fasciae in flexion and an antagonist in rotation. No case of isolated paralysis is reported, but if it does occur, the flexion of the thigh would be carried out by other flexors, the tensor, the sartorius, the rectus femoris and the pectineus. In the length rotatory direction the ilio-psoas is a strong outward rotator and its action is neutralized by the inward rotating component of the tensor. The paralysis of the latter muscle produces not only a deficiency of the active flexion of the hip joint, but chiefly an instability of the pelvis in the frontal plane with a positive Trendelenburg sign.

2. The rectus femoris

Frequency of paralysis, according to Lovett[6] was 152 among 1417 cases reported in the Vermont epidemic. As a biarticular muscle its action as a flexor of the hip joint depends upon its being stabilized at the knee by the antagonistic hamstrings. The paralysis of this muscle affects the flexion of the hip but little; so much more conspicuous is the effect on the knee joint. It is only in combination with the other hip flexor, especially the tensor, that difficulty arises in control of the hip movements. The trunk is bent forward to make gravity act as an auxiliary flexor, and in consequence, flexion contractures develop which ultimately become structural due to fibrotic changes in the paralytic or paretic muscles. The trias of hip contracture is flexion, abduction, outward rotation of the hip; associated with it is flexion, abduction and outward rotation of the knee, pelvic obliquity and exaggerated lordosis (Yount[13]).

3. The sartorius

This muscle acts on the medial aspect of the thigh similarly as the tensor does on the lateral. Being a flexor of the hip joint its paralysis should weaken the flexion power in this joint, though its effect is slight. Its action as an outward rotator becomes stronger when the hip is held in extension (Duchenne[1]), but here again the powerful outward rotators are able to make up for any deficiency arising from paralysis.

C. THE ADDUCTORS

The frequency of their being involved in lower extremity paralysis was 75% in our series (Sher[9]). Both the longus and the magnus have in upright position a flexory component which changes to an extensory at a certain degree of flexion. The pectineus also is a flexor and adductor. The adductors are considered to be outward rotators with the exception of the lower half of the adductor magnus which is a powerful inward rotator (Duchenne[1]). In general, in extension of the hip the outward rotation, and in flexion the inward rotation, prevails. This is shown in the spastic where adduction contracture is associated with flexion and marked inward rotation.

When the adductors are paralyzed one should expect an abduction contracture. This is observed occasionally in spastic paralysis after section or denervation of the adductors, but rarely in poliomyelitis, probably because of the associated abductor paralysis or because the hamstrings serve as auxiliary adductors.

D. IN- AND OUTWARD ROTATION

1. Inward rotators

Frequency of paralysis was 73% in our series (Sher[9]).

a) Tensor fasciae.

Paralysis of the tensor impairs inward rotation allowing the psoas to prevail as outward rotator when the leg is flexed; the short external rotators prevail in all positions.

b) Paralysis of the gluteus medius and minimus also causes some loss of active inward rotation but as the limb swings forward during the gait, one notices that the flail hip automatically turns into inward rotation.

2. The outward rotators

Frequency of paralysis in our series was 70% (Sher[9]).

The piriformis, the gemelli, the obturator internus, externus and quadratus are outward rotators, the piriformis being also an abductor and extensor. Paralysis of these muscles overthrows the balance in favor of inward rotation. The anterior halves of the gluteus medius and minimus, as well as the tensor and the lower half of the adductor magnus counteract the action of these six outward rotators. Patients with paralysis of the external rotators maintain the hip joint in standing or walking in strong inward rotation which gives the gait the peculiar aspect.

E. COMBINED CONTRACTURES

1. Paralysis of extensors and adductors.

This is often associated with flexion and external rotation contracture of the hip because, in this instance, the active flexors and abductors prevail.

2. Paralysis of extensors and outward rotators.

The contraction of the tensor fasciae is no longer neutralized by the gluteus maximus and the result is flexion and inward rotation contracture of the hip.

3. Paralysis of the abductors and outward rotators.

When the gluteus medius and minimus are paralyzed and an inward rotation and adduction position ensues. Because of contracture and shrinkage of the tensor and of the ilio-tibial band, the ground is laid for the paralytic dislocation of the hip. This is one of the most severe complications because, as the head loses all pelvic support, the limb becomes unfit for weight bearing. The head is forced to assume more and more a posterior position at the outer table of the os ilii and the hip goes into flexion contracture.

IV. THE MECHANICS OF RECONSTRUCTIVE PROCEDURES ON THE PARALYZED HIP JOINT
A. RESTORATION OF ALIGNMENT

1. Flexion adduction contracture

Any effort to correct the flexion contracture, whether conservative or operative, requires fixation of the pelvis against forward rotation. Dollinger's old method consisted in fixation of the pelvis by completely flexing the sound hip against the thorax so that the full amount of flexion contracture is revealed. The sound hip joint is then casted in this position and correction of the affected hip is accomplished by putting the hip in traction. A similar and more effective method based upon the same principle is that introduced by Dame Hunt.

It is obvious that a lasting correction of the contracture can only be obtained if the hip can be extended fully so that the line of gravity falls behind the hip joint center. Correction of a severe paralytic contracture can hardly be accomplished by the stripping methods of Soutter and Campbell unless there is added to it a subtrochanteric osteotomy which creates a compensatory forward kinking of the femur. Irwin[3] pointed to the inadequacy of the iliac stripping because it does not dispose of the shortening and shrinkage of the fascia lata and of the lateral intermuscular septum. He finds that Yount's[13] fasciotomy with removal of 2-3 cm. of the ilio-tibial band and the division of the lateral intermuscular septum above the knee joint followed by traction is a more efficient method.

2. The paralytic dislocation

In unreducible paralytic dislocation with paralyzed abductors and flexion adduction contracture the angulation at the osteotomy site must be such that the line of gravity falls not only behind but also lateral to the shaft so that gravity acts both as an extensor and as an abductor (Fig. 16).

Fig. 16. Paralytic dislocation, line of gravity falls behind and lateral to shaft after osteotomy.

3. The shelving operation

A shelving operation for an unstable subluxated hip for impending dislocation requires:

(1) Some remaining strength of the abductors and of the tensor;

(2) Enough extensory power to exclude the danger of flexion contracture.

B. THE RESTORATION OF MOTION

4. Substitution of abductors by tensor fasciae. Legg's operation

The shifting of the tensor to the mid crest of the os ilii deprives it of its inward rotatory and flexory ability. So does the operation by which the origin of the tensor is transplanted under the gluteus maximus. So long as these operations do not weaken the tension of the tensor which is practically indispensable to stabilize the hip joint in one leg standing, there is no objection though the last-named procedure makes such an event very likely.

5. Substitution of the abductors by the external oblique

Thomas, Thompson and Straub[12] recently proposed the use of the external oblique for the substitution of the paralyzed abductors. The operation deprives the muscle of its ability to lift the pelvis and to pull up the limb as it leaves the floor in the swinging phase. However if it provides for sufficient abductory power it should also eliminate the lateral tilt during the supporting phase. The above named authors[12] report 19 cases with only two failures. Although the operation does not entirely abolish the abductor limp, most patients showed improvement of the gait, an increased sense of security in walking, and an increased endurance.

6. Substitution of the abductor by the vastus lateralis (F. Lange[4])

The mechanical effect depends upon the lifting height of this muscle which is being elongated with silk sutures and led out through a slit in the tensor; it is attached to the iliac crest. In point of direction and also of strength it is a rational solution, but as an active abductor it is likely to fail because of insufficient contraction length. It can, however, display a considerable amount of stabilization of the pelvis.

7. The substitution of the abductor by the iliopsoas

An ingenious method by Mustard[7] is the substitution of the missing abductors by the iliopsoas. The muscle is simply detached from the lesser trochanter and then led over the femoral neck through a large cut-out made in the os ilii below the anterior inferior spine. It is then fastened to the greater trochanter. This maneuver changes this muscle from a flexor and adductor to an abductor and extensor.

8. The substitution of the abductors by the gluteus maximus (Telson[11])

Telson used for transposition the upper and lateral portion of the gluteus maximus which arises from the iliac crest. This is then implanted forward upon the crest in the frontal plane. The operation appeals from the mechanical point of view for the following reasons: (1) the muscle mass transplanted is abundant; (2) the direction of the muscle is but little changed; (3) it retains its strong attachment to the fascia lata, and (4) and finally, it seems to have sufficient lifting power. Practically the same point, however, can be made for the transposition of the tensor fascia which is certainly the easier procedure. In the absence of this muscle, the procedure of Telson is worth considering.

9. The substitution of the gluteus maximum by the sacro-spinalis

F. Lange[4] used the liberated lower end of the sacro-spinalis, elongated by silk sutures and fastened to the greater trochanter as a substitute for the paralyzed gluteus maximus. Hey-Groves, combining Legg's and Lange's operation, pulls the tensor fasciae through a tunnel at the base of the greater trochanter to be sutured to the erector spinae mass. Whether the erector is applying directly on the greater trochanter, or whether it is hooked to the tensor, it is its lifting height and the direction of its force upon which the mechanical effect of these operations depends. It is doubtful if the sacro-spinalis can substitute for the gluteus maximus as active extensor of the hip since the contraction length of the lower portion of this muscle is 8-10 cm., which is more than the erector spinae muscle can produce.

10. The substitution of the abductors by total shift of the tensor fasciae

The operation of Legg[5] consists in transferring the origin of the tensor backward to act as an abductor; that of Ober[8] adds to it the transposition of the fascia lata at the lower end. Both methods are adequate in muscle strength and direction and contractile length. They, and Mustard's transposition of the ilio-psoas, seem to be the most rational of all methods of substitution for the paralyzed abductors.

V. SUMMARY

1. There is complete neutralization by opposite sides of the horizontal component of the gravitational reactions when legs are parallel or symmetrically abducted, the force of the horizontal component being directed toward the midline as pressure against the acetabula.

2. There is likewise complete neutralization when legs are symmetrically adducted except that in this case the horizontal components are directed away from the midline acting as tension.

3. If position is asymmetrical, one leg abducted, the other adducted, the vertical gravitational component is greater on the adducted side as the line of gravity moves closer to the adducted hip.

4. Tension of the abductors decreases, that of the adductors increases the shift of the pelvis to their respective sides.

5. No rotatory moment is developed by gravity in the sagittal plane as long as the weight line falls through the center of the hip joint; this, however, is only a momentary position because of the constant back and forth swaying of the line of gravity and therefore it requires the alternating tension of the extensors and flexors to adjust the line of gravity as close as possible to the center of the hip joint.

6. Asymmetrical stance in the sagittal plane with one leg directed obliquely forward, the other obliquely backward maintains an automatic equilibrium only as long as the line of gravity bisects the angle between the two extremities.

In this case the forward horizontal component of the backward leg neutralizes the backward component of the forward leg.

7. In the horizontal plane, however, the two obliquely placed legs rotate the pelvis in the same direction; therefore they do not neutralize each other and rotation must consequently be neutralized by muscle action: by the inward rotators of the forward and the outward rotators of the backward leg.

8. In paralysis of the abductors the pelvis drops on the unsupported side, shifts to the same side and the trunk lists to the affected side developing the side sway of the gluteus medius gait.

9. If both adductors and abductors are paralyzed the adjustment of the pelvic shift is up to the contra-lateral sound ab and adductors.

10. Gravity comes to the aid of paralyzed extensors or flexors by back and forward shifting of the trunk.

11. Paralysis of the inward rotators keeps the pelvis in outward rotation on the affected side with anterior superior spine pointing forward.

12. In paralysis of the outward rotators the contraction of the inward rotators (especially the tensor) rotates the pelvis backward. Normally the pelvis undergoes forward and outward rotation in the swinging and backward and inward rotation in the standing phase. Contracture of the external rotators prevents this backward and inward rotation.

13. In paralysis of the gluteus maximus the body is thrown backward; in paralysis of the gluteus medius it tilts to the affected side; in paralysis of the flexors (ilio-psoas, rectus, sartorius) the trunk leans forward; in paralysis of the adductors abduction contracture develops rarely; in paralysis of the tensor the outward rotators prevail; in paralysis of the pelvi-trochanteric muscles the hip goes into strong inward rotation.

14. Correction of a paralytic flexion contracture requires full extensibility; Soutter's and Campbell's procedures are usually inadequate by themselves; a subtrochanteric osteotomy often becomes necessary.

15. A non-reducible paralytic dislocation is best treated by osteotomy with biplanar angulation (Schanz).

16. The prerequisites for a shelving operation for impending paralytic dislocation are sufficient residual strength of the abductors and of the tensor; and enough extensory power to prevent flexion contracture.

17. Paralyzed abductors can be substituted: (a) by shifting the tensor to mid crest of the os ilii (Legg) which weakens the latter muscle in its function of a stabilizer of the pelvis; (b) by the external oblique (Thomas, Thompson and Straub) which deprives the muscle of its ability to lift the pelvis against the trunk; (c) by the vastus lateralis (F. Lange) which has an insufficient contractile length; (d) by the gluteus maximus (Telson) which seems adequate in strength, direction and lifting power; (e) by the sacro-spinalis (F. Lange) mobilized and inserted by silk sutures to the greater trochanter; here also the lifting power is in doubt; and finally (f) by transferring the insertion of the ilio-psoas from the lesser to the greater trochanter past the neck of the femur (Mustard). This latter method seems to be the most rational of all.

BIBLIOGRAPHY

1. DUCHENNE, G. B.: *Physiology of Motion*. Translated by E. B. Kaplan. Philadelphia, Lippincott Co., 1949.
2. INMAN, V. T.: Functional Aspect of the Abductor Muscles of the Hip. *J. Bone & Joint Surg., XXIX:3.* 607, July 1947.
3. IRWIN, C. E.: The Iliotibial Band, its Role in Producing Deformity in Poliomyelitis. *J. Bone & Joint Surg., XXXI:*A. 1. 141. January 1949.
4. LANGE, F.: *Die Epidemische Kinderlähmung*. Munich, J. T. Lehmann, 1930.
5. LEGG, A. T.: Tensor Fascia Femoris Transplantation in Cases of Weakened Gluteus Medius. *New England J. Med., 209:*61, July 13, 1933.
6. LOVETT, R. W.: *Treatment of Infantile Paralysis*. Philadelphia, Blakiston & Son, 1916.
7. MUSTARD, N. T.: Iliopsoas Transfer for Weakened Hip Abductors. *J. Bone & Joint Surg., 34A:3*, 677, July 1952.
8. OBER, F. R.: Tendon Transplantation in the Lower Extremity. *New England J. Med., 209:*52, July 13, 1933.
9. SHER, J.: *Annual Report to National Foundation of Infantile Paralysis*. Iowa City, Iowa, 1941, p. 124.
10. STRASSER, H.: *Lehrbuch der Muskel und Gelenkmechanik*. III, Berlin, J. Springer, 1917.
11. TELSON, D.: Transplantation of the Gluteus Maximus for the Paralyzed Gluteus Medius. *Surg., Gynec. & Obst.,* March, 1928, p. 417.
12. THOMPSON, T. C., STRAUB, L. R., and THOMAS, L. I.: Transplantation of the External Oblique for Abductor Paralysis. *J. Bone & Joint Surg., 32A:1*, 207, January 1950.
13. YOUNT, C. C.: The Role of the Tensor Fasciae in Certain Deformities of the Lower Extremity. *J. Bone & Joint Surg., VIII:*171, January 1916.

Lecture XIX

THE MECHANICS OF THE KNEE JOINT

I. MORPHOLOGY

THERE is no joint in the body upon which locomotion has placed a heavier burden nor is there any so ready to resent abuse. Constructing a joint so exposed to mechanical stresses with any assurance of safety has been a strain on nature's ingenuity.

The joint constituents must be large because of the heavy weight stresses they have to bear; and at the same time the joint must have a wide range of motion. In the gait, for instance, it is the great shortener and lengthener of the extremity; being endowed with powerful muscles it acts with the ankle joint as a strong propeller of the body. Furthermore, placed between two long lever arms, the femur and the tibia, it receives and absorbs vigorous stresses which the lateral movements of the body impart to it in the frontal and the axial rotations in the transverse plane.

Another difficulty arose when man assumed his orthograde position. No longer could the joint be constructed as a simple "track-bound" hinge secure in its single plane track as is the case in the quadruped. The upright bipedal gait added the requirement of length rotatory motion so that the joint had to be converted from a secure single plane hinge into a trochoginglymus of two degrees of freedom of motion.

A. THE BONY CONSTITUENTS

The massive femoral condyles of the femur articulate with the condyles of the tibia. Between their extensive cartilage-covered surfaces two intra-articular discs of fibrocartilage are interposed. These interposed elastic rings fill the spaces which are left between the contacting bones because of the marked incongruency of their contours.

B. THE CAPSULE AND SYNOVIA

In order to provide these large articular surfaces with a synovial apparatus adapted for wide ranges of motion, the synovial sac had to form a large recess, the suprapatellar pouch which accommodates the sac to the different joint positions in a manner similar to that of the subdeltoid bursa in the shoulder joint. The patella is implanted into this enveloping synovial sac so that it looks into the joint through a hole in the synovial membrane. In the mid sagittal plane of the joint the synovial lining projects both in front and in back so as to form an anterior and posterior partition. The anterior is filled with fat accumulations protruding into the joint while the posterior covers the cruciates. The projecting fat pad lifts up three folds in the synovial membrane; one in the midline, the ligamentum mucosum, and two laterally, the ligamenta alaria. In this manner the knee joint cavity is almost divided into a medial and a lateral half by the two synovial protrusions (Fig. 1).

In contrast to the hip joint the effect of atmospheric pressure on the cohesion of the joint is entirely secondary; it causes the synovial membrane to fold into the dead spaces of the joint. Still, a separation of the joint surfaces in any position requires some traction to overcome atmospheric pressure (Fick[3]).

C. THE PATELLA

In order to increase the leverage of the knee extensors, nature has provided the patella as a sesamoid bone which is set into the tendon of the quadriceps. Its mechanical function is to increase the rotatory moment of the extensor. Gliding in the intercondyloid groove, the patella adjusts the oblique direction of the quadriceps so that its insertion into the tibia by means of the patellar tendon conforms with the axis of the leg. This straightening of the course of the tendon causes the quadriceps to develop a lateral directed horizontal component which tries to force the patella out of its groove.

D. THE SEMILUNAR CARTILAGES

To increase the range of motion and to make up for the geometrical incongruencies of the articular bodies nature has interposed between the condyles of the femur and those of the tibia two fibrocartilaginous bodies, one occupying the inner and one the outer half of the joint. This divides the articulation into an upper and a lower compartment. The medial cartilage is moonshape and is linked at its anterior border to the ring shaped lateral cartilage by the transverse ligament. These cartilages increase the range by moving against the tibial condyles; at the same time they deepen the tibial sockets and diminish the dead space between the articulating bodies. It is quite true that after removal of the menisci no particular disturbance of the knee joint function is observed but nevertheless they protect the capsular wall against impingement and acting as shock absorbers they protect also the articular cartilage (Fig. 2). While the total weight pressure upon the articular cartilage is the same in any position so long as the knee bears the weight, the distribution of pressure on the joint cartilage varies greatly because it covers a greater surface when the knee is in full extension. In this position also the pressure distribution over the menisci is more equable (Strasser[10]).

E. THE LIGAMENTS

Numerous ligaments support the capsular apparatus of the joint. Some are interwoven with the fibrous capsule like the lateral expansion of the extensor apparatus, or the popliteal ligaments which enforce the posterior capsular apparatus. From the viewpoint of joint mechanics two of these ligaments are particularly important: the cruciates and the collaterals.

1. The cruciates

The anterior cruciate ligament runs from the lateral portion of the inter-condyloid groove of the femur medially and forward to fasten to the anterior spine of the tibia (Fig. 3). The posterior runs from the medial portion of the intercondyloid notch to the posterior tibial spine (Fig. 4). The length ratio between anterior and posterior ligaments is 5:3. In the frontal plane their origin at the condylar notch is 1.7 cm. distant; in the sagittal plane the points of insertion are 5 cm., one behind the other. They are not exactly twisted around

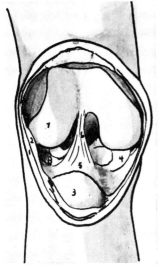

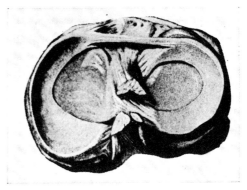

FIG. 2. Arrangement of the menisci. (Braus)

FIG. 1. Knee joint cavity from in front.
  1. Fibrous capsule.
  2. Synovia.
  3. Patella.
  4. Medial meniscus
  5. Infrapatellar fat pad.
  6. Infrapatellar fold.
  7. Condyle of femur.
  8. Lateral meniscus.

FIG. 3. Anterior cruciate ligament.

FIG. 4. Posterior cruciate ligament.

each other, but they cross at their lower half. By projecting forward they form the posterior part of the above-mentioned sagittal partition of the joint.

2. The collateral ligaments

A) THE MEDIAL OR TIBIAL COLLATERAL LIGAMENT

This ligament is, according to the newer investigations of Brantigan and Voshell[1] constituted as follows: it consists of a parallel or longitudinal anterior, and an oblique posterior portion and it is separated from the tibia by loose areolar tissue and various bursae. The anterior portion can easily be separated from the other joint structures. It inserts into the femoral epicondyle in a fan shaped fashion, while the oblique posterior portion blends intimately with the joint capsule so that it cannot be dissected from it (Fig. 5). Strasser[10] distinguishes: (1) an anterior, superficial, parallel fibered portion, and (2) a lower, short portion. The lower is behind the knee joint axis and therefore becomes tight in extension; the anterior superficial fibers are in front of the axis and they become tight in flexion. This portion arises high at the femur and attaches itself to the inner margin of the medial meniscus and to the joint surface of the tibia. It is commonly believed that in extreme extension finally

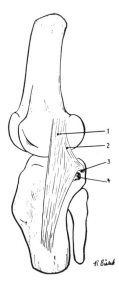

FIG. 5. Tibial collateral ligament. *1,* longitudinal anterior fibers. *2.* oblique posterior fibers. *3.* attachment of oblique fibers to capsule. *4.* semimembranosus tendon. (Brantigan & Voshell)

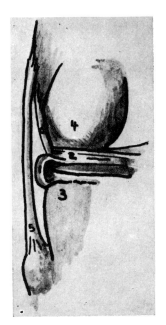

FIG. 6. Relationship of fibular collateral ligament.
1. Coronary ligament.
2. Lateral meniscus.
3. Tibia.
4. Condyle of femur.
5. Fibular collateral ligament.
6. Popliteus tendon.

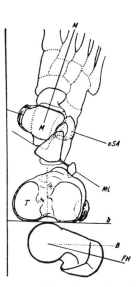

FIG. 7. The relationship of the axes of the femoral neck, the tibial condyles and the ankle joint. (Strasser)

both the anterior and posterior portions become taut and that both resist hyperextension of the joint.

B) THE FIBULAR COLLATERAL LIGAMENT (FIG. 6)

This also has a long and a short fibered portion. The principal portion is the short fibers which are attached to the capitulum fibulae. Strasser[10] finds that this ligament relaxes in flexion, first more rapidly, then more slowly but that extreme flexion causes a terminal tension of the anterior fibers. It is also relaxed in inward and it is tight in outward rotation. The short fibers of the fibular collateral ligament coming from the lateral epicondyle of the femur reinforce the lateral capsule. The entire collateral ligament is separated from the joint by the tendon of the popliteus muscle which comes from the lateral femoral epicondyle and runs underneath the ligament.

II. THE MECHANICAL ANALYSIS OF THE KNEE JOINT
A. THE AXES OF ORIENTATION AND THEIR RELATION
TO THE EXTREMITY AS A WHOLE (FIG. 7)

1. The transverse plane

The transverse axis of the knee joint stands perfectly horizontal both in extension and in flexion. The common transverse axis of both knee joints lies, furthermore, in a frontal plane, as does the common axis of the hip joints. But the axes of the neck of the femur, as explained above, are not in a frontal plane but are deflected forward. Consequently there is a torsion between the neck and the shaft of the femur to the effect that the neck is twisted outward against the shaft or, to state it conversely, the shaft is twisted inward against the neck. Comparing again the transverse axis of the ankle joint with that of the knee we find that the former also is not in the frontal plane but runs from backward and outward to forward and inward; it has, in short, an outward twist against the strictly frontally placed knee joint axis. The twist amounts to about 25° and causes the outer malleolus to lie further backward than the inner.

Comparing the axes of all three articulations in upright standing, the knee joint maintains a position of inward rotation against the head and neck of the femur as well as a position of 25° inward rotation against the lower end of the tibia.

2. The frontal plane

Here we notice the angulation of the axes of the femur and the tibia. They form an angle of 171° opening laterally. This obliquity is caused by the angulation between femoral neck and shaft as well as by the difference between the width of the pelvis and that of the supporting surface. Under normal conditions only the femur is oblique and the tibia stands perfectly vertical.

The angle of 171° between the axes of femur and tibia is divided by the horizontal axis of the knee joint so that the later forms with the axis of the femur at an angle of 81° and with the axis of the tibia an angle of 90°. Consequently the femoro-tibial angle is not strictly bisected.

In contrast the transverse axis of the elbow joint strictly bisects the angle between the axis of the humerus and that of the forearm. The result is that when the elbow is fully flexed, the upper arm and forearm cover each other completely. If, on the other hand, the knee is fully flexed, leg and thigh are not strictly superimposed but owing to the valgity of the joint the heel points medially touching the buttocks.

The difference in the angle becomes less if we consider the *mechanical axis* of the femur which goes from the center of the head to the center of the knee joint. This line has less obliquity, lacking only 3° from a straight angle with the tibial axis. It therefore makes an angle of 87° with the transverse axis of the knee joint while the angle between the knee axis and the tibia remains at 90° (Fig. 8).

Because of the obliquity of the anatomical axis of the femur the greater amount of pressure is born by the lateral condyles and, on the other hand, a greater tension stress is sustained by the soft structures of the medial condyles. This distribution is accentuated in certain pathological conditions such as the genu valgum where the mechanical axis of the extremity is displaced laterally (Fig. 9A). In the genu varum the line is deflected medially so that the longitudinal axes of femur and tibia may form a straight angle or the angle may even be reversed and open medially instead of laterally. In this case the greater pressure falls upon the internal condyles, and it is the soft structures on the lateral side which come under increased tension stress (Fig. 9B). Accord-

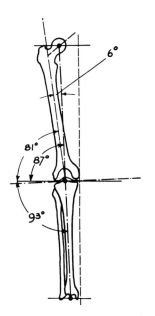

FIG. 8. The anatomical and mechanical axes of the tibia and femur.

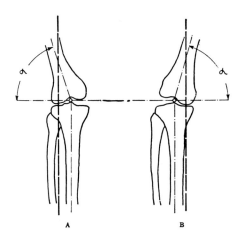

FIG. 9. Alterations in the mechanical axis of the knee joint. A. genu valgum. B. genu varum.

FIG. 10. A cardioid curve.

ing to Miculicz[9] the weight supporting line through the axis of the femur strikes the knee joint with only slight lateral deviation in 87.5%, with slight medial deviation in 12.5%. Any deviation amounting to more than 2.5 cm. from the midline must be considered as pathological.

B. THE SHAPE AND CONTOUR OF THE CONDYLES

In the sagittal elevation of condyles of the femur do not represent part of a circle as one would expect in a hinge joint but their contours describe a curve, the angular values of which constantly increase from forward to backward. In other words, the curve becomes sharper in its antero-posterior course and the radius becomes successively smaller from unit to unit of the curve so that the ratio of the radii between the foremost and hindmost part of the curve is as 9:5 (Fig. 10). Such a curve is called a cardioid. If one unites the succeeding centers of this curve, a curved line is formed which is called an evolute (Fig. 11).

If the contacting surfaces of the joint were true parts of a circle, even though each surface had a different radius, then equidistant points of the moving part would contact equidistant points of the stationary part. This is called a rocking type of motion.

As a matter of fact the knee joint represents a combination of two types of motion; one a true rocking motion (as in a rocking chair) where equidistant points of the femur contact equidistant points of the tibia or vice versa; the other is a gliding motion where the contact of one joint body is concentrated

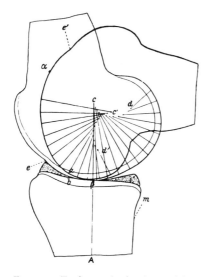

FIG. 11. Evolute of the knee joint.
(Strasser)

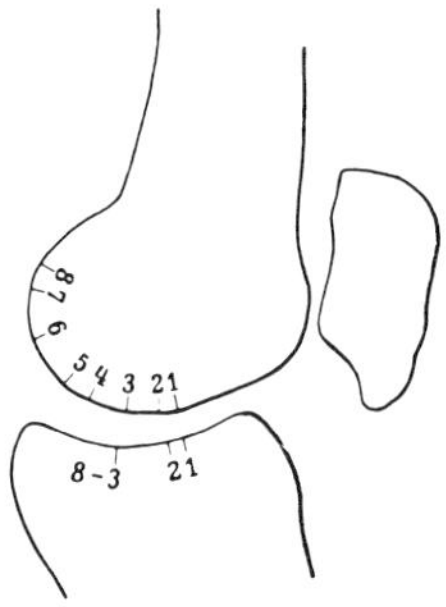

FIG. 12. Contact points, medial half of the joint. (Strasser)

to a point or a small area which sweeps over the whole contour of the other part.

To understand the mechanism of this complex motion one must consider the character of the joint surfaces. In the sagittal profile the length of the femoral joint curve is 10 cm. and that of the tibia 8 cm. But the point is that although a convex surface contact a concave one as in the elbow joint, the radii of the contacting surfaces are different; that of the tibia being much larger than those of the femoral condyle. Consequently there is neither a constant gliding as in the joint surfaces of equal radii nor is there a constant rocking as where a more curved surface contacts a more plain one, but it is rather a combination of both types of motion. Such a combination was already recognized by the brothers Weber.[11] It was left to the study of Zuppinger[12] in 1904 to make a complete and thorough x-ray study, the details of which were later corroborated by the analytical studies of Fischer[4] in 1907.

There is some difference in the motion between the medial and lateral halves of the joint (Strasser[10]). Closer analysis shows that on the medial side from 180° extension to 170° or 165°, that is over 10 to 15°, a pure rolling occurs; after this the tibial contact gradually narrows down to a point and the motion becomes a gliding one (Fig. 12). On the lateral side there is a pure rolling which occurs for 20° and then the gliding type takes over so that the rocking or rolling type in the lateral side last longer (Fig. 13) while the medial condyle does more gliding. The change between the two types of motion is not abrupt. While the motion starts with a true rocking the gliding element gradually creeps in until it completely supplants the former type of motion. In the later phases of flexion the motion becomes pure gliding, that is, one point of the tibia comes in contact successively with various points of the femur (Fig. 14).

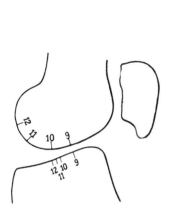

FIG. 13. Contact points, lateral half of the joint. (Strasser)

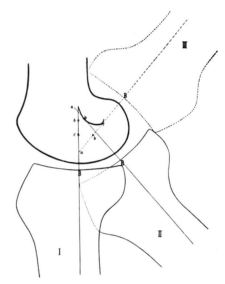

FIG. 14. Change from rocking to gliding motion of the knee. (Fick)

This change from rocking to gliding is very significant for the function of this joint in which both stability and mobility are required. The rocking motion is more compatible with stability; it is therefore predominant in the extensory phase. After 20° flexion, the knee, however, becomes looser as the arc of the condylar curves becomes smaller. The tibia moves closer to the axis of motion in the femur and therefore the ligamentous structures which are spanned between femur and tibia become relaxed. This loosening up prepares the joint then for a wider range of axial rotation.

2. The length rotation

The longitudinal axis for the external and internal rotation of the knee joint, or as it is sometimes called, pronation and supination, goes closer to the medial condyles so that the lateral condyle rotates around the medial. According to Zuppinger,[12] from 165° or 170° to full extension there is medially a pure rocking; laterally the rocking motion is combined with backward gliding of the tibial condyle which conforms with H. von Meyer's[8] terminal outward rotation of the tibia in extension. At this range $1°$ of extension corresponds to $\frac{1}{2}°$ of rotation.

The rotation about a length axis occurs mostly between meniscus and tibia. The length rotatory movement of the menisci against the femur is very slight. In inward rotation the anterior border of the lateral tibial condyle moves forward of the lateral meniscus and vice versa in outward rotation, the medial condyle of the tibia moves forward of the medial meniscus.

TERMINAL ROTATION IN FLEXION

In flexion, where the medial condyle of the tibia moves backward and the lateral forward, the movement is not as "trackbound" as the one accompanying

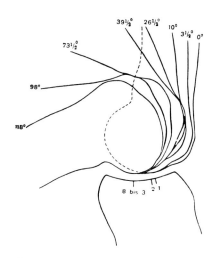

FIG. 15. Flexion of the knee; contact points of tibia farther apart (medial side). (Fick)

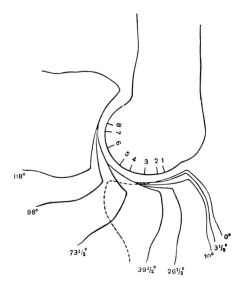

FIG. 16. Flexion of the knee. Contact points of femur closer together than on tibia (medial side). (Fick)

extension where the rotation is reversed. Comparing from degree to degree of motion the successive contacts at the medial side (Fick[3]), with those of the lateral side of the knee, one finds that on the medial side they are farther apart at the tibia (Fig. 15) than they are at the femur (Fig. 16), while on the lateral side (Fig. 17) they are closer on the tibia than they are on the femur. This means that there is inward rotation of the tibia in that respective excursion range. This inward rotation occurs between full extension and 15-$20°$ flexion at the rate of $\frac{1}{2}°$ of pronation of the tibia for each degree of flexion.

C. THE MECHANICAL STATUS OF THE PATELLA

The patella is a sesamoid bone implanted within the tendon of the extensors. It has a wide range of motion and the synovial sac into which the patella is implanted through a window has adapted itself to it by providing a large recess pouch extending upward along the anterior aspect of the femur. This recess is movable enough to follow the movements of the patella so that the latter can sweep freely up and down during the flexion-extension movements of the knee. When the knee is extended, the tightening quadriceps pulls the patella up until its upper border reaches beyond the cartilage which covers the anterior surface of the femoral condyles. The lower border is then almost in line with their lower contour. So long as the line of gravity falls behind the knee joint axis in upright standing, the quadriceps must contract in order to neutralize the rotatory effect of gravity upon the knee which would force it into flexion. As soon as the line of gravity falls within or in front of the knee joint the quadriceps can relax. Then the patella moves downward until its lower pole rests on the infrapatellar fat pad. In neutral position with the quadriceps relaxed, the lower pole is flush with the articular plane.

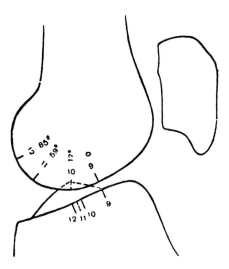

The line of pull of the quadriceps being oblique its angulation with the tibial axis produces on contraction an outward directed horizontal component. This gives rise to a tendency of the patella to slip outward over the external condyle. To offset this propensity nature has contrived to let the lateral condyle of the femur project farther forward than the medial and has also arranged that the fibers of the vastus medialis which secure the patella medially extend farthern distally on the thigh than those of the vastus lateralis.

Fig. 17. Lateral side. Contact points of tibia closer together than femur. (Fick)

Fick[3] states that the excursion of the patella from flexion to full extension is from 5-7 cm. while the distance between the lower point of the patella and the tuberosity of the tibia remains constant. Sidewise mobility is normally greater on the medial side; it becomes excessive in many pathological conditions. There is less lateral mobility in flexion. As the knee goes from extension into full flexion and the patella moves close into the intercondyloid notch the atmospheric pressure holds the patella firmly against the femur. In addition, the pressure from the quadriceps tendon presses the patella against the underlying bone so that it cannot be lifted up as it can in the position of extension.

D. THE TIBIA

It has been mentioned that in the frontal plane the angle between tibia and the anatomical axis of the femur is 171° and that between tibia and the mechanical axis is 177°.

The next point to consider is the relation of the femur and tibia in the *transverse plane*. The outward rotation of the whole tibia against the femur in full extension is 5-6°. To this must now be added the torsion angle of the tibia itself. It amounts to 20-25° so that in complete extension the total outward rotation of the tibio-astragalar joint would be between 25-30°. The so-called military position in which the feet are outward rotated amounts to 45° on each side and is therefore not a natural position because it requires an additional outward rotation of the femur. If the axis of the tibial condyles is placed strictly frontally, then the axis of the femoral condyles is inward rotated against the tibia 5-6°; in this position the axis of the malleolar mortice stands in 20-25° outward rotation against the transverse axis of the tibial condyles and in 30° outward rotation against the femoral condyles. This terminal outward rotation of the tibia against the femur which occurs in complete extension is in close relation with the mechanics of the gait. As the swinging leg puts the heel on the ground, the pelvis on this side is rotated forward which means the limb is in outward rotation. In this case the total amount of outward rotation of the feet against the sagittal midline is composed of: (1) the outward rotation in the hip joint; (2) the outward rotation of the tibia against the femur in full extension; (3) the outward torsion of the lower against the upper end of the tibia.

III. THE STRESS ANALYSIS OF THE BONES

1. Frontal plane

We recall that in discussing the form and architecture of the femur the bone seen in the frontal plane was likened to an excentrically loaded column

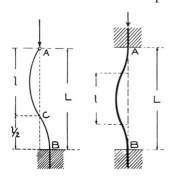

FIG. 18. Eccentrically loaded column. Deformation curve occupies 2/3 of the length.

FIG. 19. Column theory. Curve occupies ½ of length (tibia, frontal plane).

which under the column theory assumed a deformation in which the upper half was bent into a cosine curve occupying ⅔ of the length of the bone (Fig. 18). In the same plane the tibia can be considered as the column which is anchored above and below since no side motion is possible in this plane. The bone should present two curves (Fig. 19) each of which is half the length of the column. However the weight resting on the tibia is not concentric; because of the valgity of the knee it falls lateral to the axis. The inclination of the femoral axis against the tibial axis develops under gravitational stress a horizontal component which produces tension on the ligamentous structures at the medial side of the joint (Fig. 20). This horizontal component (h) depends upon the size of the femoro-tibial angle (α) (Fig. 20) and it equals the pressure force (P) times the sine of this angle.

2. The sagittal plane

A) THE FEMUR

In this plane the femur presents one long curve with forward convexity. In terms of the column theory it represents a column which in this plane is movable at both ends. The deformation curve (1) therefore occupies the entire length(L) of the bone (1 = L) (Fig. 21). The architecture of the femur is essentially conditioned by the forward convexity which is being transmitted to the upper tibia; the concave side posterior cortex is stronger (Grunewald[5]).

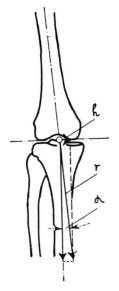

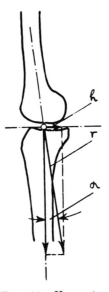

FIG. 20. Inclination of femoral axis against tibial axis producing horizontal component with resultant tension on medial joint structures.

FIG. 21. Deformation curve. Sagittal plane, 1 = L. Curve entire length of bone.

FIG. 22. Hyperextension of knee, sagittal plane. Hyperextension produces backward directed horizontal component.

The transverse trajectories go mainly to the posterior wall of the condyles, especially the lateral which is built stronger because it has the greater load.

B) THE TIBIA

There are three types of angulation (Grunewald[5]) in addition to the already mentioned torsion of the tibial shaft.

1) The retrotorsion, i.e., the backward deflection of the upper end.

2) The retroversion which is the backward slant of the tibial plateau which averages about 5.6°.

3) The retroflexion which is the backward bend of the shaft. It is believed that this retroflexion (the anterior convexity) is caused by the tension of the gastro-soleus.

The retroflexion of the tibia represents in the sagittal plane a column which is hinged at both ends. The resulting deformation is a long cosine curve (1) which is the full length of the bone (L_a).

In hyperextension gravity develops horizontal backward directed component in this plane, particularly in cases of a high patella (Fig. 22). This backward component depends upon the sagittal femoro-tibial angle (α) and is also equal to the gravitational pressure (p) times the sine of this angle. Under pathological conditions the femoro-tibial angle may be greatly increased with corresponding increase of the horizontal component. Such is the case in genu recurvatum.

IV. THE MECHANICS OF THE LIGAMENTOUS STRUCTURES OF THE KNEE

On the whole the knee joint depends for its security upon the ligamentous apparatus. The muscles rank second as stabilizers of the knee but there is always an interrelation between muscular and ligamentous tension. The ligaments, deprived of the support of the musculature sooner or later become insufficient as is seen in paralytic conditions.

The knee joint has its maximum volume capacity at between 20-30° flexion and its minimum in full flexion. Hence the joint goes into mid flexion in all pathological conditions which are associated with effusion. The capsule, having to accommodate itself to all kinds of positions with equal facility can contribute little to the stability and firmness of the joint. This is the task of the ligamentous apparatus.

A. THE COLLATERALS

In the previous description of the collateral ligaments of the knee it was mentioned that the oblique fibers of the medial collateral check extension, the straight anterior fibers check flexion, but that in hyperextension both anterior and posterior fibers become taut. The fibular collateral ligament is relaxed in all positions of flexion and only tightens when extension reaches 150°. What is to be added now is the role these ligaments play in checking lateral motion and rotation of the knee as well as the antero-posterior displacement.

1. Rotation

A) THE FIBULAR COLLATERAL LIGAMENT

1) *In extension* this ligament is already tight. It therefore checks in this position rotation as well as extension. On cutting of the cruciates the tibia cannot be rotated outward more if the tibia is extended, but it can be outward rotated on cutting of the collateral ligaments. In inward rotation of the tibia the lateral collateral ligament is at first relaxed and then becomes oblique and finally taut. Therefore inward rotation is not checked primarily by the fibular collateral ligament.

2) *In flexion.* Outward rotation of the tibia is checked by the fibular col-

lateral ligament in extension only since by cutting this ligament more outward rotation is possible. However this ligament has no checking effect when the knee is in flexion (Fick[3]).

B) THE MEDIAL COLLATERAL LIGAMENT

This ligament restrains rotatory motion of the tibia in all positions since the anterior fibers become tight in flexion, the posterior in extension.

2. Antero-posterior movement

The checking of this movement is not the exclusive function of the cruciates. The tibial collateral restrains with its oblique fibers the backward shifting of the tibia on the femur while the parallel anterior fibers do not restrain anterior or posterior gliding either in extension or in flexion. In general, most of the burden of restraining antero-posterior displacements falls upon the cruciates.

3. Abduction and adduction of the knee

The check against excessive abduction and adduction of the knee joint is taken up entirely by the collaterals, the tibial checking abduction and the fibular adduction. Again the knee is better secured against this motion in extension than it is in flexion, though in the latter the tight parallel fibers of the tibial collateral control abduction. A slight abductory and adductory range is natural for certain constitutions but any marked degree is pathological. It is significant that the safeguards against this motion are the earliest to fail. Any persistent joint effusion or even prolonged immobilization alone is apt to result in a "wobbly" knee.

B. THE CRUCIATES

For the mechanics of the cruciates their mutual relation is important. At their origin in the intercondyloid notch they are in the frontal plane 1.7 cm. apart though they twist into the sagittal plane and become 5 cm. apart at the tibia in that plane. The length ratio of the anterior to the posterior cruciate is as 5.3. These ratios determine the crossing point for the normal position only since the ligaments constantly sweep along each other with the movements of the knee.

Strasser[10] used a system of four rods to demonstrate this rather complicated relationship of the cruciates to each other (Fig. 23). His rod model shows that both cruciates cannot become taut at the same moment. In extension the anterior becomes taut; the posterior is relaxed almost to the finish of extension; it goes into tension as the knee joint proceeds with flexion.

The posterior cruciate which runs from the medial femoral condyle to the posterior tibial spine is the more medially located; the anterior which runs from the lateral femoral condyle to the anterior tibial spine is the more lateral. The posterior part of the posterior cruciate streams into the posterior border of the lateral meniscus and the posterior end of the anterior cruciate streams into the posterior portion of the medial meniscus. Thus both cruciates maintain a relationship to the menisci.

From 30° flexion on, the posterior the cruciate becomes tight, the anterior relaxed. In continued marked flexion over 90° both cruciates wind themselves around and tighten each other. The posterior particularly becomes the stabilizer.

The anterior checks the extensory gliding of the tibia on the femur but checks the rolling only at the end of the extension. The terminal, trackbound, outward rotation of the tibia at the end of extension occurs under the tension of the anterior cruciate ligament. According to Strasser[10] this terminal outward rotation is not carried out by the thigh muscles but occurs because the leg is fixed on the floor and cannot follow the inward rotation of the thigh which takes place during the standing period of the gait. He believes that the outward rotation of the extended leg, at least in standing, is a more passive

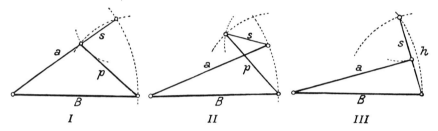

FIG. 23. Strasser's 4-rod system demonstrating relationship of cruciates. *I*. midposition. *II*. extension. *III*. flexion. (Strasser)

  a anterior cruciate
  p posterior cruciate
  B distance between insertion tibia
  s distance between insertion femur

phenomenon produced by counter resistance of the floor. This concept, however, is very doubtful; it is practically certain that the outward rotation is an active process starting with the extension of the leg and is already complete in full extension as the heel is put to the ground. It is truly "trackbound" and occurs within limits in which no checking action of the cruciate is called for.

V. THE MECHANICS OF THE MENISCI

The medial meniscus is approximately moon or sickle shaped. Its ends face each other in the middle of the joint, the anterior end being partly united with the tibia and partly its fibers continue into the ligamentum transversum to the anterior border of the lateral meniscus. The latter forms more or less a complete ring and is attached by a triangular plate to the posterior tibial spine. This betokens greater mobility of the lateral meniscus. The prime passive function of the menisci is to fill the dead spaces left by the incongruency of the articular bodies of femur and tibia; they protect the capsular wall against being impinged, between the two bones. In addition they complete and deepen the shallow tibial sockets. Virchow calls it their socket-forming function (Fig.

2). A third function of the menisci is to act as buffers which absorb pressure and concussions that reach the tibia from above.

It is well known that menisci can be removed without any particular damage to function and therefore they have been denied any importance in the kinetics of the knee. It should be emphasized, however, that the function of the menisci cannot be appraised merely by the flexion-extension range and that deficiencies may require years to become manifest. While it is true enough that upon removal of the menisci the stability of the joint remains unchanged, yet the movements are less secure, checks occur easier, and for reasons given below the extremes of motion are impaired. Gradual changes in the joint, especially in the contours of the femoral condyles, have been observed.

1. The menisci in flexion and extension

Both semilunar cartilages recede backward as the knee goes from extension into flexion. In this movement the posterior portion of the menisci more or less remain stable and it is the anterior portion which is pulled backward. In acute angular flexion the anterior halves of the semilunars may lose their contact with the femoral condyles entirely but they always remain in contact with a restricted area of the femoral condyles in the posterior portion of the joint.

As the knee joint goes into extension, the anterior halves are forced forward and the menisci are stretched; at the last phase of extension the forward slipping ceases. At the same time the pressure on the menisci increases. The forward movement continues during the gliding phase and stops with the rocking. Conversely in flexion the meniscus glides backward as soon as the gliding phase begins, at the same time putting tension upon the coronary ligaments (Strasser[10]).

2. The menisci in length rotatory motion

It was stated that the axis for longitudinal rotation falls into the medial half of the joint so that the excursions which the lateral condyles make have a greater linear value than have those of the medial condyles although their angular values naturally are the same. This is obvious since the lateral meniscus is farther from the axis of motion than is the medial. The medial meniscus in its posterior half is closely woven into the capsular apparatus and its reinforcing ligaments and, because of its C-shape, its anterior and posterior horns are farther apart. This makes for greater stability and less mobility.

In contrast the lateral cartilage is circular; its ends are closer together. It is less firmly anchored and has a greater freedom of motion. The ligaments of Humphrey and Wrisberg are strong fibrous bands which envelop the posterior cruciate and connect the posterior arc of the lateral meniscus with the medial condyle of the femur (Fig 24) (Last[7]). These ligaments have, according to Last, the function of holding the posterior arch of the lateral meniscus firmly to the rotating femur. Since the deep fibers of the popliteus insert into the posterior portion of the lateral meniscus, the contraction of the popliteus draws the meniscus backward. Last also believes that it is not so much the mobility of the lateral meniscus as the control by the ligaments of Humphrey and Wrisberg and the meniscus fibers of the popliteus which protects the lateral meniscus during rotation of the knee joint.

In inward rotation of the tibia the medial meniscus is forced backward and the lateral forward, whereas in outward rotation of the tibia the medial meniscus goes forward and the lateral goes backward, the latter always describing the greater excursion path.

Considering the mechanical function of the menisci as a whole, one finds that they modify the movement of the knee. In all positions they help to safe-

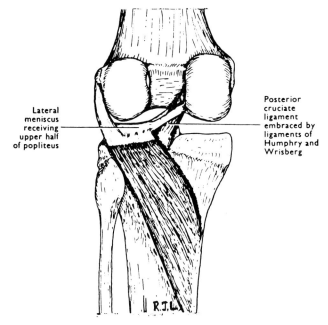

Lateral
meniscus
receiving
upper half
of popliteus

Posterior
cruciate
ligament
embraced by
ligaments of
Humphry and
Wrisberg

FIG. 24. Posterior aspect of lateral meniscus. (Last)

guard the stability of the joint. The fact that in length rotation the lateral has more freedom of motion than the medial explains why the latter is so much more exposed to traumatism, especially if it results from forced rotatory movement.

VI. THE DYNAMIC ANALYSIS OF THE KNEE JOINT

For the understanding of the dynamics of the knee joint we are indebted, in the first place, to the classical investigations of the Weber brothers,[11] to Braune and Fischer's[2] photographic methods of recording the movements of the joint, and to the x-ray studies of Zuppinger.[12] Particularly the latter's studies of the mechanism of movement in different positions has been highly instructive.

Almost in its entire range the movement of this joint is a combination of flexion-extension and rotation. Flexion starts with inward rotation of the tibia against the femur and extension ends with the outward rotation of the tibia against the femur.

A. IN THE SAGITTAL PLANE: FLEXION-EXTENSION

In the sagittal plane the normal flexion-extension range is from 40° flexion to slight hyperextension, somewhat above 140°. The hyperextension ranges according to Miculicz[9] is between 6-9°. It is often observed in children and in young individuals with constitutionally relaxed joints. One should rather speak of a pseudohyperextension because it is more often produced by slight forward obliquity of the lower femoral epiphysis. Passively flexion can be carried out until the heels touch the buttocks. Active flexion is limited by the active insufficiency of the flexors as well as by the passive tension of the extensors.

Zuppinger[12] divided the range into eight phases which he marked on the contours of the femoral and tibial condyles. He finds that the contact point at the femur moves steadily backward while at the tibia it wanders backward only to a position of 15-20° flexion and from there on the contact becomes concentrated upon a small area. This means that at the beginning of flexion the motion is a rocking one where equidistant points come in contact successively, while with continued flexion it becomes more and more a gliding motion in which one point of the tibia sweeps over the contours of the femur (Fig. 16).

B. IN THE TRANSVERSE PLANE: LENGTH ROTATION

If one lays a sagittal plane through the transverse axis of the femoral condyle and another sagittal plane through the transverse axis of the tibial condyles, one will find that when extension is completed these two planes have rotated against each other and form an angle of about 6°. If one places the transverse axis through the tibial condyles strictly in the frontal plane, then the long axes of the feet form with the plane at an angle of 20-25°. The first angle is due to the tibia rotating outward against the femur as the knee is extended; the second angle is due to the outward torsion of the tibial shaft against the condyles. The amplitude of active length rotatory motion is greatest at mid flexion and it amounts, at 90° flexion, to 75° according to Strasser[10] and to 50° according to Fick.[3] In extension the joint is locked against axial rotation by ligaments, especially the collaterals.

The ability to produce length rotation of the leg is a great factor in the human gait. In fact, it makes it possible for the leg to put the normally oblique axis of the ankle joint into a more frontal plane at the take-off. This axial rotation of the leg is also important as the foot is placed on the ground when the femur rotates inward around the fixed tibia. Furthermore, the combination of axial rotation of the knee and of the subastragalar motion is important when the feet are fixed on the ground and the knee goes into flexion as in squatting.

VII. MUSCLE DYNAMICS OF THE KNEE JOINT
A. THE EXTENSORS (FIG. 25)

The only real extensor of the knee is the quadriceps and the rectus is the only biarticular muscle. The action of the vasti is restricted entirely to the knee joint. The rectus is not able to accomplish full extension of the knee under ordinary circumstances; the final forceful extension is left to the vasti, especially the internus. The strength of the quadriceps exceeds three times that of the combined hamstrings. With a shortening distance of 8 cm., the vasti have a cross section of 148 cm.^{2} and therefor should have a working capacity of about 42 kgm. The rectus femoris represents only 1/5 of this strength. We have already mentioned that the slight deflection of the patellar tendon produces a laterally directed horizontal component and that the safeguards against lateral

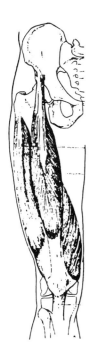

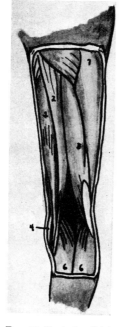

FIG. 26. Posterior thigh muscles.

1. Semimembranosus.
2. Semitendinosus.
3. Biceps femoris.
4. Gracilis.
5. Plantaris.
6. Gastrocnemius.
7. Iliotibial tract.

FIG. 25. Extensors of the knee.

displacement of the patella are the depth of the intercondylar groove and the forward projection of the lateral femoral condyle. In addition the low insertion of the fibers of the vastus internus into the extensory apparatus of the knee anchors the patella to the medial side. The contraction of the quadriceps not only tightens the tendinous insertion and pulls the patella upward, but it also puts tension on the entire anterior reinforcing apparatus of the knee.

B. THE FLEXORS (FIG. 26)

Each of the long biarticular muscles, namely the semimembranosus, semitendinosus, and the long head of the biceps are at the same time extensors of the hip joint while the sartorius is a flexor of the hip. The gracilis acts principally upon the hip joint. The combined working capacity of all hamstrings is only 15 kgm. against 42 kgm. of the extensors of the knee. Short flexors of the knee are the popliteus and the short head of the biceps. Plantaris and gastrocnemius are flexors of the knee as well as plantar flexors of the ankle joint.

With the leg free, all hamstrings flex the knee joint. But with the leg fixed the hamstrings extend the hip joint, at the same time pulling the thigh backward, especially if the pelvis is free to move upward. If the pelvis is prevented from moving upward and the leg is fixed, then the hamstrings again flex the thigh against the leg. The pairing of the flexors of the knee is significant. The biceps is synergistic with the inner hamstrings but its insertion is farther from the knee joint axis; it has a greater lever arm. The significance of this difference in lever arms is to prevent locking of the knee in full extension which might occur if the axes of the muscles all coincide with the long axis of the thigh. To prevent this further the backward projecting femoral condyles deflect the axes of the hamstring muscles. The popliteus is a flexor because it tightens when the knee goes into extension and relaxes in flexion. Its main function, however, is length rotatory.

C. THE ROTATORS

Some believe that the terminal outward rotation of the tibia on extension is caused by the anterior cruciate. However it may be pointed out that, although this outward rotation puts the anterior cruciates under tension, they can only act as a check and active muscle pull is necessary to produce rotation at least so long as the tibia is not fixed upon the ground. Fick[3] believes that the vastus lateralis, together with the biceps and the tensor, is responsible for the outward rotation.

The actual inward rotation of the leg on the thigh is produced by the internal hamstrings the tendons of which are inserted not only behind the joint center but also medial to it. In this way the muscles can combine their flexory effect with inward rotation. These muscles are: the sartorius, the gracilis, the semitendinosus, the semimembranosus and popliteus. According to Fick[3] the outward rotation has a greater range than inward rotation. The range depends upon the position: at 30° flexion it is 32°, at 90° flexion it is 42°. The importance of the terminal outward rotation lies in its stabilizing the knee in standing.

The working capacity of the rotators

Fick[3] gives the outward rotators, i.e., the biceps and the tensor, a total working capacity of 1.8 kgm., and to the inward rotators one of 2 kgm. in respect to the knee joint. This is quite a contrast to the ratio between extensors or flexors where the working capacity is 42 kgm. for the former and only 15 kgm. for the latter. It is evidence of the great preponderance which the extensors have gradually acquired with the upright position of man.

VIII. SUMMARY

1. The function of the knee joint requires massive articular constituents placed on long lever arms. The joint is a trochoginglymus, permitting 2 degrees of freedom of motion. The incongruency of its contacting surfaces is partly compensated by the interposition of the semilunar cartilages.

2. The wide range of motion requires a redundant capsular apparatus into which the patella is implanted.

3. Two curtains projecting into the joint in sagittal direction divide it into two halves which communicate through the gap between the anterior projection raised by the subsynovial fat accumulations and the posterior projection made by the bulge of the synovial covered cruciates.

4. The patella is a sesamoid bone in the quadriceps tendon with a cartilage-covered surface toward the knee joint into which it looks through a window in the synovial sac.

5. The semilunar cartilages, the inner and outer, divide each half of the joint in horizontal direction into a lower compartment between cartilage and tibia and an upper between cartilage and femur.

6. The cartilages fill the dead spaces between femoral and tibial condyles, increase the ranges of motion and provide a more even distribution of pressure.

7. From a mechanical point of view the important ligaments are the cruciates and the collaterals. The *anterior* cruciate runs from the lateral portion of the intercondyloid groove to the anterior spine of the tibia; the *posterior* from the medial portion of the intercondyloid notch to the posterior tibial spine. The medial or tibial collateral consists of an anterior longitudinal and a posterior oblique portion, the latter blending intimately with the joint capsule. The *lateral or fibular* collateral ligament is attached to the capitulum fibulæ and is separated from the joint by the tendon of the popliteus muscle.

8. The common *transverse* axis of the knee joint lies in upright standing in a horizontal and in a frontal plane. If placed in these planes the neck of the femur appears rotated outward against the shaft; the ankle joint axis, in this position, is rotated outward due to a twist of the lower against the upper end of the tibia, amounting to $25°$.

9. In the frontal plane the anatomical axes of femur and tibia form an angle of $171°$ opening laterally. The knee joint remaining horizontal, its transverse axis includes with the femur an angle of $81°$ and with the tibia one of $90°$. The mechanical axis of the femur from the center of the head to the middle of the knee joint is less oblique lacking only $3°$ from a straight angle with the tibial axis.

10. Because of the obliquity of the anatomical axis of the femur more pressure is born by the lateral condyles while, on the other hand, more tension stress is sustained by the soft structures of the medial condyles. The weight-supporting line strikes the knee joint with moderate lateral deflection (2.5 cm. from center or less) in 87.5% of the cases (Miculicz).

11. In sagittal elevation the condyles of the femur represent a cardioid line, i.e., one in which the curve becomes sharper from forward to backward and the radii commensurately decrease. Because of this changing curve a true rocking motion in which equidistant points of the moving part come in contact with equidistant points of the fixed part, cannot take place for the whole range of motion. In fact, the knee joint is a combination of two types of motion: rocking (equidistant points contacting) and gliding (a concentrated contact area glides over the contours of the other joint constituent).

12. There is some difference in motion between medial and lateral condyles. On the medial side pure rolling occurs from $180°$ to $170°$ or $165°$; then gliding takes over gradually; on the lateral side rolling occupies $20°$, from $180°$ to $160°$, and then gliding takes over. Rocking is more compatible with stability, gliding more with mobility.

13. In length rotation the axis is closer to the medial condyles so that the lateral rotates around the medial. Rotation occurs between menisci and tibia.

Extension of the knee ends with outward rotation, flexion begins with inward rotation. For the last 10°, 1° of extension corresponds to ½° of outward rotation. For the first 15-20° of flexion, 1° of flexion corresponds to ½° of inward rotation.

14. With the knee extended the tightening quadriceps pulls the upper border of the patella beyond the cartilage covering of the femoral condyles. With the quadriceps relaxed the lower pole of the patella lies flush with the level of the joint. The excursion range of the patella is 5-7 cm. Sidewise mobility of the patella is greater medially; it disappears in flexion, when the patella is pressed against the intercondyloid notch.

15. The rotation of the tibia against the femur in full extension is 5-6°. The tibia itself shows a torsion of the lower against the upper end of 25°. The terminal outward rotation of the tibia against the femur on completion of extension is of considerable importance in the gait, as the swinging leg puts the heel to the ground and the pelvis is rotated forward with the limb in outward rotation.

16. Seen in the *frontal* plane the tibia shows a curve with outward convexity which occupies half of the length of the bone and corresponds to the curve of an excentrically loaded column anchored at both ends. In the *sagittal* plane the tibia shows 3 types of angulations: a *retrotorsion* which is a backward deflection of the upper end; a *retroversion* which is the backward slant of the tibial plateau, and a *retroflexion* which is the bend of the shaft with forward convexity. It is the latter which corresponds to the curve of an excentrically loaded column free to move at both ends (in the sagittal plane) and which occupies the full length of the bone.

17. When the knee is hyperextended gravity develops a horizontal backward directed component; it tends to accentuate the femoro-tibial angle and to force the joint into a position of genu recurvation.

18. The *fibular* collateral ligament is tight in extension. In this position it also checks outward rotation. In inward rotation of the tibia the ligament is first relaxed and finally tightens again. Outward rotation is also checked in flexion. The *medial* collateral restrains rotation in all positions.

19. The checking of antero-posterior motion is not the exclusive function of the cruciates as the tibial collateral restrains it with the oblique fibers.

20. *Ab and adduction* of the knee is checked entirely by the collaterals, the tibial restricting abduction and the fibular adduction. The knee is better secured against this motion when in extension.

21. From 30° flexion on the posterior cruciate becomes tight while the anterior is relaxed. In flexion over 90° both wind themselves around and tighten each other. In extension the anterior becomes taut. Due to the fact that the posterior fibers of the posterior stream into the lateral, and those of the anterior into the medial meniscus, the cruciates maintain a stabilizing relation with the menisci. Both cruciates resist antero-posterior movements; the anterior the forward and the posterior the backward movement of the tibia.

22. The prime function of the menisci is to fill the dead spaces between the femoral and tibial condyles, to increase the range of motion, to provide a more even distribution of pressure, and to protect the capsular wall against inpingement. Both menisci recede backward in flexion; in extension the anterior halves are forced forward and the menisci are stretched; this forward movement ends, however, when the rocking type of motion sets in.

23. In length rotatory movement the axis falls into the medial half of the joint, so that the lateral condyles make greater excursions than the medial. The menisci move against the tibia and consequently the lateral must be more movable than the medial. The lateral meniscus is held firmly to the rotating femur by Humphrey's and Wirsberg's ligaments which coming from the posterior cruciate connect the posterior arc of this meniscus with the medial condyle of the femur; the popliteus, inserted with its deep fibers into the posterior portion of the lateral meniscus, draws it backward when contracting. In all positions the menisci safeguard the stability of the knee joint.

24. The normal flexion-extension range is from 40° to 180°, with occasional 6-9° overextension. According to Zuppinger the equidistant contact obtains up to 15-20° flexion; then motion becomes gliding.

25. When extension is complete the tibia has rotated outward against the femur about 6°. The torsion between upper and lower end of the tibial shaft is 20-25°. The amplitude of length rotation is 50-75° with the maximum at 90° flexion. In extension the joint is locked against axial rotation.

26. The only extensor of the knee is the quadriceps with a cross section area of 148 cm.². The vastus medialis by inserting low into the extensory apparatus is an added safeguard against lateral displacement of the patella.

27. Flexors are the hamstrings with the popliteus and short head of biceps, plantaris and gastrocnemius. With the leg free, all hamstrings flex the knee, with the leg fixed they extend the hip. There is a significant pairing of the flexors, biceps and inner hamstrings having different distances from the knee joint axis.

28. Inward rotation of the leg on the thigh is carried out by the internal hamstrings, outward rotation by the biceps and tensor. Outward rotation range is greatest at 90° flexion (42°).

29. The rotators are better matched in point of working capacity (outward rotators 1.8 kgm., inward 2 kgm.) than flexors and extensors (flexors 15 kgm., extensors 42 kgm.).

BIBLIOGRAPHY

1. BRANTIGAN, O. C., and VOSHELL, A. F.: The Tibial Collateral Ligament: its Function, its Bursae, and it Relation to the Medial Meniscus. *J. Bone & Joint Surg., 25:*121, January 1943.

2. BRAUNE, W., and FISCHER, O.: Die Bewegungen des Kniegelenkes nach einer neuen Methode am Lebenden Menschen gemessen. *Abh. Sach. Ges. d. Wiss. Math. Phys., Kl:*XVII, 1891.

3. FICK, R.: *Spezielle Gelenk und Muskelmechanik, Vol. III.* Jena, G. Fischer, 1911.

4. FISCHER, O.: *Kinematik Organischer Gelenke.* Braunschweig, 1907.

5. GRÜNEWALD, J.: Die Beanspruchung der Langen Röhrenknochen des Menschen. *Ztschr. f. Orthop. Chir.*, *39:27*, 1919, and *39:129*, 1919.

6. GRÜNEWALD, J.: Beziehungen zwichen Form und Funktion der Tibia and Fibula des Menschen und Einiger Menschenaffen. *Ztschr. f. Orthop. Chir.*, *XXXV. XXVIII:675*, 1916.

7. LAST, R. J.: The Popliteus Muscle and the Lateral Meniscus. *J. Bone & Joint Surg.*, *XXXII:*B.1. 93.

8. MEYER, H. VON: Die Mechanik des Kniegelenkes. *Arch. Anat. Phys. Wiss. Med.*, 1853.

9. MIKULICZ, H.: Über die Individuellen Formdifferenzen am Femur und an der Tibia mit Berücksichtigung der Statik des Kniegelenkes. *Arch. f. Anat. u. Entwicklungsgesch*, 1878.

10. STRASSER, H.: *Lehrbuch der Muskel und Gelenkmechanik, Vol. III.* Berlin, J. Springer, 1917.

11. WEBER, W., and WEBER, E.: *Mechanik der Menschlichen Gehwerkzeuge.* Göttingen, 1836.

12. ZUPPINGER, H.: Die Aktive Flexion im unbelasteten Kniegelenk. *Med. Habil. Schrift Zürich u. Anat. Hefte*, 1904.

THE PATHOMECHANICS OF STATIC DEFORMITIES OF THE KNEE JOINT

IN ORDER to appreciate the mechanical causes which underlie the various deformations, twists and axial deflections at the knee joint, one must follow the changes in contours and alignment as they develop when the lower extremity assumes its static functions.

I. THE STATIC DEFORMITIES IN THE FRONTAL PLANE
A. PHYSIOLOGICAL DEVELOPMENT

Of particular significance are the observations of Böhm[1] who emphasized the fact that in the normal newborn the tibia shows a varus deformity, and that

as a result the mechanical axis of the extremity, drawn from the head of the femur to the middle of the ankle joint, barely touches the medial surface of the knee (Fig. 1). In the sagittal plane a forward convexity exists, the apex of which is the knee joint, the tibia participating by a forward bend of its own (Fig. 2).

The relation of the axes of the three joints of the extremity in the newborn is as follows: the axis of the neck of the femur forms an angle of about 60° with the frontal plane due to the normal anteversion. A compensatory inward rotation of the femur with relation to the neck places in the adult the transverse axis of the knee joint strictly in the frontal plane. In the newborn, however, the tibia is still outward rotated against the femur; the knee joint axis forms an angle of 30° with the frontal plane and the axis of the lower end of the tibia is in the same plane with the knee joint axis, i.e., at an angle of 30° with the frontal plane (Fig. 3).

There are, therefore, three kinds of deviations presenting themselves in the newborn:

FIG. 1. Normal newborn. Varus position of tibia, mechanical axis medially displaced. (Bohm)

In the sagittal plane, the forward convexity.

In the frontal, a lateral convexity.

In the transverse, the rotatory deviation.

Comparing these with the adult extremity one notes: The mechanical axis of the entire extremity goes from the center of the hip joint through the middle of the knee and through the middle of the ankle joint with such physiological variations as will be discussed below. The anteversion of the neck amounts to between 12-20°; the lower end of the tibia is rotated outward against the upper end so that the transverse axis of the upper tibial condyles forms an angle of 25° with the transverse axis of the lower tibial joint axis.

Hence the normal developmental changes in post natal life are the following:

(1) The slight forward convexity of the femur with a deviation of 7 mm. in the newborn increases to 11 mm. in the adult.

(2) The retroversion of the head of the tibia which is commensurate with the flexed position of the knee in the newborn, straightens out completely at the tenth year.

(3) The anteversion of the femoral neck decreases from 60° in the newborn to 12-20° in the adult (Pitzen[15]).

FIG. 2. Sagittal plane. Forward convexity exists. (Bohm)

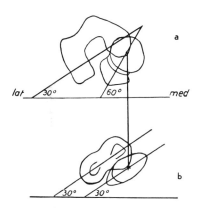

FIG. 3. *a*. Relation of femoral neck to femoral condyles. *b*. Relation of upper tibia to lower tibia. (Böhm)

According to Böhm[1] thigh and leg of the newborn form a uniform arc with outward convexity and retain it during the first year. In the second year the knees come together, the varity disappears and changes to a physiological valgity which is completed at the sixth year. Thus the axis of the extremity which in the first year barely touches the medial side of the knee joint, stays medial of the midpoint of the knee axis in the first two years and then falls into midpoint at the third and fourth and lateral to it at the fifth and sixth year.

The anteversion of the neck of the femur which amounts to 60° in the newborn diminishes to normal values during growth and the inward rotation of the femur itself decreases from 33° of the newborn to 12° of the adult. That means that the horizontal deflection between neck and shaft straightens out to adult value, apparently from both ends: a growth factor being operative at the neck and a muscle mechanical factor (especially the extensors) being operative in bring about the detorsion of the femoral shaft.

So far as the tibial torsion is concerned, which reaches 23-25° in the adult, it is, according to LeDamany,[9] also largely a muscle mechanical effect. It is the resultant of the muscle tension between the abductors of the foot, the extensors and peronei, and the opposing inward rotators of the tibia, namely the sartorius, the vastus internus, semitendinosus and popliteus.

B. PHYSIOLOGICAL VARIATIONS

There are variations within physiological limits. The inclination angle of the upper end of the tibia (frontal plane) varies in adults between 91-110°, the retroversion angle in the sagittal plane between 97-116°, and, according to Miculicz,[13] the torsion of the tibia varies from 0-48°.

It is difficult to determine the physiological border lines. We recognize from our observation of familial and hereditary cases that one factor behind the deformity is primary aberrations of the growth pattern. We also know from evidences of rickets or other conditions which weaken the bone resistance that static forces are necessary for the production of the deformity; but one is often in doubt as to how much is attributable to the growth and how much to static factors. The primary aberration of the growth pattern is less amenable to spontaneous correction than is the static element. From the clinical viewpoint it is important to be able to appraise the limits of natural correction under the laws of functional adaptation.

II. STATIC DEFORMITIES OF THE KNEE JOINT
A. THE PATHOMECHANICS OF STATIC GENU VALGUM

Bragard[2] estimates that a genu valgum exists if the displacement in relation to the line of gravity equals half the distance between the internal malleoli; 2.5cm. distance of the line of gravity from the midpoint of the knee is considered the physiological limit (Miculicz[13]). The appearance of the genu valgum deformity coincides with the two periods of rapid growth, namely from two to five years and from 14 years to maturity. In 90% the seat of the deformity is the upper metaphysis of the tibia.

1. The mechanical element
A) THE STATIC FACTOR

Already Hueber[8] recognized the relation of the deformity to the static functions of standing and walking and that the valgity as well as the outward rotation of the tibia develops under gravitational influence. The earlier observers considered the static genu valgum a pure weight bearing deformity expressing Hueter and Volkmann's theory of compression atrophy of bone.

Wolff[20] also emphasized the possibility of a genu valgum developing exclusively on static basis, but later observers, especially Schmorl[17] were of the opinion that a certain degree of pliability of bone such as exists in rickets must be operative also (J. Wolff[20]).

Conforming to the valgus deformity, the medial condyle is enlarged and the lateral is diminished and flattened as the weight bearing line proceeds toward the outer side of the knee. Another mechanical factor is transmitted strain and subsequent relaxation of the internal collateral. A pronated foot will cause the reaction from the floor to be transmitted along the tibia to the internal collateral ligament which, being sensitive to strain, becomes thickened. The internal meniscus becomes atrophic while the lateral under increased static stresses hypertrophies (Putschar[16]).

b) THE DYNAMIC FACTOR

There is a third factor operative in producing the deformity (Grunewald[7]). When the weight of the body lies vertically over the knee joint, the latter receives only pressure and no bending stresses, but if the femur is abducted, the weight line falls medially to the knee joint tending to bend the femur outward without increasing the femoro-tibial angle. Grunewald therefore concludes that it is a muscle pull rather than gravity which produces genu valgum: the quadriceps with its laterally deflected tendon, the powerful vastus lateralis which bowstrings the extended knee. While rickets may play an additional role, genu valgum is an exaggeration of the physiological valgity produced by static as well as by dynamic factors.

So far as the metaphyseal deformity is concerned, the pressure is increased at the concavity (lateral), the tension effect at the convexity (medial). The increased lateral pressure transmitted to the epiphyseal plate further deflects the normal direction of growth (Maas[11]).

2. The growth element

It is important to distinguish between actual deformation of the femoral or tibial shaft and the obliquity of the joint which is caused by the irregular epiphyseal growth. While it is certain that excessive weight may cause bony deformities in the early age period, it is usually the unequal epiphyseal growth of the femoral condyles which produces the inclination of the knee joint axis to the side of the increased growth activity (internal condyle).

As early as 1893 Ghillini[6] produced skeletal deformities, among these a genu valgum, by injuring the lateral portion of the lower femoral epiphyseal plate. Then Maas[12] obtained, by purely mechanical pressure, suppression of the lateral epiphyseal growth with subsequent deformity. It is hardly necessary to point to the ability of pathological aberrations of cartilaginous growth to produce this deformity. Exner[5] reports recently a case of an impinged lateral discoid meniscus with osteochondritis dissecans in which the lateral condyle of the femur became depressed by growth inhibition and a genu valgum developed.

From the clinical point of view one may distinguish:

(1) The purely articular aspect of the deformity resulting from the relaxation of the internal collateral ligament. Here the valgity definitely increases under weight bearing. The outlook of correction by immobilization and protection of the knee by braces is good.

(2) The osseous element of the deformity. The angulation is situated in the upper tibial metaphysis with sharp angulation or the curve is distributed over the whole length of the tibia. In the latter cases the chances of spontaneous correction under the law of functional transformation are considerably better than in cases of sharp and localized angulation, always provided that the tibia is protected against further bowing by proper appliances.

(3) A compensatory genu valgum as growth aberration when the other limb is shortened. Here the valgity develops as an effort of equalization of leg lengths under the effect of excessive growth of the medial portion of the epiphyseal plate.

B. THE PATHOMECHANICS OF THE STATIC GENU VARUM

In many instances a genu varum merely presents the continuation of the varity from the first year of life and a failure of reorientation in valgus direction during the second year. Here the genu varum really involves the entire extremity in a long outward convex arc, the summit of which lies, according to Böhm,[1] in the region of the upper tibial metaphysis. The curving is mild. A significant point is the absence of the outward torsion of the tibia or even an inward torsion with the result that the feet are held in adduction. Combination with genu recurvatum is frequent.

Similarly as in genu valgum one finds in rachitic cases the angulation usually restricted to the upper end of the tibia. A compensatory genu varum may also develop in cases of shortening of the other extremity (Putschar[16]) (Fig. 4).

Structurally the corticalis is thickened on the concave medial side and is thinned out at the convexity. The mechanical axis of the femur is shifted medially and it may even fall entirely medial to the knee joint. The pressure being concentrated at the medial condyles, functional hypertrophy develops. It is significant that in the long run the joint will resent the unequal pressure distribution much more than it does in the opposite deformity of genu valgum; it reacts with osteoarthritic degenerative changes in later life.

So far as spontaneous correction is concerned, here again the type and site of the deformity are of importance. Sharp kinks at the upper metaphysis undergo functional adaptations much more poorly than the curves distributed along the whole length of the tibia.

The deflection of the axis of the tibia in varus direction and the inward rotation of its lower end has naturally a definite effect upon the ankle joint. The astragalus shares with the tibia the varus obliquity in this joint. The subastragalar joint, on the other hand, must go into a compensatory pronation to adapt the sole to the ground. Furthermore the inward rotation of the tibia directs the body and neck of the astragalus into an adductory position. Only milder degrees of this inward rotation of the ankle joint can be neutralized by abduction of the forefoot in the mid-tarsal joint. Higher degrees take the whole foot with it and the patient stands and walks pigeon-toed.

For mechanical reasons the lateral collateral ligament does not show a similar amount of relaxation in the static genu varum as does the medial collateral ligament in the case of genu valgum. The joint is more firm and it

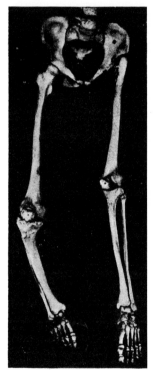

FIG. 4. Compensatory genu varum following shortening of opposite extremity. (Put-schar)

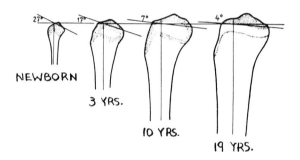

FIG. 5. Changes in slant of tibial condyle from newborn to adult. (Titze)

does not have the lateral instability. On the other hand the static stresses take more effect on the bone itself and the varus curves of the tibia often assume excessive degrees.

C. THE PATHOMECHANICS OF THE STATIC GENU RECURVATUM

It is a frequent complication of the genu valgum or varum, principally of the former. A physiological hyperextensibility of the knee to about 5° is seen in small children and it is related to the slight downward slope of the tibial epiphysis. High degrees of congenital genu recurvatum are seen, for instance, in arthrogryposis. The true static genu recurvatum ranks with the static genu valgum and varum as a deformity in which static stresses produce the deformity on the background of some insufficiencies of the joint such as constitutional ligamentous relaxation or more rarely asymmetries of epiphyseal growth. Here also belongs the group of cases in which the deformity is compensatory to a shortened opposite limb. This manner of length compensation occurs more often than by a compensatory genu valgum or varum. Another factor producing hyperextension of the knee is a short heel cord.

Gravity acts in direction of hyperextension once the line of gravity falls in front of the knee joint axis. Its effect is noted also in cases of muscular imbalance, for instance in weakness of the hamstrings where the extensors

Fig. 6. Static genu recurvatum.
(Peltesohn)

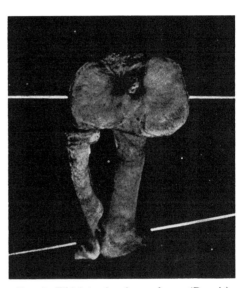

Fig. 7. Tibial torsion in newborn. (Dupuis)

of the knee operate without opposition. Furthermore, even in cases of extensor paralysis, the habitual forward flexion of the trunk which stabilizes the knee may lead to a hyperextension deformity, especially if a short tendo Achillis exercises a strong backward thrust upon the knee.

Other causes are: a downward and backward slant of the tibial condyles. This slant should decrease after birth until it reaches the adult angle of 4-6° (Titze[19]) (Fig. 5). Excluding inflammatory or destructive conditions the causes of genu recurvatum can be summarized as follows (Peltesohn[14]): (1) inequality of growth of the lower femoral epiphysis; the fibula prevents similar disalignment of the upper tibial epiphysis; (2) a static genu recurvatum with tibial angulation to shorten extremities (Fig. 6); (3) a compensatory tibia recurvata in flexion ankylosis of the knee (Patal and Cavaillon[3]).

D. THE MECHANISM OF TIBIAL TORSION (Fig. 7)

Rotatory deformity of the knee is seen both in antero-posterior (genu recurvatum) and in lateral (genu valgum and varum) deformities. Chronologically these changes occur as follows: the lateral deflection of the axis of the tibial head which in the newborn forms an angle of 76° with the transverse plane and straightens out to 86° at the age of one. The backward deflection of this axis in the sagittal plane diminishes likewise from 116° of the newborn to 96° permanent adult value; while the torsion between the upper and lower transverse axes is only 0° in the newborn, it amounts to 12° at the age of one, and finally reaches its permanent value of 23° at the age of six (LeDamany[9]). A lateral torsion about the longitudinal axis is normal, the fibula following the tibia in outward rotation. This outward rotation is seen in almost all adult tibias (Dupuis[4]). It varies in 80% of the cases between 10-30° with an average of 18%. In an earlier investigation LeDamany[10] established the average in two series at 23.5° and 20° respectively. Dupuis[4] in his interesting observations on seventy tibias of newborn found tibial rotation essentially different from the adult. In the newborn the torsion varies between 15° internal or negative and 23° external or positive rotation, and in 80% the variations were between +10 and —10 with an average of 2°11′ of external rotation. Thus, while internal rotation of the lower tibia is exceptional in the adult, it is found in 90% of the newborn.

Adult tibial torsion is therefore an acquired phenomenon. In the normal sketleton the tibia which shows only slightly outward torsion at birth (2° average) increases this torsion rapidly during the first year of life (Dupuis[4]). At the age of one it is at an average of 10°. Then the torsion remains stationary during the second and third year. Between 3½-4 years the torsion increases again to an average of 20° which approaches the adult level.

As is to be expected there is a reciprocal relationship between the torsion of the tibia and the rotation of the foot in pes plano-valgus and pes equinovarus. In the former the inward rotation and adduction of the astragalus forces the tibia into inward rotation. Inward rotation of the tibia may be ontogenetic as in congenital clubfoot or mechanical as in flatfoot with inward rotated astragalus (Sonnenschein[18]). In the clubfoot the tibial mortice may retain its normal position corresponding to the age of the infant. But as Dupuis[4] points out, the rotary force of the deformed foot in severe cases may impose upon the tibia a pathological internal rotation which may reach 55° as the static function of the gait is taken on. Consequently one finds in very severe clubfeet at birth and before treatment a slight outward torsion of the tibia in 85%; after treatment the torsion of the tibia varies according to whether the foot has a tendency to stay corrected or whether the deformity will recur. In the first case the development of the outward torsion goes on even more rapidly than normal. In the second case the development of the outward torsion is retarded or entirely arrested. This arrest of the normal outward rotation of the tibia or its reversal to inward rotation would indicate the forceful effect of the reaction from the floor upon the astragalus which, transmitting itself to the tibia, forces the latter into internal rotation (Sonnenschein[18]). In other words, the normal pattern by which external rotation of the tibia develops can be restrained or reversed by the strong derotating force of the severe and inveterated clubfoot.

III. SUMMARY

1. The curves of the tibia of the *newborn* in contrast to the adult are: a forward convexity in the sagittal plane, a lateral bowing in the frontal plane and an outward rotation of the tibia against the femur in the transverse plane.

2. The post natal developmental changes are slight increase in the forward convexity of the femur, a straightening of the retroversion of the head of the tibia and a decrease of the inward rotation of the femur corresponding to a diminished anteversion of the neck.

3. The outward torsion of the lower against the upper end of the tibia which reaches 23° in the adult (LeDamany) is also a post natal development being the result of muscle tension between abductors and extensors of the foot and the opposing inward rotators of the tibia, the sartorius, vastus internus, semitendinosus and popliteus.

4. There is a mechanical as well as a growth element operative in the production of a static genu valgum.

5. The mechanical element consists of static and dynamic factors; the valgity as well as the outward rotation of the tibia develops under gravitational influence or the stress transmitted to the internal collateral ligament by a pronated foot. The dynamic influence is represented by muscle pull rather than gravity: the quadriceps with its laterally deflected tendon, the powerful vastus lateralis bowstringing the extended knee.

6. In the growth element one must distinguish between actual deformation of the femoral or tibial shaft and obliquities of the joint caused by asymmetrical epiphyseal growth. The latter produces an inclination of the knee joint axis to the side of the accelerated growth.

7. Asymmetrical retardations of epiphyseal growth were produced experimentally by injuring the lateral portion of the epiphyseal growth plate (Ghillini) or by mechanical pressure (Maas).

8. Clinically one must differentiate between: *a)* valgity due to relaxation of the internal collateral: this valgity increases under body weight and can be corrected by mechanical means; *b)* the osseous element of the deformity; if the curve is distributed over the whole shaft the outlook for spontaneous correction is better than where there is a sharp angulation; and *c)* a compensatory genu valgum as growth aberration to equalize shortness of the other leg. This is also a growth aberration with excessive growth of the medial half of the epiphyseal plate.

9. A moderate genu varum often represents a continuation of the varity which is normal for the first year. Significant is also the retardation or failure

of the outward rotation of the lower end of the tibia so that the feet are held in adduction.

10. A compensatory genu varum may develop in cases of shortening of the other extremity.

11. Due to the medial shift of the axis functional hypertrophy of the medial condyles develops.

12. Spontaneous correction again depends upon the distribution of the curve, as in genu valgum.

13. The varus deflection of the tibia forces the astragalus in varus position in the ankle joint; the subastragalar joint goes into compensatory pronation.

14. Genu recurvatum is a frequent complication of genu valgum or varum. Here also static pressure produces the deformity upon the background of some constitutional ligamentous relaxation of the joint or on the basis of asymmetries of growth. It may also be compensatory to a short opposite limb. A short heelcord facilitates a genu recurvatum by the back thrust of the tibia. Weakness of the hamstrings, or in case of extensor paralysis, even the forward inclination of the trunk favors the development of a genu recurvatum. It may also be produced by a backward slant of the tibial condyles or appear in form of a compensatory tibia recurvata in flexion ankylosis of the knee.

15. The tibial deflection and torsion also have a developmental background: the deflection of the tibial head in the frontal plane seen in the newborn straightens out at the age of one. So does also the forward convexity. On the other hand the axial torsion develops after birth and reaches 23° (LeDamany) outward rotation of the lower end of the tibia at the age of six. According to Dupuis this rotation varies in the newborn between 15° internal and 23° external, but the average is 2°11' of external rotation; at the age of one this average is 10° and between 3½ and 4 it is 20° external rotation approaching adult level.

16. In pes planus the inward rotation of the astragalus in the subastragalar joint forces the tibia into inward rotation. In the clubfoot the tibial mortice may retain its normal position and the deformity be restricted to the foot alone. In severe cases, however, the powerful inward rotation of the foot may transmit itself to the tibial mortise and force the tibia into inward rotation. Dupuis finds slight outward rotation of the tibia at birth in 85% of clubfoot cases, but after treatment the condition varies; in those that stay corrected the outward rotation develops as normal; in those with a tendency to recurrence, the outward rotation is arrested or retarded. This arrest or even reversal to inward rotation indicates the forceful floor reaction on standing or walking which transmits itself to the tibial mortice through the astragalus and forces the tibia to follow the inward rotatory stress.

BIBLIOGRAPHY

1. Böhm, M.: Genu Varum and Valgum Infantum. *Ztschr. f. Orthop. Chir.*, *49*:321, 1927-28.

2. BRAGARD, K.: Das Genu Valgum. *Ztschr. f. Orthop. Chir., 57*, Beitr. Heft, 1932.

3. CAVAILLON and PATEL: Ankylose de la Hanche. *Revue d'Orthop*, 1904.

4. DUPUIS, PAUL V.: *La Torsion Tibiale*. Paris, Masson & Cie, 1951.

5. EXNER, Z.: Beitrag zur Pathogenese der Osteochondritis dissecans. *Ztschr. f. Orthop. Chir., 81:*3, 32, 386, 1951.

6. GHILLINI, C.: Experimentelle Untersuchungen über die Mechanische Reizung des Epiphysenknorpels. *Arch. Klin. Chir., 40:*4, 1893.

7. GRÜNEWALD, J.: Die Beanspruchungsdeformitäten. *Acta. Orth. Chir., XXXVIII, XIII:* 449, 1918.

8. HUEBER, C.: Zur Theorie und Therapie des Genu Valgum. *Arch. Klin. Chir., 9:*961, 1868.

9. LeDAMANY, P.: Le Femur, la Double Formation dans la Série Animale. *J. Anat. Phys.,* 42. 39, Paris, 1906.

10. LeDAMANY, P.: La Torsion du Tibia Normale, Pathologique, Expérimentale. *J. de l'Anatomie et Physiol. Paris,* 1904.

11. MAAS, H.: Über Mechanische Störungen des Knochenwachstums. *Arch. Path. Anat., 163:*12, 1901.

12. MAAS, H.: Über Experimentelle Deformitäten. *Ztschr. f. Orthop. Chir., XI:*122, 1903.

13. MICULICZ, O.: Die Seitlichen Verkrümmungen am Knie und deren Heilmethoden. *Arch. Klin. Chir., 23:*561, 1879.

14. PELTESOHN, S.: Zur Aetiologie und Pathologie des Genu Recurvatum und der Tibia Recurvata. *Ztschr. f. Orthop. Chir., XXII.XXVI:*602, 1908.

15. PITZEN, P.: Das X-bein Rhachitischer Kinder im Roetgenbild. *Ztschr. f. Orthop. Chir.,* Vol. 41.

16. PUTSCHAR, W.: Der Funktionelle Skeletumbau und die Sogenannten Belastungsdeformitäten Henke-Lubarsch. *Handbuch der Speziellen Pathlogischen Anatomie und Histologie, 9:*3, 19, Berlin, J. Springer, 1937.

17. SCHMORL, G.: Über Rhachitis Tarda. *Arch. Klin. Chir., 85:*203, 1905.

18. SONNENSCHEIN, A.: Zur Torsion des Unterschenkels als Ursache einer Seltenen Deformität. *Ztschr. f. Orthop. Chir., 81:*4, 56, p. 593, 1951.

19. TITZE, A.: Die Variationen der Neigung der Schienbeinkopfgelenkfläche. *Ztschr. f. Orthop. Chir., 80:*40, 436, 1950.

20. WOLFF, J.: Zur Aetiologie des Genu Valgum Adolescentium. *Ztschr. f. Orthop. Chir., 51:*98, 1929.

Lecture XXI

THE PATHOMECHANICS OF THE PARALYTIC KNEE

I. INTRODUCTION

THE knee joint is situated between two long lever arms, the thigh and the leg; of its two degrees of freedom of motion it is the one in the sagittal plane which is under the influence of gravity; it is locked anatomically only against hyperextension while the flexory range is fully exposed to gravitational stress. Hence it is the extensory force upon which the joint depends to maintain balance and to prevent collapsing.

If the extensory apparatus is eliminated or weakened by paralysis, equilibrium and weight bearing can still be maintained by the operation of two factors. The first is gravity itself; in upright standing the line falls normally in front of the joint; the knee, due to its construction being locked against hyperextension, it is possible for gravity alone to stabilize the joint in this position. The second factor is the extensory effect upon the knee of certain muscles controling motion of the hip and ankle; the gluteus maximus and the soleus (Fig. 1). Therefore, the loss of the extensory apparatus of the knee can be redeemed to a certain degree both by gravity and by substitutionary muscle action.

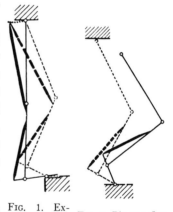

FIG. 1. Extensors of the hip and soleus extending the knee.

FIG. 2. Plantar flexors of the ankle flexing the knee.

On the other hand, gravity becomes a powerful flexor of the joint whenever its line falls behind the transverse axis of the knee joint; and, in certain mechanical situations, the biarticular hamstrings are actually aided by the plantar flexors of the ankle in flexing the knee joint (Fig. 2).

In the transverse plane, substitution of the paralyzed inward and outward rotators by rotation in the hip joint is more difficult to accomplish, although normally rotation automatically accompanies the beginning of flexion and the end of extension. This combination of rotation with flexion and extension may, in fact, cause one to question whether the initial range of knee motion can be described as a true rocking, i.e., one in which equidistant points come successively in contact. However, there should be no misgiving on that score because the amount of the accompanying rotation is, in terms of angular values, comparatively small. According to Fick,[6] and Strasser[7] for the first 5° one degree of flexion carries only 25′ of rotation; for 5-10°, 20′; and for

10-15°, only 8′; and from 15-20°, practically no rotation occurs per degree of flexion. Nevertheless, rotation of the knee is an important function in the gait and in case of failure of the rotators its substitution may become a problem in reconstruction.

A third point to be considered is the effect which the knee muscles and especially the flexors have on both the lateral and the rotatory stability of the joint. The lateral ligaments are tight in extension, the tibial in its posterior portion, the fibular in toto, and the tension of the tensor fasciae laterally and the hamstrings medially acts as a strut or reinforcement of these ligaments. Consequently, weakness of these muscles may lead to lateral instability of the knee, with outward rotators and abductors prevailing, so that a genu valgum with supination of the leg results.

II. EXTENSOR PARALYSIS

The quadriceps, which is the only extensor of the knee joint, was found involved in our series of infantile paralysis of the lower extremity in 74% (J. Sher[16]). Lovett's[13] figures for all muscles are 152 in 1,417 cases.

A. PARALYSIS OF THE VASTI

The extensory apparatus of the knee is formed by the confluence of the tendons of all four muscles constituting the quadriceps, i.e., the rectus, the vastus intermedius, and the vastus medialis and lateralis. These muscles control the movements and positions of the patella. While all muscles are synergistic in extension, the lateral balance is held between the vastus medialis and lateralis. In case of paralysis of the vastus medialis and the rectus femoris the patella might easily be dislocated laterally. Duchenne[1] states that he has observed children with atrophy of the vastus medialis in which on active extension of the leg the upward movement of the patella was associated with lateral displacement. There is, however, no spontaneous displacement to the medial side observable when the lateral vastus is paralyzed.

B. THE PARALYSIS OF THE ENTIRE QUADRICEPS

It has already been mentioned that standing and walking is possible without the quadriceps because it can be substituted by gravity and by the transmitted action of the gluteus maximus which extends the hip and of the soleus which plantarly flexes the ankle.

Still, there is a limit to this substitution. The principal requirement is that the knee be straight and not contracted. In the latter case a substitution by gravity is possible to a certain degree by bending the body forward and thus throwing the line of gravity to pass in front of the knee joint, but it becomes a difficult maneuver when the contraction exceeds 15-20°. It is common experience that in paralytic knees even a slight contracture makes a difference in standing and walking ability, and the correction of such a contracture is, in fact, a prerequisite for the substitution of the quadriceps by tendon transplantation.

The question arises what the loss of the quadriceps means in fast walking or in running, where a higher degree of flexion of the knee is required. Duchenne[4] calls attention to the fact that standing or walking becomes impossible when the thigh forms an angle with the leg opening posteriorly, and that the slightest flexion of the knee produced by contracture of the flexors makes standing on the affected side impossible. This is certainly true unless the contracture is very mild and the patient avails himself of the expedient of throwing the body forward or if he steadies the knee by pressing the thigh backward with the hand, which, in our experience, the majority of patients prefer.

But even if the knee is straight there are limitations. The patients walk slow and the standing period is shortened. When they walk fast they do not raise the affected leg as they do the sound one, but they bring it forward in a more extended position.

Under normal conditions it requires the action of the quadriceps to bring the leg forward fast enough to the position of heel contact and, at the same time, to resist the flexory component of gravity during the period of restraint. In other words, the natural swinging time of the leg is normally accelerated by quadriceps action, as every maker of artificial limbs is aware of when he uses the elastic "accelerators" over the artificial knee joint. In quadriceps paralysis this difficulty is usually overcome by the patient bending slightly forward at the moment of heel contact and thus using gravity, as it were, as an accelerator.

III. THE PARALYSIS OF THE FLEXORS OF THE KNEE
A. LOSS OF FLEXION

All of the flexor muscles of the knee have secondary functions and all of them except the short head of the biceps and the popliteus are biarticular. The tensor and sartorius are flexors of the hip joint; the gracilis is an adductor; the semitendinosus and semimembranosus as well as the long head of the biceps extend the hip joint. In addition, the flexor muscles have a secondary effect upon the length rotation. The sartorius and the long head of the biceps are outward rotators; the inner hamstrings rotate inward and so does the popliteus.

To appreciate the loss of the flexors of the knee one must first realize that biceps, semitendinosus and semimembranosus are important extensors of the hip joint. The gluteus maximus steps in where a particular extensory effort is necessary. In ordinary upright standing it is relaxed. If the patient inclines his body forward the gluteus maximus holds it in balance; if it is paralyzed the patient has the tendency to fall forward.

In the knee joint the overwhelming tension of the quadriceps, aided by the gravitational stress, forces the joint backward in direction of a genu recurvatum; but the most disturbing fact in flexor paralysis is that the limb cannot be shortened. The leg must swing forward full length and the pelvis be lifted to clear the limb off the ground. If a fixed genu recurvatum develops, this shortens the leg and pelvis lifting can be dispensed with. The heel is set to the ground with the knee fully extended or hyperextended. The restraint is firm but the take-off is markedly weakened.

B. THE LOSS OF ROTATION

The biceps forces the leg into outward rotation. *In extension,* the fibular collateral ligament of the knee which runs obliquely from the lateral epicondyle of the femur to the uppermost portion of the tibiofibular joint is relaxed when the leg is strongly outward rotated against the thigh, and it becomes tight in inward rotation. *In right angle flexion* of the knee, the fibular collateral ligament is relaxed in midrotation and its tension increases as the limit of inward or outward rotation is reached. In case of infantile paralysis, when all knee flexors are involved, the natural valgity of the knee and the normal outward rotation in extension greatly favor the development of a valgus knee with outward rotation and hyperextension, especially since the relaxed fibular collateral ligament does not resist the deformation. The foot is turned out and the lateral contour of the leg forms an obtuse angle, opening outward.

The situation is aggravated when the paralysis involves the internal rotators of the leg only and the unopposed action of the biceps results in strong outward rotation. If functioning, the popliteus, semitendinosus, gracilis as well as the tibial collateral ligament can oppose and moderate the effect of the biceps in rotating the leg outward. The sartorius has only a weak outward rotatory power.

IV. THE RECONSTRUCTIVE OPERATIONS FOR PARALYSIS OF THE KNEE FROM THE PATHOKINETIC VIEWPOINT

For reasons explained above the contracted or the distorted knee is in unfavorable position to meet mechanical stresses. Gravity accentuates disalignment because it becomes a flexory force when the knee is in flexion contracture; it becomes an abductor of the leg in genu valgum, and it acts also as an extensory force in the genu recurvatum.

The conditions which require anatomic reconstruction and which are of special kinetic interest are the genu recurvatum and the genu valgum, and, from the dynamic point of view, the deficiencies of muscular imbalance; the flexion contracture and the paralysis of the quadriceps.

A. THE METHODS OF RECONSTRUCTION OF THE GENU RECURVATUM

It is a weight-bearing deformity based on paralysis of the flexors and the relaxation of the ligamentous reinforcements of the joint. The pathological hyperextension of the knee is associated with valgus deformity and external rotation of the tibia.

The methods of reconstruction fall into three groups· the fasciodesis, the tenodesis, and the osteoplastic operations.

1. Fasciodesis

Gill's[7] method is a transposition of the posterior one-third of the fascia lata and of the collateral ligaments. These structures are sutured behind the knee joint with strips of fascia lata, both medially and laterally, so that a firm fascial support is created at the posterior aspect of the joint and in front of the vessels. It creates a very strong fascial check against hyperextension (Fig. 3). A similar type of fasciodesis is described by Heyman[8] in which both collateral ligaments are reattached to the femoral shaft posteriorly (Fig 4).

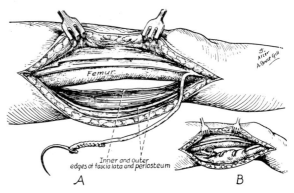

FIG. 3. Gill's fasciodesis for genu recurvatum. (Campbell—redrawn from Gill)

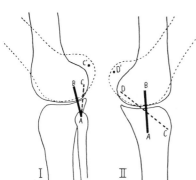

FIG. 4. Heyman's fasciodesis using collateral ligaments. *I*. The external collateral ligament is represented by the heavy line A-B. The interrupted line, A-C, represents course of the new band constructed. C′ represents the attachment C with the knee in recurvatum, which would be impossible if the band holds. *II*. The internal collateral ligament is represented by the heavy line A-B. The interrupted line, C-D, represents the course and attachments of the new band constructed. D′ represents the attachment D with the knee in recurvatum which would be impossible if the band holds. (Heyman)

The virtue of both methods lies in providing a heavy ligamentous check for the posterior aspect of the knee where the capsular apparatus is grossly distended. From what we know about fasciodesis in other situations, there is little or no give once the fascia has healed firmly into its bed and the results of the operations mentioned, which were favorably reported with observation times up to several years, should be permanent, provided always that epiphyseal growth has been completed. If it has not, a recurrence of the deformity due to asymmetrical epiphyseal growth may be expected.

2. Tenodesis

Slightly less secure is another method (Heyman[9]) in which the reinforcement comes from the tendon. Medially, the tendons of insertion of the paralyzed gracilis and semitendinosus are inserted into the posterior aspect of the internal femoral condyle; laterally, that of the biceps is fastened into the lateral condyle by means of a tube made of fascia lata (Fig. 5).

Any equinus position of the ankle joint which exerts a backward thrust

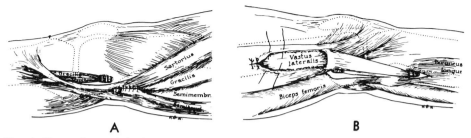

Fig. 5. Heyman's tenodesis for genu recurvatum. *A.* for medial reinforcement using gracilis and
semitendinosus; *B.* for lateral reinforcement using biceps and fascia lata. (Heyman)

of the tibia must first be corrected. Here, again, the same question arises as
does in the use of ligaments; will the tendon of a paralyzed muscle stretch
under stress? Experimental findings as well as clinical observations made by
Gallie and others on tenodesis seem to indicate that no passive elongation
occurs. The mechanical effect of the tenodesis depends more upon proper implantation and reliable anchorage.

3. Osteoplastic

The third principle applied in the reconstruction of the genu recurvatum is
that of an osteoplastic reconstruction of the joint constituents. The operations
involve three different mechanical features: the anterior bone block, the elevation of the tibial plateau, and the elimination of the joint by fusion.

a) The anterior bone block. 1) Campbell and Mitchell's[1] method implants

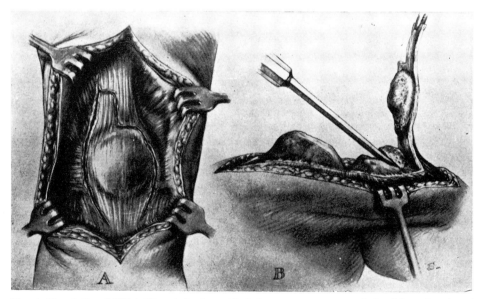

Fig. 6. Campbell and Mitchell's anterior bone block for genu recurvatum. *A.* Quadriceps lengthened
and patella and ligamentum patellae freed. *B.* Distal end of patella freed of soft tissue attachments
and denuded to spongy bone. Osseous flap pried forward from anterior aspect of tibial condyle.
(Campbell)

the patella into the tibia after lengthening of the quadriceps tendon, lodging the bone in a cleft in the tibia behind the tubercle. This forms a solid block, the object of which is to lock hyperextension in a manner similar to the action of the olecranon in the elbow (Fig. 6). These authors report satisfactory results without recurrence. One may have some misgiving about the amount of stress the knee joint will tolerate in view of the shortness of the lever arm of the newly created block.

2) To some extent this drawback is met by a modification of Mayer[14] who uses a stout tibial graft, the upper end of which is imbedded into the patella and the lower end into a slot of the tibia (Fig. 7). The advantage is that the patella is left in place, and, consequently, the lever arm abutting against the condyles is considerably longer. A disadvantage is that the graft is not as massive as the patella itself and there is greater danger that the graft will become absorbed under the pressure and bending stress. This surgeon reported 12 cases with stable knees and without recurrence of the deformity. Milgram modified the technique further by carrying out this operation in a manner which avoided entering the knee joint.

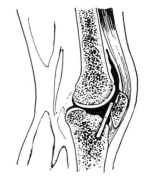

FIG. 7. Mayer's modification of bone block for genu recurvatum. Saggital section through the knee joint after the implantation of the bone graft. The deep surface of the patella has been denuded of cartilage and makes contact with the rough surface of the graft. The smooth surface of the graft faces the femur. (Mayer)

b) An entirely different principle is involved in the elevation of the tibial plateau. It was an old idea of Lexer[12] to raise the tibial plateau by a bone wedge with anterior base, thereby causing the tibial condyles to slant strongly backward. According to M. Lange,[11] this method, while satisfactory in the congenital genu recurvatum, is somewhat less reliable in the paralytic type where the hamstrings are entirely inactive.

These methods of bone block are based solely upon the mechanical effect of skeletal reconstruction and do not take into consideration the secondary relaxation of the soft capsular structures. For this reason the above mentioned soft tissue operations of Gill and Heyman may seem to offer better promise for enduring results.

Secondary ligamentous relaxation may imperil the outcome of either type of operation, however—the soft tissue as well as the osteoplastic type. If we can rely on the fact that the ligamentous or tendinous anchorage remains stable, the soft tissue procedures should be preferred.

c) The arthrodesis. Without doubt, fusion of the knee joint is the surest way to handle a genu recurvatum, but sacrificing all motion is a high price to pay. Cleveland's[2] report covers 90 follow-up cases with good results in 80%. The failures were mostly in cases operated too young, i.e., before the age of 13. There are unquestionably cases of flail knees in which this radical method is justified. From the kinetic point of view, the decision should depend on whether unsupported weight bearing, even with deformity, is possible or not. There are some knees so excessively deformed and so unstable and painful on weight bearing that this radical procedure is justified.

B. THE RECONSTRUCTION OF THE PARALYTIC GENU VALGUM

A genu valgum with outward rotated tibia is often part of the clinical picture of the paralytic genu recurvatum. The valgity can be corrected in two ways. The osteotomy can be applied at any age, except that so long as growth is not completed there is a possibility of recurrence.

If growth is not completed, a suppression of the medial half of the lower

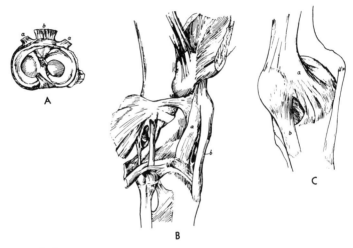

FIG. 8. Anatomy of the retinaculum (Eggers). *A.* Insertion of patellar ligament and muscular retinacula into tibia. *a:* muscular retinacula. *b:* patellar ligament. (Redrawn from Morris' *Human Anatomy,* Ed. 10, Fig. 346, p. 349.) *B.* Lateral aspect of knee joint. Retinaculum (*a*) is divided in the surgical procedure. *a:* muscular retinaculum. *b:* patellar ligament. (Redrawn from Toldt: *Anatomisher Atlas fur Studierende und Arzte,* Fig. 469, p. 231. Berlin, Urban und Schwarzenberg, 1911.) *C.* Medial side of knee joint. Retinaculum (*a*) is divided in the surgical procedure. *a:* muscular retinaculum. *b:* patellar ligament. (Redrawn from Gray's *Anatomy of the Human Body,* Ed. 24, Fig. 345, p. 334. Philadelphia, Lea and Febiger.)

femoral epiphyseal plate will correct the deformity up to a certain degree. In this respect, Blount's staple method has the advantage over the removal of the epiphyseal half in that it can be applied at an earlier age and the removal of the staples at suitable intervals can be controlled. There is at present no table available which shows the relation of asymmetrical growth suppression to the axial deviation.

C. THE RECONSTRUCTION OF THE FLEXION CONTRACTURE

Eggers[5] called attention to the limitation of extension caused by the shrinkage of the extensory retinaculum (Fig. 8). Prolonged knee flexion causes stretching of the patellar ligament but less so of the structures lateral and medial to it. Thus the extension may become restricted by the fibrous and muscular structures which extend obliquely from the sides of the patella to the condyles of the tibia. For this reason he carries out a section of the retinacula (Fig. 9). The flexion contracture which is caused by the muscular imbalance due to paralysis of the quadriceps requires surgical correction as a preliminary step to subsequent tendon transplantation.

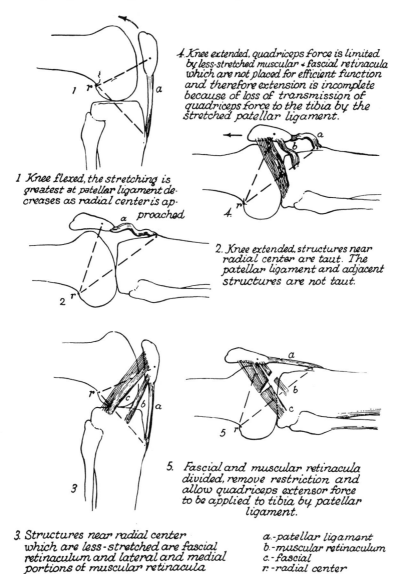

4. *Knee extended, quadriceps force is limited by less-stretched muscular & fascial retinacula which are not placed for efficient function and therefore extension is incomplete because of loss of transmission of quadriceps force to the tibia by the stretched patellar ligament.*

1 *Knee flexed, the stretching is greatest at patellar ligament decreases as radial center is approached*

2. *Knee extended, structures near radial center are taut. The patellar ligament and adjacent structures are not taut.*

5. *Fascial and muscular retinacula divided, remove restriction and allow quadriceps extensor force to be applied to tibia by patellar ligament.*

3. *Structures near radial center which are less-stretched are fascial retinaculum and lateral and medial portions of muscular retinacula*

*a.-patellar ligament
b.-muscular retinaculum
c.-fascial
r-radial center*

FIG. 9. Eggers' section of the retinaculum. (Eggers)

D. THE MUSCULAR IMBALANCE DUE TO QUADRICEPS PARALYSIS. TENDON TRANSPLANTATION

The kinetic problems are:
1. Relative strength of the substituting muscle.
2. In unilateral transplantation, the balance of the patella.
3. The rotatory moments of the transplanted muscles in respect to the tibia.

1. Transplantation of the outer hamstrings

A) Biceps Alone Into the Patella

The biceps has a shortening distance of 5.9 cm., a cross section area of 17.37 cm.,[2] and a muscle strength of 59.5 kg. (Fick[6]). The rectus femoris has a contraction length of 8.1 cm., a cross section area of 28.89 cm.,[2] and an absolute muscle strength of 104 kg. (Fick[6]). The exchange is therefore not too much out of proportion. Provided there are adequate hip muscles and there is a good gastrocnemius, the biceps, if given a straight direction of pull, should be sufficient. In fact, Crego and Fischer[3] reported 61 cases with excellent and good results in 52, and, earlier, Kleinberg[10] recorded 13 cases of transplantation of the biceps alone to the patella with good voluntary extension in all. He noted that extension after biceps transplantation is easiest with the leg in outward rotation as the direction of the pull is more straight.

B) Biceps and Tensor Fasciae

The use of both muscles for substitution of the paralyzed quadriceps, as advocated by Yount,[18] tries to meet the objection made to the use of the biceps alone, that there is danger of lateral displacement of the patella when the biceps pull is not opposed. In fact, Schwartzmann and Crego[15] observed 11 dislocations in 100 cases so treated.

If one uses both biceps and tensor, a straight pull can be effected; the gluteus maximus assists in the tension effect and the release of the lower end of the tensor from the iliotibial band helps in correction of the flexion contracture. Yount[18] reported 18 cases with good results and gives as indication cases of flexion contracture of the hip and knee and genu valgum.

C) Both Inner and Outer Hamstring Transplantations

The use of both inner and outer hamstrings disposes of the danger of lateral displacement of the patella. However, some discretion in the selection of the substituting material is necessary.

Over 20 years ago, F. Lange[11] warned that transplantation of all hamstrings should be avoided because of the great danger of developing a genu recurvatum. He also cautioned against the use of the sartorius because this muscle protects the knee against valgus deformity by reinforcing the tibial collateral ligament.

It also has been the experience of Schwartzmann and Crego[15] that in many cases of unilateral hamstring transference the transplantation had to be made bilaterally by additional operations, and that the good results of bilateral transplantation amounted to 74% against 57% of the unilateral. All in all, the hamstring transplantation for quadriceps paralysis in suitable cases has proved to be a highly satisfactory method when the gluteus maximus and the soleus fail to provide the knee with the necessary stability. According to all reports, the simultaneous transplantation of the biceps femoris and semitendinosus gives the highest percentage of favorable results.

V. SUMMARY

1. In paralysis of the extensors of the knee stabilization of the knee in extension can be effected both by gravity and by the substitutionary action of the uniarticular extensor of the hip (gluteus maximus) and the uniarticular plantar flexor of the ankle joint (soleus).

2. Paralysis of the flexors affects both rotatory and lateral stability of the joint because they normally support the collateral ligaments in their function of restraining excessive rotation and lateral motion.

3. Paralysis of the vastus medialis facilitates lateral displacement of patella.

4. Walking and standing is possible in paralysis of the quadriceps provided there is no marked degree of flexion contracture of the knee. Difficulties arise in fast walking or running.

5. Paralysis of the flexors also affects axial rotation besides predisposing to genu recurvatum. With the knee in extension the fibular collateral ligament relaxes on outward rotation; with the knee at right angle the ligament becomes tight only as the limit of inward or outward rotation is reached. In case of paralysis of all knee flexors, the natural valgity of the knee and the normal outward rotation in extension favor the development of a genu valgum with outward rotation and hyperextension. This is aggravated if only the inward rotators are paralyzed and the outward rotating biceps is unopposed.

6. Any degree of flexion contracture is unfavorable to the weight-bearing function of the knee and should be corrected. The other two deformities which require anatomic reconstruction are the genu valgum and the genu recurvatum.

7. The methods of reconstruction of the genu recurvatum fall into three groups: fasciodesis, tenodesis, and osteoplastic operations.

8. Gill's method of fasciodesis uses fascia lata and the collateral ligaments which are sutured together behind the joint and in front of the vessels. Heyman's method of reattaching the proximal ends of the collateral ligaments posteriorly to the femoral shaft is a similar procedure.

9. A later method of Heyman is a tenodesis in which the tendons of insertion of the gracilis and semimembranosus are inserted with their upper ends into the posterior aspect of the internal femoral condyle, and the tendon of the biceps into the lateral condyle.

10. The osteoplastic reconstruction method involves three different mechanical features:

a) Anterior bone block: The implantation of the patella into the tibia after lengthening the quadriceps tendon (Campbell and Mitchell). Mayer's modification of using a stout tibial graft makes the lengthening of the quadriceps tendons unnecessary and provides a longer lever arm for the block.

b) Elevation of the tibial plateau. This gives the plateau a backward and downward slant, the gap made by raising the plateau being filled by a bone wedge with anterior base. It does not provide against recurrence due to secondary ligamentous relaxation.

c) The arthrodesis is the surest way to correct a genu recurvatum but involves the sacrifice of all knee motion.

11. The operation for genu valgum after completion of growth is the supracondylar osteotomy. During growth Blount's method of inserting staples on the medial side across the epiphyseal plate is the most rational.

12. In the correction of flexion deformity the contracture of the extensory retinacula should be relieved (Eggers) before a supracondylar osteotomy is considered.

13. The restoration of active extension involves transplantation of the biceps alone, transplantation of biceps and tensor, and transplantation of both inner and outer hamstrings.

14. Transplantation of the biceps alone involves the danger of displacing the patella laterally.

15. Transplantation of biceps and tensor diminishes this danger because a straighter direction of the muscle pull can be obtained.

16. The best results are obtained by transplanting both inner and outer hamstrings though some selection of the substitutionary material is necessary to avoid a genu recurvatum. The simultaneous transplantation of the biceps and the semitendinosus gives the best results (Schwartzmann and Crego).

BIBLIOGRAPHY

1. CAMPBELL, W. C., and MITCHELL, J. L.: Operative treatment of paralytic genu recurvatum. *Am. J. Surg.*, *96*,1055, 1932.
2. CLEVELAND, M.: Operative fusion of the unstable flail knee due to anterior poliomyelitis. *J. Bone & Joint Surg., XIV:2*, 525, July 1932.
3. CREGO, C., JR., and FISCHER, F. J.: Transplantation of the biceps femoris for the relief of quadratus femoris paralysis. *J. Bone & Joint Surg., XIII:*515, July 1931.
4. DUCHENNE, G. B.: *Physiology of Movement*. Transl. E. B. Kaplan. Philadelphia, J. B. Lippencott Co., 1949.
5. EGGER, G. W. N.: Surgical division of patellar retinaculum to improve extension of the knee in cerebral palsy. *J. Bone & Joint Surg., 32A*:1, 80, Jan. 1950.
6. FICK, R.: *Handbuch d. Anatomie und Mechanik der Gelenke, III*. Jena, G. Fischer, 1911.
7. GILL, A. B.: Operation for paralytic genu recurvatum. *J. Bone & Joint Surg., XIII:*49, Jan. 1931.
8. HEYMAN, C. H.: Operative treatment of genu recurvatum. *J. Bone & Joint Surg., VI:* 689, July 1924.
9. HEYMAN, C. H.: Operative treatment of genu recurvatum. *J. Bone & Joint Surg., XXIX:*3, 644, July 1947.
10. KLEINBERG, S.: Transplantation of hamstrings. *Am. J. Orth. Surg., XV (old):*512, 1917.
11. LANGE, M.: *Orthopaedisch-Chirurgische Operationslehre*. Munich, J. F. Bergmann, 1951.
12. LEXER, E.: *Wiederherstellungschirurgie, 2d Ed. Vol. II*. Leipzig, John Ambrose Barth, 1931.
13. LOVETT, R. W.: *The Treatment of Infantile Paralysis*. Philadelphia, Blakiston Co., 1916.
14. MAYER, L.: Operation for the cure of paralytic genu recurvatum. *J. Bone & Joint Surg., XII:*4, 845, Oct. 1930.
15. SCHWARTZMANN, J. R., and CREGO, C. H., JR.: Hamstring tendon transplantation for the relief of quadriceps paralysis in residual poliomyelitis. *J. Bone & Joint Surg., XXXA:*541, July 1948.
16. SHER, J.: *Report*, National Foundation Infantile Paralysis, Iowa City, 1941.
17. STRASSER, H.: *Lehrbuch der Muskel und Gelenkmechanik III*. Berlin, J. Springer, 1917.
18. YOUNT, C. C.: Quadriceps paralysis. *J. Bone & Joint Surg.. XX:*2, 314, April 1938.

THE MECHANICS OF FOOT AND ANKLE

I. MORPHOLOGY

A. CONSTRUCTION OF FOOT AND ANKLE

THE profound developmental changes which the foot had to undergo as the orthograde type of locomotion supplanted the quadrupedal are based upon the requirement to adjust the center and line of gravity to a small supporting surface. Once gravitational stresses were so adapted to the area of support the further problem was to keep them so adjusted so that the body could not only stand erect upon it but could also be balanced by muscle action against fluctuations of the line of gravity. Furthermore, the alternating bipedal gait made it necessary to develop in the foot a propulsory ability as well as a restraining one in order that the body equilibrium could be maintained while the body is in motion.

The bipedal alternating gait of man establishes two essential postulates:

First, it becomes necessary to develop certain articulations which serve the purpose of providing for static muscular balance in standing and for dynamic propulsion in walking or running.

Secondly, the gripping ability of the foot of the primates is abandoned in favor of the ability to support. This is accomplished by changes in certain tarsal bones and by abolishing the opposibility of the big toe.

First, a joint is established between leg and foot, the tibio-astragalar articulation. This joint controls the foot in the sagittal plane and takes care of adjusting the line of gravity in standing and in providing for the propulsion and restraint of the gait in sagittal direction.

Next, a second articulation is established which permits movement in the frontal plane. It adjusts the line of gravity from side to side and, besides, has a share in the mechanism of propulsion. This is the subastragalar joint situated between the astragalus, which is the mechanical keystone of the unit, and the os calcis, which furnishes the posterior part of the support.

A third articulation, finally, interrupts the structure of the foot in the middle of the tarsus. The midtarsal joint has the function of imparting the spring necessary for propulsion. At the same time, the anterior part of the foot lying distal to the joint can adjust itself against the posterior, which gives the anterior foot plate freedom to maintain full contact with the floor independent of the position which the proximal part may assume.

1. The articulations

a) THE ANKLE JOINT

This joint is situated between the lower end of the tibia and the body of the astragalus. The latter moves in a mortise formed by the internal malleolus of the tibia and the outer malleolus of the fibula. It is a hinge joint which permits only one degree of freedom of motion, namely in the sagittal plane. This joint is capable of regulating the forward and backward fluctuations of the line of gravity so that it is kept within the limits of the supporting surface.

b) THE SUBASTRAGALAR JOINT

This joint, located between astragalus and os calcis, permits motion in a frontal plane and is therefore in position to regulate the side-swaying of the line of gravity.

c) THE MIDTARSAL OR CHOPART'S JOINT

It consists of two halves: a medial half between the head of the astragalus and the scaphoid, and a lateral half between the anterior process of the os calcis and the cuboid. This joint really allows motion in three planes, a frontal, a sagittal, and a transverse, although under ordinary conditions only that in the frontal plane, i.e., in the direction of pronation and supination, is of any consequence. It provides for compensatory movements between the front part of the foot which lies distal to this articulation, and the back part which lies proximal to it. It enables the front part to maintain close contact with the ground as the back part carries out its pronatory and supinatory or abductory and adductory movements.

d) THE METATARSOPHALANGEAL ARTICULATION

Between the metatarsal heads and the basal phalanges of the toes a joint is established which, like the midtarsal joint, also has three degrees of freedom of motion. Only one, flexion and extension in the sagittal plane, is under active voluntary control. It serves the finer adjustments between the toes and the metatarsals when the weight is thrown forward over the ball of the foot during the take-off.

These four articulations meet all the regulatory requirements necessary to maintain the line of gravity within the bounds of the supporting surface.

2. The construction of the arches

The preceding description of the location and function of these four principal joints of the foot may serve well as the basis upon which the mechanics of the human foot is to be interpreted.

Below the ankle joint the bones of the tarsus are arranged in two systems. One of these is in immediate contact with the ground through the os calcis; the other is superimposed upon it and is directed obliquely downward so that it reaches the floor only with its anterior portion. These two structures are called the outer and the inner arches of the foot.

The astragalus which receives the incumbent body weight from above rests upon the os calcis. Between the two bones the subastragalar joint is established. With its posterior process the os calcis forms the common posterior

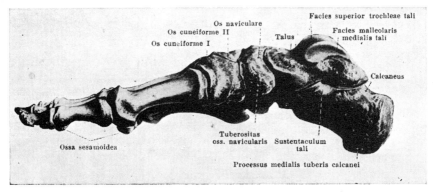

FIG. 1. Inner arch of foot, showing the three articular facets of the os calcis contacting the astragalus. (Spalteholz)

pillar of both arches. The inner arch continues forward through the body, neck and head of the astragalus to the scaphoid and through the three cuneiforms to the three metatarsals and their toes. The whole weight of the superincumbent inner arch is borne by a medially projecting process of the os calcis,

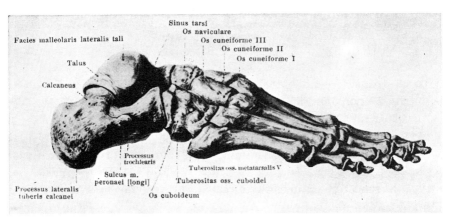

FIG. 2. Outer longitudinal arch. (Spalteholz)

the sustentaculum; three articular facets of the os calcis contact the body, neck and head of the astragalus (Fig. 1).

The outer longitudinal arch is formed by the os calcis, the cuboid, which articulates with the former at the outer half of the midtarsal joint, and by the two outer rays, namely the metatarsals IV and V and their respective toes (Fig. 2).

Due to the fact that the inner arch rests upon the outer, the translatory

weight stresses which reach the astragalus in perpendicular direction are transmitted backward and forward. Backward they go to the common pillar of the
arches, the posterior processes of the os calcis. In forward direction the
transmitting weight stresses spread apart because of the divergence of the
two arches. The inner arch touches the ground with its three medial rays
side by side with the rays of the outer arch. The result is that the anterior
support of the foot forms an oval which includes the anterior ends of both
the outer and the inner arch.

If we compare the foot to a horizontal beam supported at both ends, the
reservation should be made that the posterior support is concentrated to a
point at the heel while the anterior represents an area comprising the five

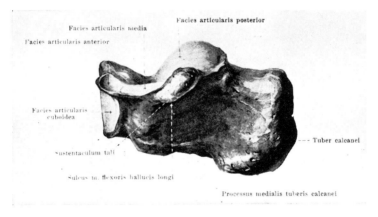

FIG. 3. Os calcis, showing articular facets. (Spalteholz)

metatarsal heads of both outer and inner arches. The inner arch which is
superimposed upon the outer does not rest perpendicularly upon the latter.
It is deflected medially and downward following the direction of the neck of
the astragalus which makes up part of the downward slope of the arch. The
immediate effect of this medial and downward divergence is that the os calcis,
carrying the body of the astragalus over the posterior facet of the subastragalar
joint, has to send forth a medially directed buttress, the sustentanculum tali,
to sustain part of the body and the deflected neck (middle facet of the subastragalar joint) and, in addition, has to make connection with the head of
the astragalus. The latter lies medially to the anterior process of the os calcis
and articulates with it by means of a small joint (anterior facet of the subastragalar joint) (Fig. 3).

B. THE AXES OF THE JOINTS OF THE FOOT

To understand the part these several joints play in the locomotor function
of the foot, a clear concept of the planes of motion and their relation to the
cardinal planes of the body is necessary.

1. The Ankle Joint

Under normal conditions of upright standing the axis of the tibio-astragalar articulation is perfectly horizontal. The knee joint axis is also horizontal, but the two axes are not parallel. When the knee joint axis is placed strictly in a frontal plane, with the patella looking straight forward, the axis of the ankle joint runs obliquely from outward and backward to forward and inward which corresponds to the outward rotation of the lower end of the tibia (Fig. 4) (Strasser[9]). Due to this tibial torsion the foot is outward rotated in relation to the knee joint, just as the neck of the femur, because its forward angulation is outward rotated in relation to the knee joint. Because of this obliquity of

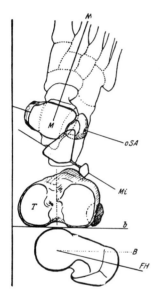

Fig. 4. Relationship of the axes of the femoral neck, the tibial condyle, and the ankle joint. (Strasser)

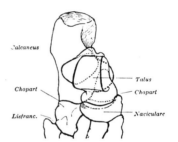

Fig. 5. Deflection of the astragalus in relation to the os calcis. (Strasser)

the ankle joint axis in upright standing position, the foot moves in the ankle joint upward and medially in dorsiflexion and downward and laterally in plantar flexion.

2. The subastragalar articulation

When the toes are planted straight forward the axis of this joint runs from backward outward and downward to forward inward and upward. Consequently, it does not point straightly in sagittal direction but is deflected about 25°, corresponding to the angle between the axis of the os calcis and that of the astragalus which is produced by the inward and downward deflection of the latter bone (Fig. 5). Motion in this joint is called pronation and supination. Here, again, because of the obliquity of the joint axis, the foot will go into abduction, i.e., away from the midline, with pronation, and into adduction with supination.

3. The midtarsal joint

This joint has three degrees of freedom of motion and the three respective axes about which the motion is carried out correspond to the orientation coordinates of the body. That is to say, the joint has a perpendicular axis, a frontal axis, and a sagittal axis, which are all approximately parallel to the respective coordinates of the body as a whole. The motion about the perpendicular axis is abduction and adduction, a motion occurring in the transverse plane; the motion about the frontal axis is dorsal and plantar flexion in the sagittal plane; and the motion about the sagittal axis is pronation and supination, which motion takes place in the frontal plan. Under normal conditions, i.e., where there is no deformation, only the motion about the sagittal axis, that is, pronation and supination, is of any consequence; it is this motion which secures the forefoot to the floor against pronatory and supinatory movements of the back part of the foot, a compensatory adjustment about which more will be said later.

C. THE INTERNAL ARCHITECTURE OF THE FOOT

The foot being an instrument of support, one must expect the inner texture of its bones to reflect the arrangement and distribution of mechanical stresses.

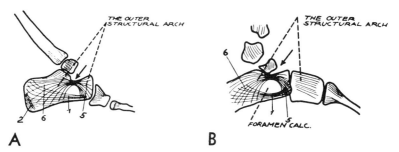

Fig. 6. Architecture of the tarsus. *A.* sagittal cut through fifth metatarsal.
B. sagittal cut through fourth metatarsal.

A series of sagittal cuts made through the foot illustrates the significance of the trajectories in this plane. The sections which go longitudinally through the outer arch, i.e., the os calcis, cuboid, and metatarsal IV and V show stress lines which go from the os calcis to the cuboid forward to the outer metatarsals with a concentration of the trabecular systems at the roof of the body of the os calcis (Fig. 6) (Steindler[s]). The section through the medial or inner arch shows trajectories rising from the heel to the center of the body of the astragalus where they meet at right angles a trabecular system coming from the tibia and converging toward the center of the astragalus, which thus becomes the mechanical center of this unit. From here these trajectories go forward in unbroken line through the neck and head of the astragalus, then through the scaphoid, the cuneiforms, into the three medial metatarsals (Fig. 7).

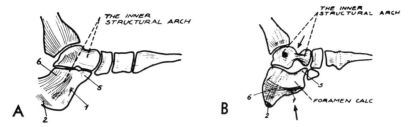

FIG. 7. Architecture of the tarsus. *A*. sagittal cut through first metatarsal.
B. sagittal cut through second metatarsal.

Manifestly, then, the lateral stress lines are directly transmitted to the os calcis so that the mechanical center where the trabeculae concentrate is located at the upper portion of the os calcis just below the subastragalar joint.

On the medial side, where the astragalus diverges medially from the os calcis, the mechanical center is in the middle of the body of the astragalus causing the area of lamellar concentration to be located below the ankle joint.

Some authors take exception to the term arch in relation to the architecture of the foot because the trajectories are not curved but run in straight lines. Engels[2] maintains that there is an arch only in lateral sections where the trajectories run

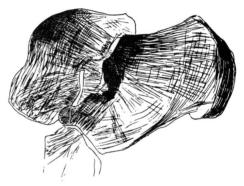

FIG. 8. The secondary trabecular system in astragalus and os calcis. (Engels)

through the os calcis and cuboid with a dorsal convexity. He maintains that the centrally loaded foot represents rather a Λ-shaped strut construction.

There are also secondary trabecular systems. One in the astragalus rises from the lower surface and crosses the long system at right angles, corresponding to the pressure lines in extreme dorsiflexion (Fig. 8). Several secondary trajectories are also described for the os calcis.

D. THE LIGAMENTOUS REINFORCEMENTS OF
THE ARTICULATIONS

From the study of the architecture it must be obvious that special provisions are necessary to protect the interposed articulations, which interrupt the trabecular system of the bones, against mechanical stresses. There is, indeed, a complicated system of ligamentous reinforcements entrusted with preventing the joints from becoming disaligned. The integrity of the joints depends in the first line upon these ligaments and only secondarily upon the muscular apparatus. These reinforcing ligaments cover all articulations and run in all possible directions. To simplify matters we shall limit ourselves to a description of the most essential ones.

1. The ankle joint

The principal ligaments are the medial collateral or deltoid and the lateral collateral. The deltoid ligament runs from the lower border of the medial malleolus, fan-shaped, to the os calcis and scaphoid in three sections: the anterior portion goes to the scaphoid as ligamentum tibiotalonaviculare; the middle portion extends to the sustentaculum tali of the os calcis as ligamentum tibiocalcaneum; the posterior portion reaches the posterior aspect of the subastragalar joint as ligamentum astragalotibiale posterius (Fick[3]). The

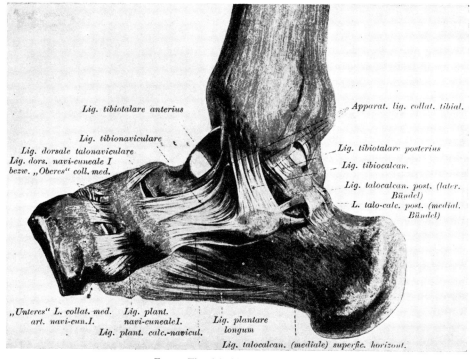

FIG. 9. The deltoid ligament. (Fick)

deltoid ligament, therefore, reinforces both the ankle and the subastragalar joints. The middle portion which goes to the sustentaculum is especially strong (Fig. 9). The lateral collateral ligament also has three portions: a ligamentum talofibulare anterius, a ligamentum talofibulare posterius, and a ligamentum calcaneofibulare. The anterior talofibular ligament runs from the middle of the lateral malleolus forward and medially to the anterior portion of the lateral surface of the astragalus; the calcaneofibular ligament runs from the point of the lateral malleolus to the outer side of the os calcis, and the posterior talofibular liagment inserts into the posterolateral portion of the subastragalar joint. This ligament also controls and checks both articulations (Fig 10).

2. The subastragalar joint

Aside from the above mentioned collateral ligaments there are others which have a checking effect on the subastragalar articulation. There is, for instance, the ligamentum talocalcaneum internum, a strong and broad ligament hidden in the narrow tarsal canal in the depth of the sinus tarsi. It is also called the ligamentum sinus tarsi anterius and it becomes tight in supination and relaxes in pronation (Fig 10). The posterior portion of the subastragalar articulation also has reinforcing ligaments: the ligamentum sinus tarsi posterius, a vertical ligament in that portion of the capsular apparatus of the joint which forms its

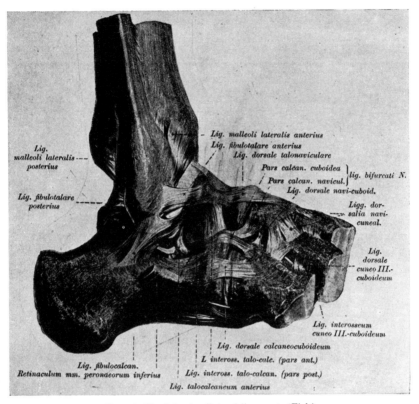

FIG. 10. The lateral collateral ligament. (Fick)

posterior wall. Finally, there is the posterior talocalaneal ligament which runs from the astragalus to the upper border of the os calcis (Fig. 10).

3. The midtarsal joint

This articulation is reinforced by a large number of separate ligaments. On the medial side there is the calcaneocaphoid or cup-shaped ligament which really is the medial and plantar portion of the astragaloscaphoid articulation. It runs from the sustentaculum tali to the scaphoid (Fig. 9). Dorsally, there is the dorsal talonavicular ligament (Fig. 11) and the ligamentum calcaneonaviculare bifurcatum (Fig. 11). On the plantar surface, there is a calcaneonaviculare plantar ligament. On the lateral side there are the dorsal and lateral calcaneocuboid ligaments, being the cuboid portion of the bifurcate ligament mentioned above (Fig. 10), and on the plantar side the long and short plantar ligaments (Fig. 12).

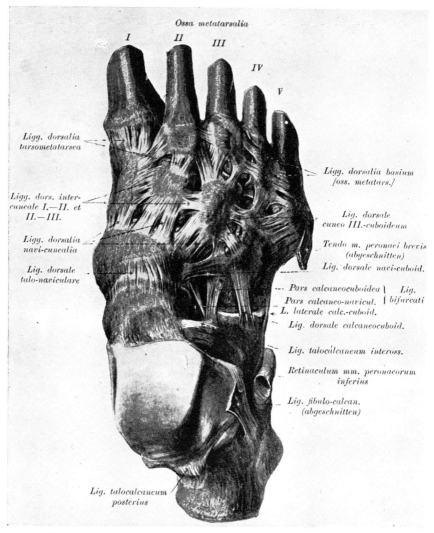

FIG. 11. Dorsal view of the foot, showing ligamentous structures. (Fick)

4. The tarsometatarsal and metatarsophalangeal joints

The tarsometatarsal joint is secured by dorsal and plantar tarsometatarsal ligaments and by the collateral intermetatarsal ligaments, medial and lateral. The bases of the metatarsals are held together by the dorsal and the plantar ligamenta basium metatarsorum.

The metatarsophalangeal joints are also amply provided with reinforcing apparatus. They are the plantar and dorsal ligaments binding the metatarsal heads and the collateral metatarsal ligaments which reinforce both sides of the joint capsule.

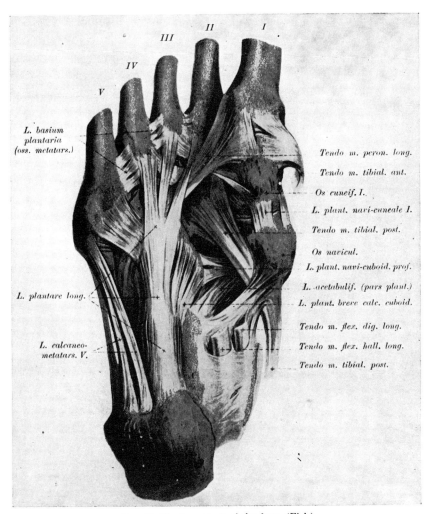

L. basium
plantaria
(oss. metatars.)

L. plantare long.

L. calcaneo-
metatars. V.

Tendo m. peron. long.

Tendo m. tibial. ant.

Os cuneif. I.

L. plant. navi-cuneale I.

Tendo m. tibial. post.

Os navicul.

L. plant. navi-cuboid. prof.

L. acetabulif. (pars plant.)

L. plant. breve calc. cuboid.

Tendo m. flex. dig. long.

Tendo m. flex. hall. long.

Tendo m. tibial. post.

FIG. 12. Plantar aspect of the foot. (Fick)

II. JOINT MECHANICS
A. THE ANKLE JOINT

The principal and, under normal conditions, the only motion in this joint is rotation around the transverse axis: flexion and extension. As we look at the articular cartilage of the astragalus we note that its borders are not strictly parallel but that the articular surfaces on the sides of the body converge slightly posteriorly. This means that the roll of the astragalus is wider in front than it is in the back, and it brings up the question whether the joint is a true hinge or whether a screw-like serpentine motion is superimposed upon it. No lateral motion can be detected if the tibia moves against the astragalus and the latter is fixed; neither can one detect any helical movement as the astragalus moves around the fixed tibia. Nevertheless, some observers maintain that there is a screw-like movement between astragalus and outer malleolus associated with flexion-extension. Langer[4] believed that on dorsiflexion the astragalus moves slightly laterally (2.5 mm.) because of the fact that the lateral facet of the roll of the talus runs forward and laterally. However, this is not likely because the medial facet has a straight sagittal direction and for a helical movement it would be necessary that both facets are directed forward and lateral. If there is any side movement at all it is minimal under ordinary circumstances, and, on the whole, the joint can be considered fully secured by the mortise.

In dorsal extension the roll of the astragalus glides backward until the anterior border disappears into the mortise, so that the neck of the astragalus approaches the anterior ledge of the tibia. Due to the fact that the roll of the astragalus is wider in front, a spring-like give of the mortise is necessary. This give is provided by the inferior tibiofibular ligaments. According to Fick,[3] the width of the posterior part of the roll of the astragalus is 2.7-3.5 cm. against the anterior width of 3.2-4 cm. Therefore, in plantar flexion, when the narrow portion of the astragalar body lies in the mortise, the tibiofibular ligament is relaxed and the tibia and fibula close up. In this position the grip of the mortise on the body of the astragalus is less firm. If, in the specimen, the malleoli are pressed closely together, the foot goes automatically into plantar flexion. In dorsiflexion the infolding capsule is saved from impingement by its adhesions to the sheath of the extensors.

Motion in the ankle joint is checked as follows: in the specimen, dorsal extension is checked by the astragalus abutting against the anterior tibial ledge; in vivo, however, it is the tension of the tendo Achillis as well as of the posterior capsule and the posterior portion of the collateral ligaments which checks dorsiflexion. Plantar flexion is checked in the specimen by the posterior process of the astragalus (Stieda's process) abutting against the posterior ledge of the tibia. In vivo, it is the strain on the anterior capsule, the tension of the extensor tendons and the anterior portions of the collateral ligament which check plantar flexion.

The excursion range of the ankle joint is from 70° dorsiflexion to 120-140° plantar flexion, a total of 50-70°.

B. THE SUBASTRAGALAR JOINT

This articulation has three facets, a posterior, a middle, and an anterior. In the posterior, the undersurface of the astragalus joins with a quadrangular facet the body of the os calcis. In the middle facet, the anterior portion of the talar body and the base of the neck contact with the articular surface of the sustentaculum; and in the anterior facet, the lateral side of the talar head joins in a small area the inner and forward end of the anterior process of the os calcis (Fig. 13). The joint is completed by a strong cartilage-covered ligament from the sustentaculum to the scaphoid, the calcaneoscaphoid or spring ligament, which also forms the plantar and medial part of the talonavicular portion of the midtarsal joint.

Normally, this joint has one degree of freedom of motion about an axis which runs obliquely from outward, backward and downward, to inward, forward and upward. Because of this obliquity the movement about this axis is, in terms of orientation planes, a combination of supination, adduction and plantar flexion in one direction and pronation, abduction and dorsiflexion in the other. The type of motion in this joint is of clinical interest. A pure gliding of the surfaces upon each other would presuppose congruency of the opposing articular surfaces. In fact, the posterior facet of the os calcis is slightly convex while the middle articular facet on the sustentaculum as well as the

anterior facets are concave. If the motion were a rocking one, at least the posterior facet would have to gape laterally in supination and medially in pronation. Since this is not the case one must conclude that the motion is not a pure rocking motion nor is it a pure gliding, but it is a compromise movement about a succession of instantaneous axes, a sort of wobbly movement or, according to Fick,[3] a so-called shift gliding.

This is particularly important for the pathomechanics of the flatfoot where the question arises whether this shift gliding may occur not only about the sag-

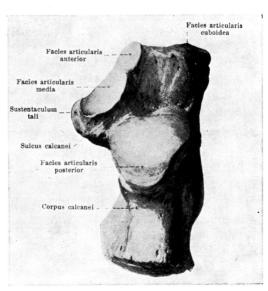

Fig. 13. Dorsal view of os calcis, showing articular facets. (Spalteholz)

ittal but also about a vertical axis. That is to say, can there be a rotatory movement of the astragalus over the os calcis about an axis which goes vertically through the astragalar body so that neck and head move around it like the hand of a clock? There is, indeed, good evidence that under pathological conditions such a rotatory gliding occurs between talus and os calcis, as is the case in the relaxed pronated foot. If there is such a length rotatory movement of the astragalus against the os calcis, what effect does this rotation have on the tibia? It has been shown that the astragalus is held firmly in the mortise and that this permits no other than flexory-extensory movement. Consequently, the astragalus cannot rotate about a perpendicular axis in the ankle joint. The motion must necessarily occur in the knee joint which, in a certain position, permits inward or outward rotation of the tibia. This point is to be discussed more fully under the pathomechanics of the static flatfoot because it is in this condition where the inward rotation of the leg becomes evident.

The range of the subastragalar joint has been given variously as from

20-45° (Fick[3]), the pronatory range exceeding the supinatory at the ratio of 3:2. The check of the pronation-abduction-dorsiflexion movement is furnished on the lateral side by the impingement of the astragalus on the os calcis and also by the strain on the posterior portion of the ligamentum talocalcaneum interosseum. The check to the supination-adduction-plantar flexion movement is effected by the impingement of the talus against the sustentaculum at the middle facet and by the tension of the anterior portion of the ligamentum talocalcaneum interosseum. At the same time, the sinus tarsi at the lateral side becomes widened.

C. THE MIDTARSAL JOINT

This joint takes care of the compensatory movements which enable the forepart of the foot to hold firmly to the ground while the back part assumes a pronatory or supinatory position or, stated differently, the forepart of the foot goes into supination relative to the back part if the latter is pronated, and vice versa. The result is a longitudinal torsion of the forefoot against the back part. This torsion is an essential part in standing and walking under normal conditions. Being dependent on the integrity of the midtarsal articulation, this compensation may be restricted or abolished by ankylosis or arthritis with detrimental effect on the mechanics of standing or walking. On the other hand, a relaxation of the midtarsal joint may lead to excessive pronation of the backfoot which again forces the forefoot in compensatory supination when it is held fixed and flush with the floor.

Aside from the pronatory and supinatory movement about the sagittal axis, there should also be an abductory and adductory movement about a vertical and a dorsal and plantar movement about a frontal axis. Both these movements are minimal under normal condition and may be disregarded, particularly the flexion-extension about a frontal axis. They become important, however, under pathological conditions where the backfoot not only goes into excessive pronation but also into adduction and plantar flexion. Here the forefoot carries out movements about the vertical and frontal axes which are compensatory in the same sense as the supination of the forefoot about a sagittal axis becomes compensatory to the pronated backfoot.

In the relaxed foot where the backfoot is in pronation and adduction and the forefoot in compensatory supination and abduction, one must assume that the astragalus has executed an axial rotation in the subastragalar joint and one must, therefore, expect some degree of inward rotation of the tibia which is commensurate with the rotation of the astragalus about a perpendicular axis.

Normally the abductory and adductory range of the midtarsal joint is very small (Fick[3]). The check to adduction is furnished by the tension of the lateral talocalcaneal ligament and that to the abduction by the spring ligament.

D. COMBINED MOVEMENTS

1. Combination with the foot free. The foot moves against the leg

The combined tibioastragalar and subastragalar motion carries the foot around the leg in a circling movement, the axis of the foot describing a cone the point of which lies in the body of the astragalus. It is the result of a simultaneous movement in both joints. The obliquity of the subastragalar axis causes the movement of the ankle joint to be combined with adduction and supination in plantar and with abduction and pronation in dorsiflexion. The midtarsal joint participates only slightly in accentuating pronation and supination.

2. Combination with the foot fixed. The leg moves against the foot

Muscle action in both the ankle joint and the subastragalar joint is necessary to maintain balance; the former in the sagittal and the latter in the frontal direction. In standing, these adjustments occur simultaneously in both planes. In the anteroposterior or sagittal direction, the astragalus moves with the rest of the foot against the tibia; in the frontal adjustment, the astragalus becomes a unit with the tibia and movement occurs between astragalus and os calcis. Supination entails a lateral, pronation medial gliding of the talus. The forefoot remains fixed on the ground and any pronatory or supinatory displacement of the talus in the subastragalar joint is taken up by the opposite and compensatory movement in the midtarsal joint, as has been explained above.

3. Combination of subastragalar and midtarsal joint movement

All movement in the subastragalar joint of the os calcis against the talus, or vice versa, is associated with simultaneous movement of the head of the astragalus against the scaphoid. There is a "trackbound" (Zwangsläufig) correlation between the two joint motions. They act as a physiological entity, both joints supplementing each other. The midtarsal or Chopart's joint is, as explained above, a double joint: an inner half between the head of the astragalus and the scaphoid and an outer between the os calcis and the cuboid. In walking the calcaneocuboid articulation is fixed against the floor as the line of gravity travels along the outer border of the foot. Any excessive passive pronation may cause a strain of the calcaneocuboid joint at that phase of the gait in which the line of gravity falls through this joint. This occurs in walking during the second or propulsion phase of the standing period. Or, both articulations become bound to the floor, as is the case in standing or in sharp stepping just before the take-off. In this case a sprain of the entire midtarsal joint may result.

4. The subastragalar joint in relation to the ankle joint

How far can pronation and supination in the subastragalar joint be forced without producing lateral motion in the ankle joint? Under normal conditions, neither a lateral (sideways) nor an axial (length rotatory) movement of the astragalus is possible in the mortise; any forced pronatory or supinatory motion of the foot will result in a strain of the collateral ligaments of the ankle joint. The very existence of these powerful collateral ligaments suggests that the ankle joint must be amply protected against lateral strain.

On the other hand, there are pathological conditions in which the ligaments are relaxed, for instance in poliomyelitis. The x-ray will then show that the joint lines of the talotibial articulation gape when pronation or supination is forced, indicating that the excessive pes valgus or varus is not entirely due to disalignment in the subastragalar joint but that the ankle joint shares in the deformity.

So far as length rotation is concerned, it has already been mentioned that, for instance, in severe pes planus, where the "break" occurs in the mid-tarsal joint and the astragalus is rotated inward against the forefoot, this rotation does not occur in the ankle joint. The astragalus forms one unit with the tibia and the rotation occurs in the knee or even in the hip joint.

E. THE MECHANISM OF THE TARSOMETATARSAL JOINTS

The movement between the first metatarsal base and the internal cunei-form is minimal. A very slight plantar flexion occurs with abduction and a dorsiflexion with adduction of the foot between the bases of the 2nd and 3rd metatarsals and the middle and lateral cuneiforms. A relatively greater dorsoplantar movement is possible between the 5th metatarsal base and the cuboid.

F. THE MECHANISM OF THE METATARSOPHALANGEAL JOINTS

The five rays constituting the arches have not only different lengths but also different heights. According to Fick,[3] the second is the longest and highest, being an average of 17-22 cm. in length, measured from the heel to the meta-tarsal head, and from $5\frac{1}{2}$-7 cm. in height, measured from the supporting surface. The lowest and shortest is the fifth with a length of 14-15 cm. and a height of 2.3 cm. The first arch has a length of 16-22 cm. and a height of 5-5.5 cm. In angular values the axis of the first metatarsal ray forms with the floor an angle of 18-26°, the second 15°, the third 10°, the fourth 5°.

The transverse metatarsal arch is highest at the second and lowest at the fifth; it is supported largely by the transverse plantar ligaments and by the tendons of the tibialis posticus and peroneus longus.

The metatarsophalangeal or the anterior arch has articulations which theo-retically should have the three degrees of freedom of motion of the ball-and-socket joints. Actually, in vivo, only the dorsoplantar movement is under voluntary control, possibly with a slight lateral movement of the lateral (4th and 5th) toes. Active dorsal extension has a range of 50-60°, passive up to 90°, and active plantar flexion 30-40°, passive up to 45-50°. The checks to these movements are the tension of the opposing muscles and the dorsal impingement of the phalanges upon the heads of the metatarsals. In upright standing the metatarsophalangeal joints of the four lateral toes are in slight dorsiflexion.

What the weight-bearing points of the transverse arch are is still a debated question. Meyer[5] believed it to be the lateral border of the fifth metatarsal, Beely[1] the head of the second and third metatarsals. According to Seitz,[7] the weight is borne on the heads of the first and fifth metatarsals, and Morton[6] estimated the ratio of heel to medial and lateral pressure as $3:2:1$.

Under weight bearing longitudinal and transverse arches flatten normally. Seitz[7] finds that in the sagittal plane the tuber of the os calcis becomes depressed under weight 1.5 mm., the anterior process 4 mm., the cuboid 4 mm., and the fifth metatarsal 3.5 mm. In the horizontal plane the head of the astragalus becomes displaced medially 1.5-6 mm. (Seitz[7]), the anterior border of the os calcis 2-4 mm.

Under weight bearing the anterior arch shows displacements also in the transverse plane (Seitz[7]). Between the first and second metatarsal heads there is a widening of 0.5 cm., between the second and third 0.2 cm., between the third and fourth 0.4 cm., between the fourth and fifth 0.15 cm., a total of 1.25 cm.

The points of pressure of the weight-bearing foot can be determined by their distance from the supporting surface. Such actual measurements of the points of pressure were carried out (Fick[3]) with the following findings. The medial portion of the posterior process of the os calcis was found in the x-ray to be 7-10 mm. from the floor, the big toe sesamoids and the fifth metatarsal head were at the same level, namely 6 mm., the third metatarsal 8.5 mm., the fourth 7 mm., the second 9 mm., thus giving the fifth metatarsal head, the posterior process of the os calcis and the sesamoid of the big toe the lowest points. This being a three-point support it has the advantage that a plane can always be laid through all three points, or, in other words, the foot can always be in position where all three points are in contact with the floor. No "teetering" is possible (the same as in a three-legged chair). It is different when the anterior arch becomes convex downward and the weight falls on the second and third metatarsal heads. The result is an unstable two-point support.

III. THE MUSCLE DYNAMICS OF THE FOOT

It is characteristic that all nine long muscles of the foot traverse the tibio-astragalar as well as the subastragalar joints and, with the exception of the gastrosoleus and plantaris, also the midtarsal joint, and that none of them inserts into the astragalus which is the keystone of the mechanical unit of the tarsus. This makes for a dynamic interdependence of the three articulations, as almost all long muscles are operative on all joints.

A. THE ANKLE JOINT

Plantar flexion is carried out by the triceps surae, the gastrosoleus and plantaris by means of the common tendo Achillis inserted into the posterior superior surface of the heel. In addition, the long flexor of the toes, the flexor hallucis, the tibialis posticus, and the peroneus longus assist in plantar flexion (Fig. 14).

The gastrosoleus is one of the strongest muscles of the human body. Its contraction length is 39 mm. for the gastrocnemius and 44 mm. for the soleus, with a cross section of 23 cm.^{2} and 20 cm.,^{2} respectively, and a combined working capacity of 6.5 kgm. This exceeds the combined strength of all

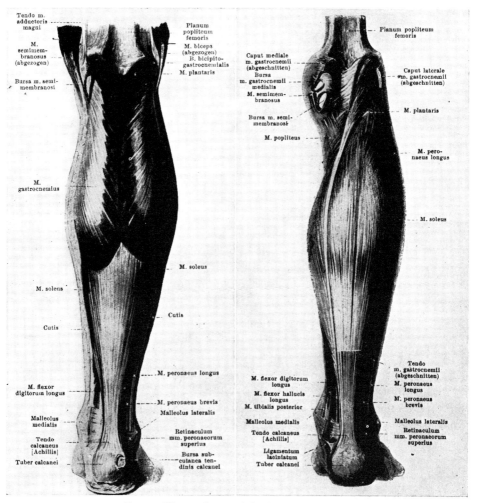

FIG. 14. Plantar flexors of the foot. (Spalteholz)

the other plantar flexor muscles of the ankle joint, i.e., the flexor of the toes, the flexor hallucis, the tibialis posticus, the peroneus longus and brevis, which have a total working capacity in respect to the ankle joint of only 0.5 kgm.

The dorsiflexors of the ankle joint are the tibialis anticus, extensor of the toes, extensor hallucis and peroneus tertius (Fig. 15). The combined working capacity of the dorsiflexors is only about 1.4 kgm., or hardly 1/4 to 1/5 of the plantar flexors. The explanation of this gross discrepancy is partly developmental and partly mechanical. The development of the calf muscles was a prerequisite of the upright position and of the elevation of the body center of gravity. The fact that the weight line usually falls in front of the ankle joint made the rotatory component of gravity a powerful dorsiflexor of the foot against which the gastrosoleus must be equipped to hold the equilibrium.

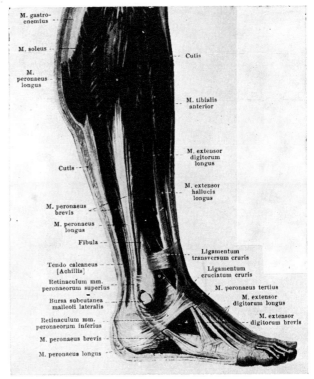

FIG. 15. Extensors of foot and toe. (Spalteholz)

B. THE SUBASTRAGALAR JOINT

The principal *supinator* is the tibialis posticus which, running behind the internal malleolus and under the sustentaculum, reaches the tuberosity of the

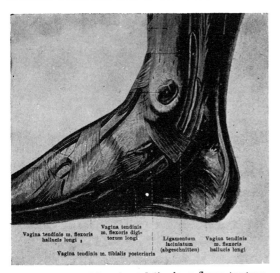

FIG. 16. The relationship of the long flexor tendons of the toe at the ankle joint. (Spalteholz)

scaphoid and the cuneiforms. It, therefore, crosses all three articulations: the ankle joint, the subastragalar joint, and the midtarsal joint. It is assisted by the flexor of the big toe. This muscle crosses the long flexor tendons of the toes at the sole of the foot and goes to the base of the end phalanx of the big toe (Fig. 16). The common flexor of the toes further assists supination. The tibialis anticus can pronate or supinate to midposition, i.e., it becomes a pronator in the pronated and a supinator in the supinated foot, always retaining its function of dorsiflexion.

The principal *pronators* in the subastragalar joint are the peroneus longus and brevis; they are further assisted by the common extensors of the toes and the peroneus tertius (Fig. 15).

Comparing the working capacity of the supinators and the pronators relative to this joint, we find that of the supinators to amount to 2.82 kgm., whereas the pronators have a working capacity of 1.16 kgm. only. Here, again, the difference in strength is made up by the fact that weight acts as a pronatory force.

C. THE MIDTARSAL JOINT

In this joint only pronation and supination occurs under normal conditions, except for negligible amounts of abduction and adduction. The pronators of this joint are the peroneals, the extensor digitorum, and the peroneus tertius. The combined pronatory working capacity is only 0.52 kgm. The supinators are the tibialis anticus (to midposition only), the posticus, the flexor longus digitorum and hallucis, and, to a degree, the extensor hallucis longus. The combined strength of the supinators is 0.5 kgm. almost equalling that of the pronators.

D. THE METATARSOPHALANGEAL JOINT

Abduction of the big toe in this joint is carried out by the abductor hallucis, a short muscle running from the tuber of the os calcis and the plantar fascia to the medial side of the basis of the first phalanx (Fig. 17).

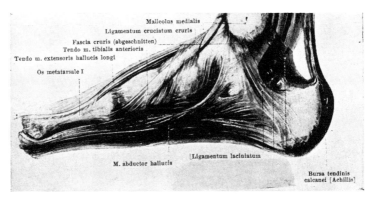

Fig. 17. Medial aspect of the foot, showing relationship of abductor hallucis. (Spalteholz)

The flexion of the toes is carried out by the long flexors, hallucis and digitorum, which are inserted into the bases of the end phalanges, and the short flexors which apply to the middle phalanges of the toes. The short flexor action is similar to that of the flexor sublimis of the hand (Fig. 18).

The action of the intrinsic muscles is similar to that in the upper extremity. They flex the toes in the metatarsophalangeal articulation and extend them in the interphalangeal joints; the interossei, acting as abductors and adductors of the toes, are arranged similarly as they are in the hand except that voluntary abduction and adduction is minimal save for the fifth, which has its own abductor. The plantar interossei as well as the flexor digitorum longus and the flexor hallucis pull the toes toward the midline at the third metatarsal, while the dorsal interossei spread the toes apart. An opponens hallucis, corresponding to the opponens pollicis, is absent.

The working capacity of the long flexors of the toes is about 0.8 kgm., that of the extensors, however, only 0.3 kgm. Thus the flexors are almost three times as strong as the extensors.

In vivo, the toes often show a claw position without paralysis of the interossei but simply due to insufficiency of the interossei and the overwhelming pull of the extensors coincident with the depression of the anterior arch.

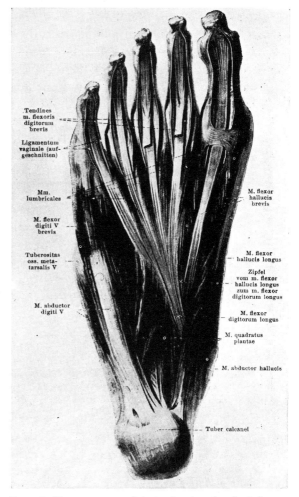

FIG. 18. Plantar aspect of foot, showing the short flexors. (Spalteholz)

IV. THE CONDITIONS OF EQUILIBRIUM OF THE JOINTS OF THE FOOT

A. THE ANKLE JOINT

We assume that the foot represents a beam concentrically loaded by the superincumbent body weight and supported at the end at ball and heel. The translatory component of the gravitational force is equilibrated as long as the sum of the two reactions (at the supports) equals the load. Let us first suppose that the line of gravity goes through the center of the ankle joint. In this event, so far as the ankle joint is concerned, both translatory and rotatory components are in equilibrium because in this case no rotatory component of the force of gravity develops (the lever arm which is the distance of the line of gravity from the center of motion is zero).

In this case there is, therefore, no necessity for the display of muscle forces to maintain the equilibrium in respect to the ankle joint. Both extensors and flexors of the joint are relaxed. The conditions of equilibrium in this case can be formulated as follows: for the *translatory component:* reaction A times its distance from the line of gravity is equal and opposite to reaction B times its distance from the line of gravity b or Aa $=$ $-$Bb.

For the *rotatory component:* the gravitational force R times the distance (d) from the center of motion when d is zero, Rd $=$ 0 (Fig. 19).

Now let us suppose that the line of gravity falls in front of the ankle joint.

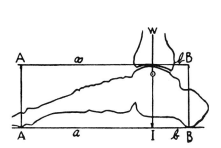

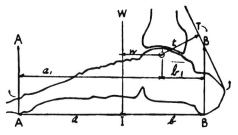

Fig. 19. Force analysis of ankle joint. Sagittal plane. Equilibrium in ankle joint. The line of gravity falls through the center O. Condition of equilibrium: Aa $=$ $-$Bb.

  W superincumbent weight
  A anterior reaction
  B posterior reaction
  I intersection point of Line of Gravity
      with Base Line
  a distance of A from I
  b distance of B from I
  O center of ankle joint

Fig. 20. Force analysis of ankle joint. Sagittal plane. Equilibrium in ankle joint. The line of gravity falls in front of center O. Condition of equilibrium. 1. Translatory: Aa $=$ Bb. 2. Rotatory in respect to ankle joint: Ww $=$ $-$Tt.

  W superincumbent weight
  A anterior reaction
  B posterior reaction
  I intersection gravity and base line
  a distance of A to I
  b distance of B to I
  w distance line of gravity to center
  O center of ankle joint
  T tension in force of calf muscles (tendo
      Achilles)
  t distance of T from center O
  a_l distance of A from center O
  b_l distance of B from center O

The gravitational stress develops a rotatory moment R (gravitational rotation component) times d (distance from the center of motion). This moment tries to move the tibia against the foot in the direction of dorsiflexion. It is resisted by the tension of the posterior muscles, especially the tendo Achillis muscles (Fig. 20). The amount of the gravitational rotatory moment depends upon the weight (P) and the distance (d) of the weight line from the center of the ankle joint. The calf muscle must furnish an equal and opposite moment of resistance to hold the equilibrium. This moment is the product of the tension of the tendo Achillis (T) times its distance (t') from the center of the joint. Consequently, the condition of equilibrium is Pd $=$ $-$ Tt'.

Let us further assume that the line of gravity moves farther forward until it falls into the ball of the foot. As it moves forward the tension in the tendo Achillis steadily increases, simply because, as the rotatory moment of gravity increases with the lengthening of its distance from the center of motion in the ankle joint, the moment of the tendo achillis must also increase proportionally in order to hold the equilibrium. Since the distance of the tendon from the center of motion is constant, it is the tension of the heel cord which must increase. By the time the line of gravity falls into the ball of the foot, all weight is borne by the ball and none by the heel.

The rotatory moment of gravity now is the product of the entire superincumbent body weight (since the heel bears no weight) times the perpendicular distance of the ball from the center of the ankle joint. In other words, the two equilibrated moments are P (entire weight) times d (its distance) and T (tension of the tendo Achillis) times d_1 (its distance from the center of motion). These two products must be equal if equilibrium is to be maintained ($Pd = -Td_1$). Since the distance of the tendo Achillis from the center of motion is only $\frac{1}{2}$ to

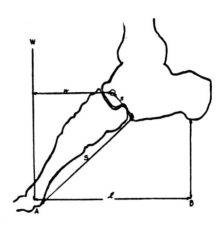

FIG. 21. Force analysis of Chopart's joint. Sagittal plane. Effect of raising on toes. Equilibrium in respect to midtarsal joint. The line of gravity goes through center O. A = W, Condition of Equilibrium Ww = Ss. As w becomes smaller the moment Ww and consequently also Ss becomes smaller. When in standing on toes, line of gravity goes through center O as well as through A, w is o and Ss is also o, consequently there is then no tension in the sole of the foot.

$\frac{1}{3}$ of that of the ball of the foot from the center, it follows that the tension of the heel cords must equal two to three times the body weight to hold the balance against the gravitational rotatory moment. Thus if a man weighing 75 kg. stands on the balls of both feet, the tension in each tendo Achillis must be $\frac{1}{2} \times 75 \times 2$, or 75 kg.; and if standing on one foot, it would be 75×2, or 150 kg. A similar computation can be made for the midtarsal joint (Fig. 21).

V. SUMMARY

1. The orthograde type of locomotion necessitated for the purpose of maintaining balance in standing and walking the special development of the ankle joint, the subastragalar, the midtarsal, and the metatarsophalangeal articulations.

2. Of these, the ankle joint with one degree of freedom of motion provides for the anteroposterior balance regulating the fluctuations of the line of gravity in the sagittal plane; the subastragalar joint, also having one degree of freedom of motion, takes care of the lateral balance by regulating the fluctuations of the line of gravity in the frontal plane; the midtarsal joint, with three degrees of freedom of motion, provides for compensatory movements between the back- and front part of the foot; and the metatarsophalangeal joints,

with only one of their three degrees of motion, namely, flexion and extension under voluntary control, serve the finer adjustments of the toes as the weight is thrown on the ball of the foot.

3. The tarsal bones are arranged in two systems, interrupted by the above-named articulations: the outer and inner arch. The outer is in immediate contact with the ground through the posterior process of the os calcis. The inner arch which receives the superincumbent body weight rests on the outer; in forward direction it continues through astragalus, scaphoid and cuneiforms to the three medial rays of the foot, while the outer is formed by the os calcis, the cuboid and the two lateral rays. Thus there is a posterior common support concentrated at the heel, while anteriorly the five rays through the heads of their metatarsals form an oval area of support.

4. In resting upon the os calcis, the astragalus forms with the former the subastragalar articulation consisting of three distinct facets: a posterior one, between the body of the astragalus and the os calcis; a middle one, between the sustentaculum and the astragalar body and neck; and an anterior one, between a small area of the astragalar head and the anterior process of the os calcis.

5. The axis of the ankle joint in upright standing is directed from outward and backward to forward and inward, corresponding to the outward torsion of the lower end of the tibia; the axis of the subastragalar joint runs from backward outward and downward to forward inward and upward, corresponding to the angle of deflection of the axis of the astragalus. Because of this obliquity, pronation is combined with abduction and supination with adduction of the foot. The three axes of the midtarsal joint correspond more closely to the sagittal direction for pronation and supination, to the frontal for flexion and extension, and to the perpendicular for abduction and adduction.

5. The inner architecture of the foot reflects in its trajectory systems the weight distribution and other stresses. The trajectories of the medial or inner arch rise from the heel to the body of the astragalus, converging toward its center, and are there crossed by trajectories which go forward through neck and head, scaphoid and cuneiforms, to the three medial metatarsals; those of the outer arch run through the os calcis, the cuboid, to the outer two metatarsals.

7. All articulations are heavily reinforced by ligaments. The medial collateral of the ankle joint runs in three sections from the medial malleolus, respectively to the scaphoid, to the sustentaculum, and to the posterior aspect of the subastragalar joint. The lateral collateral also has three portions: one forward to the astragalus, one to the outer side of the os calcis, and one posterior to the posterolateral portion of the subastragalar joint. The subastragalar joint is checked in addition by the lig talocalcaneum interosseum hidden in the narrow tarsal canal, and a posterior ligament in the sinus tarsi. The midtarsal joint has numerous ligaments, the more important of which are the dorsal talonaviculare and calcaneonaviculare bifurcatum as well as the cup-shaped ligament from the sustentaculum to the scaphoid. The lateral side of

the joint is reinforced by the dorsal and lateral calcaneocuboid ligaments and the plantar surface by the long and short plantar ligaments. Reinforcing the metatarsophalangeal joints are plantar and dorsal ligaments binding the metatarsal heads and the collateral metatarsal ligments on both sides of the joint capsule.

8. Motion in the ankle joint is practically a pure hinge movement. The anterior part of the roll of the astragalus being wider than the posterior, a spring-like give of the tibiofibular ligament is necessary in dorsiflexion. Dorsiflexion in this joint is checked by the tendo achillis, the posterior capsule, and the posterior portion of the collateral ligaments; plantar flexion is checked by the anterior capsule, the tension of the extensors and the anterior portion of the collateral ligaments.

9. Because of the incongruency of its facets, motion in the subastragalar joint is neither pure rocking nor pure gliding, but a combination, a sort of shift gliding which permits movement of the astragalus against the os calcis about a vertical axis. This assumes significance under pathological conditions such as the relaxed flatfoot. No length rotation is normally possible between astragalus and tibia, and any inward rotation of the malleolar mortise is due to inward rotation in the knee or even in the hip joint, or to torsion of the tibial shaft. Pronation-abduction-dorsiflexion is checked by the talocalcaneal interosseous ligament and the astragalus impinging on the body of the os calcis; supination-adduction-plantar flexion is checked by the astragalus impinging against the sustentaculum and by the anterior portion of the talocalcaneal ligament.

10. The function of the midtarsal joint is to place the forefoot in positions compensatory to that of the backfoot so that the forefoot may always retain its full contact with the ground. Under normal conditions, this compensation involves only the pronatory and supinatory range. Under pathological conditions, compensatory abduction and adduction, dorsiflexion and plantar flexion come into play.

11. With the foot fixed and the leg moving against the foot, as in standing, the astragalus moves with the rest of the foot against the tibia in the sagittal plane; in the frontal plane, the astragalus becomes a unit with the tibia and movement occurs between astragalus and os calcis. This is associated with simultaneous movement of the head of the astragalus against the scaphoid. The outer half of the midtarsal joint between os calcis and cuboid is fixed against the floor in walking, and any excessive pronation causes a strain of this joint as the line of gravity passes through it. Since the mortise of the ankle joint prevents both lateral and length rotatory motion, any forced pronation or supination results in strain and injury to the collateral ligaments.

12. The height and length of the longitudinal arches formed by the tarsus and the metatarsal rays varies. The second is the longest and highest, the fifth the shortest and lowest. The transverse metatarsal arch is highest at the second and lowest at the fifth ray. Relative to the points of support, the

lowest distance from the floor is found at the posterior process of the os calcis (7-10 mm.), the sesamoids of the big toe (6 mm.), and the fifth metatarsal head (6 mm.). Under weight bearing the transverse arch spreads a total of 1.25 cm.

13. The gastrosoleus is the strongest plantar flexor of the ankle joint; its strength exceeds many times that of the other plantar flexors. The dorsiflexors have hardly more than 1/4 to 1/5 the strength of the plantar flexors, but gravity acts as an auxiliary dorsiflexor.

14. In the subastragalar joint the supinators have more than twice the strength of the pronators, but here, again, gravity acts as a pronatory force. In the midtarsal joint the pronators and supinators are more evenly matched. In the metatarsophalangeal joints there is also a preponderance of the toe flexors over the extensors.

15. The conditions of a static equilibrium in the joints of the foot are based on the principle that translatory gravitational stresses alone are at work in respect to a certain joint only if the line of gravity falls into the center of the joint; otherwise, rotatory components become operative which have to be neutralized by muscle action. If, for instance, the line of gravity falls a certain distance in front of the ankle joint, the rotary effect of gravity must be equalized by a tension of the calf muscle of such degree that the product of this tension times the perpendicular distance of the muscle from the center of motion of the ankle joint equals the gravitational reaction applying to the forepart of the foot times its distance from this center. If the line of gravity falls so far forward that it strikes the ball of the foot (which is the anterior support of the beam represented by the foot), then the anterior gravitational reaction (R_1) equals the entire body weight since no weight is borne by the heel. Because the distance of this reaction from the center of motion is twice that of the tendo Achillis muscle, it follows that the tension of the latter must be twice the body weight if equilibrium against gravitational rotation stress is to be established in the ankle joint.

BIBLIOGRAPHY

1. BEELY, F.: Zur Mechanik des Stehens (Fussgewölbe). *Langenbeck's Arch. f. Chir.*, 27, 1882.
2. ENGELS, W.: Über den Normalen Fuss und den Plattfuss. *Ztschr. f. Orthop. Chir.*, *XII.XXXVI:*461, 1904.
3. FICK, I.: *Handbuch d. Anatomie und Mechanik d. Gelenke.* Jena, G. Fischer, 1911.
4. LANGER, C.: Über das Sprunggelenk der Säugetiere und des Menschen. *Sitzungsber. d. Akad. d. Wiss., Wien,* 1856.
5. MEYER, H. VON: *Studien über den Mechanismus des Fusses.* Jena, 1883-1888.
6. MORTON, D. J.: *The Human Foot.* New York, Columbia University Press, 1935.
7. SEITZ, L.: Die vorderen Stützpunkte des Fusses unter normalen und pathologischen Verhältnissen. *Ztschr. f. Orthop. Chir., VIII.III:*37, 1901.
8. STEINDLER, A.: The Architecture of the Tarsus. *Am. J. Orth. Surg.*, Oct. 1914.
9. STRASSER, H.: *Lehrbuch der Muskel- und Gelenkmechanik III.* Berlin, J. Springer, 1917.

Lecture XXIII

THE PATHOMECHANICS OF THE STATIC
DEFORMITIES OF FOOT AND ANKLE

I. INTRODUCTION

DEVELOPMENTAL FACTORS

THE study of the mechanical forces which underlie the static disalignment of the foot directs our attention to certain developmental features which serve as a philogenetic background for the inability to resist normal mechanical stresses. Frequently, one finds predisposing anatomical variations in cases of static flatfoot which are nothing else but an arrest in the developmental cycle which leads from the ancestral primate to the perfect foot of the bipedal human race. According to Morton,[6] the principal earmarks of this cycle are the following:

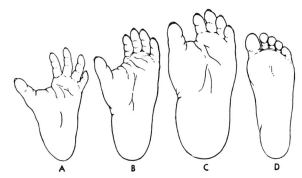

FIG. 1. Anthropoid and human feet, showing loss of opposition of the big toe. A. Chimpanzee B. Lowland gorilla (G. gorilla). C. Mountain gorilla (G. Beringei). D. man. (Morton)

In contrast to the anthropoids, the human foot is characterized by the complete loss of the opposibility of the hallux (Morton[6]) (Fig. 1). The metatarsal heads are no longer rotated toward each other as the gripping function of the toes requires, but they are directed straight anteroposteriorly (Fig. 2). The axis of leverage has shifted from between the second and third to be-

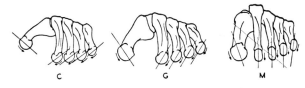

FIG. 2. Loss of rotation of the metatarsals. C. Chimpanzee. G. Gorilla. M. Man. (Morton)

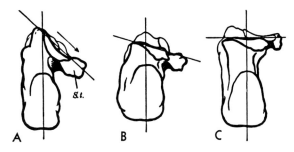

FIG. 3. Horizontalization of the sustentaculum tali (S.t.). A. Gorilla. B. Neanderthal man. C. Modern man. (Morton)

399

tween the first and second metatarsals and, as a result, the latter is increased in length. A most important point is that the orthograde locomotion causes the sustentaculum tali to become more massive and to assume a more horizontal position (Fig. 3). In the anthropoids, the adduction of the great toe gives the

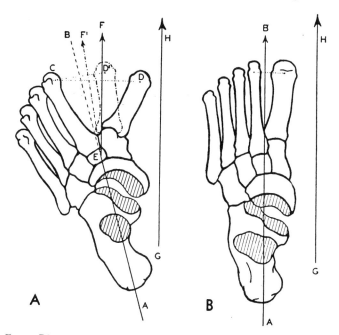

FIG. 4. Disappearance of medial angulation of the great toe. (Morton)

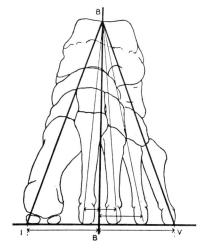

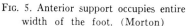

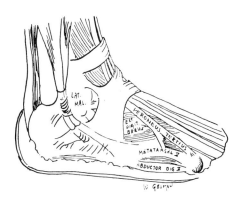

FIG. 6. Lateral aspect, foot and ankle.

FIG. 5. Anterior support occupies entire width of the foot. (Morton)

forefoot a medial deflection. This angulation disappears in the human (Fig. 4). The axis of the entire foot follows a straight line so that the first and second metatarsal bones run more parallel. The so-called metatarsus primus varus in man may well be considered as an atavistic phenomenon. With the plantigrade gait, the os calcis changes its downward position to an upward direction, and it becomes at the same time more massive. As the axis shifts medially to the interval between the first and second metatarsal, the inner border of the foot becomes more flattened and depressed. With the shift of balance from lateral to medial, the transverse arch develops and the supporting ball occupies the entire width of the foot (Fig. 5).

The plantigrade locomotion of man which has forced the foot into a right-angle position to the leg has also produced some changes in the arrangement of the musculature. For instance, the tibialis anticus has, in the pro-anthropoid, still a split tendon; in man, its insertion into the hallux disappears and the muscle has only a single tendon, losing all its ability to abduct the big toe. Nevertheless, according to Morton,[6] in 10% of the human feet there is still a split of this tendon noticeable. The extensor digitorum longus as well as the extensor hallucis split off from the primitive extensor plate as separate units, and a new

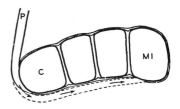

FIG. 7. Migration of the peroneus longus across the plantar aspect of the foot. (Morton)

unit is formed by the peroneus tertius; according to Morton,[6] this muscle is only found in the gorilla Berengei. It is a muscle which aids in the pronation of the formerly supinated foot and it helps to hold the planta firmly to the ground; in man, it is absent in 10% (Fig. 6).

A characteristic feature is the development of the peroneus brevis and longus. In the first place, the tendons running behind the axis of the ankle joint now change from dorsiflexors to plantar flexors and assume a propulsive function. Then, because of the increased stress that is put on the anterior arch with the plantigrade locomotion, the longus has migrated clear across the sole of the foot. It now acts, by virtue of its insertion into the base of the first metatarsal, as a tiepiece which maintains the arch of the foot against depression (Fig. 7).

Certain functional disabilities may arise from the arrest of development and the reversal to an atavistic state of both the skeleton and the musculature. For instance, a persisting short first metatarsal may predispose the anterior arch to depression. A failure of the peroneus tertius to develop additional pronatory power may be the anatomic basis of a congenital clubfoot; or, the tibialis posticus may fail to develop its terminal fanshaped insertion which deprives the longitudinal arch of its principal support and facilitates pronation deformity and flattening of the arch.

II. THE PATHOMECHANICS OF THE STATIC DEFORMITIES OF THE FOOT

A. PREMISES

A static so-called pes planus or planovalgus develops when for some reason or other the foot has lost is normal resistance to gravitational stresses. Starting on the proposition that the normal foot is one which maintains a perfect equilibrium under weight bearing, one may state by contrast that a foot which becomes deformed under the effect of gravity has lost the ability to maintain an active equilibrium by means of muscular tension, and that as a result a disalignment of the constituent parts of the foot occurs.

In the case of the static pes valgus the disalignment involves not only the joints of the foot but the entire extremity as well. The disturbed relationship can be expressed for the parts concerned in reciprocal terms. Any inward rotation of the leg against a stabilized foot means that the latter is outward rotated against the leg. Any adduction in the back part against the forepart of the foot may be designated as abduction of the forefoot in the midtarsal joint. Plantar flexion of the backfoot against the forefoot is tantamount to dorsiflexion of the forefoot in the midtarsal joints; the basic mechanical condition is that one part being immobilized, usually the peripheral one, it cannot follow the movement of the other and, therefore, a disalignment or a break occurs in the articulation which joins the two portions.

The mechanical analysis of static foot disorders is, therefore, confronted with a series of compensatory distortions in joints in which under the effect of gravity muscular or ligamentous resistance has become inadequate.

B. THE PATHOLOGICAL EQUILIBRIUM

The normal alignment of the different joints of the foot represents the stable physiological equilibrium in which the function of all structures entrusted with the maintenance of this equilibrium is unimpaired. Conversely, the disalignment seen in the static flatfoot betokens the overthrow of this equilibrium. Instead, a pathological equilibrium is established which is unstable. It is constantly overthrown and lasts no longer than the intervals between the intermittent sallies of the progressing deformity. The point of interest is to investigate certain morphological features which may predispose to this loss of equilibrium.

1. The pronated foot

a) Does any independent pronation ever occur in the ankle joint? Except for the paralytic flail foot, we have not been able to observe such a pronatory motion. The mortise holds the astragalus strictly to one degree of freedom of motion, i.e., to flexion-extension.

b) Does the axis of the ankle joint ever become oblique from pathological causes, thereby producing a pronatory inclination of the entire foot? This is, indeed, the case in disturbances of the lower tibial epiphysis which result in asymmetry of growth, i.e., a thinning of the lateral half of the epiphysis. Under this effect the transverse axis of the ankle joint becomes inclined or oblique. This is observed, for instance, after forcible correction of clubfoot. In the traumatic pes valgus the pronation is not the effect of deflected growth but is due to the deformation of the tibia itself, following fracture.

c) Does pronation occur as a compensatory disalignment? Such compensation is seen, for example, in the case of genu varum, as a compensatory pronation of the foot; it may also be caused by a short heel cord (Harris and Beath[3]) (Fig. 8), which does not permit the foot to be placed flat on the ground except in pronation, and which forces the os calcis into equinus position. This compensatory pronation which occurs in the subastragalar joint may become a contributory, if not the primary factor producing a hypermobile foot.

2. The instability of the subastragalar joint

The subastragalar joint becomes unstable primarily because of certain morphological anomalies which concern the manner in which the talus is supported by the os calcis.

Normally, the os calcis carries three joint facets which correspond with

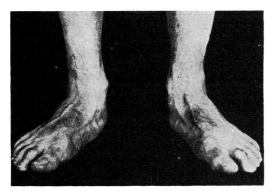

Fig. 8. Flat foot due to a short heel cord. (Harris & Beath)

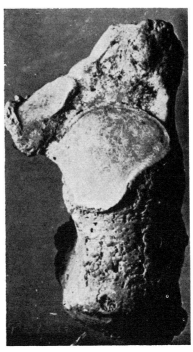

Fig. 9. Os calcis, superior view showing absence of anterior facet and narrow articular facet of sustentaculum. (Courtesy of Dr. John Cooper)

similar facets at the lower surface of the astragalus. A large posterior quadrangular facet at the undersurface of the body of the astragalus articulates with the facies articularis posterior of the os calcis; a facies articularis media covers the sustentaculum and is separated from the posterior facet by the sulcus calcanei. It fits into the concave middle facet of the astragalus. An anterior facies articularis of the calcaneus occupies a small area medially at the upper surface of the anterior process and articulates with a convex anterior facet of the head of the talus. Harris and Beath,[3] who made a comprehensive study of the unstable foot, emphasize the importance of the anterior facet which in the normal foot supports the head of the talus. In the weakly supported foot this supporting facet between the end of the calcaneus and the head of the talus may be absent.

The signs of firm support are the broad rounded sustentaculum which runs forward to the anterior margin of the calcaneus. In the poorly supported foot the sustentaculum is a narrow tongue which springs from the medial side of

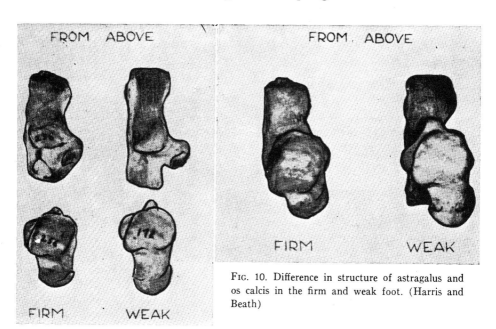

FROM ABOVE   FROM ABOVE

FIRM   WEAK

FIRM   WEAK

FIG. 10. Difference in structure of astragalus and os calcis in the firm and weak foot. (Harris and Beath)

the os calcis with a narrow base and the facet at the anterior end of the os calcis is missing (Fig. 9). A firmly supported foot has a short and thick head and neck of the astragalus superimposed over the anterior end of the os calcis, which is also thick and short. A weak support is an elongated head and neck of the astragalus projecting forward and medially much beyond the anterior border of the os calcis. A firmly supported foot is one in which the head of the talus lies over the center of the anterior end of the os calcis and flush with it. In a weakly supported foot, the center of the head lies medially to the center of the anterior end of the os calcis and projects forward from it. In these cases there is no contact between head and anterior border of the os calcis, but there is rather a diastasis (Fig. 10). The absence of this anterior facet is not necessarily a sign of hypermobile foot, since Harris[3] found it also in the normal foot; while, on the other hand, the facet may be present in some hypermobile feet. However, in the majority of the hypermobile feet he noticed that the facet was absent. He, therefore, believes that this absence is at least a contributory factor, especially since it facilitates the downward displacement of the astragalar head.

How does this instability of the subastragalar joint affect the function? Does it mean only an increased pronatory range? Or, is there, in addition, some other pathological motion? There is, associated with the pathological displacement which separates the head of the talus from the anterior process of the os calcis, a horizontal gliding of the astragalus against the os calcis, the result of which is that the angle between the axes of the two bones, normally 30-35°, is increased (Fig. 11). In severe cases of the hypermobile foot this increase occurs regularly and may reach as much as 55° or 60° (Harris[3]). This increase of the angle between the axes of astragalus and os calcis was already recognized by Fick.[2]

Strasser[7] points out that the horizontal inward rotation about a vertical axis must be carried out by the talus in conjunction with the leg, the talus being held firmly in the mortise; the fibular side of the body of the astragalus rotates forward, the tibial side backward, and the head inward. As the leg rotates inward with the astragalus, the outer malleolus comes to lie more forward, the inner more backward. It is this horizontal rotation of the talus which increases the talocalcaneal angle. With this horizontal rotation is also associated an increase of the slope of the sustentaculum, thereby approaching the increased sustentacular slant of the anthropoid.

Fig. 11. Angle between os calcis and astagalus which is increased in flat foot. (Harris and Beath)

The astragalar head, which no longer overlaps the anterior process of the os calcis, articulates only with its lateral half with the scaphoid. The inward rotation of the talus causes the external malleolus or the styloid of the fibula to approach the lateral surface of the anterior process of the os calcis. In short, there is an inward rotation of the whole ankle mortise together with the astragalus, with the result that the medial or inner arch does not simply flatten down but actually glides off from the supporting calcaneus. The break in the whole structure of the foot is in the midtarsal joint. The forefoot is dorsiflexed in this joint against the plantar flexed backfoot, abducted against the backfoot, and supinated against it (Henke[4]).

III. THE PATHOMECHANICS OF THE SOFT STRUCTURES

In the foregoing certain skeletal anomalies were interpreted as developmental retardations which predispose the foot to become mechanically insufficient and, under favorable conditions, prepare the way for the development of the static flatfoot.

The actual skeletal disalignment of the foot must be preceded by a deficiency of the soft structures, because it is the latter which safeguard the

normal relationship of the tarsal bones. Their functional impairment ultimately causes them to become relaxed and elongated so that they are unable
to prevent skeletal deformation. Fortunately for the clinician, all these
functional strains make themselves known early by aches, pains, and by
definite pressure points. The discussion of their clinical significance falls outside the scope of this work. It shall be mentioned only that, due to the fact that
these structures are richly endowed with sensory fibers, pain appears early

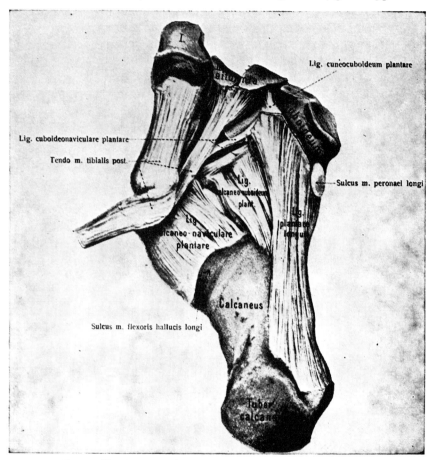

FIG. 12. The plantar calcaneonavicular ligament. (Rauber-Kopsch)

and is of prime importance for the proper localization of the injured structures. It is also significant that deformities which follow the functional impairment of the soft structures make their appearance at first only under the effect
of gravitational stresses. Only later, when the bones have adapted themselves
secondarily to the abnormal static conditions, do they become stationary.

An analysis of the normal function of the soft structures safeguarding the
different joints and of the effect of their functional impairment follows:

A. THE SUBASTRAGALAR JOINT

1. The principal ligamentous structure is the calcaneonavicular ligament
(Fig. 12). When this ligament comes under strain, a discrete pressure point

develops at the astragaloscaphoid articulation and under the head of the astragalus.

2. The long flexors of the toes and the tibialis posticus running under the sustentaculum sustain the longitudinal arch against gravitational stress. When they are under tension, they develop tenderness along their course. At the same time, these muscles running under the sustentaculum sustain the pronatory and supinatory equilibrium and guard the normal relationship between os calcis and astragalus. The latter has the tendency to slip off the os calcis, increasing thereby the calcaneo-astragalar angle (Hohmann[5]). This

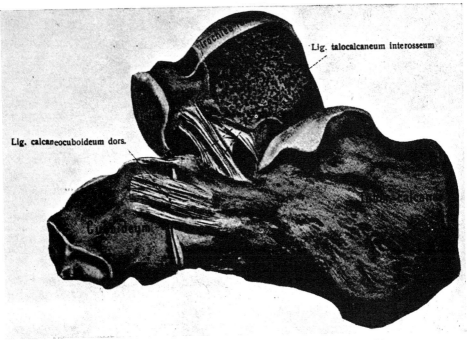

Fig. 13. The talocalcaneo interosseous ligament. (Rauber-Kopsch)

produces additional strain in the talocalcaneo interosseous (Fig. 13), in the deltoid ligaments and in the calcaneoscaphoid ligament mentioned above.

B. THE MIDTARSAL JOINT

This is an articulation with three degrees of freedom of motion: dorsi- and plantar flexion, ab- and adduction, and pro- and supination. Pro- and supinatory movements are the only ones occurring under normal conditions; they serve the purpose of adapting the foot plate to the floor.

Ligaments and other soft structures protect the arch of the foot by preventing pathological dorsiflexion of the forefoot. The plantar aponeurosis prevents a flattening of the arch. Of the muscles, the most important ones sustaining the plantar arch are the short muscles, especially the short flexor of the toes, which is intimately connected with the plantar fascia. In addition, the deep ligaments, especially the long plantar ligament and the calcaneonavicular ligament, help sustain the arch. Additional to the flexor brevis the muscles maintaining the longitudinal arch in its normal height are the abductor hallucis as well as the tibialis posticus and the long flexors of the toes. The strain of these structures causes definite trigger points. One of these is found along the medial side of the plantar ligament, and there is not infrequently another in front of the posterior process of the heel which is often mistaken for a heel spur.

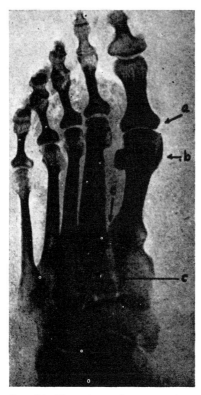

FIG. 14. Morton's syndrome. *a.* short first metatarsal; *b.* proximal displacement of sesamoids; *c.* diastasis between first and second cuneiform. Also note hypertrophy of first and second metatarsal. (Morton)

FIG. 15. The route of the peroneus longus across the plantar aspect of the foot. (Rauber-Kopsch)

C. THE ANTERIOR ARCH

We have already referred to the significance of a short first metatarsal such as constitutes a part of the so called Morton's syndrome. This shortness causes the axis of balance to rest more upon the head of the second metatarsal instead of lying between the first and second. The second metatarsal, in consequence, becomes thickened (Fig. 14).

The deficiency of the structures maintaining the arch can be recognized first by the relaxation of the intermetatarsal ligaments and, secondly, by the strain of the ligaments of Lisfranc's joint. Both of these structures, when under strain, produce areas of tenderness at the tarsometatarsal junction. In addition, we know that the tibialis posticus as well as the flexors of the toes are essential aids in maintaining this arch; their strain produces further tenderness located along the course of these tendons at the underside of the foot. It is evident, therefore, that a relaxation of the anterior arch will produce tenderness not only directly behind the metatarsal heads but also farther upward, along the sole of the foot.

The interossei, due to their ability to flex the metatarsophalangeal joints, further assist in maintaining the arch; strain of the latter often causes tenderness of these muscles on pressure at the dorsum of the foot. An essential maintainer of the arch is the peroneus longus. Its tendon crosses the cuboid at the planta and inserts into the base of the first metatarsal. It constitutes a tiepiece holding the transverse arch together. It is for this reason that sometimes the flattened anterior arch produces tenderness along this muscle which can be traced to its course at the sole as well as at the lateral border of the foot (Fig. 15).

IV. SECONDARY SKELETAL CHANGES FOLLOWING RELAXATION OF SOFT STRUCTURES

A. THE ANKLE JOINT

While the joint itself does not suffer any real displacement, being held firmly in the malleolar mortise, the downward slant of the os calcis raises the heel into an equinus position. The back part of the foot, therefore, is in position of plantar flexion and the raised heel causes the tendo Achillis to become shortened. It is, therefore, impossible in the fully developed valgus to obtain a right-angle position between foot and leg when the foot is held in supination.

B. THE SUBASTRAGALAR JOINT

The greatest changes occur in this joint. The disalignment is not confined to one plane as one might suppose from the fact that this joint normally has only one degree of freedom of motion. There is first the pronatory position of the subastragalar joint as a result of which the line of gravity is strongly deflected laterally so that gravitational stresses tend to increase the pathological pronation.

In addition to this, the astragalus becomes depressed and slides forward and downward. It takes with it the anterior portion of the os calcis and thereby allows the posterior process of the heel to rise up, causing the already described shortening of the tendo achillis and the plantar flexed position of the foot in the ankle joint (Fig. 16).

Of greatest importance, however, is the displacement which occurs in the transverse plane, namely, the inward sliding of the astragalus away from the os calcis. This increases the angle between astragalus and os calcis which normally is about 35°, to as much as 60° in the flatfoot. Due to the pronatory tilt of the os calcis, the sustentaculum projects farther medially (Harris and Beath[3]). The horizontal rotation of the talus, in which, as explained above, tibia and astragalus rotate together as one unit against the os calcis, can be observed in all cases of advanced and fully developed static flatfoot. It causes the head of the astragalus to project sharply medially, making the inner

FIG. 16. Depression of the astragalus with compression of the scaphoid and plantar flexion of the os calcis.

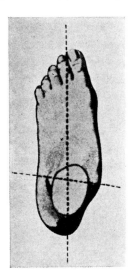

FIG. 17. Flat foot, inward rotation of the leg, outward rotation of the foot. (Hohmann)

border of the foot convex. The scaphoid retains its relation to the forefoot and subluxates toward the lateral half of the astragalar head.

Hohmann[5] and, lately, Starr, have pointed out that even in the normal foot the inward rotation of the leg causes pronation and outward rotation of the foot (Fig. 17) the outward rotation of the leg produces supination in the subastragalar joint and inward rotation of the foot. This latter position can be observed in cases of clubfoot where the leg is rotated outward against an inwardly rotated foot, thereby accentuating the latter's supinatory position.

C. THE MIDTARSAL JOINT

This joint has three degrees of freedom of motion. In the pes planovalgus, the forefoot is dorsiflexed about its frontal axis against the back part of the foot after the ligamentous and muscular safeguards have given way. In relation to the sagittal axis, the forefoot is supinated against the back part of the foot; the forepart of the foot, being held firmly to the ground by the weight of the body, is not able to follow the pronatory twist of the back part of the foot. Retaining its original position, it becomes relatively supinated against the hindfoot. In relation to the perpendicular axis, the forefoot undergoes an abductory movement. This also is relative only and it is due to the fact that the forefoot remains immobilized by the body weight and is, therefore, not able to follow the adductory movement of the back part of the foot which, as mentioned above, is the result of the inward and forward movement of the astragalus.

Thus we find compensatory adjustment of the forefoot against the backfoot about all three axes: a compensatory supination about the sagittal, a compensatory dorsiflexion about the frontal, and a compensatory abduction about the perpendicular axis.

V. CLINICAL ANALYSIS

Securing the forefoot firmly to the ground against the distortion of the back part requires not only considerable muscular effort, but the muscle governing the movement of the forefoot must also be in perfect control of the compensatory position. Otherwise, the forefoot will follow the backfoot distortion, as one observes, for instance, in the spastic flatfoot.

It is particularly the task of the peroneus longus and the flexor hallucis longus to produce plantar flexion of the first metatarsal and internal cuneiform. Tibialis posticus has a similar effect. Both muscles press the forefoot forcibly against the head of the talus and thereby they prevent the stretching of the plantar ligament and oppose the abduction of the forefoot. Furthermore, the flexor hallucis longus forces the scaphoid medially against the astragalar head, opposing the tendency of the scaphoid to glide laterally off the head of the astragalus.

According to Duchenne,[1] the tibialis posticus is a strong adductor causing the medial border of the foot to become concave. Of particular importance is the synergism which exists between tibialis posticus and peroneus brevis. The latter muscle alone produces abduction of the foot which is more powerful even than that of the peroneus longus. It elevates the lateral border of the foot, turning the plantar surface outward. It also dorsiflexes the plantar-flexed foot, but if the foot is in strong dorsiflexion, then it becomes a plantar flexor until the foot reaches the right angle. It is this synergism of the muscles which is designed to maintain the foot firmly on the ground in standing and to prevent its turning either laterally or medially.

In the spastic flatfoot one often sees a forward dislocation of the peroneus longus, and the tibialis anticus becomes converted from a partial supinator to a pronator. In this type of flatfoot both tibialis anticus and peroneus longus are contracted. They may neutralize their respective flexory and extensory action, but they are synergistic pronators. The strong pronatory and abductory action of the peroneus longus forces the foot into pronation and abduction, and if the peroneus brevis is also contracted, then the abductory effect becomes much accentuated. If the contracture includes the extensors of the toes and the peroneus tertius, the dorsiflexory effect is still more pronounced, as is the associated abduction and pronation. The strong valgus position of the subastragalar joint crowds the structures in the sinus tarsi; they become tender to pressure.

The effect of these muscles on the plantar arch is as follows: When the medial border of the foot is lowered, i.e., plantar flexed, the plantar arch is heightened. This is effected by the peroneus longus which, at the same time, tightens the anterior arch and increases its plantar concavity. In this effect the peroneus longus is supported by the flexors of the toes which, together with the interossei, flex the metatarsophalangeal joints. By doing so they impart a pronatory movement to the big toe and a supinatory movement to the little toe; in other words, they hollow out the anterior arch, automatically supporting and heightening it. In the flatfoot the contracture of the toe extensors weakens the flexors of the toes and impedes the action of the interossei with the result that the anterior arch becomes depressed and assumes a plantar convexity.

VI. SUMMARY

1. Developmental factors incident to the development of the orthograde locomotion in man are: loss of opposibility of the great toe, massiveness and upward direction of the os calcis, development and horizontalization of the sustentaculum, loss of medial deflection of the first metatarsal, single tendon of insertion of the tibialis anticus, appearance of the peroneus tertius as a new unit, and the wandering of the peroneus longus across the planta of the foot.

2. In the static flatfoot a pathological equilibrium exists which is unstable and changes with the progression of the deformity.

3. No independent pronation takes place in the ankle joint proper but asymmetry of epiphyseal growth or fracture disalignment can tilt the axis of the ankle joint in a pronatory sense (traumatic ankle valgus). A compensatory pronation in the subastragalar joint is observed in cases of supinatory deflection of the ankle joint axis, as in tibia vara.

4. Certain morphological variations favor the instability of the subastragalar joint: a narrow sustentaculum, an elongated head and neck of the astragalus projecting beyond the midtarsal joint line, absence of the anterior articular facet with diastasis between head of astragalus and anterior process of the os calcis.

5. The instability of the subastragalar joint not only leads to an excess of pronation but it also engenders a horizontal gliding of the astragalus against the os calcis, about a perpendicular axis, whereby the angle between the length axes of the two bones is increased. This rotation is carried out by the talus in conjunction with the tibia, so that the outer malleolus rotates forward and the inner backward. The break in the whole structure of the foot is in the

midtarsal joint; here the forefoot is dorsiflexed against the backfoot, and abducted and supinated against the latter.

6. Actual skeletal disalignment is preceded by deficiency of the soft structures, which become relaxed and elongated. Strain of these structures develops characteristic pressure points.

7. The sequence of events which leads to the formation of the flatfoot from its very incipiency to its fully-developed form is as follows: The first stage is the strain of the supporting soft structures which safeguard the normal relationship of the joint constituents. This is followed by a relaxation of these structures which imparts to the joints abnormal mobility. Deformation at first occurs only when the joints are put under static stress. In the beginning, the static flatfoot remains movable and becomes deformed only when sustaining the body weight. A gradual adaptation of the joint to the abnormal position takes place; it has its expression in reactive changes in the form of exostoses, spurs, or degenerative arthroses.

8. The fully-developed flatfoot presents the following features:

a) Inward rotation of the leg against the foot in the ankle joint.

b) Pronation of the backpart of the foot and a compensatory supination of the frontpart.

c) A spreading of the angle between the axes of os calcis and astragalus, with forward and inward gliding of this bone into an adductory position; the backfoot is, therefore, adducted while the frontpart shows a compensatory abduction.

d) Depression of the os calcis by the astragalus, elevation of the heel in a plantar-flexed position in the ankle joint.

e) The plantar-flexed position of the backpart of the foot is not being followed up by the forefoot; the latter appears, therefore, in dorsiflexion in respect to the backfoot.

f) Likewise, the forefoot does not take part in the adductory position of the backfoot since it is fixed to the ground. Therefore, the forefoot is abducted in relation to the backfoot.

9. Since these deformities follow a very typical and stereotype pathological course, it should be a simple matter to apply appropriate measures. It seems obvious that the impending or the actual strain demands first the passive support by appropriate mechanical devices, and secondly, the rehabilitation of the strained muscles through rest and physiotherapeutic measures. A foot, when it is still pliable, will yield to conservative measures so long as there is a possibility of the ligaments recovering their tone and tension and the muscles are capable of rehabilitation. Relaxed muscles no longer capable of recovery because of constitutional conditions or chronicity or age will require such operative measures as will restore the stability of the joints. The less extensive the operation, i.e., the less it interferes with the normal function of the joints, the better will be the ultimate functional result.

The foot that has already assumed permanent pathological disalignments in the different articulations which no longer disappear when the superincumbent weight is relieved, e.g., the so-called spastic or the rigid flatfoot, requires above all restoration of the normal joint relationship. This is a prerequisite to any further procedures. In extreme cases the most radical of these measures must be used where the joints are persistently disaligned, and it consists of the fusion of the joints in corrected position.

BIBLIOGRAPHY

1. DUCHENNE, G. B.: *Physiology of Motion*. Transl. by G. B. Kaplan. Philadelphia, J. B. Lippincott, 1949.
2. FICK, R.: *Handbuch der Anatomie und Mechanik der Gelenke III*. Jena, G. Fischer, 1911.
3. HARRIS, R. I., and BEATH, T.: *Army Foot Survey, Vol. I*. National Research Council of Canada, Ottawa, 1947.
4. HENKE, W.: Die Kontroversen über die Fussgelenke. *Ztschr. f. Med., 2*, 1857.
5. HOHMANN, G.: *Fuss and Bein*. München, J. F. Bergmann, 1948.
6. MORTON, DUDLEY J.: *The Human Foot*. New York, Columbia University Press, 1935.
7. STRASSER, H.: *Lehrbuch der Muskel- und Gelenkmechanik III*. Berlin, J. Springer, 1917.

THE PATHOMECHANICS OF THE PARALYTIC FOOT AND ANKLE

I. INTRODUCTION

HIGHER demands for stability are made on the articulations of the foot and ankle than on any other group of joints in the human body. A completely rigid foot is more serviceable than a flail foot with all joints intact. In fact, under the prevailing concept the requirement of stability overshadows that of mobility so much that operative measures to procure active mobility have receded into the background in favor of stabilizing procedures. This is almost a reversal of ideas which prevailed not too long ago when tendon transplantation was accorded a wide field in the treatment of paralytic disabilities of foot and ankle.

The reasons for this change of policy were largely mechanical: the transplantation did not stand up against the static demands, and deformities were either not corrected at all or they recurred under weight-bearing conditions. It was only when a system of combining stabilization with tendon transference was adopted for the majority of situations that sound and enduring results were obtained. What is still lacking in many cases is, first the recognition of the fact that not all joints are depending on stability in the same measure or are equally independent of mobility to perform their function. Secondly, it is not always appreciated that when one joint is eliminated for the sake of stability, this elimination is bound to have its effect upon the other joints of the extremity and, in fact, upon the whole body, and that far-reaching readjustments become necessary. It is only after analyzing the locomotor situation of the entire extremity, if not of the whole body, that any evaluation of an operative procedure is acceptable.

Relative to the above mentioned remote effect of stabilizing operations, one must remember that whenever a muscle is prevented from acting upon its own joint because of ankylosis, its effect upon the neighboring joint increases. Its contractile strength is no longer being spent on its own articulation. For example, the soleus becomes an extensor of the knee when ankylosis of the ankle joint prevents plantar flexion. The rectus femoris increases its flexory effect upon the hip joint when the knee is ankylosed, and so also the hamstrings become more forceful extensors of the hip when movement at the knee joint is eliminated.

II. THE CONDITIONS OF EQUILIBRIUM FOR THE JOINTS OF FOOT AND ANKLE

A. THE ANKLE JOINT. FLEXION-EXTENSION EQUILIBRIUM

The principal flexor is the powerful gastrocnemius, one of the strongest single muscles of the human body. It is the primary plantar flexor of the ankle joint and is assisted by the common flexors of the toes, the tibialis posticus and the peroneus longus, all of which take effect also upon the more distal subastragalar and midtarsal joints. According to Biesalski and Mayer,[2] the gastrosoleus is also a considerable adductor and supinator to the amount of 13° adduction and 12° supination. The peroneus brevis is also a plantar flexor but its strength in this direction is less than one-half that of the longus.

The extensors, or dorsiflexors, are the tibialis anticus, the extensor of the toes, the extensor hallucis and the peroneus tertius, whose combined working capacity is hardly more than $\frac{1}{4}$ to $\frac{1}{5}$ of the strength of the plantar flexors. In standing position gravity assists dorsiflexion so long as the line of gravity falls in front of the ankle joint. The tension of the gastrosoleus, which counteracts dorsiflexion, increases with the forward displacement of the line of gravity until it reaches for each leg one-half of the body weight in standing on both legs, and full body weight in standing on one leg.

Biesalski and Mayer[2] find that in dorsiflexion the tibialis anticus develops a considerable adductory component, especially if acting against resistance. This adductory component of the muscle seems to be missing when the foot drops into plantar flexion. This would indicate that normally any adduction or supination in plantar flexion comes from the tibialis posticus and not from the anticus. In paralysis of the extensors the joint is relaxed and drops automatically into plantar flexion and adduction, while dorsiflexion is more "trackbound" due to the greater width of the anterior portion of the roll of the astragalus.

This opens the question whether, under condition of paralytic relaxation, an abnormal mobility of the ankle joint can develop in pro- and supinatory or in length rotatory sense. In the horizontal plane of length rotation the kidney-shape medial articular surface of the astragalus remains the fixed point, while motion occurs on the lateral side of the mortise. In the frontal plane, a pronatory separation of the astragalus from the tibia can be demonstrated in cases of paralysis.

The preponderance of the plantar flexors over the extensors easily leads to equinus deformity in recumbency. This occurs even when the extensors of the ankle joint are not paralyzed, as can be seen in arthritis or in cases where an equinus deformity develops from an improperly applied cast; but this discrepancy becomes much more potent in cases of weakened or paralyzed extensors (dorsiflexors) of the ankle joint, and special precautions are necessary to prevent equinus deformity.

B. THE SUBASTRAGALAR JOINT

All muscles, pronators as well as supinators, cross all three joints of the foot—the ankle joint, the subastragalar, and the midtarsal—and, consequently, have a rotatory effect on all.

The principal supinator is the tibialis posticus, which is also a plantar flexor of the ankle and an adductor of the forefoot. The total supinatory range of the forefoot against the leg is 52° (Biesalski and Mayer[2]), while it is only 34° between the astragalus and the os calcis. This difference indicates the additional supination in the midtarsal joint.

The tibialis posticus is assisted in plantar flexion and adduction by the flexor of the big toe and the common toe flexor. They are neutralized by the pronators, primarily the peronei, and the common extensors of the toes, assisted in standing by gravity.

The peroneus longus is a strong abductor as well as a pronator, while the extensors of the toes and the peroneus tertius are principally dorsiflexors of the ankle joint. The latter is also a strong abductor and pronator in the subastragalar and midtarsal joints. According to Biesalski and Mayer,[2] its abduction range is even greater than the flexion—extension range.

In working capacity, there is a discrepancy in favor of the supinators (2.82 kgm. versus 1.16 kgm. for the pronators) which is made up by gravity in upright position. In resting position, when not bearing weight, the foot has the tendency to assume a position of supination.

C. THE MIDTARSAL JOINT

The pronators are abductors and dorsiflexors; the supinators are adductors and plantar flexors of the midtarsal joint within the limits of the ranges of this articulation.

In pronation and supination in the midtarsal articulation the two muscle groups are more evenly matched (0.52 kgm. pronatory and 0.50 kgm. supinatory capacity). In addition, the short flexors flex the joint plantarly. In the sagittal plane there exists between the short flexors, the tibialis anticus and the gastrosoleus, a sort of triangular balance.

D. THE METATARSOPHALANGEAL ARTICULATION

In this articulation the equilibrium depends on the balance between the long flexors, the long extensors and the intrinsic muscles of the foot. Because all of these long muscles pass the proximal joints from the ankle joint down, they may be already in a state of muscular imbalance in any of the proximal articulations before they reach the metatarsophalangeal joints. The extensor of the toes may become stretched by an equinus position of the ankle joint; an imbalance of the midtarsal joint causing a cavus deformity may produce passive tension of the flexors and this, together with the hyperextension in the metatarsophalangeal joints, causes the interphalangeal joints of the toes to go into flexion and produce hammertoes. So it is well in cases of paralytic claw—and hammertoes to look for the more proximal joints as the original seat of the muscular imbalance.

III. WHAT DEFORMITIES DEVELOP FROM SPECIFIC MUSCLE DEFICIENCIES?

A. THE ANKLE JOINT

Since there is only one degree of freedom of motion, namely in the sagittal plane, one has to consider only two contracture positions: the talipes calcaneus, resulting from a deficiency of the plantar flexors; and the equinus, due to the disability of the dorsal extensors.

1. Talipes calcaneus

The main plantar flexors are the gastrosoleus and the peroneus longus. According to Duchenne,[8] the gastrosoleus is also an adductor of the foot in the subastragalar joint; the peroneus longus is an abductor as well as a plantar flexor.

a) In *isolated paralysis of the gastrosoleus,* the foot assumes a calcaneus position in the ankle joint. At the same time, the tricornered equilibrium between triceps surae, tibialis anticus and the short flexors is overthrown in relation to the midtarsal joint. The peroneus longus, then, abducts the forefoot, and there is no opposition to its pronatory and plantar flexory effect on the first metatarsal.

Irwin[13] distinguishes four component deformities in the paralytic calcaneocavus. The loss of the gastrosoleus causes the flexors to draw the os calcis forward, the talus is forced into dorsiflexion, and the plantar fascia contracts. Then, there is an increased effort of both peronei to plantar flex the forefoot with a resulting cavus deformity. The angle of the midtarsal joint is sharpened and, finally, because of this dorsal kink, the dorsum of the foot is elongated, tightening the long extensors and causing a depression of the transverse arch and a clawfoot.

b) The paralysis of the peroneus longus. In this instance, all three joints are involved. The ankle joint has lost some of its plantar flexory power, the subastragalar joint yields to the supinatory effect of the unopposed tibialis posticus and goes into supination, and the midtarsal joint goes into adduction.

The sole of the foot is flat rather than hollow. The transverse arch is relaxed, having lost its tiepiece, and the ball of the foot cannot be pressed against the ground; in the subastragalar joint, the foot goes into adduction and supination and callouses develop on the lateral side of the foot and under the heads of the lateral metatarsals.

c) In combined paralysis of the gastrosoleus and the peroneus longus the resulting deformity is a pes calcaneus, cavus, adductus, supinatus, in contrast to the calcaneocavus and abductus which develops when only the gastrosoleus is paralyzed (Fig. 1).

2. Talipes equinus

The principal dorsiflexors are the tibialis anticus and the long extensors of the toes; the former has an adductory or supinatory, the latter have an abductory or pronatory effect on the subastragalar and midtarsal joints. The peroneus tertius is an auxiliary to the long extensor, the extensor hallucis to the anterior tibial (Duchenne[8]). The adduction produced by the tibialis anticus in the midtarsal joint is negligible and its supinatory effect on the subastragalar joint varies. Isolated paralysis of the tibialis anticus is frequent. Dorsiflexion of the foot can still be accomplished by the extensors of the toes and the peroneus tertius but with more difficulty and with the foot carried in abduction, that is, elevation of the lateral border (Fig. 2).

On the other hand, if the extensor digitorum i͜ aralyzed, the foot can be dorsiflexed by the tibialis anticus but this time it goes into adduction, so that in walking and standing the lateral border of the foot touches the ground.

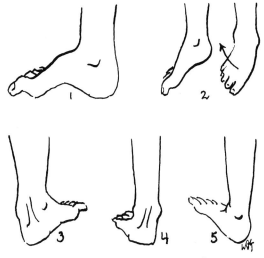

B. THE SUBASTRAGALAR JOINT

The principal muscles controlling pro- and supination in this joint are the peroneus brevis and the posterior tibial. Loss of the *peroneus brevis* deprives the foot of its ability to move in abduction without being dorsiflexed. The

FIG. 1. Calcaneocavus adductus deformity.
FIG. 2. Talipes equinus; if dorsiflexion attempted with tibialis anticus paralysis, foot goes into abduction.
FIG. 3. Varus deformity.
FIG. 4. Valgus deformity.
FIG. 5. Calcaneoplanus deformity.

foot, therefore, inclines to varus position under the influence of the unopposed action of the posterior tibial (Fig. 3).

The loss of the *posterior tibial* muscle produces the opposite deformity, namely, a valgus in the subastragalar joint, although the pure abductory component of the now unopposed peroneus brevis is limited in the midtarsal joint (Fig. 4).

On the other hand, if *both muscles are paralyzed*, the deformity is generally light and there is less functional disturbance than in paralysis of one muscle alone. Duchenne[8] states even that "It is better to lose all the leg muscles moving the foot than have a certain number of them saved."

C. THE MIDTARSAL JOINT

The long muscles of the foot insert distally to this articulation after traversing the subastragalar and the ankle joints. In case of muscular imbalance, a disalignment is established in these proximal joints and the question is merely what the midtarsal joint contributes to the deformity. This joint has three degrees of freedom of motion: one about a frontal axis, i.e., dorsi- and plantar flexion, which shares with the ankle joint; one about a sagittal axis, i.e., pro- and supination, which shares with the subastragalar joint. Therefore, it adds to the deformations produced by these articulations by accentuating the equinus or calcaneus deformity of the ankle joint by a plantar or dorsiflexion of its own. It can also increase a pro- and supinatory deformity of the subastragalar joint by pro- and supination of its own. But, in addition, it has a third degree of motion about a perpendicular axis, i.e., ab- and adduction. The calcaneus or equinus of the ankle joint and the pes supinatus or pronatus of the subastragalar joint then becomes a calcaneocavus abductus or an equino-cavus adductus.

If all the muscles which flex the forefoot plantarly, i.e., the *peroneus longus* and the *flexors of the toes* are paralyzed, a *pes planus* develops. If there is, in addition a paralysis of the gastrosoleus, the result is a *calcaneoplanus*. This is an unusual deformity in infantile paralysis, although weak toe flexors combined with weak gastrosoleus are often seen in the congenital calcaneus (Fig. 5). If there is, in addition, a deficiency of the long extensors of the toes, the sole of the foot turns medially and the result is a *calcaneoplanovarus*.

If there is paralysis of the gastrosoleus or tibialis anticus, or both, with good extensors and, especially, a good peroneus longus, the picture is changed; there is, of course, the calcaneocavus, produced by the paralysis of the gastrosoleus, but, in addition, the long peroneus flexes the first metatarsal plantarly and at the same time abducts it. The deformity is then a *pes calcaneocavus abductus*.

A simple calcaneocavus means loss of the triceps surae only, although a slight degree of pronation should accompany dorsiflexion, just as a slight degree of supination accompanies plantar flexion in the ankle joint.

D. THE METATARSOPHALANGEAL ARTICULATION

It has been indicated that the normal position of the toes depends upon a three-cornered equilibrium established between the long extensors, the long flexors, and the intrinsic muscles of the foot. Any disturbance of this equilibrium will overthrow the normal mechanism.

Atrophy or paralysis of the intrinsic muscles produces a foot which is a combination of *cavus, clawfoot* and *hammertoes*. The three are mechanically interrelated.

IV. THE ARTHRODESES AND ARTHRORISES OF PARALYTIC JOINTS FOR THE ESTABLISHMENT OF EQUILIBRIUM

A permanent and stable equilibrium of the joints of the foot by muscle transference alone is possible only in a small number of selected cases. In fact, the only situation where such a transference would be self-sufficient is the case of an isolated paralysis of the tibialis anticus. Here the muscles can be supplanted by the peroneus longus, which is pulled through the sheath of the tibialis anticus to its point of insertion after the technique developed by Biesalski and Mayer.[2]

The more frequent situation is that the joint must be eliminated by arthrodesis because it is impossible to establish active muscular balance. The question arises what effect the arthrodesis has upon the equilibrium of the neighboring joints. A great many of the operative failures arise from the fact that the secondary effect of the arthrodesis on neighboring joints is not taken into consideration.

A. STABILIZATION OF THE ANKLE JOINT

1. The arthrodesis

As this operation is still performed occasionally, it is well to investigate the kinetic effects which arise from it. It is certain that securing the lateral balance of the foot in the more peripheral joint is more essential for the act of standing and walking than is the maintenance of the anteroposterior balance by arthrodesing the tibioastragalar articulation.

Of the flexor group, the gastrosoleus becomes entirely eliminated by this procedure, but the peroneus longus still retains its effect on both the midtarsal and the subastragalar articulations. The tibialis posticus is also a flexor of the ankle joint. If its function as flexor is likewise eliminated, it becomes a powerful adductor, well capable to neutralize the abductory component of the peroneus longus.

On the extensor side, the ankylosis of the ankle joint eliminates the common extensors of the toes, the peroneus tertius, and the tibialis anticus. While the abductory component of the extensors is by far greater than the adductory of the tibialis anticus, this deficiency is made up by the adductory power of the tibialis posticus.

The conclusion is that when the arthrodesis of the ankle joint is done for paralysis of the extensors, as is usually the case, one must rely upon the remaining peroneus longus and the tibialis posticus to hold the lateral balance against each other. If the arthrodesis is done for paralysis of the gastrosoleus with calcaneus deformity, the preservation of the lateral balance, both in the subastragalar and the midtarsal joints, relies upon the posterior tibial to neutralize the pronatory effect of both peronei as well as the abductory effect of the peroneus longus.

One can readily see the hazards of this operation if one overlooks any degree of imbalance in either the subastragalar or the midtarsal articulation, or in both; lateral deformities in either pro- or supinatory, or in ab- or adductory sense are almost certain to follow.

2. The arthrorisis

An arthrorisis is a checking operation which produces, by means of osteoplastic procedures, a mechanical block against certain movements. It does not eliminate motion; it only restricts its range. In the ankle joint, arthrorises are performed both for the equinus and for the calcaneus. Two principles are applied: either the bone block is so constructed that in a certain position it strikes against the tibia and no further motion is possible, or the check is based on a compensatory principle where two adjacent joints are placed in opposite terminal positions. For instance, an extreme plantar flexion of the astragalus in the ankle joint is compensated by an opposite dorsiflexion of the os calcis in the subastragalar joint.

The first type of procedure is represented by the so-called bone block methods. In Campbell's bone block and in its modification (Fig. 6) for the correction of paralytic foot drop, a bone peg is placed behind the tibia the end of which strikes against the posterior tibial wall, preventing plantar flexion in the ankle joint.

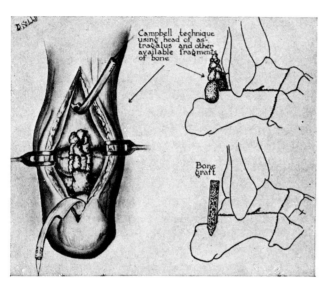

FIG. 6. Campbell's bone block and modification with use of bone graft for paralytic foot drop. (Steindler)

In the anterior bone block of Putti for paralytic calcaneus, the peg is inserted in front of the tibia and, in striking against the latter, it prevents dorsiflexion of the ankle joint.

It seems that early reports on the end results of this posterior bone block were favorable. Reports from the literature speak of 84.5% to 93.8% of good results either with the original Campbell procedure or with its numerous modifications (Wagner[23]). Ingram and Hudley,[12] reporting on 477 operative cases of which more than half were followed up sufficiently, found satisfactory permanent correction in a high percentage; but the operation should be done between 10-20 years of age and should be combined with triple arthrodesis, and a tendon transplantation should be added to procure active dorsiflexion.

However, the question is, what is the effect of the strain on the bone peg? Striking at one point, the situation is comparable to a cantilever beam where the maximum bending moment is at the tip and the maximum shear at the point of implantation. Experience shows that in the long run the bone atrophies under stress, principally at the free end, and the bone peg often fails to provide the desired check. This occurred in 66% of our cases.

The other principal used in arthrorisis is the tilting of the articulation. In the method devised by Snow[20] for paralytic dropfoot, a wedge of bone is taken from the lower end of the tibia with anterior base so that when the cut surfaces are adapted the joint tilts forward and upward in an exaggerated calcaneus position and further plantar flexion is thereby restricted. For the calcaneus deformity, a wedge is taken with posterior base which puts the joint in exaggerated equinus position and further dorsiflexion beyond the right angle is restricted. Gill's[9] operation is devised upon the same principle. He uses the astragalar joint surface for the tilting, which is done by placing a heavy bone block with posterior wedge into the gap.

A somewhat different principle of joint tilting is applied in the operation of Lambrinudi.[14] It is a combination of ankle joint arthrorisis with arthrodesis of the subastragalar and the midtarsal joints. Here the idea is that the strong

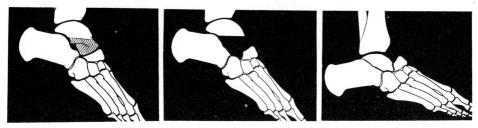

Fig. 7. Principle of Lambrinudi operation for foot drop. (Hart)

forward tilt of the astragalus places the joint in extreme plantar flexion so that no further flexion is possible. This position is then neutralized by an extreme calcaneus position of the os calcis in the subastragalar joint (Fig. 7). The point is that the terminal plantar flexion in the ankle joint is neutralized by a terminal calcaneal position in the subastragalar articulation.

For the adult joint this argument seems valid enough, but for the juvenile joint there are two points to consider. First, there is the possibility that the joint will, under the effect of the flexor muscles, adjust itself by structural transformation and acquire an increased flexory range. Secondly, the long flexor muscles will exert their full strength upon the more distal joints, the midtarsal and even Lisfranc's joint, and produce a flexion deformity of the forefoot. In fact, both events are observed after this operation unless: (1) the operation is done closer to puberty age, and (2) the accompanying subastragalar arthrodesis is extended to and beyond the midtarsal joint to the scaphocuneiform articulation.

The Whitman[24] astragalectomy is also an arthrorisis which is effective in the pes calcaneus or calcaneocavus. It has two main features: 1) the astragalus is removed, and 2) the foot is placed backward against the leg until, when the tibial mortise is set into the hole left by the removed astragalus, the outer malleolus lies against the calcaneocuboid junction and the inner malleolus strikes against the scaphoid (Fig. 8).

Both of these features are sound from the mechanical point of view. The astragalus is the keystone of the structure of the foot in respect to mobility; it transmits movement in both anteroposterior and lateral directions. Removal of the astragalus contacts the tibia directly with the os calcis and cuboid and thereby reduces the range of anteroposterior motion.

The first to devise the backward displacement was G. G. Davis,[7] with his horizontal transverse section. This is also mechanically sound because it produces a more concentric loading of the arch by the superincumbent weight. Together with the restriction of the anteroposterior motion, it makes the reactions of the ball and heel more equal in standing and greatly adds to the stability of the joint.

The mechanics of Whitman's[24] operation meet the requirements for antero-

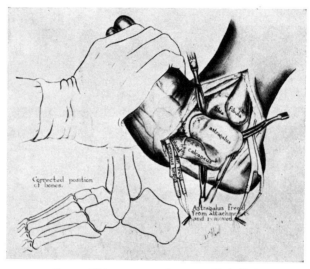

FIG. 8. Whitman astragalectomy. (Steindler)

posterior balance but lateral imbalance is not secured by this operation, which, therefore, would be inadequate for the paralytic varus or valgus. On the other hand, we have found the astragalectomy with posterior displacement of the foot an excellent method for the pure calcaneus or the calcaneocavus, and have followed up patients up to 28 years who had used the foot without any discomfort. Thus, when the indications were held strictly to the calcaneus and the calcaneocavus foot, our good results reached 88%.[21] For the cavus deformity, the operation has the added advantage that the concentric load is placed just at the summit of the deflection and weight bearing counteracts at this point the buckling of the foot

B. THE STABILIZATION OF THE SUBASTRAGALAR AND MIDTARSAL JOINTS. MECHANICAL PRINCIPLES

Lateral balance is more difficult to establish for the reason that two joints are involved, the subastragalar and the midtarsal, being controlled by the same long muscles which, in addition, act as plantar and dorsiflexors as well as ab- and adductors in the midtarsal joint. The question is, then, to what extent the arthrodesis of the tarsus must be carried out to control the lateral balance which cannot be reestablished by tendon transference alone.

1. Single joint arthrodesis

A) SIMPLE SUBASTRAGALAR ARTHRODESIS

Up to the time of the introduction of the triple arthrodesis by Ryerson,[19] it was believed that the arthrodesis of the subastragalar articulation alone suf- fices to secure the foot against lateral imbalance. The arthrodesis of the sub- astragalar joint alone leaves the balance in the midtarsal joint unprovided for. The peroneus brevis produces pronation and abduction, seconded by the peroneus longus, while the tibialis posticus aided by the flexor of the big toe supinates and adducts in the subastragalar joint. After the latter is fused, these muscles still exercise their effect upon the midtarsal joint, and unless they are perfectly balanced, deformity will occur in this articulation.

B) THE TRANSVERSE HORIZONTAL SECTION OF DAVIS[7]

This operation consists in denudation of the subastragalar articulation by complete transverse section which separates entirely the outer from the inner arch of the foot so that the latter can be displaced backward a distance of 2 cm. This combines with the obliteration of the subastragalar joint the advantage of lengthening the posterior arm of the angulated weight-bearing beam repre- sented by the foot.

C) MIDTARSAL ARTHRODESIS WITH WEDGE RESECTION (BICK[1])

With removal of the midtarsal wedge, this procedure corrects a cavus de- formity and, while useful in nonparalytic conditions, produces no lateral sta- bility.

2. The two-joint arthrodesis

A) SUBASTRAGALAR AND ASTRAGALOSCAPHOID ARTHRODESIS OF HOKE[11]

This procedure provides for the talocalcaneal arthrodesis, the resection of the head and neck of the astragalus with architectural restoration of the foot by trimming the neck of the astragalus and replacing it, as well as for an as- tragaloscaphoid arthrodesis and backward shift of the foot (Cole[5]) (Fig. 9). The great merit of this procedure lies in the restoration of the architecture of the tarsus by osteoplastic trimming and in the backward displacement of the foot. Miller,[16] reporting 125 cases of paralytic clubfeet and 17 of paralytic calcaneus with almost uniformly good results, states that the correct position of the head of the astragalus is the secret of a normal looking, stable, well- balanced foot. Because of the fact that the calcaneocuboid joint is left free, which opens the possibility of a forefoot disalignment, Miller[16] resects the calcaneocuboid joint, adding to it the transplantation of the anterior tibial to the outside to ensure balance against the adduction of the forefoot. This makes it a triple arthrodesis.

In a report on 25 cases, Brown[4] states that the functional results were good in 70%, but good mechanical alignment was obtained in only 40%, which

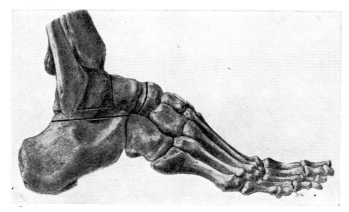

Fig. 9. Hoke's subastragalar and astragaloscaphoid arthrodesis. Diagram shows bone incisions. (Cole)

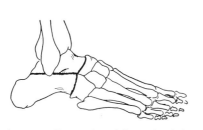

Fig. 10. Ryerson's triple arthrodesis. (Steindler)

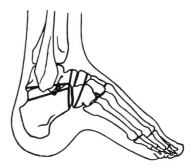

Fig. 11. Naughton Dunn's triple arthrodesis with posterior displacement of the foot. (Dunn)

shows that in cases of forefoot imbalance even this generally accepted stabilizing operation needs further provisions against forefoot disalignment.

3. The three-joint arthrodesis

A) THE TRIPLE ARTHRODESIS OF RYERSON[19]

The operation provides for: 1) the arthrodesis of the astragaloscaphoid joint; 2) the arthrodesis of the calcaneocuboid joint; 3) the arthrodesis of the subastragalar joint, and 4) in selected cases, arthrodeses of the scaphocuneiform, the cuneiformmetatarsal, and the cubometatarsal joints (Fig. 10).

However, the usual procedure is to confine the arthrodesis to the subastragalar and the two portions of the midtarsal joints.

B) THE TRIPLE ARTHRODESIS WITH POSTERIOR DISPLACEMENT OF THE FOOT

The operation of Naughton Dunn[18] combines the advantages of the triple arthrodesis with that of Davis' and Whitman's posterior displacement of the foot and provides for a more equal weight distribution between heel and ball. Since the line of gravity now falls close to the ankle joint center, the rotatory moment applying to this joint is lessened and the stance is more stable (Fig. 11).

C) THE TRIPLE ARTHRODESIS WITH ANTEROPOSTERIOR ARTHRORISIS OF THE ANKLE JOINT

This is represented by the already mentioned procedure of Lambrinudi[14] (Fig. 7) and by that of Brewster and Larson.[3]

Lambrinudi's principle is that of tilting the joint surface of the astragalus

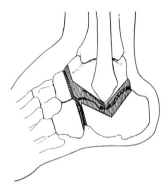

FIG. 12. Brewster's countersinking operation. (Brewester & Larson)

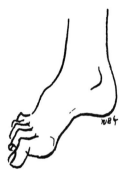

FIG. 13. Paralytic claw foot.

into extreme equinus position against the dorsally tilted os calcis. The wedge removed for this purpose from the neck and head of the talus serves as arthrodesis of the subastragalar joint. The pointed head of the astragalus is lodged in a niche made in the denuded scaphoid. It is an ingenious device from the mechanical point of view. The only weak point, in our opinion, is the fact that because of the reduction of the contacting surfaces the fusion of the astragaloscaphoid joint is jeopardized.

The difficulty with this method, as we have found it, is not so much the actual recurrence of the equinus deformity in the ankle joint but rather the development of an equinus in the midtarsal joint, produced by the plantar flexors of this joint.

The countersinking operation of Brewster and Larson[3] is one which combines a simple arthrodesis with a check against plantar and dorsiflexion of the ankle joint. The principle is that the astragalus is countersunk into a cavity chiseled out from the superior surface of the os calcis, which produces a subastragalar fusion and at the same time locks flexion and extension in the ankle joint. To this the midtarsal arthrodesis is added (Fig. 12).

4. The panarthrodesis of foot and ankle

The operation was designed by Lothioir[15] for the flail ankle without deformity, and has been used by the writer under this indication in flail and nondeformed feet with good or fairly preserved hamstrings (Steindler[22]). The operation provides for arthrodesing the subastragalar, midtarsal, as well as the tibiotarsal articulations. The ankle joint is placed in 20° flexion to accommodate the heel of the shoe.

The mechanical effect is a firm foot, but there is no takeoff. The stiff ankle transmits a powerful backward thrust to the knee, thereby stabilizing it, but at the same time the knee must be protected by good hamstring muscles against a genu recurvatum deformity.

Statistics of Miltner[17] from our clinic give 61% good functional results and, later, Hamsa[10] reports 74.1% good results in 85 properly selected cases.

From a mechanical point of view, the danger of an ill-advised or poorly performed panastragalar arthrodesis can easily be appreciated; too much equinus lengthens the leg and compels flexion of the knee; weak hamstring muscles involve the danger of genu recurvatum.

From the mechanical point of view alone, the indications should be restricted, therefore, to a flail, nondeformed foot with straight knee and well-preserved hamstrings.

5. The restoration of the metatarsophalangeal balance

The tricornered equilibrium between the long extensors, the long flexors, and the intrinsic muscles can be overthrown by paralysis of the long flexors of the toes, paralysis of the intrinsic muscles, or by contracture of the extensors of the toes against passively insufficient flexor muscles in cases of pes cavus.

The result of this overthrow is the depression of the transverse arch, hyperextension in the metatarsophalangeal joints, and the flexion contracture of the interphalangeal joints, all features of the clawfoot (Fig. 13). Restoration of the normal muscle mechanics, therefore, requires not only the correction of all passive disalignment but also the active lifting of the depressed transverse arch by transference of the toe extensors to the metatarsals. The result depends largely upon the return of function of the intrinsic muscles, i.e., whether or not an active equilibrium between the flexory and extensory apparatus as a whole can be obtained.

V. WHAT MECHANICAL CAUSES ARE RESPONSIBLE FOR FAILURE AND RECURRENCE AFTER STABILIZATION?

There are two principal causes of failure:

1. The operation selected does not provide for sufficient stability of the joints because it is mechanically inadequate.

2. The operation provides sufficient stabilization at the time but a persisting muscle imbalance produces secondary deformities either by overcoming previous fusion or, more often, by its effect on nonfused peripheral joints.

That muscle disequilibrium can force an apparently fused joint into a contracture position is nothing new. An example is the progressive deformity of the fused hip joint in cases of tuberculosis, where the process is still active and the muscles still continue in spasm. It must be remembered that stability, even in fused joints, can be depended upon only if there are no constant deforming forces at work.

We have already discussed inadequate fusion as cause of postoperative deformity. A case in point is the plantar flexion of the foot in the midtarsal joint after Lambrinudi's operation, where no special precautions are taken to secure the peripheral articulations.

Crego and McCarroll,[6] analyzing 1100 arthrodesed feet, found recurrences in 212, or almost 20%; and, of the latter, 155 recurrences, or 73%, were due to abnormal muscle pull. They found in 88 cases that valgus and abduction deformity was due to the unopposed action of the peronei, even from the peroneus brevis alone if the longus had been transplanted. A pes varus and adductus was caused by the unopposed pull of the tibialis anticus alone in 30 cases, and by the combined unopposed action of the tibialis anticus and postiticus in 27 cases.

This is a strong argument that arthrodesis should be combined with muscle transplantation in such a way as to establish, as much as possible, a lateral balance. It is especially important to secure the pro- and supinatory balance for the forefoot by careful selection of the muscles and of the points of their implantation. This means that the strength of the muscles as well as their direction and the lever arm must be carefully considered. One can "unbalance" a forefoot by injudicious transplantation just as badly as the original paralysis did in converse direction. An example is the pes varus and adductus which occurs after transplantation of the long peroneus into the anterior tibial in those cases in which the peroneus brevis and tertius are not fully adequate opponents.

VI. SUMMARY

1. High demands for stability have relegated tendon transplantation into the background in the treatment of the paralytic foot and ankle. They have their place as auxiliary procedures.

2. In eliminating joints by arthrodesis for the purpose of stabilization, the remote effect of such elimination on other joints and parts of the extremities must be taken into consideration.

3. Of the muscles maintaining the flexion-extension equilibrium in the ankle joint, the flexors greatly preponderate. All of the muscles take effect on peripheral joints, most of them both on the subastragalar and midtarsal joints; the gastrosoleus is also an adductor and supinator; the tibialis anticus an adductor, the peroneus brevis and the extensors of the toes are pronators and abductors, etc.

4. Under paralytic relaxation, some pronatory separation of the astragalus from the tibia can be demonstrated.

5. The preponderance of the flexors over the extensors of the ankle joint leads to equinus deformity in recumbency, even without paralysis.

6. In the subastragalar joint the supinators are the tibialis posticus, the flexor of the big toe, and the common flexor. Their range is 34° in this joint, but there is another 18° or 20° of supination effected by these muscles in the midtarsal joint. Pronators are the peroneus longus and brevis. The total strength of the supinators is more than twice that of the pronators.

7. In the midtarsal joint the pronators are also abductors and dorsiflexors, the supinators adductors and plantar flexors. Here the pro- and supinator muscle groups are more evenly matched.

8. In the metatarsophalangeal joint the balance depends on the equilibrium between flexors, extensors, and the intrinsic muscles. Since the long muscles pass the ankle, subastragalar and midtarsal joints before they reach the toes, any deformity in these joints may overthrow the balance, causing contractures in the metatarsopholangeal and, secondarily, hammertoe deformities in the interphalangeal joints.

9. Talipes calcaneus is caused by isolated paralysis of the gastrosoleus. In addition, the increased effort of the peronei flexes the forefoot plantarly, causing a cavus deformity. The resulting tightness of the long extensors produces a depression of the transverse arch and a clawfoot.

10. In paralysis of the peroneus longus alone, the subastragalar joint goes into supination by the action of the unopposed tibialis posticus. The sole of the foot becomes flat and the transverse arch is relaxed. Combined gastrosoleus and peroneus longus paralysis produces a pes calcaneus, cavus, adductus and supinatus.

11. Paralysis of the dorsiflexors, especially the tibialis anticus and extensors of the toes, results in pes equinus. If the tibialis anticus alone is preserved, the foot can be dorsiflexed by it but goes also in adduction and supination.

12. The midtarsal joint adds to the deformity produced in the proximal articulations by paralysis of the long muscles. Paralysis of the peroneus longus and the toe flexors results in pes planus or, with paralysis of the gastrosoleus, in calcaneoplanus. Additional deficiency of the long extensors of the toes produces a calcaneoplanovarus; paralysis of gastrosoleus and tibialis anticus, a calcaneocavus adductus.

13. In stabilization of the ankle joint by arthrodesis for paralysis of the extensors, one must rely principally upon the peroneus brevis and the tibialis posticus to hold the lateral balance against each other. The same is true if the arthrodesis is done for paralysis of the gastrosoleus.

14. The arthrorises are checking operations designed to restrict (but not to abolish, as do the arthrodesing operations) the range of the tibioastragalar articulation. Campbell's posterior bone block restricts plantar flexion, Putti's anterior block the dorsiflexion. Another method of arthrorisis is the tilting of

the articular surface by resection of a bone wedge above it, with anterior base for the equinus and posterior base for the calcaneus deformity.

Early reports on block operations were very favorable but long-range observations show a high percentage of cases in which the stress on the block caused atrophy with recurrence of the deformity. A different principle is that of Lambrinudi's operation which accomplishes restriction of talotibial movements by placing ankle and subastragalar joints in two opposite terminal positions. Whitman's operation, consisting in astragalectomy and backward displacement of the foot, is also an arthrorisis. While these operations control the anteroposterior balance, they have no effect on the lateral. They are, therefore, em-employed best where there is no lateral imbalance, or, if there is, it is taken care of by additional surgical procedures.

15. The stabilization of the subastragalar and the midtarsal joints takes care of the lateral balance. Arthrodesis of the subastragalar joint alone still leaves the peroneus longus and brevis, the tibial posticus and the flexors of the toes free to act on the midtarsal joint. Unless these muscles are perfectly balanced against each other, this operation is inadequate. The same can be said of Davis' horizontal transverse section.

16. The combined subastragalar and astragaloscaphoid arthrodesis of Hoke has the advantage that by temporary removal and reshaping of the astragalar head and neck skeletal deformities in the sense of ab- or adduction, pro- or supination can be corrected, but, here again, restoration of muscular balance requires additional procedures.

17. The triple arthrodesis of Ryerson stabilizes both subastragalar and midtarsal joints, and that of Naughton Dunn combines the advantage of a triple arthrodesis with a shortening of the anterior lever arm and better distribution of the superincumbent weight. Lambrinudi's operation is a combination of triple arthrodesis and wedge resection of the subastragalar joint which permits the maximally plantar flexed talus to contact the strongly dorsally flexed os calcis. This results in a right-angular position of the foot against the leg from which no further plantar flexion of the astragalus is possible in the ankle joint, since the bone is already in terminal position of flexion. Brewster's countersinking operation likewise combines a subastragalar fusion with an arthrorisis of the ankle joint by creating an anterior and posterior block in the recess of the os calcis into which the astragalus is countersunk.

18. The panastragalar arthrodesis stabilizes all joints of which the astragalus forms a part. It secures the foot against anteroposterior as well as pro- or supinatory imbalance; it involves, however, the danger of a genu recurvatum and does not take care of ab- or adductory imbalance. It is, therefore, restricted to cases of flail nondeformed feet with good hamstrings. Under proper selection we obtained good and lasting results in 74%.

19. The restoration of metatarsophalangeal imbalance with transverse arch depression and clawfoot and hammertoe deformity requires correction of all disalignment and transference of the toe extensors into the metatarsals.

20. The principal causes of failures after stabilization are inadequate arthrodesing methods and persisting musclar imbalance not provided for operatively. Recurrences in arthrodesed feet were found by McCarroll and Crego in not less than 20% of a large series. A valgus and abduction deformity resulted from the unopposed action of the peronei, a varus and adductus from the pull of the tibialis anticus. A judicious tendon transplantation in addition to the arthrodeses must be resorted to in order to avoid these recurrences.

BIBLIOGRAPHY

1. BICK, E. M.: Tarsal wedge arthrodesis. *J. Bone & Joint Surg.*, *XX*:3, 726, July 1938.
2. BIESALSKI, N., and MAYER, L.: *Die physiologische Sehnentransplantation.* Berlin, J. Springer, 1916.
3. BREWSTER, A. H., and LARSON, C. B.: Cavus feet. *J. Bone & Joint Surg.*, *XXII*:1, 361, April 1940.
4. BROWN, L. T.: Stabilizing operation on the foot. *Am. J. Orth. Surg.*, *XXII (old)*:839, Oct. 1924.
5. COLE, W. H.: Subastragalar arthrodesis. *J. Bone & Joint Surg.*, *XII*:289, April 1930.
6. CREGO, C. H., and McCARROLL, H. R.: Recurrent deformation in stabilized paralytic feet. *J. Bone & Joint Surg.*, *XX*:609, July 1938.
7. DAVIS, G. G.: The treatment of the hollow foot. *Am. J. Orth. Surg.*, *XI*:231, 1913.
8. DUCHENNE, G. B.: *Physiology of Motion.* Transl. E. B. Kaplan. Philadelphia, J. B. Lippincott, 1949.
9. GILL, A. B.: Operations to make a posterior bone block at the ankle. *J. Bone & Joint Surg.*, *XV*:166, Jan. 1933.
10. HAMSA, W. R.: Panastragaloid arthrodesis. *J. Bone & Joint Surg.*, *XVIII*:732, July 1936.
11. HOKE, M.: Stabilizing paralytic feet. *J. Bone & Joint Surg.*, *XIX (old)*:494, Oct. 1921.
12. INGRAM, A. J., and HUDLEY, J. M.: Posterior bone block of ankle for paralytic equinus. *J. Bone & Joint Surg.*, *33A*:3, 679, July 1951.
13. IRWIN, C. E.: Correspondence letter, Nov. 16, 1950.
14. LAMBRINUDI, C.: New operation in dropfoot. *Brit. J. Surg.*, *XV*:103, 1927.
15. LOTHIOIR: *J. de Chir. et Anat., Soc. Belge de Chir.*, *XI*:184, 1911.
16. MILLER, O.: Two hundred cases of paralytic foot stabilization after the method of Hoke. *J. Bone & Joint Surg.*, *XXIII (old)*:85, Jan. 1925.
17. MILTNER, L. J.: Stabilization of the foot. *J. Bone & Joint Surg.*, *XIII*:502, July 1931.
18. NAUGHTON DUNN: Reconstructive surgery in paralytic condition of the leg. *J. Bone & Joint Surg.*, *XII*:2, 799, April 1930.
19. RYERSON, E. N.: Arthrodesing operations on the foot. *J. Bone & Joint Surg.*, *XXI (old)*:425, July 1923.
20. SNOW, L. C.: Mechanical and anatomical principles for footdrop. *Surg., Gynec. & Obst.*, *LI*:252, Aug. 1930.
21. STEINDLER, A.: *Orthopedic Operations.* Springfield, Illinois, Charles C Thomas, Publisher, 1940.
22. STEINDLER, A.: Panastragalar arthrodesis. *J. Bone & Joint Surg.*, *XXI (old)*:384, April 1928.
23. WAGNER, L. C.: Modified bone block (Campbell) of ankle for paralytic dropfoot. *J. Bone & Joint Surg.*, *XXIX*:141, Jan. 1931.
24. WHITMAN, R.: Astragalectomy and backward displacement of the foot. *J. Bone & Joint Surg.*, *XX*:206, 1922.

Lecture XXV

THE LOWER EXTREMITY AS A WHOLE

I. INTRODUCTION

EQUILIBRIUM AND STABILITY

TAKEN as a mechanical unit, the extremity can be considered as a system consisting of four links movable against each other. They are represented by the pelvis, the thigh, the leg, and the foot, articulating in the hip, the knee, and the ankle joints. Hip and knee joints move in more than one plane and for each of these the conditions of equilibrium have been investigated in a former lecture. In addition, the analysis of the equilibrium in the frontal plane include the subastragalar joint, which plays an important part in maintaining the body in balance.

A. THE SAGITTAL PLANE

1. The hip joint

It was the opinion of H. von Meyer that the balance is a passive phenomenon affected by the tension of the iliofemoral ligament. This idea is not acceptable because in the usual stance the line of gravity runs in front of the hip joint; therefore, the iliofemoral ligament is not under tension and active contraction of the extensors of the hip becomes necessary. Only in the slouched position the line of gravity falls behind the hip joint axis and the balance in the hip joint can be considered as "passive."

2. The knee joint

In normal stance the line of gravity falls in front of the knee joint axis and gravity becomes an extensory force in this joint; action of the quadriceps is superfluous in this position and the muscle is relaxed. The question arises whether in complete extension of the knee the outward rotation of the tibia against the femur is a factor in stabilizing the whole extremity. It is a fact that inward rotation of the femur tightens the iliofemoral ligament and that outward rotation of the tibia against the femur produces tension of the medial collateral ligament of the knee joint. But the tension in these ligaments is not enough to lock the joint in the sagittal plane.

3. The ankle joint

In normal stance, the line of gravity passes in front of the ankle joint axis. Consequently, balance must be maintained by the tension of the calf muscles. In this upright position, the axis of the ankle joint is oblique. It points from

backward and outward to forward and inward and is rotated 30° against the
frontal plane.

Here the problem presents itself of how movements can be carried out simultaneously in both ankle and knee joints, as in squatting. This is only possible
if motion in the ankle joint is combined with motion in another joint in such a
way that the resultant of the axis of both joints conforms with the axis of the
knee joint. The hip joint offers no difficulties because it has three degrees of
freedom of motion and, therefore, it can adjust itself to any position of the knee
or ankle. This, however, does not solve the difficulty which lies in the discrepancy between the axis of knee and ankle joints. Another joint is required
which combines its movement with that of the ankle joint in such a way that
the resultant parallels the knee joint axis. This joint is the subastragalar
articulation.

4. The subastragalar joint

The axis of this joint is oblique from backward, outward and downward
to forward, inward and upward. Although the joint has no motion in the sagittal plane in combination with the ankle joint an adjustment is effected which
makes it possible to carry out simultaneous bilateral motion in the ankle and
knee joints (Baeyer[1]).

5. The combined bilateral ankle and knee motion

The obliquity of the plane of the ankle joint movement causes: 1) dorsiflexion to be associated with abduction and plantar flexion with adduction
of the foot; 2) dorsiflexion to be combined with pronation in the subastragalar
joint and plantar flexion with supination. It is this pro- and supinatory motion
in the subastragalar joint which is able to make the adjustment of the divergence between the axes of knee and ankle joints, and which makes it
possible for both ankle joints to move at the same time when the foot is resting on the floor bearing weight and when the patella looks straight forward
and both knee axes are in the same frontal plane. It is this pro- and supinatory adjustment in the subastragalar joint accompanying dorsi- and plantar
flexion in the ankle joint which permits squatting in this position (Fig. 1).

The situation is different if in squatting the knees are spread apart. This involves the outward rotation of the femur, with the tibia relatively inward
rotated. The axes of knee and ankle joint are now more parallel. The feet
follow the ankle joint axis and go into abduction, which again causes the
tibia to rotate more inward. On the other hand, inward rotation of the feet
makes squatting difficult, if not impossible, because the composite action
of the ankle joint and subastragalar joint does not permit adduction of the
feet.

What happens in the case of a subastragalar arthrodesis where the joint
is eliminated from combining in motion with other joints of the extremity?
Since the tarsal plate cannot carry out pro- and supinatory motion, it cannot

adjust itself to the dorsal and plantar flexion of the ankle joint which occurs in an oblique plane. Consequently, squatting with closed knees is impossible. The patient can only squat with his knees widely abducted (Fig. 2).

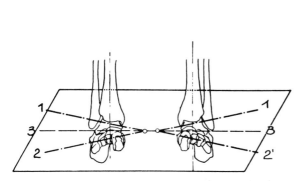

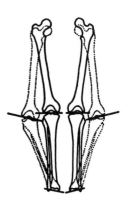

FIG. 1. Bilateral movement in the ankle joint. Flexion in the ankle joint is combined with supination in the subastragaloid articulation; extension in the ankle joint, with pronation in the subastragaloid articulation.

FIG. 2. Bilateral movement in knee and ankle joints. In flexion of the knees, the thighs go in outward rotation in the hip and knee joints, aligning the knee joint axes with those of the ankle joints.

B. RELATION OF ANKLE, KNEE, AND HIP JOINT IN THE STANCE

1. Relationship in normal position

The hip joint, having three degrees of freedom of motion, can adjust itself to any position of the knee or ankle joint axis. By rotating inward it can place the ankle joint axis in a strictly frontal plane. It cannot, however, by any rotatory movement, affect the discrepancies between the knee and ankle joint axes; this can only be equalized through the cooperation of the subastragalar articulation. The important function of the hip joint in the stance is to regulate the fluctuations of the line of gravity, both in the sagittal and frontal planes, so that it falls between the supporting points of heel and ball.

2. Relationship in malposture

The relationship of the four principal articulations of the lower extremity when the body is balanced in malposture depends essentially upon the position of the hip joint.

a) The hip joints are in hyperextension (static malposture): the femur goes into outward rotation; the ilio-femoral ligament is under tension since the line of gravity now passes behind the hip joint. The knees go into flexion with an inward rotated tibia to neutralize the outward rotation of the femur so that the axis of the tibial condyles remains frontal.

Fig. 3. Pronation of the hindfoot with adduction of forefoot. Knee in flexion and tibia rotated internally.

b) The hip joints are in flexion (arthritic or spastic malposture): the knee likewise goes into flexion, at the same time rotating the tibia inward. This inward rotation is then transmitted to the astragalus, similarly as we see in advanced static flatfoot. In this type of malposture the foot assumes a pronatory position with its posterior portion, while the forefoot goes into abduction (Fig. 3).

C. THE TRANSVERSE PLANE

Rotatory motion in the transverse plane is possible in the hip and the knee joints. There is no length rotatory motion in the ankle joint. The foot depends upon the hip and knee joints for the inward rotation and upon what adductory movement the astragalus can perform in the subastragalar joint. A length rotatory impulse coming from the upper part of the body is thus transmitted through the entire extremity to the toes. These rotatory impulses, generated from the trunk or from the arm, partially spend themselves in visible motion in the hip joint; the surplus of force is transmitted to the knee joint where the narrower rotatory range is soon checked, causing sprain, and the ankle joint with no transverse rotatory range at all bears the full brunt of the unspent rotatory force.

II. DYNAMICS. MUSCLE INTERRELATIONSHIP. REMOTE EFFECT IN KINETIC CHAIN

In the standing body the lower extremity is a closed kinetic chain, i.e., a system of articulations joined to an external resistance. This resistance is the superincumbent weight or, rather, the gravitational reaction of the floor which it produces. These reactions are transmitted from below to the subastragalar, the ankle, the knee, and the hip joints.

In such a closed kinetic chain, where the limb is not moving freely because of the external resistance, the action of the muscles is quite different from what it is when the muscle operates on a free lever arm which is unencumbered except by its own weight. In a closed kinetic chain the muscle often develops a rotatory effect upon a remote joint which lies outside of the muscle. Pluriarticular muscles, under changed mechanical conditions, sometimes reverse their motor effect on the same particular joint.

A. UNIARTICULAR MUSCLES

They are distributed about different articulations as follows: in the hip joint, the gluteals, the adductors and the iliopsoas; in the knee joint, the vasti, the popliteus and the short head of the biceps; in the ankle joint, the soleus.

Examples of remote action under external resistance are as follows:

1. The soleus

a) With the knee joint straight, the whole sole fixed to the floor, contraction of the soleus causes the whole leg, tibia and femur, to rotate backward with the pelvis (Fig 4).

Fig. 4. Knee joint straight. Sole fixed to floor. Contraction of soleus causes backward rotation of the entire leg.

Fig. 5. Knee straight. Heel free. Contraction of soleus increases plantar flexion, elevates whole body.

Fig. 6. Knee straight. Pelvis fixed. Standing on ball of foot. Plantar flexion of ankle joint causes flexion of knee.

b) With the knee joint straight and the heel free to rise, contraction of the soleus increases plantar flexion, elevating the whole body (Fig. 5).

c) With the knee flexed and the pelvis prevented from rising and the ball of the foot contacting the floor, the plantar flexion of the ankle joint increases the flexion of the knee (Fig. 6).

d) With the knee flexed, the pelvis free to rise, and the whole sole fixed to the floor, the contraction of the soleus produces an extension of the knee (Baeyer[2]) (Fig. 7).

Thus, in one instance (Figs. 4 and 5) the soleus stabilizes the knee in extension, while in another (Fig. 6) it flexes the knee, and in a third it extends it (Fig. 7), according to how the external resistance is arranged in this closed kinetic chain.

2. The adductors

Inversion of muscle action (Baeyer[2])

a) When the hip joint is free to swing and there is no external resistance, the effect is pure adduction.

b) When the hip and knee are extended and the ball of the foot is fixed,

Fig. 7. Knee flexed. Pelvis free. Sole fixed to floor. Soleus contraction causes extension of knee.

Fig. 8. Hip and knee extended. Ball of foot fixed. Adduction is combined with outward rotation.

Fig. 9. Hip and knee flexed. Ball of foot pivoting on ground. Adduction is combined with inward rotation.

pivoting on the floor, adduction is combined with outward rotation (Fig. 8).

c) When hip and knee are slightly flexed, the ball of the foot pivoting on the ground, adduction is associated with inward rotation (Fig. 9).

B. BIARTICULAR MUSCLES

Swinging, shortening and lengthening of the limb are combinations of rotatory motions in the several joints.

One can say in general that the contracture of all muscles of the anterior surfaces of the limb results in a forward swing; all rotation occurs in the same direction, i.e., clockwise or counterclockwise (isotropic rotation). Similarly, the contraction of all muscles on the posterior aspect of the limb results in a backward swing. The anterior muscle group consists of the flexors of the hip, the extensor of the knee, and the dorsiflexors of the ankle joint; the posterior group, of the extensors of the hip, the flexors of the knee, and the plantar flexors of the ankle.

Conversely, alternating contraction of the muscles of the anterior and posterior surfaces results in shortening or lengthening of the extremity. In shortening the contracting muscles are: the flexors of the hip joint, the flexors of the knee joint, and the dorsiflexors of the ankle joint. In lengthening of the limb the contracting muscles are: the extensors of the hip, the extensors of the knee, and the plantar flexors of the ankle joint.

In the case of lengthening or shortening, the rotations do not take place in the same direction but alternate between clockwise and counterclockwise (heterotropic rotation) (Strasser[3]). It is in this type of movement where a true translatory effect of lengthening or shortening of the extremity is accomplished.

The biarticular muscles of the lower extremity are: anteriorly, the rectus femoris; posteriorly, the semimembranosus, the semitendinosus, the gracilis, the long head of the biceps, and the gastrocnemius. There are, furthermore, biarticular muscles which change sides in their course; the tensor fasciae and the sartorius, from an anterior flexor of the hip to a posterior flexor of the knee. Thus there are three kinds of muscles: uniarticular muscles at the anterior and posterior aspects of the limb; biarticular muscles, either at the anterior or at the posterior aspect; and biarticular muscles crossing from the anterior to the posterior side.

When the extremity carries out a straight backward or forward swing, the biarticular muscles act on both ends, that is, at insertion and origin; or, in other words, origin and insertion points are drawn together. Thus, when the leg is swung forward, the rectus femoris flexes the hip and extends the knee at the same time; the opposing hamstrings are passively distended. When the leg swings backward, the hamstrings act on both ends, extending the hip and flexing the knee, while the rectus femoris is now passively distended.

On the other hand, when the extremity is shortened or lengthened, the biarticular muscles act on one end only while the other is stabilized by the action of the opponent. Thus, the muscle retains its original length, its contraction is isometric, which makes it possible for the muscle to display greater tension than in the case of swing where the muscle shortens and its contraction is isotonic.

To illustrate the situation: When the leg is shortened by drawing the heel against the buttocks, the rectus femoris acting at the upper end flexes the hip joint; at the lower end, however, the muscle is prevented from extending the knee by the contraction of the hamstrings, which flex the knee.

At the upper end the rectus prevails over the hamstrings. At the lower end the hamstrings, in flexing the knee, prevail over the rectus femoris. Both muscles retain their original length but they slide against each other, the rectus femoris shifting downward with the anterior inferior spine as the hip is flexed, while the hamstrings shift upward with the tuber ossis ischii. This is called a parallel or concurrent shift (Baeyer[1,2]) (Fig. 10).

When the limb is extended and the heel moves away from the buttocks, the same pattern applies. The hamstrings, extending the hip cooperating with the gluteus maximus, shift downward with the tuber ossis ischii, while the rectus, extending the knee, shifts upward with the anteroinferior spine of the os ilii (Strasser[3]) (Fig. 10).

In contrast in for- and backward swing, origin and insertion approach each other in the contracting muscle, while in its antagonist they are drawn apart. In forward swing the contracting muscle is the rectus; in backward swing, the hamstrings. The muscles no longer shift parallel but in opposite direction. This is called countercurrent shift (Baeyer[1,2]) (Fig. 11).

III. INTERPRETATION OF PATTERNS. ACTIVE AND PASSIVE INSUFFICIENCY

The whole significance of concurrent and countercurrent action of the biarticular muscles, and why the uniarticular muscles join their action with the biarticular synergists will become apparent when we consider what shortening or distending of a muscle beyond its natural length means for its efficiency.

It has been mentioned before that with continued contraction the muscle

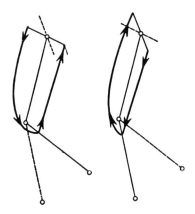

FIG. 10. Concurrent shift of rectus and hamstrings.

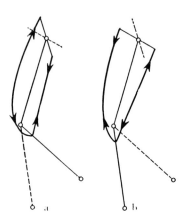

FIG. 11. Counter current shift.

tension rapidly decreases. A muscle which is at maximum contraction no longer develops any tension. This is called active insufficiency; a muscle which is distended maximally and opposes further elongation by the increasing elastic resistance is called passively insufficient. It can easily be seen that the bi-articular muscles will rapidly become actively insufficient if they shorten from both ends, such as the rectus femoris in the forward and the hamstrings in the backward swing; likewise, in the forward swing the hamstrings will soon become passively insufficient being distended on both sides, while in the back-ground swing, for the same reason, the rectus becomes passively insufficient. It is because of this that all biarticular muscles have a uniarticular companion which, though also subject to active or passive insufficiency, yet is so arranged that it steps in when the biarticular muscle reaches the state of active in-sufficiency; the iliopsoas in the flexion range and the gluteus maximus in the extension of the hip joint support the rectus and the hamstrings, respectively; the soleus comes to the assistance of the gastrocnemius when the knee is flexed.

The last forceful extension of the knee is then procured by the vasti, especially the medialis; the extreme flexion of the hip is carried out by the iliopsoas, as one can observe on the football field.

Active and passive insufficiency of muscles plays a prominent role in the development of contractures. For instance, an equinus contracture of the ankle joint produces by stretching of the extensor tendons of the toes a passive insufficiency of the extensors which gives rise to a retraction of the toes in the metatarsophalangeal joints. This hyperextension produces, in turn, a passive tension of the toe flexors with the result that the interphalangeal joints are forced into flexion (Fig. 12).

The monarticular muscles are arranged closer to the joint so that per unit contraction they cover a greater angular excursion than do the biarticular. This makes the uniarticular muscles more adequate in covering the entire excursion field of the joint and less apt to become actively insufficient. The excursion of the biarticular muscles in terms of angular values is not equal to that of its accompanying uniarticular muscle. Samples of active and passive insufficiency of the biarticular muscles are as follows:

1. Inability to extend the knee fully when the hip is maximally flexed (active insufficiency of the rectus femoris, passive insufficiency of the hamstrings).

2. Inability to flex the knee fully when the hip is maximally extended (active insufficiency of the hamstrings and passive insufficiency of the rectus femoris).

3. Inability to flex the hip fully with the knee maximally extended (insufficiency the same as in 1).

4. Inability to dorsiflex the foot beyond 90° when the knee is extended (passive insufficiency of the gastrocnemius).

IV. SPECIAL MECHANICAL SITUATIONS OF THE LOWER EXTREMITY AS A KINETIC CHAIN

A. STANDING ON THE TOES

In a previous lecture it was explained that the weight line of the body must fall into the ball of the foot before the heel can leave the floor. The situation, so far as the ankle joint is concerned, is still that of a lever of the first order but the lever arms are changed. While the whole sole was still on the ground, the two reactions applying at the ball and heel, respectively, were holding each other in equilibrium in the ankle joint because the moment of rotation of the anterior lever arm was equal and opposite to that of the posterior. As the heel is lifted off the ground there is no posterior reaction. The tension of the tendo Achillis must furnish a rotation moment equal to that of the gravitational reaction acting upon the ball of the foot, which reaction now equals the full body weight.

The arrangement has changed only insofar as the lever system is now standing perpendicularly instead of horizontally. The upper lever arm is the entire body to the ankle and the lower arm is the foot to the ankle joint. The forces to be held in equilibrium are now moments of the gravitational force acting from the center of gravity of the body, which is equal to the moment of the reaction from the floor. This rotatory moment in respect to the ankle joint forces the foot into dorsiflexion. The condition of equilibrium is that the moment of the gravitational reaction applying to the ball be neutralized by the tension of the heel cord.

Let us now suppose that the individual rises on his toes so much that the

FIG. 12. Equinus deformity of foot with clawing of toes.

FIG. 13. Line of gravity falls directly through ankle joint.

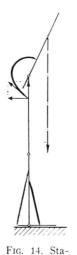

FIG. 14. Stabilization of the straight knee by forward bending of the body. Soleus and gluteus stabilizing knee. Body flexed slightly forward.

FIG. 15. Knee flexed. Soleus and gluteus essential to stabilize flexed knee. Body thrown farther forward.

line of gravity falls through the ankle joint. Then, gravity (or its reaction) develops no rotatory moment in respect to the ankle joint and, consequently, no tension is required of the calf muscles to hold the equilibrium. All gravitational force (or its reaction) becomes a translatory compression force which is absorbed by the rigidity of the skeleton. The toe dancer, therefore, works her calf muscles while she is rising upon the toes, but once this extreme plantar flexion has been accomplished, the heel cord tension is diminished (Fig. 13).

People with paralysis of the extensors cannot stand on their toes. They can rise on them but they cannot maintain themselves because the action of the antagonist which regulates the tension of the calf muscles is missing. On the other hand, people in whom the foot is fixed in a definite equinus position are well able to rise on their toes.

B. GRAVITY AS THE FLEXOR OF THE HIP AND EXTENSOR OF THE KNEE

It is well known that even without the quadriceps the knee joint can be kept from collapsing by the simple expedient of throwing the body forward so that the line of gravity falls in front of the knee joint and gravity then becomes an extensory force. It is not the only force, however, which can secure the knee in extension. The tension of the gluteus maximus rotates the thigh backward, provided the pelvis is fixed against backward rotation; and the tension of the soleus rotates the tibia backward, provided again that the ankle joint is fixed against plantar flexion.

There are, however, certain limits to this maneuver. In the first place, a forward bent position of the body is impossible in the absence of the extensors of the hip. The hamstrings can stabilize the pelvis only up to a certain angle. As long as the knee is straight the tension of the soleus can secure it in cooperation with the extensors of the hip (Fig. 14).

In case of moderate flexion contracture of the knee, the line of gravity must be thrown farther forward to fall in front of the knee joint axis (Fig. 15), and both glutei and soleus become necessary for the stabilization of the knee. These patients are able to walk but they are unable to run.

C. GRAVITY AS AN EXTENSOR OF THE HIPS

Extension of the hips can be procured by throwing the body backward, which causes the line of gravity to fall behind the common hip joint axis. While this relieves the tension of the gluteals it increases the tension in the quadriceps, opposite to what occurs when the common center of gravity is thrown forward.

Under normal conditions, changes in stance and posture alternatingly invoke these gravitational stresses which release the tension of certain muscles; the slouch relieves tension in the gluteals, whereas the fatigue gait with the body bent forward eases the quadriceps tension.

V. SUMMARY

1. The normal stance, with the line of gravity running in front of the hip joint, relaxes the iliofemoral ligament and calls for tension of the extensors of the hip. Only in the slouched position is the iliofemoral ligament tight.

2. In the knee joint, also, the line of gravity falls in front of the knee joint axis in normal stance. Outward rotation of the tibia against the femur tightens the medial collateral but this is not sufficient to stabilize the joint.

3. In the ankle joint, anteroposterior balance is maintained by the tension of the calf muscles because here, likewise, the line of gravity falls in front of the ankle joint axis. The problem of how both ankle joints can move as one unit, as for instance in squatting, when their axes are not in a straight line is explained by the fact that: 1) inward rotation of the tibia bring the common joint line of the ankles more into a frontal plane corresponding with the knee

joint axis, and that 2) the cooperation of the subastragalar joint with its pro- and supinatory range further adjusts the divergence between knee and ankle joint axis. This makes squatting with knees close possible. In squatting with knees apart, the ankle and knee axes are more in line; the feet go into abduction and the tibia rotates inward. In contrast, adduction of the feet makes squatting difficult, if not impossible.

4. The change of relationship of the four principal articulations of the lower extremity which occurs in malposture depends essentially upon the position of the hip joint. If hyperextended, the iliofemoral ligament is tight, the knees are flexed, the tibiae are inward rotated, the backfoot assumes a pronatory position, the forefoot goes into abduction.

5. In standing or walking, the lower extremity acts as a closed kinetic chain, being attached to an external resistance. Both uni- and biarticular muscles develop remote effects which are not observed in the open chain where the peripheral lever arm is free to move.

6. Examples of uniarticular muscle action in the closed chain are as follows:

a) The soleus, with knee straight and sole fixed to the floor, rotates the whole leg backward; with the heel free to rise, it elevates the whole body; with knee flexed and ball contacting the floor, it increases the flexion of the knee; with knee flexed and the whole sole fixed to the floor, it extends the knee.

b) The adductors in free swing with no external resistance adduct the hip; with hip and knee extended and the ball of the foot pivoting on the floor, adduction is combined with outward rotation; with hip and knee flexed and ball of foot pivoting, adduction is associated with inward rotation.

7. Contraction of all muscles at the anterior surface results in forward swing (isotropic rotation); that of the posterior, in backward swing. Alternating contraction of the muscles of the anterior and posterior surface results in shortening and lengthening of the extremity (heterotropic rotation). In the for- and backward swing, the biarticular muscles, moving both joints, contract countercurrently; in the shortening and lengthening mechanism, one end contracts, the other is stabilized. The muscles shift parallel, their action is concurrent. Thus, in forward swing, the rectus, in backward swing, the hamstrings contract countercurrently. In shortening and lengthening of the extremity, rectus and hamstrings operate concurrently; in lengthening, the rectus shifting upward, the hamstrings downward; in shortening, the rectus shifting downward, the hamstrings upward.

8. In for- and backward swing the biarticular muscles, contracting at both ends, easily become actively insufficient, i.e., they lose their tension with increasing contraction. For this reason the biarticular muscles are paired with uniarticular ones: the rectus with the iliopsoas for flexion of the hip and with the vasti for extension of the knee; the hamstrings with the gluteus maximus for extension of the hip and with the short head of the biceps and the popliteus for flexion of the knee.

9. Samples of active and passive insufficiency are: inability to extend knee

fully when hip is maximally flexed; or to flex knee fully when hip is maximally extended; inability to dorsiflex foot fully when knee is extended.

10. When the heel is lifted off the ground all weight is borne by the ball, and the tension of the tendo Achillis which prevents the foot being forced into dorsiflexion by the gravitational stress must be proportionally increased. With continued rising on the toes the line of gravity approaches the center of the ankle joint; because of this shortening of the gravitational lever arm the rotation moment of gravity decreases and so does, commensurately, the tension in the tendo Achillis. When the toe raising is continued until the ankle joint lies perpendicular over the ball of the foot, there is no rotation moment produced by gravity and the tendo Achillis is relaxed. All gravitational stress has become translatory.

11. Gravity acts as flexor of the hip and extensor of the knee if the line of gravity falls in front of both joints; this prevents the knee from collapsing in case of quadriceps paralysis. Gravity also stabilizes the soleus in the ankle joint so that it rotates the tibia backward; and it, likewise, stabilizes the gluteus maximus and prevents it from rotating the pelvis backward.

12. When the body is thrown backward, gravity acts as extensor of the hip; this relieves the tension of the gluteals but increases that of the quadriceps.

BIBLIOGRAPHY

1. BAEYER, H. VON: Natürlicher Ausgleich von Bewegungsstörungen. *Ztschr. f. Orthop. Chir., 50:*3/4, 458, 1928.
2. BAEYER, H. VON: Synapsis in der allgemeinen Gliedermechanik. *2d Internat'l Orth. Cong., London,* 1933.
3. STRASSER, H.: *Lehrbuch der Muskel- und Gelenkmechanik, Vol. III.* Berlin, J. Springer, 1917.

Lecture XXVI

MECHANICS OF SHOULDER-ARM COMPLEX

I. INTRODUCTION

IN ADAPTING itself for its highly specialized function the upper extremity has, to a large extent, followed its own way in the philogenetic development. Nature's goal was to produce a motor unit which would satisfy high demands on motility and freedom of action and which, at the same time, would be stable enough to give its movement force and precision. This general trend to produce an organ of free mobility becomes singularly manifest when we compare its relationship to the trunk with that of the lower extremity.

Shoulder blade, clavicle and shoulder joint form a mechanical unit called the shoulder-arm complex; they correspond to the combination of the os ilii, the os ischii, the os pubis, and the hip joint in the lower extremity. But, in contrast to the massive pelvic ring, only one flimsy articulation connects the complex with the skeleton of the trunk: the sternoclavicular joint. The clavicle interposes itself between thorax and shoulder blade, articulating with the latter in the acromioclavicular joint. The scapula joins the humerus in the glenohumeral articulation; thus a summation of mobility is established by three individual joints, all mutually interdependent. This makes the ultimate range of the shoulder joint itself far greater than that of the hip joint, which does not enjoy the benefit of having its mobility increased by auxiliary articulations. The clavicle rotates around the sternum, the scapula around the clavicle, and the humerus around the scapula. The latter, in addition, carries out rotatory and translatory movements against the thoracic wall in what is metaphorically called the scapulothoracic joint. Such an arrangement greatly favors the mobility of the shoulder-arm complex but it makes the problem of stabilizing it against the thorax much more difficult. The main burden of this stabilization rests upon the powerful musculature which secures the shoulder girdle against the thorax. Because the tone of these muscles varies so much with different attitudes and postures it is difficult to speak of a normal relationship of the scapula to the thorax without conceding rather wide physiological fluctuations.

One might consider as the normal position of the clavicle one in which it is retracted backward and forms with the frontal plane an angle of 30°. In this position the scapula is closely attached to the posterior thoracic wall (Fig. 1) (Fick[6]). When the shoulders are carried high and retracted, the outer end of the clavicle is elevated and points obliquely backward. When the posture relaxes, however, the clavicle approaches more the frontal plane or may even point slightly forward with its acromial end.

II. MORPHOLOGY

A. RELATION OF THE SCAPULA

Many refer to the relation of the scapula to the thoracic wall as the scapulothoracic joint. While this may not be correct from the strictly physical point

of view, it is a convenient concept to describe the movements which the scapula carries out around the thoracic wall. It is, of course, obvious that all these movements, whether rotatory or translatory, produce a combined movement of the sternoclavicular and acromioclavicular articulation, the former being the center of motion. But, after all, motions in the intervertebral articulations are also mostly

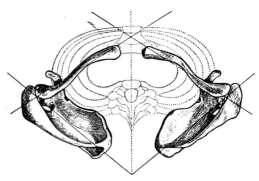

FIG. 1. Relation of scapula to thoracic wall. (Fick)

gliding movements of the articulating facets about a center outside of them; so one should not, we believe, be too pedantic about using the term scapulothoracic joint.

In military posture the shoulder blades are drawn backward so that the medial border of the scapula is less than 5-6 cm. from the midline. In normal

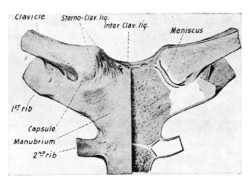

FIG. 2. Sternoclavicular articulation. (Redrawn from Braus)

position the distance is somewhat increased; the upper end of the scapula lies at the level of the 2nd rib and the lower at the level of the 7th to 8th rib.

So far as the firm attachment of the scapula to the thoracic wall is concerned, a number of facts are to be considered. The principal force which holds the shoulder blade to the thorax is the atmospheric pressure. We assume the average size of the triangular surface of the scapula to be 15 cm. long and 10 cm. wide,

which would give it an area of 75 cm.^{2} The atmospheric pressure is 760 mm. Mercury per cm.^{2}, or 760×14 mm. of H_2O. Thus the atmospheric pressure acting upon the scapula would be $75 \times 76 \times 14$ cm. of H_2O, or 79,800 cm.^{3} H_2O equal to 79.8 kg.

Secondly, the scapular motion is controlled by the scapulothoracic muscles which likewise hold the bone in close contact with the trunk and which restrict its movements to the contours of the thoracic wall.

B. THE STERNOCLAVICULAR ARTICULATION

1. Anatomy

This is more a saddle than a ball-and-socket joint. In the frontal plane the articular end of the clavicle is rounded, almost spherical, and articulates with a concavity of the sternum from which it is separated by a constant intraarticular meniscus (Fig. 2). While the articular surface of the sternum is concave in the frontal plane, it is somewhat convex in the sagittal plane. The sternal end of the clavicle is convex in the frontal but shows a concave depression in the sagittal plane. The joint has three degrees of freedom of motion which permit rotation of the scapula on its length axis, forward and backward movement about a perpendicular, and upward and downward movement about a sagittal axis. Only the length rotatory axis goes through the center of the articular facets of the joint. This axis runs from lateral upward to medial downward approximately in the frontal plane. The perpendicular and the sagittal axes lie outside of the joint, somewhere in the costoclavicular ligament. Therefore, the movement about these two axes is a gliding one between clavicle and meniscus and meniscus and sternum. The perpendicular axis for the forward and backward gliding runs from above medial to below and lateral. In this forward and backward movement the sternum furnishes the convexity and the clavicle the concavity of the articulation. The sagittal axis for upward and downward movement also goes through the costoclavicular ligament. In this movement the sternum furnishes the concavity and the clavicle the convexity.

The backward movement of the sternal end of the clavicle is combined with elevation and the forward movement with depression of the clavicle; similarly, the upward movement of the clavicle is combined with backward and the downward movement with forward movement of the clavicular end. In backward and forward movement the convexity of the sternal fossa offers more stability than the upward and downward movement which occurs about a sagittal axis around the concavity of the sternal socket. The intraarticular disc shows many variations in thickness which can be explained by the individual differences in the contours of the joint constituents. In addition, there are synovial protrusions into the joint which can cause indentations in either the disc or the joint facets. The disc is woven into the joint capsule, but it sends loose fibrous bundles to the sternal socket and to the 1st costosternal junction.

Ligaments. An anterior sternoclavicular in front and a posterior sternoclavicular ligament behind reinforce the capsular wall. An interclavicular ligament extends from clavicle to clavicle over the sternal notch. The anterior checks the forward and the posterior the backward movement of the head of the clavicle, while the interclavicular checks the downward movement (Fig. 2).

The pivot of these movements, except the axial, is the costoclavicular ligament. It consists of an anterior triangular and posterior, more rhombic portion. This very strong ligament restricts excessive elevation as well as excessive forward and backward movement of the clavicle. It does not check downward movement nor does it seem to have much effect on length rotation, in contrast to the coracoclavicular ligament.

2. The excursion ranges

A) LENGTH ROTATORY

Isolated movements are not possible in the living; they always accompany movements in the other planes. The excursion is small when the clavicle is in the frontal plane but it increases considerably (to 30° according to Fick[6]) when the acromial end of the clavicle is carried backward. At any rate, of the three degrees of freedom of motion in the joint, this is the most restricted and is especially in need of the supplementary increase furnished by the acromioclavicular joint.

B) SAGITTAL AXIS: ELEVATION AND DEPRESSION

Waldeyer,[15] the first to examine the motion in vivo, found that when the clavicular head is depressed the acromial end rises up to 8-9 cm., or to 10 cm. according to Mollier,[12] and Albert[1] estimates the total excursion range at 60° Elevation of the clavicle is checked by the lower portion of the capsule and the costoclavicular ligament; depression by the upper portion of the capsule and the interclavicular ligament. In addition, the subclavius muscle resists elevation of the clavicular head.

C) PERPENDICULAR AXIS: THE ANTEROPOSTERIOR MOVEMENT

Here the range is less; only 25-30°, and the excursion of the acromial end amounts to 6-7 cm. However, the data regarding this excursion field vary greatly from Albert's[1] 60° to Strasser's[13] 35° and Steinhausen's[14] 25°. It is most likely that in the living it amounts to not more than 25-30°.

D) CIRCUMDUCTORY MOVEMENT IN THE STERNOCLAVICULAR JOINT

The longitudinal axis of the clavicle describes the periphery of a cone; the acromial end describes a circle which is the base of a cone; Waldeyer[15] modified this by stating that this base is not a circle but an elliptical figure which is more in accord with the fact that, as mentioned above, the forward and backward excursion of the acromial end of the clavicle amounts to 6-7 cm. while the upward and downward excursions are 8-9 cm.

The maximum excursion is upward and forward in which the posterior border of the clavicle rotates upward around its longitudinal axis in pronatory direction. On the other hand, the maximum excursion backward and downward is associated with a backward rotation of the posterior clavicular border in the sense of a supinatory length rotation. The point is that the movements carried out in this articulation are combinations of frontal and sagittal plane movements with axial rotation.

C. THE ACROMIOCLAVICULAR ARTICULATION

1. Anatomy

In some respects the role which the acromioclavicular joint plays in the movement of the entire complex surpasses that of the sternoclavicular joint.

This is true at least for the motion in or close to the sagittal plane.

This joint which connects the clavicle with the shoulder blade has small elliptical facets, the larger diameter of the ellipse pointing from forward lateral to backward medial so that the calvicular facet looks lateral and backward and the acromial medial and forward. The convexity of the joint is represented by the clavicle and the concavity by the acromion.

An intraarticular meniscus completely or incompletely developed is found occasionally filling out a triangular space in the joint. It rarely reaches the stage of a fibrocartilaginous ring which would be comparable to the meniscus of the knee joint. According to Krause,[10] only in 1% of the cases separate joint spaces are formed by an intraarticular disc.

The joint is secured by strong ligaments which extend between the adjacent surfaces of the acromion and the acromial end of the clavicle. These ligaments are the capsule reinforcing ligamenta acromioclavicularia superius and inferius. Both ligaments restrain the clavicle from being displaced posteriorly. Medial to the joint, the coracoclavicular ligament secures the coracoid process to the clavicle and thereby gives stability to the acromioclavicular articulation (Fig. 3). The lateral portion, the trapezoid ligament, runs in almost sagittal direction from the coracoid to the undersurface of the clavicle and restrains its forward movement; it also prevents the clavicle from gliding off the slanting facets of the acromion. The medial and less powerful portion is the conoid ligament. Also coming from the medial surface of the coracoid process, it runs upward more in a frontal plane to the undersurface of the clavicle. Both ligaments also restrain backward movement of the shoulder blade (Fick[6]). They unite the clavicle so strongly to the coracoid process that a fracture of the clavicle lateral to the ligaments or even going through between the two ligaments causes no displacement. In addition to these ligaments there is also a firm fascial band (Henle's ligamentum coracoclaviculare anterius) running from the coracoid process upward to the acromial end of the clavicle. This fascial bend blends below with the muscle fascia of the pectoralis minor and above with that of the subscapularis. It is, according to Fick,[6] sometimes strongly developed and comes under tension when the arm is pulled laterally.

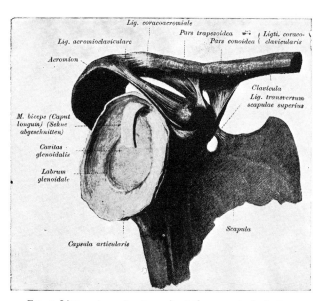

FIG. 3. Ligamentous structure about the acromioclavicular articulation. (Fick)

2. Ranges of motion

This joint also has three degrees of freedom of motion about three axes which are better described in terms of their relation to the shoulder blade than in terms of the cardinal planes of the body since the joint constantly changes its relation to the trunk.

A) MOVEMENT ABOUT A VERTICAL AXIS OF THE JOINT (FORWARD AND BACK-WARD CIRCLING OF THE SCAPULA)

The axis goes vertically through the acromial end of the clavicle midway between joint and coracoclavicular ligament. The scapula circles the joint about this axis in a forward and backward movement. As the acromion glides backward the angle between clavicle and shoulder blade is decreased; as it glides forward it is increased. A compensatory movement in the sternoclavicular joint accompanies these movements.

In normal position the angle between clavicle and shoulder blade projected unto a horizontal plane is 30° (Fig 1). As the acromion glides backward and forward against the clavicle about the vertical axis, the backward movement is checked by the joint capsule and the trapezoid, the forward movement by the trapezoid and conoid ligaments.

B) MOVEMENT ABOUT THE SAGITTAL AXIS (ABDUCTION AND ADDUCTION OF THE SCAPULA)

This motion has a limited range. In abduction both coracoclavicular ligaments become tight (the conoid first), while in adduction the coracoid impinges against the clavicle. Fick[6] finds that this range amounts to only 10°.

C) MOVEMENT ABOUT A FRONTAL AXIS (FORWARD AND BACKWARD SWING)

This is a pendulum movement which allows the shoulder blade to swing forward and backward. It permits full vertical elevation of the forward flexed arm and its amplitude is estimated at 60-70° in the living. The checks of motion are as follows: the backward swing of the scapula is checked by the tension of the anterior portion of the joint capsule, then by the medial portion of the trapezoid ligament; and the forward swing is checked by the posterior portion of the capsule, the lateral portion of the trapezoid and the impingement of the coracoid process against the clavicle.

In summary, the acromioclavicular articulation can be considered as a ball-and-socket joint with three perpendicular axes of motion which has its center of motion at a point midway between the coracoclavicular ligament and the middle of the joint facet. All motions except that about the length axis are therefore gliding motions and the coracoclavicular ligaments are far enough away from the center of motion to act as restraint.

The most important function of the joint is that it furnishes an additional range of motion for the shoulder complex after the ranges of the sternoclavicular joint are exhausted. And here again by far the greatest addition is that to flexion and extension in the sagittal plane. (Fick[6])

D. STATIC STRESSES OF THE CLAVICLE

Both pressure and tension forces are operative on the clavicle; they are absorbed by the resistance of the structure but they become manifest when the integrity of the shoulder girdle ring is destroyed, by fracture, dislocation, muscular imbalance, etc. In contrast to the pelvic ring which is a complete structural unit, the pectoral ring is incomplete as a bony structure. Consequently, intrinsic stresses require for their equilibration the cooperation of the muscles of the shoulder-arm complex.

These intrinsic stresses operate as follows:

1. Pressure stress

A pressure stress is produced along the longitudinal axis of the clavicle toward the sternoclavicular joint by the musculature; for example, the trapezius or pectoralis minor forces the clavicle against the sternum. This is increased by pressing the shoulders together or by lying on one shoulder. In fracture of the clavicle it becomes manifest by the overriding of the fragments. If one, resting upon the dorsiflexed arms, tries to let the trunk sink backward between the shoulders, such a pressure effect is produced on both clavicles.

2. Tension stress

Tension stresses acting upon the clavicle are produced if the arm is abducted in the direction of the clavicle and is pulled out. In this case deltoid and pectoralis major provide the tension stress. If one rests on the forward flexed arm or if one, hanging by the arms, swings forward, or in many other common situations such tension stresses are produced and they may even lead to discontinuity; here dislocations are more likely to occur than fractures.

3. Torsional rotational forces

Torsional rotational forces likewise tax the resistance of the clavicle. For instance, a forceful outward rotation of the arm is transmitted through the scapula to the clavicle. This force also takes more effect on the joints and the reinforcing ligaments than on the continuity of the bone.

4. Bending stresses

A downward force acting on the acromial end produces bending stresses which may assume proportions leading to discontinuity, especially if the depressed clavicle becomes impinged against the coracoid process after rupture of the acromioclavicular reinforcing ligaments.

E. THE SCAPULOHUMERAL JOINT

1. Anatomy

A) SKELETAL

The ball-shaped head of the humerus and the shallow concave glenoid fossa form the scapulohumeral articulation or the shoulder joint proper. The most striking feature is the incongruity of the joint constituents. The head represents almost a half sphere or at least 2/5 of it, having an angular value of 153°; the glenoid fossa of the scapula on the other hand, which is more shallow and therefore has a larger radius, has an angular value of only 75°.

In the adult the center of the head of the humerus is about 2.5 cm. from the periphery; because of an angulation between humeral head and shaft, the center is about 1 cm. medial to the length axis of the shaft. There is some slight difference in the curving of the head in the sense that the radii become shorter from laterally to medially (Strasser[13]), but for all practical purposes the head can be considered as a half sphere (Fig. 4).

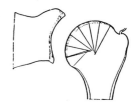

FIG. 4. The contour of the head of the humerus. (Strasser)

The neck and head of the humerus are angulated against the shaft 45-50° in the frontal plane (Fig. 5). In the transverse plane the head and neck of the humerus are twisted backward or internally against the shaft so that the axis of the elbow joint becomes oblique in relation to the axis of the head (Fig. 6). This angle between the two axes is, according to Hultkranz,[8] 43-49° but it varies with age and race in such wide

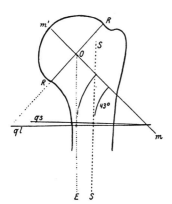

FIG. 5. Angulation of head and neck of the humerus. (Strasser)

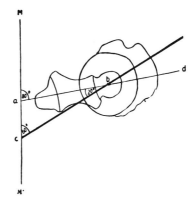

FIG. 6. Angulation of the humerus in the transverse plane. (Fick)

limits that it is impossible to speak of a standard value. According to DePalma,[4] the torsion angle increases as the orthograde stage is approached and as the anteroposterior flattening of the thoracic cage and the dorsal displacement of the scapula causes the glenoid fossa to be directed more laterally.

The glenoid is more uniformly curved and has a radius of 24-26 mm. in the adult. Even with the limbus glenoidalis its angular value in the frontal plane does not exceed 75° (against 153° of the humerus), and its greatest diameter is 3.5-4 cm. (against 6.5-7 cm. of the head). This difference of 78° corresponds in linear length to 4 cm. In the sagittal plane the diameter of the glenoid is 2.5-3 cm. and its angular value is only 50°.

The *atmospheric pressure*. In contrast to the hip joint, the atmospheric pressure plays only a minor role in the stabilization of the shoulder joint. The contact area of the head with the glenoid is small, occupying only a portion of the lower medial quadrant of the head and of the corresponding area of the glenoid. With the labrum, the joint surface of the glenoid is substantially increased. On the other hand, the capsular apparatus intrudes under atmospheric pressure between head and socket, thus filling a considerable space of the cavity and leaving only a small portion of the glenoid in direct contact with the humeral head. It is estimated that the pressure amounts to not more than 6.5 kg. Therefore, it has little effect in holding the head tightly to the glenoid

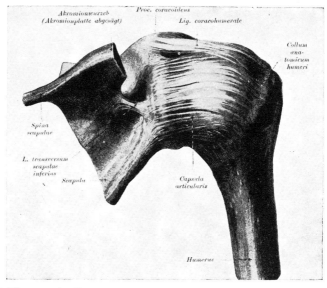

Fig. 7. The posterior ligamentous and capsular structures of the shoulder. (Fick)

against the weight of the extremity. In the living the tendons of the rotatory cuff which are woven into the capsule restrict somewhat the penetration of the capsular apparatus between the joint surfaces. It is believed that the musculature acts like a suction valve so that the effect of the atmospheric pressure may be of greater importance for the stabilization of the joint in the living than it would appear from the study of the anatomic specimens.

In paralysis of the shoulder muscles, especially the deltoid, the relaxation permits fringes of the synovia to be drawn into the joint, filling the empty recesses very similar as in the knee joint when a relaxed quadriceps fails to keep the synovial villi out from between the joint constituents. However, the effect of the atmospheric pressure can still be demonstrated if the deltoid is removed and if the interposition of the capsule is prevented by rotating the arm so that the posterior muscles are tightened and the fatty synovial fringes are kept out of the joint. In the absence of the deltoid the humerus is still held in the socket in perpendicular position, but as soon as the arm is abducted or the scapula rotates relaxing the coracohumeral ligament, the head immediately leaves the socket (Fick[6]).

B) CAPSULE AND LIGAMENTS

The ligaments supporting the capsule of the shoulder joint have principally a suspensory function. The fibrous capsule itself is loose, redundant, particularly anteriorly where it arises at various distances from the anterior side of the neck of the scapula. Here recesses are established between the capsular insertion and the labrum glenoidale which at times are of definite pathological significance (DePalma[4]).

Two ligaments reinforce the anterior aspect of the capsule: the coracohumeral and the glenohumeral. The former comes from the basis of the coracoid and runs underneath the coracocromial ligament obliquely downward and outward to the transverse ligament of the intertubercular sulcus (Fig. 7). It is not so much a suspensor as it checks rotation of the arm. The other ligament is the glenohumeral, which consists of a superior, a middle, and an inferior portion (Fig. 8). The superior is deep-seated, coming from the base of the coracoid as well as from the upper pole of the glenoid to the lesser tuberosity. The middle ligament is a smaller bundle, arising below the former and going also to the lesser tuberosity as does the inferior

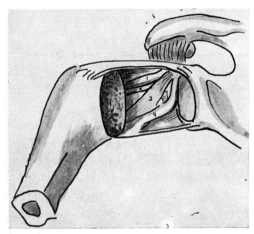

FIG. 8. The anterior ligamentous and capsular structures of the shoulder (seen from behind). After Grant.
1. Long head of biceps.
2. Superior glenohumeral ligament.
3. Middle glenohumeral ligament.
4. Inferior glenohumeral ligament.
5. Cut edge of subscapularis muscle.

glenohumeral ligament which arises from the lower part of the anterior glenoidal rim, going over the surgical neck to the lesser tuberosity. All portions check outward rotation of the arm.

The fibrous capsule is intimately related to the so-called rotator cuff which consists of the tendons of the subscapularis, the supra- and infraspinatus, and the teres minor. They are all woven into the fibrous capsule. The long head of the biceps, coming from the supraglenoid tubercle, passes through the capsular sac and then enters the intertubercular sulcus in which it glides up and down in a synovial sheath, an extension of the synovial membrane.

Movements of this joint successively tighten different portions of the capsule (Fig. 9). In midposition all portions of the capsule are equally relaxed. When the humerus is carried forward, the posterior portion of the capsule becomes tight; when it is carried backward, the anterior tightens. With the arm hang-

ing down, the glenohumeral and particularly the coracohumeral ligaments become tight.

c) BURSAE

There are numerous bursae about the shoulder joint, some of them inter-
communicating and some simple prolongations of the synovial sac. The subacromial, subdeltoid and subcoracoid bursae overlie the scapulohumeral joint
between the rotator cuff below and the deltoid and acromion above. They present the gliding mechanism between these two muscle groups. The infraspinatus and subscapularis bursae and the bicipital sheath for the long head
are all synovial extensions which serve the same purpose. Some of these
bursae become of clinical significance by virtue of adhesions which prevent
the free gliding mechanism. This is especially true of the subdeltoid bursa and
of the bicipital sheath.

The vault of the shoulder joint is formed by the acromion and by the
coracoacromial ligament. This triangular structure forms a strong arch between
the outer edge of the coracoid and the opposing anterior edge of the acromion,
lying just forward of the acromioclavicular articulation. Under it lies part
of the subacromial bursa.

2. Orientation planes and joint ranges

In contrast to the hip joint, the incongruity of the shoulder joint surfaces
necessitates that motions in all directions except axial rotation are a combination of gliding and rocking, as it is in the knee joint. A line drawn from the
center of the head of the humerus to the midpoint of the transverse elbow
joint axis represents the mechanical axis of the humerus. A plane laid through
the elbow axis and through the center of the head is the determination plane
of the humerus in relation to the trunk.

The mechanical midposition of the scapulohumeral joint is one in which
the center of the articular surface of the head coincides with the center of
the glenoid. In this midposition the humerus is abducted from the thorax 45°
and forward flexed 45°. The lesser tuberosity looks straight forward and the
anterior surface of the arm is slightly inward rotated.

This midposition is impractical for measuring the relationship between arm
and thorax. It is more serviceable to use the normal or resting position of the
arm. In this position the arm hangs down vertically; its volar surface is
directed anteriorly and slightly inward; the elbow axis is no longer frontal but
runs diagonally from forward lateral to backward medial, forming with the
frontal plane an angle of 10° and with the sagittal plane an angle of 80°
(Fig. 6).

As to the position of the plane of the scapula in relation to the humerus
in this rest position, it forms an angle of 30° with the frontal plane (Fig. 1).
This angle changes constantly with the forward and backward movement of
the scapula, becoming smaller in backward and larger in forward movement.

One can best study the ranges of motion in different planes with the use of a wire net globe system indicating the excursion in meridians and parallels (Fig. 10). If one takes into account the obliquity of the scapular plane with the frontal plane (Strasser[13]), which is 30°, one must designate the meridians of the net globe so that the frontal plane is in meridian 30-150° instead of 0-180°.

a) The Abduction and Adduction Range

Fischer[7] found the abduction range of the shoulder joint in this plane between 71-80°, while Strasser[13] gives it only 64°.

If in the wire net globe the pole axis is placed anteroposteriorly (Fig. 10),

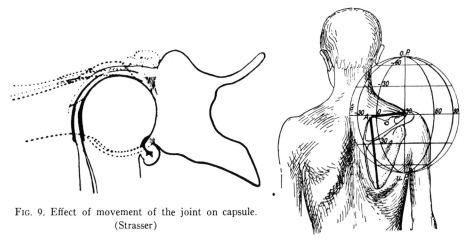

FIG. 9. Effect of movement of the joint on capsule. (Strasser)

FIG. 10. Wire net globe system for indicating excursions. (Strasser)

one finds that the closer the abductory and adductory motion occurs to the poles, i.e., the more it is combined with forward flexion and backward extension, the less it is accompanied by longitudinal rotation; and, the farther from the pole axis, i.e., the closer it is to the equator, the more length rotation occurs (Strasser[13]). In other words, if one flexes the elbow of the hanging arm at right angles so that the forearm looks straight forward and then elevates the arm strictly in the sagittal plane until it gradually reaches the vertical position, no length rotation of the humerus occurs during the entire range and the flexed forearm remains strictly in this plane. But, if one starts out from the same position of the hanging arm holding the elbow at right angle flexion and then abducts the arm strictly in the frontal plane, then the elbow axis which is almost frontal at the outset becomes sagittal at 90° abduction and the forearm points upward in the frontal plane. In further abduction it even points backward of the frontal plane. There has occurred, in plain words, in the course of abduction an outward rotation of the humerus of over 90°.

This outward rotation of the humerus which is associated with abduction in the frontal plane is of greatest importance for the function of the shoulder joint. It makes it possible for the greater tuberosity to dive under the anterior edge of the acromion as the arm is abducted and so to complete its abduction range of about 90°. Any injury to the outward rotators of the humerus, therefore, is a great obstacle and causes painful impingement symptoms.

In view of later investigations on the relations between shoulder blade and humeral movement, it is of interest to note that Fick[6] had already stated definitely that in the living (as can be seen by x-ray) a lateral movement of the lower angle of the scapula (rotation of the scapula) occurs with the very beginning of abduction, so that in spite of the fact that the glenohumeral joint has its own abduction range of 90°, no abduction ever occurs in the joint which is not combined with movement in the other joints of the shoulder girdle.

b) The flexion Extension Range

The range of pure ventral and dorsal elevation is variously given as between 100-150° for the glenohumeral joint alone, of which range most falls upon the forward elevation.

From Braune and Fischer's[2] observations on the cadaver it appears that the forward elevation of the arm in a strictly sagittal plane is possible in this joint only to 58-62°, while elevation in a forward diagonal plane, especially in the plane of the scapula, has a much wider range (up to 104°).

Ventral elevation is checked by the posterior capsule, the posterior portion of the coracohumeral ligament, the teres minor, and the infraspinatus. Dorsal elevation is much more restricted. This is especially so in the strictly sagittal plane where no backward elevation is possible without rotation of the shoulder blade taking the greater part. Even in the more diagonal plane the range of dorsiflexion is comparatively small. The dorsal elevation is checked by the superior and anterior portions of the capsule, the anterior portion of the coracohumeral ligament, the supraspinatus, and the subscapularis.

c) The Inward and Outward Rotation Range

To understand the mechanism of certain strains which involve the shoulder joint, one should appreciate that in pronation or inward rotation of the humerus the head glides backward in the socket, in supination or outward rotation it glides forward, and that horizontal forward elevation causes a downward gliding of the head. The excursion range of length rotation is greatest in 90° abduction, where it amounts to 120°, while in the forward flexed position it is only 50° (Fick[6]).

In normal position with the arm hanging down, inward rotation is checked by the impingement of the lesser tuberosity against the anterior, and outward rotation by impingement of the greater tuberosity against the posterior rim of the glenoid. But when the arm is abducted 90° in the frontal plane, there is no impingement in either inward or outward rotation, and the check is produced by the capsule, the coracohumeral ligament, and the muscles of the rotator cuff.

d) The Circumpolar Movement in the Shoulder Joint

So far only the motion in single planes, i.e., in the frontal, sagittal and transverse planes, has been considered. It now remains to analyze arm movements in which the elementary movements of these three planes are combined.

Let us again visualize the humeral head placed in a wire globe as the center of motion. If the arrangement is such that the pole axis of this globe runs from forward to backward in sagittal direction, then, clearly, abduction and adduction in the frontal plane are sweeps over successive meridians, and forward and backward flexion in the sagittal plane sweep over successive parallels of the globe. Suppose next that flexion-extension occurs in abduction. The extremity sweeps again along to reach a certain meridian, not necessarily the meridian of 90° but, according to the degree of abduction, some intermediate one. If this position is maintained and the extremity then proceeds with forward flexion, it crosses a number of parallels but remains in the same meridian plane. But, let us assume that the two movements are not carried out successively but simultaneously. Then, there will be a continuous instantaneous change of meridians and parallels during the combined abduction-flexion movement. The arm no longer moves in one plane but describes the surface of a cone. On the surface of the globe the arm traces a circle which may or may not be a parallel but at any rate is smaller than the equator or any meridian. Furthermore, in describing this circle there is not only a momentaneous change in abduction or adduction, flexion or extension, but also in length rotation.

When we describe with the outstretched arm a circle in the air or when we draw a circle on a blackboard in front of us, the succession of movement is as follows: starting from a position in which the humerus is adducted and forward flexed, the arm goes into forward flexion, abduction, supination, backward extension, adduction, pronation, and forward flexion again.

These movements about the three axes occur simultaneously. Since the shoulder girdle movements accompany all movements of the glenohumeral joint, the same principle of combined instantaneous movement in all three planes applies to the movement of the shoulder-arm complex as a whole.

III. THE DYNAMICS OF THE SHOULDER-ARM COMPLEX
A. THE MOVEMENTS OF THE SCAPULA

Through its connection with the sternum, the clavicle holds the acromion at a certain distance from the thoracic wall. The scapula forms a bridge between the outer end of the clavicle and the thorax so that the vertebral border of the shoulder blade is closely attached to the thoracic wall.

The movements of the shoulder blade are of two types: they are either translatory, where all points of the bone move in straight parallel lines, up and down or forward and backward; or they are rotating about a perpendicular axis near the upper outer or upper inner angle of the bone. Usually the movements are combinations of both the translatory and the rotatory types. In fact, since most of the muscles reach the shoulder in oblique direction, so that their line of pull does not coincide with the centers of motion of the entire shoulder complex, they necessarily develop rotatory as well as translatory components. For example: the anterior serratus pulls the shoulder blade forward and downward and at the same time rotates it about a center lying close to the acromial angle; the upper trapezius pulls the scapula in the opposite direction, namely upward and backward, at the same time it rotates the scapula about a center lying closer to the upper inner angle. The result is that to give the movements of the scapula a specific direction the secondary functions of the operating muscles have to be neutralized by antagonistic action before a resultant in the desired direction is obtained.

1. The translatory movements of the shoulder blade

Although straight translatory movement, i.e., without rotation, can occur in many directions, it suffices to describe the muscle mechanism in two planes only, the frontal and the horizontal, that is, up and down, and the forward and backward excursions of the scapula.

A) IN THE FRONTAL PLANE (Lanz and Wachsmuth[11])

1) Upward movement (Fig. 11). Levator and trapezius are the principal muscles. Yet, neither muscle acts strictly translatory. The trapezius has a slightly rotatory effect around the inner scapular angle, and the levator also pulls the scapula upward and forward with a slight rotatory effect about the outer angle. Secondary elevators are the rhomboids; but, here again, their rotatory effect which pulls the lower angle of the scapula closer to the vertebral spines must be neutralized by the upper fibers of the anterior serratus. Upward movement is thus a combined effort of levator and upper trapezius with the help of the rhomboids, and with mutual neutralization of all rotatory components of these muscles. Upward elevation amounts to 2-3.5 cm., according to Duchenne.[5]

2) Downward movement (Fig. 12). Gravity becomes a factor. Under its effect, the shoulder blade assumes the resting position of complete relaxation. Correct posture is an antigravity position as the backward pull of the shoulder entails some elevation. Active depressors of the shoulder blade are: directly, the lower trapezius and the anterior serratus; indirectly, the pectoralis minor by pulling the coracoid process downward and forward; the pectoralis major with its lower fibers and the latissimus dorsi by pulling the humerus down; also the subclavius which pulls the clavicle down against the first rib. Obviously, all these muscles must have their rotatory components as well as their translatory components in forward and backward movements mutually neutralized before a true resultant in downward direction is achieved. In this neutralization scheme the muscles are paired as follows: to suppress rotation of the scapula: the lower trapezius versus the lower serratus; to suppress forward and backward movement: the latissimus versus pectoralis major and pectoralis minor.

B) IN THE TRANSVERSE PLANE

1) Forward movement (Fig. 13). The principal muscles are: the anterior serratus, the pectoralis major with its midportion, and the pectoralis minor.

Translatory movement of the scapula

a. subclavius
b. pectoralis minor
c. pectoralis major
d. serratus anterior
e. trapezius
f. lattissimus dorsi
g. sternocleidomastoid
h. levator scapulae
i. rhomboideus minor
j. rhomboideus major

FIG. 11. Cranial elevation of the scapula. (Redrawn from Lanz & Wachsmuth)

FIG. 12. Caudal depression of the scapula. (Redrawn from Lanz & Wachsmuth)

FIG. 13. Forward movement of the scapula. (Redrawn from Lanz & Wachsmuth)

FIG. 14. Backward movement of the scapula. (Redrawn from Lanz & Wachsmuth)

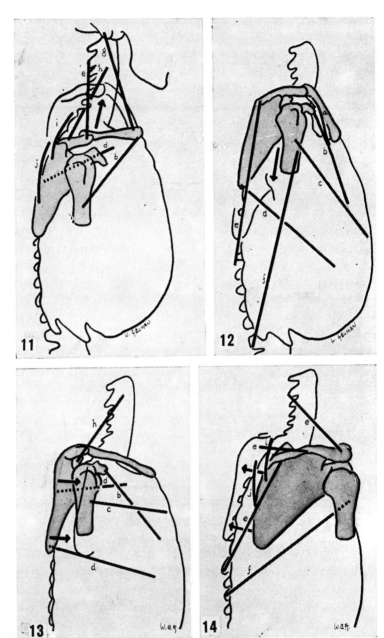

All these muscles act more nearly in forward translatory direction, although a tendency to downward pull by the lower fibers of the serratus anterior must be neutralized by the upper trapezius. The forward rotatory effect of the pectoralis minor is neutralized by the lower fibers of the serratus. The forward translatory range of the scapula is 2-4 cm.

2) Backward movement (Fig. 14). This is somewhat more complicated because of the more oblique direction of the axes of the participating muscles. The direct backward movers are principally the middle and lower trapezius and the rhomboids; indirectly, they are assisted by the latissimus through its action on the humerus. The backward rotatory action on the lower angle of the shoulder blade by the rhomboids and by the latissimus is neutralized by the upper and lower trapezius. The entire backward excursion from the rest position is, according to Duchenne,[5] 2-4 cm.

It must be realized that all these translatory movements of the scapula transmit themselves to the acromioclavicular and to the sternoclavicular joint. In the strictest sense, these movements are not precisely translatory. The acromial end of the scapula is always held to the thorax by the constant length of the clavicle, which is the radius for all motion in the sternoclavicular joint. However, the freedom of the acromioclavicular joint allows the scapula to adjust itself to the thoracic wall both in up- and downward and in for- and backward movements. The translatory movements of the scapula are, therefore, so to speak, incorporated in a rotatory system involving the entire shoulder girdle.

In rest position, the scapular plane forms with the clavicle an angle of 30°. This angle increases with forward and decreases with backward movement of the shoulder blade. Similarly, in up- and down movement, the angle decreases on elevation and increases on depression of the scapula.

2. The rotatory movement of the shoulder blade

A) THE FORWARD ROTATION OF THE LOWER ANGLE (Fig. 15)

Forward rotators of the scapula are the upper trapezius and the lower serratus. The overwhelming strength of the latter is the reason that, at the end of rotation, the lower angle is always pulled out laterally. Rotation of the lower angle always accompanies abduction of the arm in the shoulder joint.

In addition, the upward component of the upper trapezius must be equalized by the downward pull of the lower trapezius. These two muscle portions constitute a pair of forces rotating the scapula upon an axis vertical to its plane, around a center which lies midways between the acromion and the vertebral end of the scapular spine. The acromioclavicular portion of the trapezius inclines the head to its side and pulls it backward. Only when this inclination has reached its maximum or when the head is fixed in neutral position by opposing muscles does the elevation and forward rotation of the shoulder blade follow. It is then that the lower trapezius acts to neutralize the elevation.

As can be seen, the action of the trapezius varies in its several sections. The upper portion elevates the shoulder blade and brings its upper inner angle closer to the midline; of the midportion, the lateral half also elevates the acromion, while the medial half approaches the scapular border to the midline. The lower portion pulls the scapula down, lowering its inferior angle, and, at the same time also, pulls the scapular border toward the midline.

The anterior serratus, coupled with the trapezius, is a powerful forward rotator of the shoulder blade. The rotation produced by both muscles is around an axis lying closer to the upper inner angle, which, therefore, remains relatively stationary while the acromion moves strongly upward (Duchenne[5]).

The question when and how the rotation of the scapula becomes auxiliary to the movements in the shoulder joint has become of vital interest. The maximal abduction in the glenohumeral joint is, according to Duchenne,[5] 90°; Mollier[12] gives it 112°; Strasser,[13] 110°. This would make the angle between the lateral border of the scapula and the humerus in maximal abduction about

The rotatory movements of the scapula.

a. trapezius
b. serratus anterior
c. levator scapulae
d. rhomboideus minor
e. rhomboideus major
f. pectoralis minor

Fig. 15. Forward rotation of the lower angle. (Redrawn from Lanz & Wachsmuth)

Fig. 16. Backward rotation of the lower angle. (Redrawn from Lanz & Wachsmuth)

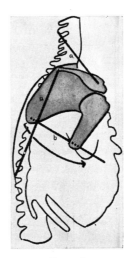
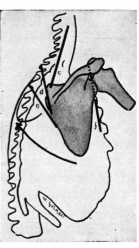

Fig. 15   Fig. 16

140-150°. As this maximum angle is reached in the shoulder joint the scapula has already rotated 25° (Strasser[13]), or 30° (Mollier[12]). As abduction proceeds, scapular rotation gradually increases and at full perpendicular abduction of the arm it amounts to 60°. At the same time, the lower angle of the scapula has moved forward 1-2 cm. (Duchenne[5]).

b) The Backward Rotation of the Lower Angle (Fig. 16)

The principal muscles are: the levator anguli scapulae, the rhomboids, and the pectoralis minor. The latter suppresses the elevatory action of the levator and of the rhomboids. In this backward movement of the lower angle, rotation is carried out principally by the rhomboids and the pectoralis minor acting as a force couple. An auxiliary force is provided by gravity. The movement occurs about an axis lying between the acromion and the vertebral end of the scapular spine; the lower angle of the scapula, therefore, has a considerable excursion, amounting to 2-4 cm. (Duchenne[5]).

B. THE MOVEMENTS IN THE STERNOCLAVICULAR ARTICULATION

This articulation is not moved by one muscle alone. With the exception of the subclavius, the effect of which upon the joint is negligible, all muscles whether they insert into the clavicle or the scapula engage the acromioclavicular joint simultaneously with the sternoclavicular. Furthermore, the scapulothoracic muscles as well as the scapulohumeral and the thoracohumeral group, while not prime movers of this joint, exert a constant indirect effect on that articulation by virtue of their action on the acromioclavicular joint, which maintains a constant reciprocal relation with the sternoclavicular articulation.

It will be recalled that the center of motion for the sternoclavicular articulation lies within the costoclavicular ligaments and that a forward movement of the clavicle in the sternoclavicular joint produces a backward movement of the acromioclavicular and a backward movement in the former joint produces a forward movement in the latter. Only the axial movement about the longitudinal axis of the clavicle occurs in the same direction in both joints. The clavicular portion of the trapezius, in moving the outer end of the clavicle backward and upward, forces the sternal end downward and forward. Acting upon the clavicle indirectly through the shoulder blade, the pectoralis minor is forcing the scapular end forward and downward, causing the sternal end to move upward and backward; the lower trapezius and the latissimus, which move the shoulder downward and backward, in transmitting this movement through the clavicle cause the sternal end to move upward and forward.

C. THE MOVEMENTS IN THE ACROMIOCLAVICULAR JOINT

This articulation also supplies additional ranges of motion to the other joints of the shoulder girdle. Of these additions, the one for forward and backward swing of the scapula about a frontal axis is by far the most important for the mobility of the entire shoulder-arm complex. Rotation of the scapula in elevation of the arm depends largely upon this additional range. But the range in the acromioclavicular joint is not enough. The clavicle itself must rotate to allow for the full amplitude of forward flexion. Inman, Saunders and Abbott[9] found that beyond a certain height further elevation of the scapula was only possible because of the length rotation of the clavicle itself which relaxed the straining coracoclavicular ligament.

D. THE MOVEMENTS IN THE GLENOHUMERAL JOINT

1. Ab- and adduction

A) ABDUCTION

The principal abductors are the *deltoid* and the *supraspinatus* muscles. The latter fixes the arm against the glenoid so as to enable the deltoid to start its abductory function. According to Duchenne,[5] the deltoid action is limited to the horizontal, at which point the tension of the teres minor restrains further elevation of the arm by the deltoid. The maximum abduction is carried out by the anterior fibers and it is facilitated by the outward rotation of the humerus. In this respect, the findings of Inman, Saunders and Abbott[9] are of special interest because they furnish proof that the shoulder blade rotates during the entire abduction range. It was known to earlier observers (Fick[6]) that rotation of the scapula starts at the very beginning of abduction and not after abduction has reached 90°, as some, ignorant of previous findings, seem to have believed. To full perpendicular abduction the glenohumeral joint contributes 120°, the rotating scapula, 60°.

Duchenne[5] emphasizes that the most important function of the supraspinatus is to maintain the head of the humerus firmly in contact with the glenoid cavity in which it is assisted by the tension of the long heads of triceps and biceps. Whether or not it is an actual abductor is still debatable. The action current curve of this muscle is, according to Inman[9] and others, almost a pure sine wave which reaches its peak at 90° or 100°, the same as the deltoid. In forward flexion its peak is reached somewhat sooner. Inman[9] believes, therefore, that it is not merely an initiator of abduction but that it acts together with the muscle all the time. At any rate, the deltoid is powerless as an abductor without the supraspinatus, as one can see in cases of complete rupture of the supraspinatus tendon when active abduction becomes impossible.

B) ADDUCTION

Principal adductors are the pectoralis major, the latissimus dorsi and the teres major. The clavicular portion of the pectoralis major is a forward flexor and inward rotator of the outward rotated arm. In normal position, the sternal portion is also a forward flexor. The adductory function of the muscle naturally depends on the relation of the muscle axis to the center of the head. As long as this axis runs below the center, the muscle remains an adductor. But with abduction increased beyond the horizontal, some of the clavicular portion of the muscle comes to lie above the center and, therefore, becomes abductory, while the fibers of the sternal portion always remain adductors.

The teres major has little power as an adductor (Duchenne[5]). In order to carry out adduction of the humerus in the glenohumeral joint the scapula must first be stabilized against the rotatory effect of the rhomboids and the other backward rotators of the lower angle. Only when the shoulder blade has become a fixed point, the adductory effect of the teres major appears. Both in the teres major and latissimus dorsi the adductory component increases, at the expense of the inward rotatory, as abduction increases. Both are also dorsal extensors of the arm and their dorsal extensory effect also increases with abduction.

2. In- and outward rotation

A) OUTWARD ROTATION

The rotator cuff consists of the three principal outward rotators, namely, the supraspinatus, the infraspinatus, and the teres minor.

The supraspinatus has already been mentioned as a stabilizer of the humeral head. Its outward rotatory ability seems to have been overlooked by Duchenne[5] who ascribes to it the power of active abduction of the arm even without the deltoid. The muscle is an outward rotator in all positions together with the infraspinatus and the teres minor. These latter muscles are called by Duchenne[5] the posterohumeral rotators, and he assigns to them a maximum outward rotation range of 90°.

B) INWARD ROTATION

The principal inward rotators are the teres major, the latissimus, and the subscapularis. All three portions of the subscapularis rotate the humerus strongly inward but the upper portion is also a slight forward flexor and the other two are slight dorsal extensors. With increasing abduction, the inward rotatory effect of the subscapularis diminishes in favor of forward flexion. Vice versa, with diminishing abduction, its power of inward rotation increases.

3. Flexion-extension

A) FORWARD FLEXION

It is particularly in this plane where the muscle action changes with the ab- or adductory positions. One observes that forward flexion and inward rotation are interrelated. The deltoid in normal position is an inward rotator with its anterior fibers. With increasing abduction, they become forward flexors and the same is the case with the long head of the biceps and the coracobrachialis. The pectoralis major is likewise a forward flexor with its upper portion. Additional forward flexors are the long and short heads of the biceps and the coracobrachialis. The long head of the biceps is really an adductor of the shoulder, but with increasing abduction, its forward flexory component increases. The short head of the biceps and the coracobrachialis are forward flexors, adductors, and inward rotators of the arm; they also increase their forward flexory component as abduction increases.

B) BACKWARD EXTENSION.

Principal backward extensors are the latissimus, the teres major, and the long head of the triceps. All three are also adductors. In normal position, their principal action is adduction and inward rotation. As the arm goes into abduction, their dorsal extensory component increases. The posterior fibers of the deltoid, which are outward rotators in normal position, become backward extensors with progressive abduction.

E. THE WORKING CAPACITY OF THE SHOULDER GIRDLE MUSCLE (FICK[6])

Figures reduced to the Recklinghausen index of 3.6 kgm. give the absolute muscle power per cm.² as follows: 1) The inward rotators, 5.374 kgm.;

2) the outward rotators, 2.561 kgm.; 3) the forward flexors, 4.375 kgm.; 4) the backward extensors, 4.465 kgm.; 5) adductors, 4.320 kgm., and 6) abductors, 2.074 kgm.

F. ANALYSIS OF THE MOVEMENTS OF THE WHOLE SHOULDER-ARM COMPLEX

Codman[3] calls the cooperation of the clavicular joints with the movement in the glenohumeral articulation the scapulohumeral rhythm which he considers an index for the normal function of the complex. We have long parted with the idea that the movements in the clavicular joints are successive to those of the glenohumeral articulation. Fick[6] and others have taught us that they are concomitant. The question is now how much do they add to the range of the shoulder joint and in what ratio?

1. The increments

The share which the sternoclavicular joint has in the movements of the entire arm was studied by Braune and Fischer.[2] The sternoclavicular increment is especially great in the lateral abduction, while in the sagittal elevation the acromioclavicular joint adds most materially (47°) to the range.

It is evident that elevation in the sagittal plane is more free than abduction in the frontal plane. There are reasons for this aside from the amplitude of the auxiliary joints. A limit is set to the rotation of the scapula in frontal plane abduction by the lumbodorsal fascia and by the muscles covering the lower angle, especially the upper border of the latissimus. This obstacle does not exist in forward elevation which also has the advantage of the greater amplitude of the acromioclavicular joint in sagittal direction.

Another factor already mentioned by Strasser[13] is the length rotation of the clavicle. Steinhausen[14] states that this rotation of the clavicle begins at the

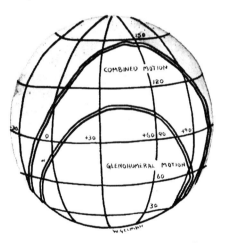

FIG. 17. Combined ranges of the shoulder arm complex.

horizontal position and lasts until 150° is reached. As mentioned above, this point has been studied lately by Inman and Saunders.[9] The clavicle is held to the scapula by means of the coracoclavicular ligament. Elevation of the clavicle in abduction or forward elevation of the arm is first carried out in the sternoclavicular joint. This movement begins early and is almost complete during the first 90°, during which time there are 4° of clavicular elevation (about a sagittal axis) for every 10° of abduction or elevation of the arm. Then, the tension of the ligament prevents further rotation of the scapula unless in some way the ligaments can become relaxed. This is accomplished by the longitudinal rotation of the clavicle. Inman[9] demonstrated the existence of such clavicular rotation about a length axis in the living subject and found it to be up to 50°.

The *combined ranges of the entire shoulder-arm complex* in the living are as follows (Fick[6]) (Fig. 17):

A) FORWARD ELEVATION

The range amounts to 155-195°, with an average of 180°, of which 120° originates in the glenohumeral joint and 60° in the clavicular joints. Straight forward elevation does not change the position of length rotation of the humerus. The bi-epicondylar axis retains its frontal position. Only beyond the horizontal there is a forced inward rotation or pronation caused by the ligamentous tension.

B) LATERAL ELEVATION

Lateral elevation amounts to 150°. Movement about a sagittal axis, i.e., in the frontal plane, causes, in contrast to the straight forward elevation, an outward or supinatory rotation of the humerus.

C) BACKWARD ELEVATION

In backward elevation of the arm, the shoulder blade first remains in its usual position. Then the vertebral border of the scapula begins to move medially and the scapula goes into a frontal plane entirely upon the posterior surface of the thorax. The range of elevation in extreme backward extension of the arm does not exceed 30-40°.

2. The pairing of the muscles of the shoulder-arm complex

If we examine the patterns of combinations in the movement of the complex from the purely functional point of view, we find some definite groupings of muscles which represent stereotypes and frequently-used motor performances. The following patterns may be recognized (Mollier[12]).

A) THE UP- AND DOWNWARD SWING (GOLF SWING, THE SWING OF THE LUMBERJACK)

The upper trapezius and the pectoralis minor form a sling for the swing from upward and backward to downward and forward. The upper trapezius raises the shoulder blade and pulls it backward. The pectoralis minor then pulls it forward and downward with force.

B) THE OBLIQUE LATERAL SWING (BASEBALL BATTING)

The swing is from lateral upward to medial and downward. The midportion of the trapezius and the upper and middle serratus perform the proper rotation of the scapula. The sternoclavicular joint is the center of motion. Middle trapezius and upper and middle serratus, in rotating the scapula, provide for the initial abductory position of the arm. Then, the adductors and particularly the pectoralis minor, restrained by the rhomboids, produce the forceful adductory and forward flexory movement which follows.

c) The Straight Downward Stroke (Post Hammering)

The shoulder joint is stabilized; motion is principally for- and backward rotation of the scapula. The latter is neutralized against translatory gliding; rhomboid and levator form the force-couple with the trapezius and serratus for the rotatory movement of the scapula. The centers of rotation are both acromioclavicular and sternoclavicular joints. The arm is poised in forward flexion, the shoulder blade being aligned with it. The lower angle of the scapula is rotated forward by the trapezius and serratus. With the glenohumeral joint fixed, the downward blow is accomplished by the forceful action of the levator and rhomboids, assisted by the pectoralis minor. This motion is carried out principally in the acromioclavicular joint.

d) Postural Alignment of the Scapula

1) Shoulder shrugging. The axilla is brought to the level of the first rib. It is a pure translatory movement, the excursion of which, according to Fick,[6] amounts to 5-8 cm. The clavicle becomes elevated at its outer end. Seen in the frontal plane, the movement in the acromioclavicular joint causes the angle between clavicle and length axis of the scapula to become smaller. Toward the end of the shrugging motion a rotation of the scapula occurs by which its lower angle turns laterally and upward. It is probably the tension which the elevation of the scapula places on the serratus which is responsible for the rotatory effect.

2) The scapula in military position. In the for- and backward movement around the thoracic wall, the scapula shifts its vertebral border a distance of about 8 cm. When pulled maximally backward, the border is about 6 cm. from the vertebral spines. The movement occurs around the sternoclavicular joint, but the acromioclavicular articulation provides the adjusting of the scapula to the thoracic wall. Seen in the transverse plane, the angle between clavicle and spine of the scapula changes during this movement. When the shoulder blade is carried backward the angle decreases; in maximum backward position it amounts to 30°. The levator scapulae and the lower trapezius, while they neutralize each other in rotation of the scapula, combine their action in the translatory movement which pulls the scapula backward and upward, giving the patient a military posture. Contrarywise, in the slouched position, the shoulders fall forward by gravity and are also pulled downward and forward by the pectoralis minor. Motion occurs in both acromioclavicular and in sternoclavicular joints in reciprocal direction.

IV. SUMMARY

1. Shoulder blade, clavicle, and shoulder joint form a mechanical unit, the shoulder-arm complex, containing three individual joints—the sterno-clavicular, acromioclavicular, and the glenohumeral. All having three degrees of freedom of motion, this combination provides for an unusual degree of mobility, but it makes the problem of stabilization of the shoulder-arm complex more complicated.

2. In normal position, in which the scapula is closely attached to the thoracic wall, the clavicle forms with the frontal plane an angle of 30°. In military position, the shoulder blade is drawn backward and its medial border stands less than 5-6 cm. from the midline; in normal position, this distance is somewhat increased.

3. The atmospheric pressure on a scapula of 15 cm. length and 10 cm. width amounts to 79.8 kg. In addition, the muscles hold the bone in close contact with the trunk.

4. The sternoclavicular joint is more a saddle than a ball-and-socket joint, each of its facets being convex in one and concave in the other plane. Only the length rotatory (frontal) axis goes through the centers of the articular facets. The perpendicular and sagittal axes lie outside of the joint, somewhere in the costoclavicular ligament, so that the movements about these axes become gliding movements. An intraarticular disc divides the joint and increases its range. The joint is reinforced by the anterior and posterior sternoclavicular ligaments checking for- and backward movement of the head of the clavicle, and by the interclavicular ligament which checks the downward movement.

5. The length rotatory excursion ranges of the sternoclavicular joint is, according to Fick, 30° (50° according to Inman); that about a sagittal axis (elevation and depression), 60°, and the anteroposterior range about a perpendicular axis is 25-30°.

6. In the circumductory movements, the long axis of the clavicle describes the periphery of a cone with elliptical base. The maximum excursion is upward and forward accompanied by a length rotatory movement in pronatory sense, while the downward and backward rotation is associated with a supinatory length rotation (the posterior border of the clavicle rotating downward).

7. In the acromioclavicular articulation, the movement of the scapula in or close to the sagittal plane has the greatest range. Only in 1% an intraarticular meniscus is found. The joint is reinforced by the superior and inferior acromioclavicular ligaments checking posterior displacement, and by the coracoclavicular ligament preventing forward movement of the scapula and keeping the clavicle from gliding off the acromion. Both trapezoid and conoid portions of the ligament restrain backward movement of the shoulder blade.

8. For- and backward circling of the scapula about a vertical axis is checked by the joint capsule and the coracoacromial ligament. Ab- and adduction about a sagittal axis, amounting to only 10°, is checked by the coracoclavi-

cular ligament and the impingement of the coracoid against the clavicle. The most important movement is for- and backward swing of the scapula about a frontal axis. It amounts to 60-70° and is checked by the posterior and anterior portions of the capsule, respectively, by the trapezoid ligament, and by the coracoid process impinging against the clavicle.

9. The clavicle sustains pressure stresses along its longitudinal axis produced by the push of the trapezius or the pectoralis minor against the sternum. This is, in case of fracture, the cause of overriding. Tension stress on the clavicle is produced by deltoid and pectoralis major upon abduction and pull on the arm. Torsion stresses result usually in discontinuity of the joints. Bending stresses are produced by blow upon or depression of the clavicle which impinges against the coracoid process.

10. The scapulohumeral or glenohumeral joint consists of the humeral head with 153° and the glenoid with only 75° angular value; neck and head are angulated against the shaft 45-50° (Hultkrantz) of backward rotation in the transverse plane. In contrast to the hip joint, the atmospheric pressure plays only a minor role, amounting to not more than 6.5 kg. However, its effect can still be demonstrated after removal of the deltoid if interposition of the capsule is prevented by rotating the humerus.

11. The ligaments reinforcing the anterior capsule, the coracohumeral as well as the glenohumeral with its three portions—a superior, middle, and inferior—check outward rotation.

12. The rotator cuff is intimately interwoven with the fibrous capsule. The latter is relaxed in midposition but movements tighten the capsule in different portions: backward movement the anterior, forward movement the posterior, etc. Of the numerous bursae of the shoulder joint, only the subscapularis and the bicipital sheath communicate regularly with the joint. All bursae serve the gliding mechanism between muscle groups.

13. A plane laid through the elbow axis and through the center of the humeral head is the determination plane of the humerus in relation to the trunk. In midposition, the centers of head and glenoid coinciding, the humerus is abducted 45° and forward flexed 45°; in normal position, the arm hanging down, the elbow axis runs from forward lateral to backward medial, forming with the frontal plane an angle of 10°. The scapula forms in this rest position an angle of 30° with the frontal plane.

14. The abduction range in the glenohumeral joint is variously given from 64-90°. The closer abduction moves to the sagittal plane, i.e., the more it is combined with forward flexion or backward extension, the less is it associated with longitudinal rotation. There is no length rotation in sagittal elevation; there is a 90° rotation associated with abduction in the frontal plane by the time the point of perpendicular abduction has been reached.

15. Abduction in the glenohumeral joint is associated from the beginning with scapular rotation.

16. The flexion-extension range is variously given from 100-150°, most of

which is in the flexion range. Forward flexion is greatest in a forward diagonal plane, amounting up to 104°. Ventral elevation is checked by the posterior capsule, the coracohumeral ligament, the teres minor, and infraspinatus; dorsal elevation is checked by the anterior capsule, the supraspinatus, and the subscapularis.

17. The humerus glides backward in inward and forward in outward rotation, and downward in forward elevation. Length rotation is greatest in 180° abduction. Inward rotation is checked by impingement of the lesser tuberosity against the anterior, outward rotation by impingement of the greater tuberosity against the posterior, rim of the acetabulum.

18. The circumpolar movement is a combination of simultaneous movements in all three planes, as in describing a circle in the air or drawing a circle on the blackboard; the sequence of this motion from a position of adduction and forward flexion is abduction, supination, backward extension, adduction, pronation, and again forward flexion.

19. The movements of the scapula are translatory, up and down, for- and backward, or rotatory about axes going perpendicularly through the scapular plate, or combinations of the two. In all movements secondary muscle actions must be suppressed by antagonists if motion in a certain direction is to be accomplished.

20. Translatory up and down movement in the frontal plane is carried out by the levator, upper trapezius, and rhomboid for the upward movement; by gravity, the lower trapezius, the pectoralis minor, pectoralis major, and latissimus for the downward movement; in the transverse plane, forward movement is carried out by the anterior serratus, the pectoralis minor, and the mid-portion of the pectoralis major; the principal backward movers are the middle and lower trapezius, the rhomboids, and, indirectly, the latissimus.

21. Forward rotation of the lower angle of the scapula is accomplished by the upper trapezius and the lower serratus. The acromioclavicular portion of the trapezius inclines the head to its side and pulls it backward. The head must, therefore, be fixed by opposing muscles before the action of the trapezius on the shoulder blade can be realized.

22. Scapular rotation becomes auxiliary to the glenohumeral motion from the beginning. By the time the abduction in the shoulder joint reaches 90°, the scapula has already rotated 25-30°, and as abduction proceeds scapular rotation gradually increases. In perpendicular abduction this rotation amounts to 60°.

23. The backward rotation of the lower scapular angle is carried out by the levator, the rhomboids, and the pectoralis minor.

24. Movements in the sternoclavicular articulations are always associated with reciprocal movements in the acromioclavicular joint. Forward movement in one joint produces backward movement in the other; elevation in one, depression in the other. Only the axial movements (longitudinal) occur in the same direction in both joints.

25. The movement in the acromioclavicular joint, as those in the sterno-clavicular, supply auxiliary ranges of motion for the shoulder joint. Of these, the rotation about a frontal axis, i.e., for- and backward swing of the scapula, has the greatest range. To allow full amplitude of forward flexion, the clavicle must undergo a length rotation to relax the coracoclavicular ligament.

26. In the glenohumeral joint the principal abductor is the deltoid. The supraspinatus fixes the head against the glenoid to facilitate the abductory function of the deltoid. To full perpendicular abduction of 180°, the gleno-humeral joint contributes 120°, the rotation of the scapula 60°. The principal adductors are the pectoralis major, the latissimus, and the teres major. In ex-treme abduction, some parts of the pectoralis major come to lie above the center of the shoulder joint and, therefore, act in an abductory sense. The principal outward rotators are the supra- and infraspinatus and the teres minor. The maximum inward and outward rotation range is 90°. The main inward rotators are teres major, latissimus, and subscapularis. Forward flexors are the anterior deltoid and the pectoralis major, with additional help of the long and short heads of the biceps and the coracobrachialis. Forward flexion and internal rotation are interrelated insofar as with increasing abduction the inward rotatory action of the anterior deltoid decreases in favor of its forward flexing component, and the pectoralis major, likewise, increases its forward flexory effect with abduction. Backward extensors are the latissimus, the teres major, and the long head of the triceps. They also increase their dorsal extensory component with abduction, as do, likewise, the posterior fibers of the deltoid.

27. So far as working capacity is concerned, the inward rotators exceed the outward rotators 2 to 1, while flexors and extensors are evenly matched. On the other hand, the adductors have more than twice the working capacity of the abductors.

28. So far as the additional range of motion furnished by the auxiliary joints is concerned, the sternoclavicular increment is especially large in lateral abduction, while in forward elevation the acromioclavicular joint adds most materially (47°). The rotation of the clavicle, which is necessary for the free abduction range, begins at 90° abduction. Up to this point there are 4° of clavicular elevation to every 10° of abduction. Then, the tension of the coraco-clavicular ligament prevents further elevation unless a longitudinal rotation of clavicle occurs which relaxes the ligament (Inman).

29. The combined ranges of the entire complex are: forward elevation 180°, of which 120° are in the glenohumeral joint and 60° in the clavicular joints; lateral elevation 150°, with supinatory (outward) rotation of the humerus; backward elevation does not exceed 30-40°.

30. For stereotype motor performances, there are certain patterns of com-bined muscle action: a) up and down swing (golf swing): upper trapezius and pectoralis minor; b) oblique lateral swing (baseball batting): middle trapezius and upper and middle serratus, then adductors and pectoralis minor; c) straight downward stroke (post hammering): arm poised in forward

flexion, trapezius and serratus rotate lower scapular angle forward; the downward blow is accomplished by the forceful action of levator and rhomboids, assisted by pectoralis minor; d) shoulder shrugging: a purely translatory movement of trapezius and levator, with their rotatory component neutralized; e) military stance: levator and lower trapezius neutralize their rotatory components and combine their translatory effect to pull the scapula back; f) slouched position: the scapula is allowed to fall forward by gravity and is pulled down by the unopposed (serratus and trapezius) pectoralis minor.

BIBLIOGRAPHY

1. ALBERT, A.: Zur Mechanik des Schultergelenks des Menschen. *Wien. Med. Jahrb.* 1877.
2. BRAUNE, W., and FISCHER, O.: Über den Anteil den die einzelnen Gelenke des Schultergürtels an der Beweglichkeit des menschlichen Humerus haben. *Abh. d. Kgl. Sächs. Ges. d. Wiss.*, 1888.
3. CODMAN, A.: *The Shoulder.* Boston, T. Todd Co., 1934.
4. DE PALMA, A. F.: *Surgery of the Shoulder.* Philadelphia, J. B. Lippincott, 1950.
5. DUCHENNE, G. B.: *Physiology of Motion.* Transl. E. B. Kaplan. Philadelphia, J. B. Lippincott, 1949.
6. FICK, R.: *Specielle Gelenks- und Muskelmechanik,* Vol. III. Jena, J. Fischer, 1911.
7. FISCHER, O.: *Kinematik organischer Gelenke.* Braunschweig, Vierweg, 1907.
8. HULTKRANTZ, J. W.: *Das Ellbogengelenk und seine Mechanik.* Jena, 1897.
9. INMAN, V. T., SAUNDERS, J. B. DEMC, ABBOTT, L. C.: Observations on the function of the shoulder joint. *J. Bone & Joint Surg., XXVI:1,* 1, Jan. 1944.
10. KRAUSE, W.: Skelett der oberen und unteren Extremität. *Bardeleben's Handb. d. Anat. d. Menschen XVI:3,* 1909.
11. LANZ, T. V., and WACHSMUTH, W.: *Praktische Anatomie, Vol. I.3,* Berlin, J. Springer, 1935.
12. MOLLIER, S.: Über die Statik und Mechanik des Menschlichen Schultergürtels unter normalen und pathologischen Verhältnissen. *Festschr. f. C. v. Kupffer,* Jena, 1899.
13. STRASSER, H.: *Lehrbuch der Muskel- und Gelenkmechanik, IV.* Berlin, J. Springer, 1917.
14. STEINHAUSEN: Beiträge zur Lehre von dem Mechanismus der Bewegungen des Schultergürtels. *Arch. anat. Phys., Phys. Abt. Suppl.,* 1899.
15. WALDEYER, W.: De Claviculae articulis et functione. *Diss.* Berlin, 1861.

THE PATHOMECHANICS OF PARALYSIS
OF THE SHOULDER

I. INTRODUCTION

SINCE the sternoclavicular articulation represents the only skeletal attachment of the shoulder-arm complex to the trunk and since the clavicle interposed between trunk and scapula participates in movements carried out simultaneously in both the sternoclavicular and acromioclavicular articulations, it becomes obvious the loss of any of the principal muscles which act on either the scapula or the clavicle must have a detrimental effect on the shoulder-arm complex as a whole.

Even in upright position, without any movement of the extremity, active muscle force is necessary to hold the arm by the side of the body because, in contrast to the hip joint, the atmospheric pressure alone is unable to do so. In this position the clavicle can still be depressed which proves that in vivo the arm complex is, at least partially, carried by the thoracoscapular muscles (Strasser[15]). The equilibrium of the shoulder-arm complex in any position depends not only upon the muscles which produce the position but also upon their antagonists which stabilize it; to make the motor performance direct and precise depends largely upon antagonistic control. For example, the elevation of the acromion by trapezius and serratus lacks precision and guidance without the controlling and restraining influence of the opposing rhomboids and pectoralis minor.

The shoulder girdle is able to carry out its full movements independent of those of the arm; but the full range of arm movement requires the supplementary ranges of the sternoclavicular and acromioclavicular joints. To obtain the best abduction range in the scapulohumeral joint the scapula aligns itself with the plane of motion. The abduction field of the humerus is greatest in the forward oblique plane, and it is the function of the scapulothoracic muscle to place and to hold the shoulder blade in this plane; in fact, movement in the shoulder joint is facilitated in any plane by the ability of the scapula to align itself to this plane.

For this reason one may expect difficulties in the function of the shoulder-arm complex from the paralysis of almost any of the important muscles. Paralysis of the rhomboids has an effect on the abduction of the arm, even though in this performance the muscle plays only a passive role as stabilizing antagonist.

In general, it can be stated that the loss of any of the scapulothoracic muscles, quite aside from the effect it has on the movement of the scapula itself, necessarily impedes and restricts the movement of the glenohumeral articulation.

II. THE PARALYSIS OF THE THORACOSCAPULAR MUSCLES OF THE SHOULDER GIRDLE COMPLEX

A. PARALYSIS OF THE TRAPEZIUS

In the Vermont epidemics of 1894-1922, the trapezius was found paralyzed in 49 of 1452 cases, and it was second in frequency of the shoulder girdle muscles affected.

It will be recalled that the clavicular portion of the muscle inclines the head toward the side and pulls it backward; only when the limit of this movement is reached or when the head is held stabilized can this portion of the muscle elevate clavicle and shoulder blade. The lateral half of the middle portion elevates the acromion and moves the inferior angle of the scapula away from the midline; the medial half of the middle portion pulls the scapula toward the midline and the inferior portion of the muscle pulls the scapula down.

1. At rest

The paralysis of the trapezius has the following effect: when the shoulder is at rest, the scapula is depressed and the acromion droops forward. The sternum deviates to the unaffected side. With the forward drooping of the acromion the vertebral border of the scapula moves away from the midline 10-12 cm., compared with the normal distance of 6 cm. between the spinous processes and the vertebral border of the scapula. If the lower portion is also paralyzed the scapula rotates, its lower angle moves medially, and its upper outer angle downward and laterally (Duchenne[1]).

2. In motion

a) When the arm is elevated in a forward plane the usual backward rotation of the acromion is missing. It is normally the function of the middle trapezius to see to the scapular rotation. If it is lost, the patient fatigues easily on forward elevation of the arm. Since only the serratus is left to rotate the lower angle of the scapula forward, the total rotatory range is no more than half of normal.

b) When the arm is elevated in the frontal plane, the difficulties are still greater. In total paralysis of the trapezius the unopposed rhomboids and the levator become hypertrophic. The arm cannot be elevated much beyond 75°.

c) Straight forward movement of the shoulder blade in absence of the trapezius is carried out principally by the pectoralis major, assisted by the pectoralis minor, which is neutralized by the serratus against downward rotation of the scapula.

B. THE PARALYSIS OF THE SERRATUS ANTERIOR

This muscle acts with the trapezius as a couple of forces rotating the lower angle of the shoulder blade outward and forward when the arm is elevated. As the inferior portion of this muscle rotates the scapula, the upper inner angle remains fixed while the acromion is elevated and the lower angle is carried forward and outward. The rhomboid and the levator restrict this movement and as they come under tension, the scapula moves upward under the effect of the serratus and the midportion of the trapezius (Duchenne[4]). In addition, the anterior serratus cooperates with the pectoralis major in forward movement of the shoulder.

1. At rest

The loss of this muscle makes the position of the scapula insecure. The acromion remains elevated by the upper and middle trapezius. Unopposed by the serratus, the rhomboid and the levator rotate the shoulder blade so that its

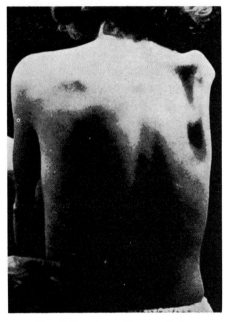

Fig. 1. Serratus anticus paralysis. Winged scapula. (Steindler)

Fig. 2. Combined paralysis of trapezius and serratus. The inferior angle (1) is almost at the level of the lateral angle (2). The superior angle (3) is carried outward and upward. (Redrawn from Duchenne)

lower angle approaches the vertebral spines and its vertebral border becomes downward oblique. Since the muscle no longer holds the scapula firmly to the thorax, a "winged" shoulder develops, especially when the arm is carried forward in a sagittal plane.

2. In motion

On forward elevation of the arm, the absence of the muscle is definitely disturbing because only the trapezius is left for the rotation of the scapula. "Winging" of the scapula becomes very marked and the arm cannot be elevated above 130-140°; the scapula itself rotates no more than 20°. Further elevation of the arm has to be accomplished by backward and side bending of the trunk (Fig 1).

In frontal elevation less difficulty is encountered because the scapula is held backward closer to the thoracic wall by the trapezius, which becomes hypertrophic (Strasser[15]).

C. THE COMBINED PARALYSIS OF TRAPEZIUS AND SERRATUS

In this combination the acromion drops so low and the lower angle rises so high that both may lie in the same horizontal level. The acromion is carried far outward. This position becomes greatly aggravated when attempting to elevate the arm (Duchenne[4]). The intact rhomboids and the levator become hypertrophied. If the anterior portion of the trapezius has escaped paralysis, there is no drop of the shoulder blade but it is sharply rotated about its upper outer angle and the vertebral border approaches horizontally (Fig. 2).

D. THE PARALYSIS OF THE RHOMBOIDS

This muscle shows considerable force in drawing the shoulder back and, in connection with the lower trapezius, maintains the vertebral border of the scapula firmly against the posterior thoracic wall. Loss of the muscle causes the lower angle of the scapula to rotate outward under the action of the serratus, which is no longer opposed by the rhomboids. With the absence of this muscle, the pectoralis minor loses its antagonist and is free to pull the scapula forward; and trapezius and serratus lose a stabilizer of the shoulder blade so that they also are handicapped in the upward rotation of the scapula as the arm is elevated. In combination with levator paralysis, the shoulder blade also loses an important elevator and medial adductor. In combination with the pectoralis minor, the rhomboids rotate the scapula downward lowering the acromial angle. If both muscles are paralyzed, the scapula assumes a position of abduction and upward rotation. It is, therefore, no longer able to add to the abduction range of the shoulder point by rotation.

III. THE PARALYSIS OF THE SCAPULOHUMERAL MUSCLES

A. PARALYSIS OF THE DELTOID (FIG. 3)

This muscle is the principal abductor of the arm. From the hanging position, the middle portion of this muscle abducts in straight frontal plane; the anterior portion abducts the arm in a forward oblique, and the posterior in a backward oblique plane. The maximum elevatory range, however, is that of the anterior portion, namely 90°, while the posterior abducts the arm hardly to 45°. The question whether other muscles start abduction and the deltoid merely continues it has been discussed above. Suffice it to say that the stabilizing effect of the supraspinatus is necessary for abduction. This is shown by the fact that a complete tear of the supraspinatus tendon makes active abduction entirely impossible, while passive abduction is completely unrestricted.

When, in the course of the abduction, does the rotation of the scapula start? It was already known to Duchenne[4] that the instant the arm is abducted from the thorax the acromion rises, instead of descending. Were the deltoid left alone to accomplish this abduction, its contraction would have to pull the acromion against the humerus precisely in inverse proportion to the masses represented by the arm and the shoulder girdle, respectively. The rotators of the shoulder blade, i.e., the trapezius and serratus, not only add to the abduction range by rotating the scapula, but in earlier phases of abduction, when their rotatory effect is still secondary, they must, together with the antagonistic rhomboids and pectoralis minor, stabilize the scapula against the thorax so as to make it a fixed point against which the deltoid can exert its abductory effect. It is, therefore, not correct to say that the action of the outward rotators gradually increases with abduction; it is only the rotatory effect of these muscles which becomes dominant as abduction progresses. The stabilizing effect these muscles have on the scapula in cooperation with their dynamic antagonists must precede or be simultaneous with any abductory action of the deltoid. It is for this reason that paralysis of the rhomboid and pectoralis minor, although they are downward rotators of the scapula, actually interferes with abduction, especially if carried out against considerable external resistance.

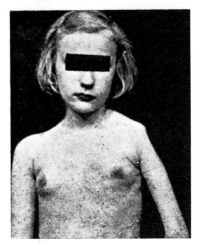

FIG. 3. Paralysis of the deltoid.
(Steindler)

Of all the muscles of the upper extremity, the deltoid is the one most frequently paralyzed. The frequency is given for the Vermont epidemics (Lovett[10]) as 57 in 1417 cases; the relation of partial to total paralysis was found to be 3.7 to 1.

Most frequently paralyzed is the midportion. Yet, even in absence of this portion, the arm can still be elevated in the frontal plane by the synergism of the anterior and posterior portions which mutually neutralize their forward and backward flexory components. If the anterior portion is paralyzed, the arm can still be elevated in the frontal and oblique backward, but not in the forward or oblique forward planes. When the posterior portion is paralyzed, the patient has difficulty in elevating the arm backward; in placing his hand in the hip pocket, for instance, or moving it toward the back above the gluteal region, etc. (Duchenne[4]).

B. PARALYSIS OF THE SUPRASPINATUS

As an auxiliary muscle for the elevation of the arm its main function is to stabilize the humeral head against the glenoid. For active abduction its lever arm is too small to be of great assistance to the deltoid. As a stabilizer of the head, on the other hand, its function is, as we have pointed out, indispensable for the abductory effect of the deltoid. When the deltoid is paralyzed but the supraspinatus acting, there is very little displacement of the humeral head;

but if both muscles are paralyzed, the head subluxates. The supraspinatus, therefore, has a definite role in maintaining the equilibrium of the arm against gravitational forces (Inman[7]). The supraspinatus is also an outward rotator, though less so than the infraspinatus and the teres minor. According to Strasser,[15] it is likewise a considerable backward extensor of the arm in all positions.

C. THE PARALYSIS OF THE INFRASPINATUS AND THE TERES MINOR

These are the principal outward rotators of the humerus, irrespective of the position of the arm. Paralysis of the infraspinatus is rare (17 in 1417 cases of infantile paralysis, Vermont epidemic[10]), but when they are paralyzed the lack of outward rotation becomes very disturbing. This is particularly apparent in writing. Duchenne[4] observes that when outward rotation is absent the hand in writing can trace out the first letters but as soon as one or two words are written the patient cannot continue and is forced to stop. To continue writing he must pull his paper from right to left with the opposite hand.

D. THE PARALYSIS OF THE SUBSCAPULARIS

We can dispense with the secondary function of this muscle which is forward flexion of the upper and extension of the middle and lower portions because these are negligible. On the other hand, all three portions are strong inward rotators. Still, the loss of this muscle causes little difficulty because of the strong inward rotatory effect of the thoracohumeral muscles, namely, the pectoralis major, the latissimus dorsi, and the teres major.

E. PARALYSIS OF THE CORACOBRACHIALIS AND SHORT HEAD OF THE BICEPS, LONG HEAD OF THE BICEPS, AND THE TRICEPS

Both coracobrachialis and short head of biceps have a slight adductory and forward flexory component, the loss of which has very little effect upon the mobility of the humerus. On the other hand, the stabilizing effect which the long head of the triceps and the long head of the biceps exert on the shoulder joint is not to be underestimated. Together with the supraspinatus, these last two muscles may press the humerus against the glenoid fossa sufficiently so that humerus and scapula move as one unit. The arm can then be abducted to some degree by the rotators of the scapula.

IV. THE PARALYSIS OF THE THORACOHUMERAL MUSCLES

These muscles move not only the humerus against the scapula but indirectly also, by means of their humeral insertion, the entire shoulder girdle against the thorax. To do this, the humerus must be first fixed by mutually neutralizing all components which produce movement of the humerus in the glenohumeral articulation. Only two muscles need to be considered: the pectoralis major and the latissimus dorsi.

A. PARALYSIS OF THE PECTORALIS MAJOR

It is a powerful adductor, forward flexor and inward rotator of the arm. Paralysis of the muscle is not infrequent (29 in 1417 cases with the ratio of partial to total as 3.1 to 1). In the adduction of the arm from an elevated position, gravity plays a definite role so long as the rate of adduction does not exceed the gravitational acceleration and is not more forceful than the weight of the falling arm would provide. Consequently, the patients are not handicapped in ordinary tasks involving the dropping of the elevated arm. But when the arm is brought to the side sharply and with force, such as in swinging a club or an axe, then the loss of the muscle if definitely noticed; nor are the patients able to hold a briefcase tightly under their arm. If the clavicular portion is paralyzed, there is also some difficulty in bringing the arm forward.

B. THE PARALYSIS OF THE LATISSIMUS DORSI

The upper portion of this muscle draws the arm backward, adducting it at the same time, while the lower portion pulls the shoulder down; in depressing the shoulder it keeps the vertebral border of the scapula parallel to the spine, maintaining the trunk in a straight position. Duchenne[4] observed that when the latissimus was paralyzed the military position could not be maintained without elevating the shoulders because the elevatory action of the trapezius and of the levator was unopposed.

C. THE COMBINED PECTORALIS MAJOR AND LATISSIMUS PARALYSIS

Both muscles depress the shoulder by means of pull on the humerus. Both are adductors and inward rotators of the shoulder joint. Consequently, when they are both paralyzed and the upper and middle thirds of the trapezius are intact, the shoulders are elevated and the trapezius, contracting progressively, strongly resists any depression of the shoulder. Adduction of the arm can not be carried out with force but inward rotation of the humerus can still be accomplished by the subscapularis, though it lacks force and leads to early fatigue.

V. THE REVIEW OF RECONSTRUCTIVE OPERATIONS FOR PARALYSIS OF THE SHOULDER GIRDLE SYSTEM FROM THE KINETIC POINT OF VIEW

A. OPERATIONS FOR PARALYSIS OF THE TRAPEZIUS

There are two objectives: the fixation of the shoulder blade and the active rotation of the scapula. The material on hand for transplantation consists of the rhomboids and the levator. The drooping shoulder must be pulled upward and backward. This is accomplished by transplanting first the levator from the upper inner angle to behind the acromion, thereby elevating the acromial angle of the scapula. Secondly, the rhomboids, which rotate the scapula downward, are transplanted laterally into the shoulder blade to give them a more transverse direction and to accentuate their backward pulling of the scapula (Lange[8]) (Fig. 4). The levator acts as rotator only to a moderate extent because it has less strength and a lesser contractile distance than the trapezius. Both of the muscles, however, stabilize the scapula which is of great advantage for the action of the serratus anterior. Lange[8] reports that this gives excellent and lasting results in controlling the displacement of the shoulder blade. It is primarily a stabilizing operation.

Another method is one recently published by Dewar and Harris,[3] who add to the transplantation of the levator scapulæ the fasciodesis of the upper inner angle to the second and third dorsal spines. From the mechanical viewpoint,

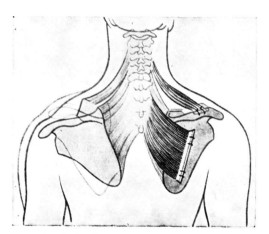

Fig. 4. Paralysis of trapezius. Transplantation of rhomboids and levator scapulae. (Max Lange)

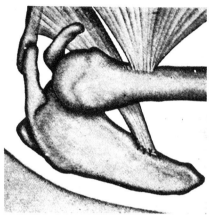

Fig. 5. Paralysis of the anterior serratus. Substitution by transplantation of clavicular portion of pectoralis major to lower angle of scapula. (Steindler)

the advantage is that the position of the scapula is an abductory one from the beginning which facilitates the effect of the forward transplanted levator. Satisfactory results are reported.

B. THE OPERATIONS FOR THE PARALYSIS OF THE ANTERIOR SERRATUS

An isolated paralysis of the anterior serratus is unusual in anterior poliomyelitis, in contrast to the traumatic type due to injury of the long thoracic nerve. Both for anatomical and kinetic reasons, the situation is complicated. One must distinguish between the requirement of stabilizing the scapula and that of substituting the power of upward rotation. The first objective is comparatively easier to obtain. Whitman[16] secured the vertebral border of the scapula against the spinous processes of the fourth to seventh dorsal vertebræ by fascia lata, which was passed through drill holes at the vertebral scapular border and was fastened to the spinous processes. Others immobilized the scapula by fastening its lower angle to the ninth to tenth ribs. The benefit of the fasciodesis is to stabilize the scapula without preventing its being upward rotated by the trapezius, an advantage which is not shared by the methods of scapulopexy where the shoulder blade is firmly anchored to the ribs. The only disadvantage of the fasciodesis seems to be in forward elevation of the arm since the shoulder blade is kept from aligning itself to the plane of the arm.

The second requirement is the substitution of the active upward rotator of the scapula by muscle transference. It would seem that the pectoralis muscle, especially the major, should be the most suitable, and F. Lange,[9] among others, has used this muscle, separating it from its insertion at the humerus and implanting it into the lower angle of the scapula. A similar method is that of Hass who transplants the clavicular portion of the pectoralis major to the lower angle of the scapula (Fig. 5). Chavez[2] isolates the pectoralis minor from the coracoid process and implants it by means of fascial sutures near the vertebral border of the scapula at the junction of the upper and middle thirds. As a result of the stabilization of the scapula, the winging disappears when the arm is carried into the forward elevation.

C. THE OPERATION FOR THE PARALYSIS OF THE DELTOID

Before entering into the discussion of substitutionary operations for paralysis of the scapulohumeral muscles it should be emphasized that the stabilization of the scapula as obtained by the above-mentioned methods is a major prerequisite. Failing this, the action of the muscle, for instance, the deltoid, must take effect on both ends, rotating the scapula against the humerus as well as the humerus against the scapula. Provided the rotation moments (depending on angle of application and lever arm) are equal at both ends, the rotatory acceleration so produced will be inversely proportional to the respective masses of the humerus and the shoulder girdle. In other words, so long as the proximal arm of the lever system, in this case the scapula, is not fixed, a great deal of the motor effort will be lost on moving the proximal portion. It is, therefore, important to see to it that the scapula is duly secured, either by its own muscles if they are intact or by proper substitutionary or stabilizing methods.

1. Substitutionary methods

It has been pointed out that in paralysis of the middle portion the anterior and posterior acting synergistically can abduct the arm. Furthermore, a strong contraction of the long heads of biceps and triceps may secure the head of the humerus so firmly against the glenoid that, for a short excursion at least, humerus and scapula may act as one unit and abduction can be brought about by the scapulothoracic muscles rotating the scapula. In our experience the latter maneuver is usually inadequate for practical use. Since there is no other muscle capable of abducting the shoulder joint, the muscle, if completely paralyzed, must either be supplanted or the scapulohumeral motion must be eliminated by arthrodesis.

A) Substitution by the Pectoralis Major

The long history of this type of transplantation goes back to Hildebrand.[6] He used the entire pectoralis major which he folded over and transplanted to the lateral one-third of the clavicle and to the acromion. Because of certain anatomical difficulties, principally connected with the nerve supply, the method has gradually fallen into disuse, although using the clavicular part only there is less danger of injuring the anterior thoracic nerves and better results should be obtainable.

The question is, however, whether this clavicular portion is adequate for substitution either in point of strength or in contractile length. The clavicular part of the muscle is the weakest; it has a cross section of only 3 cm.[2] and a contractile distance of 4 cm. The acromial portion of the deltoid which it supplants has a cross section area of 22 cm.[2] and a contractile length of 6 cm.

B) The Substitution by the Trapezius

This idea goes back to Hoffa (1901) and was used by F. Lange[9] in cases of birth palsy. It was later revised by Mayer[12] who greatly refined the method, using a fascial cuff and silk sutures to elongate the tendon. Since the contractile length of the midportion of the deltoid is only 6 cm., and the midportion of the trapezius is of adequate strength to supplant the deltoid, the prospect of this muscle exchange should be good from the kinetic point of view, provided always that a straight line of pull can be obtained. The difficulty lies in providing a smooth gliding apparatus and in preventing adhesions. If this fails and the pull transmitted by the muscle is intercepted by adhesions, the operation acts merely as a tenodesis.

C) The Substitution by the Latissimus Dorsi and Teres Major

The method of Bastos Ansart[1] uses for substitution of the posterior portion of the deltoid the latissimus dorsi and the teres major which are implanted into the greater tuberosity. There seems to be insufficient myokinetic basis for the transposition of these two muscles from the crest of the lesser tubercle to the greater except in cases where the middle and anterior portions of the deltoid still have some power left, although the author reports 11 cases, all with good results.

D) The Substitution by the Short Head of the Biceps and the Long Head of the Triceps

The operation of Ober[13] uses the long head of the triceps and the short head of the biceps which are both inserted in the acromion. It is a combined transplantation. The idea of a double substitution of the deltoid can be traced back to Spitzy[11] and to Mau[11] who used the pectoralis minor and the trapezius, and to Bastos Ansart[1] who used the pectoralis major, the teres major, and the latissimus dorsi. So far as the method of Ober is concerned, the long head of the triceps which is an auxiliary adductor of the humerus (Duchenne[4]) has a maximum contraction length of 5.2 cm., so that in this respect there can be no objection to their use. Of all the double substitution methods, this appears to be the most rational from the kinetic point of view.

Multiple transplantations for the paralysis of the deltoid with paralysis of the external rotators are reported by Harmon.[5] External rotation is furnished by transplantation of the latissimus and teres major to the lateral bicipital groove; abduction is provided for by shifting the origin of the posterior deltoid, if it is preserved, anteriorly and of the clavicular portion of the pectoralis major to the acromion; in addition, the long head of the triceps and the short head of the biceps are transplanted to the acromion following Ober's technique. There seems to be enough slack of both the auxiliary and the anterior thoracic nerves to permit the shifting of the respective muscles.

2. The arthrodesis of the glenohumeral joint for paralysis of the deltoid

Investigating the different substitutionary operations for deltoid paralysis, a committee of the American Orthopedic Association, in 1942, arrived at the conclusion that the functional results of the arthrodesis are more satisfactory on the whole than those of muscle transplantation, including the methods of Mayer,[12] and Ober[13] which from the kinetic point of view are the most promising.

Before discussing the several methods of arthrodesis of the shoulder joint from the kinetic point of view, it must be made clear what the mechanical conditions are under which the arthrodesis can work as a substitute for the paralyzed deltoid.

1) Because transplantation methods have proved practical in cases of partial paralysis of the deltoid, the arthrodesis should be reserved for severe or total paralysis of this muscle.

2) Because the rotators of the scapula become the real abductors of the arm when the glenohumeral joint is fused, the power of these muscles, especially the trapezius and anterior serratus, should be unimpaired or at least sufficient power should be left for the upward rotation of the scapula.

3) Because every upward rotation of the scapula requires for guidance and control the antagonistic action of the rhomboids and the levator, it is desirable though not indispensable that these muscles also be intact.

4) Because the acromioclavicular and sternoclavicular articulations take over the function of the shoulder joint when the latter is arthrodesed, their combined excursion ranges must be adequate.

5) Finally, because these combined excursion ranges, while adequate, naturally add up to less than that of the entire shoulder girdle, the optimum position for the arthrodesed shoulder must be carefully selected. This should be the midposition of the normal range of the two auxiliary joints, the sternoclavicular and the acromioclavicular articulations.

A number of modifications have been devised to improve on the simple intraarticular arthrodesis by denudation and wire or chromic sutures, although we have found the latter to be entirely adequate.

From the kinetic point of view, the most important aspects are:

1) The position of the arm.

2) The possibility of increasing the range of the two auxiliary joints.

ad 1) To be able to bring the arm strictly to the side of the body is a necessary requirement. An arm which stands off because it cannot carry out com-

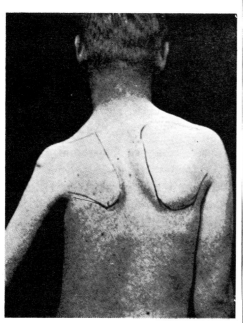

Fig. 6. Arthrodesis of shoulder in excessive abduction. Inability to adduct completely. (Steindler)

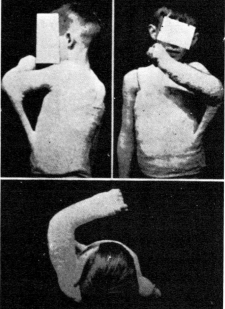

Fig. 8. Position of arm in cast following arthrodesis of shoulder. (Steindler)

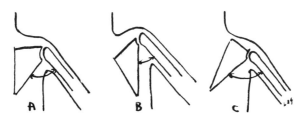

Fig. 7. Position of scapula and humerus in arthrodesis of the shoulder joint. The scapula must be held in midposition (A). The arm is the same distance from the thorax in each instance but in (B) the scapula is elevated, and in (C) the scapula is depressed.

plete adduction denotes a failure (Fig. 6). It follows, therefore, that the joint must not be in too much abduction; 70° for the young and 60° for the adult. In judging the degree of abduction which is to be obtained, the scapula should be held in midposition between elevation and depression. The angle of abduction must be measured between the axis of the humerus and the vertebral border of the scapula and not the side of the thorax. When the scapula is elevated this angle is actually smaller, and when it is depressed it is actually greater than the angle formed between the side of the thorax and the humerus (Fig. 7).

For practical purposes, the favored plane of motion of the arm is not the frontal but a forward diagonal plane with the humerus in slight inward rotation. This imposes a position of about 45° forward flexion and about 20-25° inward rotation. The positions in these three planes combine themselves to what is called the position of salute (Fig. 8).

ad 2) Can the range of the two auxiliary joints, the acromio- and the sterno-clavicular, be increased without damage to the stability of the shoulder girdle? The only increase obtainable is by resection of the acromioclavicular joint up to, and exclusive of, the coracoclavicular ligaments. To go beyond this, or even to take out the whole clavicle, imperils, we think, the stability of the shoulder girdle which is so necessary to carry out movements of the arm with force. Since we have made it almost a routine to add the acromiclavicular resection to the arthrodesis of the shoulder joint, we find that it adds to the range and that the arthrodesed humerus can, therefore, be given a more abductory position.

VI. SUMMARY

1. The equilibrium of the shoulder girdle complex depends not only on the action of the muscles producing motion, but even more so upon their antagonists. The very lability of this system of three articulations joined to the thorax by a single small joint makes the problem of providing fixed points of application for the muscles carrying out visible motion a difficult one. For this reason, the performance of a muscle is always impeded when its antagonist is paralyzed, although the muscle itself may be intact.

2. To obtain the most favorable position for movements in the shoulder joint, the scapula must align itself with the humerus in the plane in which movement is contemplated. Such alignment requires full control of all muscles needed to hold the scapula immobilized in this particular plane. Hence the loss of any of the scapulothoracic muscles in itself impedes and restricts motion in the glenohumeral joint.

3. Paralysis of the trapezius causes the shoulder to be lowered and the acromion to droop forward. In forward elevation the backward rotation of the acromion is missing. Only the serratus is left to rotate the lower angle forward and the patient fatigues easily. The difficulties are increased on elevation of the arm in the frontal plane.

4 Paralysis of the anterior serratus causes the acromion to remain elevated by the upper and middle trapezius. On carrying the arm forward in the saggittal plane, a winged shoulder develops. Elevation of the arm is not possible beyond 130-140°. Less difficulty arises on elevation in the frontal plane because the scapula is held closer to the thoracic wall by the trapezius. In combined paralysis of the trapezius and the serratus the acromion drops low, the lower

angle rises, so that both lie almost in a horizontal plane.

5. In paralysis of the rhomboids, the lower scapular angle rotates outward; the unopposed pectoralis minor pulls the scapula forward. If both pectoralis minor and rhomboids are paralyzed, the scapula assumes a position of abduction and upward rotation.

6. In paralysis of the deltoid, active abduction is gravely impaired although tension of the biceps and triceps may secure the humerus so firmly in the socket that it can move with the scapula as one unit for a short distance, provided the rotators of the scapula as well as their stabilizing antagonists are intact. Even then, the abduction is not adequate for practical purposes. If only the mid-portion is paralyzed the arm can be abducted by the anterior and posterior portions acting together.

7. Paralysis of the supraspinatus interferes with the abduction by the deltoid because of lack of fixation of the humerus in the glenoid fossa.

8. Paralysis of the outward rotators, infraspinatus and teres minor, is disturbing for the sweep of the hand, for instance, in writing; that of the inward rotating subscapularis is less disabling because this muscle is amply substituted by the inward rotating thoracohumeral muscles, the pectoralis major and the latissimus dorsi.

9. In paralysis of the pectoralis major, the arm loses a powerful adductor, forward flexor, and inward rotator. Gravity comes to the aid, but forceful pressing of the arm against the body is impossible. When the latissimus is paralyzed, a military position cannot be maintained without elevating the shoulders by the trapezius. In combined paralysis of pectoralis major and latissimus the shoulder is kept elevated by the trapezius, which contracts and strongly resists depression of the shoulder.

10. The operation for paralysis of the trapezius has two objectives: fixation of the scapula and procurement of active rotation. Transplanting the levator to behind the acromion and the rhomboids laterally into the scapula provides elevation and backward pull of the shoulder blade as well as a moderate amount of rotation (M. Lange). To this method Dewar and Harris add the fasciodesis of the upper inner angle to the second and third dorsal spines.

11. To stabilize the scapula in paralysis of the serratus, the vertebral border can be secured to the fourth to seventh dorsal vertebrae by fasciodesis, which is preferable to fastening the lower angle to the ninth or tenth rib because it does not prevent rotation of the scapula by the trapezius. For substitution of motion, the best choice is the pectoralis major, separated from its insertion and anchored to the lower angle of the scapula (F. Lange, Hass). Chavez uses the pectoralis minor, which is detached from the coracoid process and sutured to the vertebral border.

12. Prerequisite for operative procedures in deltoid paralysis is stabilization of the scapula. As substitutes have been used the clavicular portion of the pectoralis major (Hildebrand), the trapezius (Hoffa, Mayer), the latissimus (Bastos Ansart), and the short head of the biceps and the long head of the

triceps (Ober). Of these, the substitution by biceps and triceps seems to be the most rational. The substitution by the trapezius, thoroughly acceptable from the kinetic viewpoint, often fails because of lack of proper gliding facilities; the method then acts as tenodesis.

13. The arthrodesing operation is to be reserved for cases of severe or total deltoid paralysis. It requires well-functioning trapezius and serratus muscles and, preferably, intact rhomboids and levator. A number of modifications of the simple denudation arthrodesis have been advanced, but the simpler operation proves entirely adequate provided the indications are properly met. The operation should not be carried out before the age of 10.

From the mechanical angle, the position of the arthrodesed joint is an essential point. It must be the midposition of the normal range of the shoulder joint in all three planes, as follows: abduction, 70° for children, 60° for adults; 45° forward flexion and 20-25° inward rotation: these positions combine themselves to what is called the position of "salute." An increase of the range of motion, especially abduction, can be obtained by resecting the acromial end of the clavicle up to the coracoclavicular ligament.

BIBLIOGRAPHY

1. ANSART, BASTOS: Die Myoplastik bei der Paralyse des Deltoideus. *Ztschr. Orth. Chir.,* *48:*57, 1927.
2. CHAVEZ, J. P.: Pectoralis minor transplanted for paralysis of the serratus anterior. *J. Bone & Joint Surg., XXXIII:*B.2, 228, May 1951.
3. DEWARD, F. P., and HARRIS, R. I.: Restoration of function of the shoulder blade following paralysis of the trapezius by fascial sling fixation and transplantation of the levator scapulae. *Am. J. Surg., 132:*6, 111, Dec. 1950.
4. DUCHENNE, G. B.: *Physiology of Motion.* Transl. E. B. Kaplan, Philadelphia, J. B. Lippincott, 1949.
5. HARMON, P. H.: Surgical reconstruction of the paralytic shoulder by multiple muscle transplantation. *J. Bone & Joint Surg., XXXII A:*3, 583, July 1950.
6. HILDEBRAND, A.: Über eine neue Methode der Muskeltransplantation. *Arch. klin. Chir., 78:*75, 1906.
7. INMAN, V. T., SAUNDERS, J. B. DE CM, and ABBOTT, L. C.: Observations on the function of the shoulder. *J. Bone & Joint Surg., XXVI:*1, 1, Jan. 1944.
8. LANGE, M.: *Orthopaedisch-Chirurgische Operationslehre.* München, J. F. Bergmann, 1951.
9. LANGE, F.: Die Entbindungslähmung der Arme. *Münchner Med. Wchnschr., 1421,* 1912.
10. LOVETT, R. W.: *The Treatment of Infantile Paralysis.* Philadelphia, P. Blakiston & Co., 1916.
11. MAU, C.: Kombinierte Muskelplastik bei Deltoideus Lähmung. *Verh. dtsch. orth. Ges., 22:*236, 1927.
12. MAYER, L.: Transplantation of the trapezius for paralysis of the abductors of the arm. *J. Bone & Joint Surg., IX:*412, July 1927.
13. OBER, F.: An operation to relieve paralysis of the deltoid. *J.A.M.A., 99:*2182, Dec. 24, 1932.
14. SPITZY, H.: Aussprache zur Deltoideuslähmung, Muskelplastik. *Verh. dtsch. orth. Ges., 22:*239, 1927.
15. STRASSER, H.: *Lehrbuch der Muskel- und Gelenkmechanik, Vol. IV,* Berlin, J. Springer, 1917.
16. WHITMAN, A.: Congenital elevation of the scapula and paralysis of the serratus magnus muscle. *J.A.M.A., 99,* 1932.

Lecture XXVIII

THE MECHANICS OF THE ELBOW JOINT

THE joint is subservient to the function of the hand in the sense that it enables hand and fingers to be properly placed in space. Two types of mobility are necessary to comply with this requirement. The first and principal one takes care of shortening and lengthening of the extremity so that the hand can be brought in the sagittal and frontal planes of the limb. The other movement adjusts the position of the hand in relation to a transverse plane by pro- and supination of the forearm. If any external resistance immobilizes the extremity so that the forearm and hand are the fixed part of the lever system, then the joint undergoes correspondingly reciprocal movements in which the upper arm and rest of the body present the movable portion of the lever system. Both situations must be given attention in the analysis of the elbow motion; free movement of the forearm against the humerus (open kinetic chain), and the movement of the body against the fixed forearm (closed kinetic chain).

I. THE ANATOMICAL CONSTRUCTION OF THE ELBOW

The joint has two degrees of freedom of motion. It is a trochoginglymus, i.e., a joint which permits a hinge motion in one plane and axial rotation in another. The proximal constituents of the joint are, on the ulnar side, the trochlea; on the radial side, the capitulum of the humerus. The distal constituents are the bifurcated upper end of the ulna and the head of the radius.

A. JOINT CONSTRUCTION

1. The trochlea

The trochlea is a hyperboloid which means a body which is generated by the rotation of a hyperbola about a tangent or a parallel to a tangent as axis. The

FIG. 1. Trochlea and capitulum of humerus. (Fick)

surface produced thereby has the shape of an hourglass and is concave in the frontal and convex in the sagittal plane, where it forms almost a complete circle. Only about seven-eighths, i.e., 320°-330° of the trochlea is covered by cartilage because the anterior and posterior cartilaginous surfaces are separated by a bony wall which forms the floor of the olecranon fossa. The medial free border of the trochlea is, according to Fick,[5] not a true circle but it is a helical line having a slant directed radially. The roll of the trochlea is separated from the capitellum of the humerus by a groove of various depths (Lanz and Wachsmuth[8]) (Fig. 1).

2. The capitulum humeri (eminentia capitata)

This is a completely spherical body (Fig. 1) although óf not quite constant radius since its curve increases from proximally to distally. It is covered by hyaline cartilage, the center of which reaches, in the adult, 5 mm. in thickness.

3. The upper end of the ulna (cavitas humeralis ulnae)

Coronoid and olecranon processes hug the trochlea in almost perfect fit, forming approximately a semicircle (Fig. 2). A projecting ledge divides the articulating surface of the ulna and fits into a corresponding groove formed

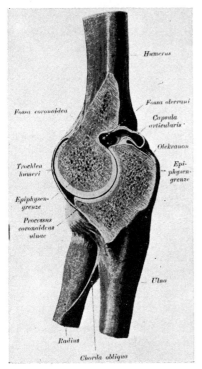

FIG. 2. The ulnotrochlear articulation.
(Fick)

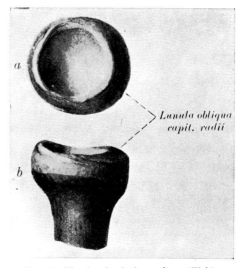

FIG. 3. The head of the radius. (Fick)

by the waist of the trochlea. The lesser sigmoid cavity is a cutout at the proximal end of the ulnar shaft which faces the head of the radius. It has a concave cylindrical surface of the angular value of 60-90° and a radius of 12-15 mm. (Strasser[12]).

4. The head of the radius

Articulating with the capitulum humeri, the head of the radius carries a shallow saucer representing a cavity of 70-80° angular value (Fig. 3). The cartilage of this saucer continues over the head to articulate on the ulnar side with the ulna. The hyaline cartilage covering of the articulating surface is thickest in the middle.

B. THE AXES AND PLANE RELATIONS

The length axis of the shaft of the ulna lies dorsally to the length axis of the radius. Consequently, the transverse axis of the elbow is so placed that in the hanging arm the medial condyle of the humerus points somewhat more dorsally and the lateral somewhat more volarly.

1. The transverse axis of the elbow joint

THE POSITION

Going approximately through the middle of the trochlea, this axis bivalves the angle between the long axis of the humerus and that of the forearm (Fischer[6]). It is not entirely constant but it oscillates, so to speak, between extreme flexion and extreme extension, thereby describing the surface of a cone with medial base. The angle of this cone, according to Fick,[6] is 10°, and it has a base of 2 mm.

The question may, therefore, be asked whether the elbow joint is a pure hinge joint or a helical one. This point is still debatable but if a helical motion exists it is certainly minimal and has no practical importance. Meissner[9] believed that the right elbow represents a right-handed and the left elbow a left-handed screw. Hultkrantz[7] also considers the lateral and length rotatory deviations of the ulna very small, and he states that in the middle ranges of the flexion-extension excursion they amount to only 1-2°, but that in extreme position they reach 5-10°, with the result that the stability of the joint is less in terminal position, a factor which has some bearing on the strains of the elbow joint. In fact, it is only in midposition that the fossa olecrani is in close contact with the trochlea, i.e., at a position of about 120°. In extreme extension, the entire medial border of the olecranon surface gapes because the trochlea is smaller posteriorly. Conversely, in extreme flexion, it is the lateral border of the olecranon which shows a gap because this border of the trochlea is smaller anteriorly (Fig. 4).

A longitudinal displacement between ulna and radius occurs in extension; the radius shifts distally. This is due to the spiral shape of the contour of the capitellum. Furthermore, the small ab- and adductory movements of the ulna are necessarily associated with an upward shift of the radius in radial and a downward shift in ulnar abduction. Such shifts occur also in daily tasks, for instance in resting the hand upon a table or in pushing a load with the hand.

One further observes that free flexion and extension cause automatic outward and inward rolling of the radius. The rotation occurs in the lesser sigmoid cavity which is fortified front and back by the annular ligament. The latter forms a ligamentous ring which, due to its insertion into the ulna, leaves the head of the radius free to secure a point of anchorage for the lateral collateral ligament of the joint. The closest attachment of the radial head to the capitellum is at the position of 90-120°.

2. The axis of length rotation of the elbow joint

The radiohumeral joint which forms the lateral half of the elbow joint has two axes. One is part of the common transverse axis of the elbow joint and coincides with the ulnohumeral axis. The other is vertical to the former and is part of the common radioulnar axis of the upper and lower radioulnar articulations. This pro- and supinatory axis goes through the convex head of the radius in the upper radioulnar joint and through the convex articular surface of the ulna in the lower radioulnar joint. The axis, therefore, is oblique to the length axes of both the radius and the ulna.

3. The relations between ulnohumeral and radiohumeral joints in combined motion (Fischer[6])

If flexion and rotation are carried out simultaneously at a certain constant ratio, then the combined axis of both motions is given by the hypotenuse

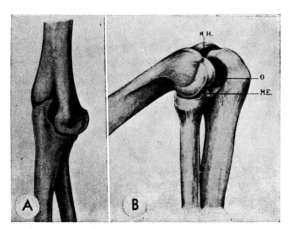

FIG. 4. A. Gap of the medial border of the olecranon surface with elbow in extreme extension. B. Gap of the lateral border of the olecranon in extreme flexion. (Fick)

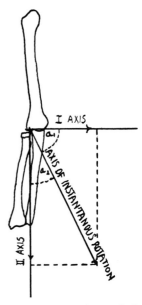

FIG. 5. The resultant of the two axes in combined flexion and rotatory motion. (Redrawn from Fischer)

of a right angle triangle, the acute angles of which represent the deflection of the rotatory from the transverse axis (a_1 and a_2). Let us assume that, starting from complete extension and supination to complete flexion and pronation, the angular values of rotation to that of flexion up to right angle be 2:1 (Fischer[6]), or flexion of 90° be combined with 180° of rotation. In this case of combined movement the integrated axis describes a cone, the axis of which is the resultant of the two axes (Fig. 5) (the hypotenuse with angles

α_1 and α_2) which forms an angle of 63° 26′ with the flexion extension axis, and 26° 34′ with the pro- and supination axis. A similar complication may apply for any position of the joint so long as the flexion and rotation angles retain their relation of 1:2. The same relationship exists if the forearm is fixed and the humerus moves around it from full extension and supination to full flexion and pronation (Fischer[6]).

4. The cubital or carrying angle

The deviation between the axes of the humerus and forearm is called the cubital angle. It amounts to an average of 10-15° in men and 20 to 25° in

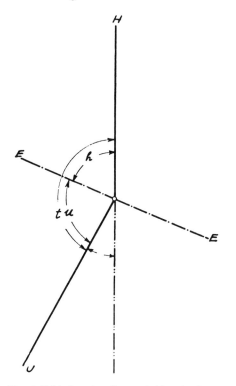

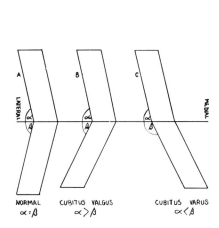

Fig. 7. Cubital angles. *A*. Normal cubital relation; flexion of elbow results in super-imposition of forearm and upper arm. *B*. Cubitus valgus; flexion of elbow results in medial deflection of forearm. *C*. Cubitus varus; flexion of elbow results in lateral deflection of forearm.

Fig. 6. Cubital angle; elbow axis bisecting it.

   EE elbow axis
   H   axis of humerus
   U   axis of ulna
   t   carrying angle

$$u = h = \frac{t}{2}$$

women. When the elbow is acutely flexed, forearm and arm cover each other and there is no apparent deviation. This is due to the fact that under normal conditions the axis of the elbow joint almost exactly bisect this angle (Fig. 6). In other words, the angle between the forearm axis and elbow joint axis, or the ulnar cubital angle, is the same as the angle between the axis of the humerus and the transverse axis of the elbow, or the humeral cubital angle.

The total cubital angle is between 155-170°. Fick's table shows that humeral and ulnar cubital angles are not always alike, that of the humeral angle fluctuated between 79.5° and 88°, and that of the ulnar angle between 81° and 90°. But so long as the difference between the cubital angles is small, any angulations should disappear in full flexion and forearm and upper arm should cover each other completely (Fig. 7A). However, if there is an appreciable difference, no such superposition occurs, as can well be demonstrated by a simple cardboard pattern. If the ulnocubital angle is considerably larger (as in the cubitus varus), then when the elbow is fully flexed the hand points laterally to the shoulder joint instead of lying directly over it (Fig 7C). If,

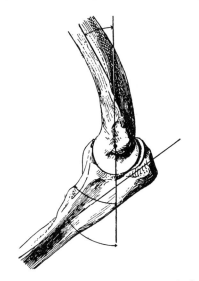

FIG. 8. Dorsal convexity of lower end of humerus. (Strasser)

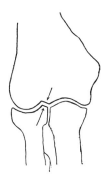

FIG. 9. Ulnar border of radial head fitting into groove between trochlea and capitellum. (Fick)

on the other hand, the angle is smaller (as in the cubitus valgus), the hand points medially to the shoulder joint when the elbow is in full flexion (Fig. 7B).

5. The dorsal curving of the lower end of the humerus

The distal end of the humerus is bent forward causing a slight dorsal convexity, the deflection amounting to about 20° (Fig. 8). Consequently, the fully extended elbow is, in fact, in a position of hyperextension.

C. THE ANGULAR VALUES AND CONTACT AREAS OF THE JOINT CONSTITUENTS

1. The humerus

The trochlea has an angular value of 330°, almost a complete circle, being interrupted only by the thin bony wall separating the coronoid from the olecranon fossa of the humerus. The angular value of the capitellum is 180°, forming a complete half sphere.

2. The forearm

On the other hand, the angular value of the joint constituents of the fore-arm are much smaller, leaving a considerable portion of cartilaginous surface of the lower end of the humerus exposed. The angular value of the olecranon fossa of the ulna is 190° against 330° of the opposing trochlea; and the angular value of the head of the radius is 40° against 180° of the capitellum. It will be noted that the differences in angular values are the same for the ulnohumeral and for the radiohumeral portions of the joint, namely 140°, which represents the flexion-extension range.

3. The contact

We have already mentioned the gap which opens in extreme extension at the medial and in extreme flexion at the lateral border of the olecranon (Fig. 4). There is, besides, considerable gaping on extreme flexion between the coronoid process and the trochlea because the latter is smaller at the upper than at the lower circumference. The bifurcated end of the ulna seems too large to hold the trochlea snugly. Conversely, in extreme extension, the olecranon impinges posteriorly into the fossa olecrani, although more often than not a soft tissue buffer comes between humerus and ulna and prevents further extension.

Because of the flatness of the radial concavity, only a small portion of the radiohumeral joint is in contact with the capitellum, leaving a gap in front and back. In extreme flexion, Fick[5] believes that the anterior margin of the radial head may impinge against the humerus. The obliquely sloped ulnar border of the radial head (Fig. 9) fits into the groove which separates the trochlea from the capitellum and serves as a sort of "guiding crest" in flexion movement of the elbow.

D. THE CAPSULAR LIGAMENTS AND LIGAMENTOUS CHECKS

1. The capsule

In extension, the anterior capsular wall tightens over the anterior surface of the trochlea, especially at its medial aspect. In flexion, the posterior wall tightens over the posterior surface of the trochlea while the anterior wall folds up. With these movements the small gaps between humerus and ulna at the posterior aspect of the joint allow fat pads to enter the joint when the capsule becomes redundant in extension; but in flexion, when the posterior capsule tightens, the fat pads are forced out from behind the trochlea.

2. The collateral ligaments

The strongly developed collateral ligaments, the internal as well as the external, safeguard the joint in ab- and adduction and in axial rotation. The radial collateral ligament comes from the lateral epicondyles and splits in front and behind the head of the radius in two halves which run to the anterior and posterior borders, respectively, of the lesser sigmoid cavity.

Between them, the annular ligament runs around the neck of the radius (Fig. 10). The medial collateral ligament comes from the internal epicondyle and runs to the medial border of the ulna. It consists of a posterior thin por-

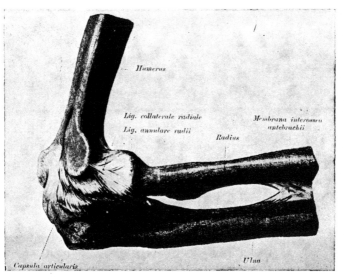

FIG. 10. Annular and lateral collateral ligaments of elbow. (Fick)

tion which forms the lateral wall of the osseofibrous canal for the ulnar nerve; and a superimposed, strong, parallel-fibered anterior strand which goes to the base of the coronoid process (Fig. 11). The radial collateral ligament checks the radius against movement toward the ulna, and the medial collateral prevents movement of the ulna toward the radial side.

3. The interosseus ligament

Almost along their entire length, radius and ulna are united by the interosseous membrane, a flat, shining, tough ligamentous structure (Fig. 12). Its course is upward from the ulna to the radius. Only in the upper and lower interspaces between

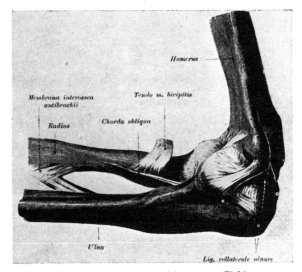

FIG. 11. Medial collateral ligament. (Fick)

the two bones there are two ligamentous structures, the upper and the lower chorda obliqua which run their fibers in opposite direction, namely, upward from the radius to the ulna. They are separated from the main portion of the membrane

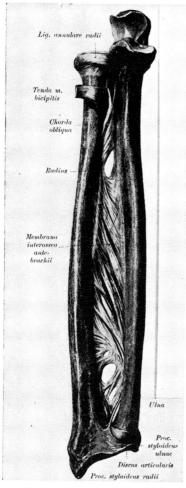

Fig. 12. Interosseous ligament.
(Fick)

by wide foramina for the passage of nerves and vessels. The upper chorda obliqua is a more rounded band arising between the biceps insertion at the tuberosity of the radius and it runs upward to a roughened area at the lateral side of the ulna (tuberositas ulnae). As can be seen from the direction of the fibers, the purpose of the membrane is to prevent longitudinal displacement of the two bones. The main portion, the interosseous membrane proper, prevents the upward pushing of the radius as might occur, for instance, by bracing the palm of the hand against a table. The upper and lower portions, which have their fibers running in opposite directions, prevent downward displacement of the radius against the ulna, such as might occur by a forceful pull on the hand.

It is of special interest to know the relation of this interosseous membrane to pro- and supination of the forearm bones. In both positions, the ligament relaxes, as one can observe not only in the ligamentous specimen but also when the other soft structures are preserved. The two forearm bones are farthest apart in midposition. This position is especially important in cases of fracture of both forearm bones, when synostosis between the radius and ulna is to be avoided.

E. TENSION RESISTANCE

Fessler[3] found the tension resistance of the ulnar collateral ligament to be from 100 -120 kg.; and that of the radial from 80-160 kg. The interosseuos membrane resists tension up to 65-70 kg. in longitudinal and to 91 kg. in transverse direction. On the other hand, the radiohumeral articulation sustains a tension force of only 15-40 kg., while the tension resistance of the ulnohumeral joint was found at anywhere from 85-230 kg. The anterior capsule breaks under a tension load of 46-70 kg. at 9 cm. lever arm, or a moment of up to 630 kgcm., while torsion with a lever arm of 17 cm. a load of 14 kgm., or a torsion moment of 235 kgcm. always produces a tear (Fessler[3]). In children, a dislocation of the head of the radius may occur from pulling because the head slips out easily from the loosely attached annular ligament. In the adult, a strong pull is more likely to be followed by a tear of the ligament. It is obvious that dislocation of the ulna must occur easier in extension, where the point of the olecranon digs into the posterior olecranon fossa of the humerus and acts as hypomochlion. At the same time the width of the joint and the strong collaterals make lateral dislocation more unlikely.

II. THE TYPES OF MOTION

Because of the difference of the joint constituents, the type of motion is different in the ulnohumeral and the radiohumeral portions of the articulation.

A. THE ULNOHUMERAL ARTICULATION

Since there is almost complete congruency of the articular surfaces the motion in this joint is a gliding one, but, in contrast with the knee, the gliding is evenly distributed over the entire surface of the articulation. It is a hinge motion and, as pointed out above, the lateral displacement which accompanies it is too insignificant to be taken into account.

The flexion-extension range averages 135-140°, exactly the difference between the angular value of the trochlea and that of the upper end of the ulna. Hyperextension of 10-20° is not uncommon in loose-jointed individuals, and the most acute flexion angle is about 20°, so that an extreme range of 160° or more can be observed.

From the viewpoint of joint mechanics, the movement between ulna and humerus which occurs in extreme flexion and extension is of interest. H. von Meyer[10] observed that these terminal ranges are accompanied by certain axial rotations. The ulna carries out a longitudinal inward rotation at the end of the extension and a longitudinal outward rotation at the end of flexion. This falls in with the gaping at the outer olecranon border in flexion and the inner one in extension, which has been mentioned above. Terminal extension is also associated with abduction and terminal flexion with either ab- or adduction. This terminal abduction in extension is attributed to the action of the anconeus (Duchenne[2]).

B. THE RADIOHUMERAL ARTICULATION

The incongruency between the capitellum and the shallow cavity of the radial head is responsible for a sort of point gliding. It differs from that of the ulnohumeral portion in that the contact area of the radius is small. For a small distance the medial edge of the head of the radius is in contact with the intermediate crest between trochlea and capitellum (Fig. 9). The propensity of this articulation, particularly the head of the radius, to undergo degenerative changes is probably due to the lesser stability of the joint. The angular value of the capitellum being 180° and that of the radius only 40°, the natural range of motion is from 180° to 40°, or in comformity with the range of motion of the ulnohumeral joint. The radiohumeral articulation is really a ball-and-socket joint. However, its mobility is restricted to two planes, the sagittal and the transverse, because there is no rotation in the frontal plane and the axial rotation is controlled by the common axis of the upper and lower radioulnar articulation.

C. THE RADIOULNAR ARTICULATION

In this articulation the radius rotates about the ulna in both the upper and the lower radioulnar joints, provided that the forearm is free and the upper arm with the rest of the body is the fixed part of the system. In free rotation of the forearm the thenar border of the hand describes the greatest excursion. The interosseous ligament can be considered to belong to both articulations. If, on the other hand, the hand and forearm are the fixed part of the system, then pro- and supination is carried out by the ulna. Since the ulna cannot rotate in the elbow joint, the center of rotation must be the glenohumeral articulation.

1. The upper radioulnar joint

The head of the radius moves in the lesser sigmoid fossa; the ligamentum annulare radii is fastened front and back to the edges of the incisura ulnae. It hugs the neck of the radius, becoming more tight distally and forming part of the elbow joint capsule (Fig. 13).

2. The lower radioulnar joint

The capitulum ulnae articulates with an excavation of the lower end of the radius from which the triangular cartilage spreads ulnarly to be attached to the styloid process of the ulna. The lower and broader end of the radius rotates like a swinging door around the head of the ulna as a center. A thin joint capsule is spanned from radius to ulna.

3. The combined radioulnar motion

The movements in both radioulnar joints are, except for minimal longitudinal shifts mentioned above, simply excursions about an axis which goes through the convex constituents of the two joints, namely, the head of the radius and the capitulum of the ulna. In the proximal radiohumeral joint, the interosseous ligament acts really as an articular reinforcement since it prevents upward dislocation of the joint, while the chorda obliqua prevents its downward displacement. In the proximal joint the radius moves strictly about the axis of the head since any movement to or from the ulna is prevented by the annular ligament which encloses the neck of the radius. In the distal joint, the lower end of the radius moves in a circle about the capitulum ulnae. When the radius is immobilized a reversion of the pro- and supinatory movement takes place in the sense that the ulna now moves around the radius together with humerus, the center of motion being the shoulder joint. This "reversal" of movement which has been referred to above requires some closer consideration. It occurs often in daily locomotor performances such as by bracing the extended arm against a table or by any other fixed resistance.

The humerus follows the movement of the ulna around the head of the radius. If the elbow joint is in extension, then the humerus rotates in the shoulder joint, together with the ulna, about the common longitudinal axis of the upper extremity (Fig. 14). If, on the other hand, the elbow is flexed, then the axis of motion of the forearm does not go through the humerus. The latter does not carry out an axial rotation but its lower end describes a circle. The

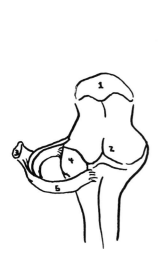

Fig. 13. Upper radioulnar
      articulation.
1. Olecranon.
2. Coronoid process.
3. Lateral ligament of elbow.
4. Radial notch of ulna.
5. Annular ligament.

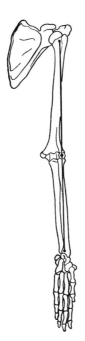

Fig. 14. Elbow
joint in exten-
sion. Humerus
r o t a t e s   i n
shoulder   joint
with ulna about
t h e   c o m m o n
longitudinal axis.
(Fick)

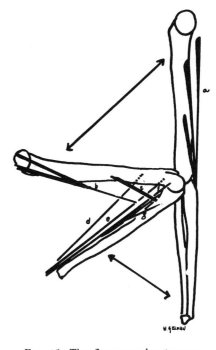

Fig. 15. The flexory and extensory
      muscles of the elbow joint.
a. triceps
b. biceps
c. brachialis
d. brachioradialis
e. extensor carpi radialis longus and
   brevis
f. flexor carpi radialis, and palmaris
   longus
g. pronator teres

radius of this circle is greatest at right angle flexion and becomes successively smaller as 180° are reached.

Pro- and supinatory ranges. In the living the checks of pro- and supination do not come from ligaments alone but also from the passive resistance of the stretched antagonists. Braune and Flügel[1] found in the cadaver that removal of the muscles increases the excursion range more than the resection of all the checking ligamentous structures. The impingement of tissues between the bones restrains pronation, especially the flexor pollicis longus which is being forced against the deep finger flexors. The entire pro- and supinatory excursion range in the living, as mentioned above, is, therefore, no more than 120-140°, whereas, when the muscles are removed in the cadaver, Braune found the range to be up to 185-190°, and when cutting the ligaments also he found it increased to 205-210°.

III. JOINT DYNAMICS

A. FLEXION-EXTENSION

1. The flexors

The brachialis and the biceps are the two principal flexors. Stimulation of the brachialis produces powerful flexion of the elbow joint without any pro- or supinatory effect (Duchenne[2]). Stimulation of the biceps produces supination of the pronated forearm and simultaneously flexion of the elbow.

Auxiliary flexors are: the brachioradialis and the extensor carpi radialis longior. The brachioradialis, or supinator longus, is a flexor of the elbow which at the same time maintains the joint in midposition between pro- and supination. It supinates the pronated and pronates the supinated forearm. The extensor carpi radialis longior has a minimal flexory effect. It never produces extension of the elbow because its axis runs in front of the epicondyle (Fig. 15).

2. The extensors

All extension of the elbow is supplied by the three heads of the triceps and the anconeus, although some slight extensory effect is attributed to the humeral head of the supinator brevis. So far as extension is concerned, the isolated action of all three heads is identical. The anconeus participates in all extensory movements of the elbow and, due to the obliquity of its fibers, it also produces a lateral movement of the ulna. Of all the three heads, the long one is the least powerful extensor but it is also an auxiliary adductor in the shoulder joint (Fig. 15).

3. Comparative strength

Expressed in kilogrammetrical values the flexors are one and one-half times stronger than the extensors. This is philogenetically a step forward since the preponderance of the flexors over the extensors is much greater in the higher apes. It should also be noted that the flexory effect is greater when the forearm is in pronation than when it is in supination because in the former position the distance of the muscle axes from the center of motion is greater. The maximum work of the elbow flexors with the forearm in midposition is, according to Fick,[5] 5.185 kgm., while that of the extensors is 3.345 kgm.

B. PRO- AND SUPINATION

1. Supination

The principal supinators are the biceps and the supinator brevis.

A) THE BICEPS

The insertion of the biceps into the tuberosity of the radius lies behind the longitudinal axis of the latter so that the tendon winds around the ulnar border of the radius. The supinatory power of the biceps decreases rapidly as supination increases and the last phases of supination must be taken over by the supinator brevis. The supinatory rotation movement of this muscle is greatest at right angle flexion (A. E. Fick[4]). This is the position one choses instinctively when using a screw driver with considerable force. The tension of the biceps can then still be increased by carrying the arm backward in the shoulder joint. An example of reversal occurs when the radius is fixed (for instance by a resistant doorknob) and the flexed elbow is elevated to the horizontal. If it is then allowed to drop toward the side of the trunk, the ulna this time carries out a supinatory movement by means of outward rotation and adduction (Fig. 20) in the shoulder joint. This movement is transmitted to the radius, permitting the latter to add its own supinatory effort. This is a common situation in which supination meets with considerable external resistance.

B) THE SUPINATOR BREVIS

This muscle is capable of performing almost half of the superinatory work of the biceps when the elbow is extended. When the elbow is flexed, however, the capacity of this muscle amounts to scarcely one-third of that of the biceps. At the same time, the tension of the latter greatly decreases with flexion and, therefore, it is in flexion of the elbow (contrary to Duchenne[2]) where one must depend more upon the supinator brevis.

C) THE BRACHIORADIALIS

The brachioradialis is, according to Fick,[5] a considerable supinator beyond the midposition; also, the extensor carpi radialis longior acts as a supinator when the elbow joint is in extension.

2. The pronators

a) The principal muscle is the *pronator radii teres*. It displays its greatest power in midflexion position of the elbow. It is this muscle which establishes the pro- and supinatory equilibrium against the biceps and the supinator brevis. In fracture of the radius above the insertion of the pronator teres, the proximal fragment goes into full supination. With increasing extension of the elbow, the pronatory effect of this muscle is diminished.

b) The *pronator quadratus* is a forceful pronator in all positions and, therefore, a welcome synergist to the pronator teres, especially when the elbow is extended. Both pronators, particularly the quadratus, try to approach the forearm bones to each other, which is especially noted in fractures of the middle of both forearm bones.

c) The *flexor carpi radialis,* when stimulated, first flexes the wrist and then pronates it (Duchenne[2]).

3. The comparative strength of pro- and supination

The maximum work of the supinators is in right angle flexion and amounts, according to Fick,[5] to 0.604 kgm.; that of the pronators is 0.568 kgm. The supinatory power becomes considerably smaller with the arm in extension. The working capacity of the pronators in right angle flexion of the elbow amounts to 0.6 kgm.; with the arm in extreme flexion it falls to 0.5 kgm.; with the elbow in extension it is only 0.32 kgm.

The question of the effect of the position of the elbow on pro- and supinatory strength has been investigated recently by Salten and Darcus.[11] They found a linear relationship between position and strength. The pronatory moment increased as the forearm goes into supination. Of all positions tested the right angle flexion was the optimum for both pro- and supination. This advantage decreases as the position of the elbow changes in either direction. At midpoint of the flexion-extension range, pronation is stronger than supination. These findings are in accord with the above-mentioned data of Fick.[5]

Compensatory pro- and supination can be furnished by movement in the shoulder joint when such movements are prevented in the forearm by rigidity or by muscle paralysis. With the elbow flexed, abduction plus inward rotation of the shoulder joint places the forearm in pronation. Compensatory supination is much more difficult to obtain because it requires the adduction and outward rotation of the humerus, the former being impeded by the arm striking against the side of the thorax.

IV. SUMMARY

1. The elbow joint serves the shortening and lengthening of the extremity as well as the movement of the hand in pro- and supinatory direction.

2. It has two degrees of freedom of motion, hinge motion as well as axial rotation, i.e., it is a trochoginglymus.

3. The surface of the trochlea is generated by a hyperbola rotating about its tangent or a parallel to it; it is hourglass shape, concave in the frontal and convex in the sagittal plane. The trochlea is separated from the capitulum of the humerus by a groove. This capitulum is completely spherical. The upper end of the ulna carries two large processes, the olecranon and the coronoid, which enclose the greater sigmoid fossa (cavitas humeralis ulnae). A lesser sigmoid cavity is a cutout at the proximal end of the ulna which accommodates the head of the radius. The latter articulates with the capitellum by a shallow saucerlike cavity.

4. The transverse axis of the elbow joint runs in the hanging arm from medial dorsal to lateral volar and, going approximately through the middle of the trochlea, it bisects the angle between the long axes of humerus and forearm. It is debatable whether flexion-extension is a pure hinge motion

because the axis oscillates somewhat, describing a cone. However, the angle of this cone is minimal, being only 1-2° in midranges, but reaching 5-10° in extreme positions. In extreme extension, the medial, and in extreme flexion the lateral surface of the olecranon gapes. Also, in extension the radius shifts slightly distally and there is an upward shift of the radius in radial and a downward shift in ulnar abduction. Furthermore, free flexion and extension is associated with an automatic outward and inward rolling of the radius in the lesser sigmoid cavity. These displacements, however, are insignificant and for all practical purposes the ulnohumeral joint is a pure hinge.

5. The length rotatory axis for pro- and supination goes through the convexities of the upper and lower radioulnar joints. It is, therefore, oblique to the length axes of both radius and ulna.

6. If flexion and rotation are carried out simultaneously the combined axis depends upon the ratio between rotation and flexion. If this ratio is 2 to 1, i.e., if for each two degrees of rotation there is one degree of flexion, this combined axis forms an angle of 63°26′ with the transverse and 26°34′ with the longitudinal axis.

7. The cubital or carrying angle amounts to 15° average in men and to 25° in women. Under normal conditions the transverse axis bisects the angle between humerus and forearm into an equal humerocubital and ulnocubital half. The result of this equality is that in complete flexion of the elbow forearm and upper arm cover each other. If the ulnocubital angle is larger, as in cubitus varus, then on flexion of the elbow the hand points laterally to the shoulder joint; if it is smaller, as in cubitus valgus, it points medially. The distal end of the humerus is bent forward, causing a deflection of about 20°.

8. The angular value of the trochlea is 330°, that of the articulating ulna 190°; the angular value of the capitellum is 180°, that of the head of the radius 40°. The differences are, therefore, equal for both halves, namely 140°, which represents the flexion-extension range. The ulnohumeral contact is not perfect; aside from the lateral gaping of the olecranon in full flexion and the medial in full extension, there is also some dehiscence of the coronoid process anteriorly on extreme flexion. Only a small portion of the radial head is in contact with the capitellum, leaving a gap in front and back.

9. The capsule tightens in front on extension and in back on flexion. Strong collateral ligaments reinforce the elbow joint. The medial collateral consists of a posterior thin portion which forms the lateral wall of the osseofibrous canal for the ulnar nerve, and a strong anterior portion going to the base of the coronoid process; the lateral collateral splits in front and behind the head of the radius; between the two halves the annular ligament runs around the radial neck. The medial prevents movement of the ulna radially, the lateral movement of the radius medially.

10. The interosseous ligament running from the ulna in proximal direction to the radius prevents the radius from being pushed upward; the su-

perior and inferior chorda obliqua, which run in opposite direction, resist the downward pulling of the radius. The interosseous membrane is relaxed in pronation and supination and tight in midposition, where the forearm bones are farthest apart.

11. The tension resistance of the collaterals is 100-120 kg. for the ulnar and 80-160 kg. for the radial. The anterior capsule breaks under a load of 46-70 kg. and a lever arm of 9 cm. Dislocation of the head of the radius occurs in children from pulling on the forearm.

12. Motion in the ulnohumeral articulation is a gliding with a range of 135-140°. Hyperextension of 10-20° is not uncommon. Terminal flexion and extension is accompanied by axial rotation; inward rotation of the ulna at the end of extension and outward rotation at the end of flexion. There is also abduction on terminal extension.

13. Because of incongruency of the joint surfaces the movement in the radio-humeral articulation is a point-gliding, where a concentrated area. of the head glides over the capitellum of the humerus in flexion-extension. In pro- and supination a small area of the radius rotates about a small area of the capitellum.

14. The common axis of the upper and lower radioulnar joint restricts motion to the transverse plane. When the forearm is free the radius swings around the ulna in a circle. If the radius is immobilized a reversion of the pro- supinatory movement takes place. The ulna now moves around the radius, the center of motion being the shoulder joint. If the elbow is straight the axis of motion is the length axis of the extremity. If the elbow is flexed the elbow describes the surface of a cone by a circular motion in the shoulder joint.

15. Pro- and supinatory range in the living is 120-140°. Principal checkers are the muscles of the forearm.

16. Flexors of the elbow are biceps and brachialis with the brachioradialis and the extensor carpi radialis as auxiliary flexors; the brachioradialis moves the forearm to midposition. Principal extensors are the triceps and anconeus. The flexors are one and one-half times as strong as the extensors, with 5.185 kgm. against 3.345 kgm. working capacity. The flexory effect is greater with the forearm in pronation.

17. Biceps and supinator brevis are the supinators. The supinatory rotation moment of the biceps is greatest at right angle flexion. Its tension can still be increased by carrying the arm backward in the shoulder joint. If the radius is fixed and the flexed elbow elevated to the horizontal, and then made to drop to the side of the body, the ulna carries out a supination by rotation and adduction in the shoulder joint. This being transmitted to the radius, the latter can then continue the supinatory effort, which may or may not be entirely blocked by the external resistance. The supinator brevis is particularly called upon when the elbow is in flexion.

18. The pronator teres and quadratus are the main pronators; the former holds the equilibrium against biceps and supinator brevis. The pronator

quadratus is particularly needed when the elbow is extended. The force of the pronators increases as the forearm goes into supination. Right-angle flexion is the optimum position both for pro- and for supination. Compensatory pro- and supination can be obtained by movement in the shoulder joint. Abduction and inward rotation for pronation, and adduction and outward rotation for supination. The latter is more difficult to obtain because the side of the thorax blocks adduction.

19. In working strength, pro- and supinators are evenly matched. Maximum strength at right-angle flexion is 0.604 kgm. for the supinators and 0.568 kgm. for the pronators.

BIBLIOGRAPHY

1. BRAUNE, W., and FLÜGEL, A.: Über Pronation und Supination des menschlichen Vorderarms und der Hand. *Arch. Anat. Physiol. Anat. Abt.*, 1882.
2. DUCHENNE, G. B.: *Physiology of Motion.* Transl., E. B. Kaplan. Philadelphia, J. B. Lippincott, 1949.
3. FESSLER, J.: Festigkeit der menschlichen Gelenke mit besonderer Berücksichtigung des Bandapparates. *München*, 1893.
4. FICK, A. E.: Über die Methode der Bestimmung von Drehungsmomenten. *Arch. Anat. Physiol. Anat. Abt., Suppl.* 1889.
5. FICK, R.: *Spezielle Gelenk- und Muskelmechanik.* Jena, G. Fischer, 1911.
6. FISCHER, O.: *Kinematik organischer Gelenke.* Braunschweig, F. Vierweg und Sohn, 1907.
7. HULTKRANTZ, J. W.: *Das Ellbogengelenk und seine Mechanik.* Jena, 1897.
8. LANZ, T. V., and WACHSMUTH, W.: *Praktische Anatomie.* Berlin, J. Springer, 1935.
9. MEISSNER, G.: Lokomotion des Ellbogengelenkes. *Ber. üb. d. Fortschr. d. Anat. u. Physiol.*, 1856.
10. MEYER, H. V.: Das Ellbogengelenk. *Arch. Anat. Physiol. Anat. Abt.*, 1866.
11. SALTEN, N., and DARCUS, H. D.: The effect of the degree of elbow flexion on the maximum torques developed on pronation and supination. *J. Anat., 86:2*, 197, April 1952.
12. STRASSER, H.: *Lehrbuch der Muskel- und Gelenkmechanik.* Berlin, J. Springer, 1917.

Lecture XXIX

THE PATHOMECHANICS OF THE
PARALYTIC ELBOW

INTRODUCTION

IN THE elbow joint the requirement for mobility takes preference over that of stability because the principal function of the joint is to shorten and lengthen the extremity in order to place the hand at any point which lies within the length of the extremity. The fact that the joint depends less on stability than on mobility does not mean that the stability is to be dispensed with entirely. A flail joint fails to provide the muscle with a stable proximal part of the lever system against which the distal part is to be moved. Consequently, the rotatory effect of muscle action becomes insufficient. The paralysis of any group of muscles not only abolishes their primary function and makes movement in this direction impossible, but it also deprives the opposing muscles of antagonistic control. For example, paralysis of the elbow flexors not only eliminates active flexion but it also interferes with the control and grading of the extensory effort.

The second function of the joint is pro- and supination in the radiohumeral and radioulnar articulations. If pronation is lost, it can to a degree be substituted by movement in the shoulder joint. Supination, on the other hand, is difficult to supplant in this manner and often calls for operative reconstruction.

I. THE PARALYSIS OF THE EXTENSORS OF THE ELBOW JOINT: TRICEPS AND ANCONEUS

Statistics of the Vermont epidemics give the frequency in which the triceps is involved in poliomyelitis as 28 in a total of 1417 muscles affected. The ratio of partially to totally paralyzed muscles is given as 3.6 to 1 (Lovett[5]). Since the long head divides its strength between extension of the elbow and adduction of the shoulder joint, its extensory power lags behind the two short heads, the action of which is identical. The anconeus participates in all movements of extension and has about the same extensory force as the long head.

If the extensors are paralyzed, passive extension can be procured easily enough by gravity, but such extension is uncontrolled and lacks stability. Besides, patients with paralysis of the triceps are severely handicapped if paralysis of the lower extremities imposes the use of crutches.

As substitute for this muscle, F. Lange[4] used the latissimus elongated by silk strands. The triceps has a cross section area of 15 cm.[2] and when it extends the arm its average shortening is 5 cm. (Fick[2]) (Fig. 1). The latissimus seems adequate both in point of strength and shortening distance. The question remaining is that of the direction. Another technique used by the same surgeon is the direct transplantation of a portion of the trapezius which is separated from the acromion, elongated by silk strands, and sutured directly to the olecranon. Here, requirements of direct pull seem to be met; the question remains whether the transplanted portion has the necessary strength and, especially, whether it commands the required excursion range comparable with the contractility of the triceps.

One can readily see that, quite aside from technical difficulties, none of these operative plans fully satisfies the mechanical requirements of transplantation.

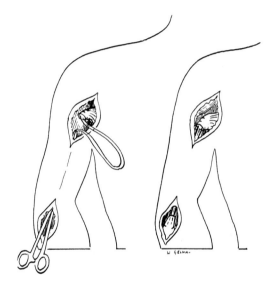

FIG. 1. Use of the latissimus dorsi for paralysis of triceps. (Redrawn from Max Lange)

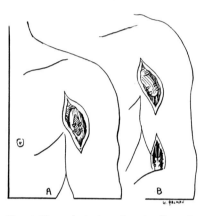

FIG. 2. Transplantation of pectoralis major for paralysis of elbow flexors. *A.* Direct, into belly of biceps. *B.* Indirect, into tendon of biceps using silk sutures. (Redrawn from Max Lange)

II. THE PARALYSIS OF THE FLEXORS

They are the biceps, the brachialis, and, to a lesser extent, the brachioradialis. The frequency of the paralysis of the biceps in poliomyelitis is given as 31:1417, and the ratio of partial to total paralysis is 3.4:1 (Lovett[5]). Isolated paralysis of the biceps is rare. Operative substitution is indicated when both muscles and possibly also the brachioradialis are paralyzed. Bunnell[1] mentions four methods of substitution:

1. By the triceps.
2. By the pectoralis.
3. By the sternocleidomastoid.
4. By the author's method of transposition of the forearm muscles arising from the internal condyle of the humerus.

A. SUBSTITUTION BY THE TRICEPS

This muscle is adequate in point of strength; it has a cross section area of 15 cm.[2] and a contractile length of 4-6 cm. (Fick[2]).

The operative methods cover both partial and total transposition of the muscle. If only the lateral head is used, the difficulty lies in the insufficient length of the muscle. Biesalski used the long head, crossing it to the flexor side and suturing it to the biceps tendon. The method of partial transplantation of the triceps does not seem sound in the physiological sense because, innervationally, the triceps acts as one unit and its parts cannot be made to work as antagonists against each other.

The total transplantation of the triceps to the flexor side is more sound physiologically but it entails the sacrifice of all active extension and leaves the extension entirely to gravity which, as mentioned above, is not a satisfactory substitute. The triceps must be elongated by fascial or tendon graft as it is too short even to reach the tendon of the paralyzed biceps. Occasional good results notwithstanding, the method can hardly be recommended on kinetic grounds.

B. SUBSTITUTION BY THE LONG RANGE TRANSFERENCE: PECTORALIS MAJOR (FIG. 2)

This method was first recommended by Hohmann[3] who, with F. Lange,[4] elaborated the technique. It consists essentially in the mobilization of the pectoralis tendon and its implantation into the biceps, elongating the implanted tendon by silk strands.

The great problem, unsolved by these procedures, is that of direction and transmission of pull. Whether one uses the pectoralis major or the sternocleidomastoid (see below), the question of direction takes precedence over strength and excursion range of the donor muscle. In transplanting the pectoralis tendon into the muscle belly of the biceps it must be lead through a buttonhole made in the latter and lengthened with silk strands to reach the tendon of the insertion of the biceps to which it is anchored (Hohmann,[3] F. Lange[4]). The axis of the pectoralis muscle is still at an angle (α) of at least 45° to the axis of the biceps. That means that of the force displayed by the pectoralis major only a portion (F $\times$ cosine α) is actually pulling upward, while the rest (F $\times$ sine α) is pulling sideways, trying to dislodge the muscle inward. As for the contraction length, the excursion range is restricted to the same proportion of the total excursion (L $\times$ cosine α). This is all the lifting height which the elongated pectoralis muscle can dispose of.

C. SUBSTITUTION BY LONG RANGE TRANSFERENCE: STERNOCLEIDOMASTOID

So far as we know, the method is original with Bunnell.[1] Separated from the sternum and clavicle, the muscle is passed subcutaneously down the arm and prolonged by a long and free graft of the fascia lata. The latter is attached with its lower end to the bicipital tubercle of the radius. Thus the muscle become a flexor of the elbow as well as a supinator. Looking at it from the kinetic point of view, the method has much to recommend it. Strength and excursion range seem to be adequate. The advantage is the more direct transmission of pull and the avoidance of fascial slings such as are used in the substitution by the pectoralis. Cases demonstrated by Bunnell[1] show indeed surprising results so far as strength and excursion range of the transplanted and elongated muscle is concerned. The only possible handicap is that the head must be stabilized in symmetrical position or even in contralateral inclination. By rotating the head to the affected side the sternocleidomastoid muscle tightens so that flexion of the elbow can still be carried out under the control of the eye.

D. THE METHODS OF TRANSPOSITION OF THE FOREARM MUSCLES

It will be recalled that the wrist flexors, i.e., the flexors carpi radialis and ulnaris, have by virtue of the epicondylar origin a certain flexory effect on the elbow joint, although the perpendicular distance of these muscles from the axis of the joint is short and the rotation moment accordingly very small. Of the muscles arising from the internal epicondyle, the pronator teres has the greatest contraction length in respect to the elbow joint, 4 cm., i.e., almost twice as much as the flexor carpi radialis with 2.4 cm., and four times as much as the palmaris longus with 1.1 cm. The combined working capacity of these three muscles (i.e., contraction length times cross section area times a constant of 3.6 kgm. per cm.² of cross section) is 0.72 kgm. against 3.6 kgm. of the combined working capacity of the brachialis, the two heads of the biceps, and the brachioradialis (Fick[2]).

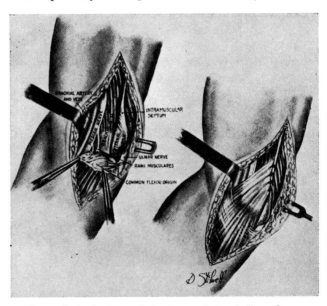

Fig. 3. Author's flexor plasty for paralysis of elbow flexors.
(Steindler)

1. The author's transposition of the flexor group at the elbow (Fig. 3)

By increasing the lever arm of the wrist and finger flexors as well as that of the pronator teres in respect to the elbow joint, a satisfactory degree of active flexion can be obtained. The operation of flexor transposition at the elbow is based on this principle. It is accomplished by transposing the common origin of the group from the medial epicondyle to a point higher up in the humerus. This increases the rotation moment so that the elbow can be flexed actively against gravity and against some external resistance.

2. The modifications of the author's[6] flexor plasty

The one drawback to the above method is that the transposition causes a pronatory tendency. For this reason a number of modifications were devised. Bunnell[1] tried the lengthening of the freed end of the flexors by means of a strip of fascia which, according to him, gives a moderate but not complete

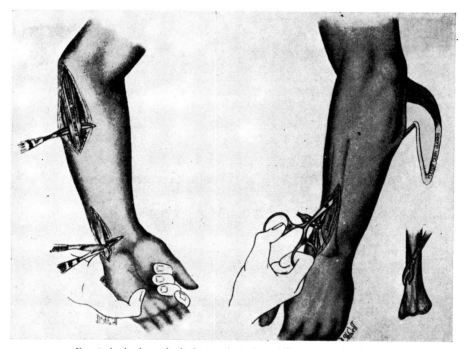

FIG. 4. Author's method of transplantation of flexor carpi ulnaris for supinatory paralysis of forearm. (Steindler)

correction of the pronatory tendency. It seems better to provide for an active supinator to counteract the pronator teres. For this purpose we have used the flexor carpi ulnaris (Fig. 4). Of all the epicondylar flexor muscles this has the least effect upon flexing the elbow but it has a definite pronatory action. If this muscle is freed from its insertion and mobilized through its entire length, it can be led diagonally over the dorsal surface of the forearm and fastened by drill holes into the lower end of the radius. This converts the muscle from a pronator into a supinator which not only counterbalances pronation but makes possible active supination considerably beyond the midposition of the forearm.

3. The pathomechanics of the flail elbow

In case of a completely flail elbow with no muscles available for transplantation or transposition, it is debatable whether one should perform an arthrodesis or whether one should be content with a splint which holds the joint in desirable position.

Usually, the arthrodesis is not performed for the flail elbow alone but as part of the reconstruction program for an entire flail arm, especially where a flail shoulder is associated with the flail elbow. If the arthrodesis of the shoulder is justified by the preservation of sufficient power of the rotators of the shoulder blade, the fusion of the completely flail elbow in proper position may still result in a fairly useful member. Where elbow flexion can be secured by transposition, the ability of

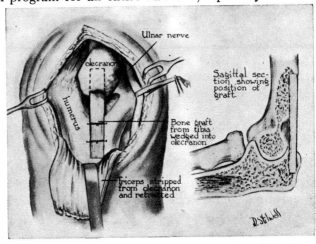

FIG. 5. Author's method of arthrodesis of the elbow. (Steindler)

abducting the arm in the shoulder joint is of definite advantage because the elbow can be flexed so much easier in the horizontal than it can in the perpendicular position. Therefore, the greater the weakness of the rotators of the shoulder blade, the less is the advantage which is offered by the flexor plasty of the elbow.

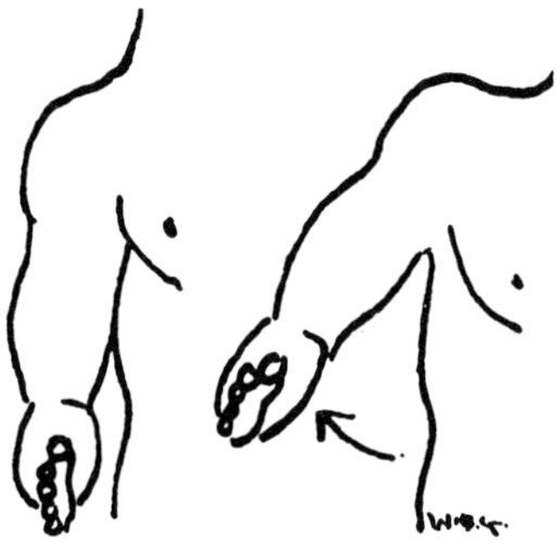

FIG. 6. Pronation of hand substituted by abduction of shoulder joint.

On the other hand, the arthrodesis of the elbow has a definite place in cases in which all three joints are hopelessly paralyzed. The combination of wrist and elbow arthrodesis makes it possible to use the extremity like a hook, and if there is some shoulder movement left, this combination of operations can be of considerable benefit (Fig. 5).

What position should be given to the arthrodesed elbow? That the flexion position should be close to the right angle or less is generally understood. However, the length rotatory position requires some consideration. It depends to a great extent on the abductory ability of the shoulder joint. If the shoulder joint is flail and there is neither active abduction nor can it be provided for by glenohumeral arthrodesis, the elbow joint should be ankylosed in midposition. If, however, abduction is possible, the position of the elbow joint should be in slight, supination. The reason is that a pronatory position of the hand can be procured by abduction in the shoulder joint while, conversely, supination cannot easily be produced by abduction which is limited by the thoracid wall (Fig. 6).

This correlation between the motion of the shoulder girdle and the position of the elbow joint must be further supplemented by procuring a functional position of the wrist, i.e., one fo 15-20° extension.

III. SUMMARY

1. While the elbow joint depends for its function more on mobility than stability, the latter is an important prerequisite for movement carried out with strength because it provides the acting muscle with a firm point of application.

2. If the extensors are paralyzed, passive extension is possible by gravity but such movement is uncontrolled and lacks stability. Efforts of substituting the paralyzed extensors have been made by use of the latissimus (F. Lange) or of the trapezius elongated by silk strands and fastened to the olecranon. Quite aside from technical difficulties, neither of these operations seems to meet the mechanical requirements.

3. For paralysis of the flexors, the muscles recommended for transplantation are: the triceps, the pectoralis, the sternocleidomastoid, and the transposition of the flexor group (author's method).

4. Partial transplantation of the triceps is unphysiological. Total transplantation to the flexor side, while more sound physiologically, entails sacrifice of all active extension.

5. Long range transference of the pectoralis major (Hohmann, F. Lange) into the biceps by means of elongating silk strands offers the problem of direction and transmission of pull. A considerable portion of the tension is spent in side pull; the longitudinal pull is commensurate with the cosine of the angle of application; if the latter is 90°, there is no longitudinal effect.

6. Long range transference of the sternocleidomastoid (Bunnell) gives both adequate strength and excursion range and is also more favorable in point of direction. The head must be stabilized to permit the muscle to take effect on the elbow.

7. Transposition of the flexor group from the internal epicondyle to a higher point at the humerus is based on the proposition that this maneuver greatly increases the rotatory moment of these muscles in respect to the elbow joint. A disadvantage is that it produces a pronatory tendency. This is being met by Bunnell's modification which lengthens the freed end of the flexors by means of fascia lata. Active supination counteracting the pronatory pull of the pronator teres can be obtained by transposition of the flexor carpi ulnaris, a muscle which can be spared from the flexor group because it has the least flexory moment of all. The muscle is mobilized through its entire length and is led diagonally over the dorsal surface of the forearm and fastened to the

lower end of the radius by drill holes. This changes the muscle into a supinator.

8. Arthrodesis of a completely flail elbow is indicated especially in combination with arthrodesis of the shoulder. The latter also facilitates the effect of the flexor plasty by providing abduction of the arm, which makes flexion of the elbow by the transposed muscles easier.

9. The position of the arthrodesed elbow should be right-angle flexion. If abduction in the shoulder joint is provided for, the forearm should be in slight supination; if no abduction is possible, the forearm should be in midposition.

BIBLIOGRAPHY

1. BUNNELL, S.: Paralysis of elbow. Paper, *Am. Soc. Surg., of Hand*, Jan. 26, 1952.
   BUNNELL, S.: *Surgery of the hand*. Philadelphia, J. B. Lippincott, 1944.
2. FICK, R.: *Handbuch of Anatomie*. Mechanik der Gelenke III. Jena, G. Fischer, 1911.
3. HOHMANN, G.: Ersatz des gelähmten Biceps brachii durch den Pectoralis major. *Münch. Med. Wchnschr., 1240*, 1918.
4. LANGE, F.: Die epidemische Kinderlähmung. *München*, J. F. Lehmann, 1930.
5. LOVETT, R. W.: *The Treatment of Infantile Paralysis*. Philadelphia, P. Blakiston & Co., 1916.
6. STEINDLER, A.: Tendon transplantation in upper extremity. *Am. J. Surg., 44:*260, 1939.
   idem: *Orthopedic Operations*. Springfield, Illinois, Charles C Thomas, Publisher, 1940.

Lecture XXX

MECHANICS OF HAND AND FINGERS

INTRODUCTION

THE anatomical construction of this motor unit is so complex and its function so intricate that it becomes necessary to go into morphological and mechanical aspects in greater detail than almost any other part of the human body would require. Here is a chain of articulations, none of which operates in ordinary usage by itself; all depend for their mechanical effect upon intimate correlation.

This system includes: three joints in the formation of which the carpus participates: the radiocarpal, the triquetropisiform, and the intracarpal joint. It further includes three joints involving the five rays: the carpometacarpal, the metacarpophalangeal, and the interphalangeal joint. The ratio of the relative length of carpus, metacarpus, and fingers is $2:3:5$; eight-tenths of the length of the hand are occupied by the five rays. In the foot, this ratio is reversed, namely, $5:3:2$; only one-half of the length of the foot is occupied by the five rays.

I. MORPHOLOGY

A. THE RADIOCARPAL ARTICULATION

1. Relationship and angular values

In this articulation, the proximal carpal row, consisting of the scaphoid, the lunate, and the triquetrum, furnish the convexity, while the articulating surface of the radius together with the triangular cartilage attached to it forms the concave portion of the joint. The lower end of the radius, being deflected slightly forward, causes the joint to point not strictly distally but distally and somewhat forward (Rauber-Kopsch[9]) (Fig. 1).

The oval joint surface of the radius has a longer radioulnar and a shorter dorsovolar diameter. It is so poised that it reaches farther distally on the radial side, so that the joint also locks slightly ulnarly. There are two facets on the radius: one is triangular and articulates with the scaphoid, the other is more quadrangular and articulates with the lunate. A triangular cartilage between the ulnar edge of the radius and the styloid of the ulna contacts in midposition the ulnar half of the lunate and the triquetrum (Fick[5]) (Fig. 2).

The joint is curved both anteroposteriorly and from side to side. The anteroposterior curve is sharper, the radioulnar flatter. Consequently, one must expect the excursions to be greater in the flexion-extension range than in that of radial or ulnar abduction. Scaphoid and triquetrum show a sharper radio-

ulnar curve than the corresponding concavity of the radius and the triangular
cartilage, the radius of the former being 3 cm. against 4 cm. of the latter.
The total angular value of the proximal carpal row in radioulnar direction
is 109°, while the angular value of the radius is only 75°. In dorsovolar direction, the curve of the articular surface of the radius has an angular value of
only 67° against 113° of the convex part of the joint. Specifically, the angular

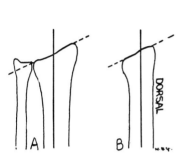

FIG. 1. Distal view of radius. A, ulnar
slope. B, volar slope.

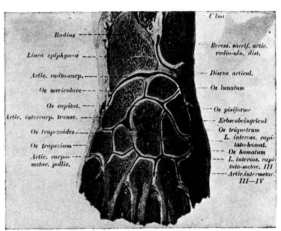

FIG. 2. Sagittal section through wrist joint showing
triangular cartilage. (Fick)

value of the scaphoid is 130°, of the lunate 115°, and of the triquetrum 108°,
averaging 113°.

2. The axes, planes, and excursions

This joint has two degrees of freedom of motion, one about an antero-
posterior or dorsovolar, and another about a transverse or radioulnar axis.
There is no movement about the longitudinal axis; pro- and supination occur
exclusively in the upper and lower radioulnar articulations, at least under
normal conditions. All these movements are supplemented by the intracarpal
articulation, which also has two degrees of freedom of motion about corresponding axes.

The midposition of the joint is not 180°, but slight volar flexion and slight
ulnar abduction. But when one considers the midposition of the intracarpal
articulation together with that of the radiocarpal joint, then the combined
midposition of both articulations is, according to Recklinghausen,[8] 12° extension and 3° ulnar abduction.

The axis of the dorsovolar movement between radius and proximal carpal
row corresponds to approximately the greatest diameter of the articular surface of the radius. The axis for the lateral movement is perpendicular to the
former, going through the convexity of the carpus and striking in antero-
posterior direction the head of the capitate, approximately parallel to the
shortest diameter of the articular surface of the radius.

Because the axes for both anteroposterior and lateral motion are more or less constant and go through the convexity at some distance from the articular surfaces, the gliding element prevails in the movements of this joint. Thus, when the hand is abducted radially, the proximal carpal row glides over to the ulnar side (Lanz and Wachsmuth[7]) (Fig. 3); when the hand is abducted ulnarly, it glides to the radial side (Fig. 4). Similarly, in dorsiflexion, the

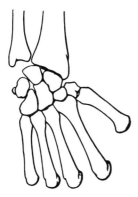

FIG. 3. Radial abduction.
(Lanz & Wachsmuth)

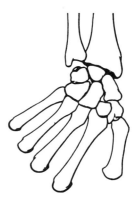

FIG. 4. Ulnar abduction.
(Lanz & Wachsmuth)

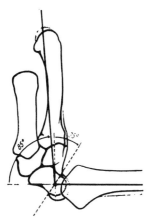

FIG. 5. Dorsiflexion of wrist.
(Lanz & Wachsmuth)

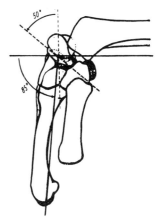

FIG. 6. Volar flexion of wrist.
(Lanz & Wachsmuth)

proximal row glides to the volar side (Fig. 5) and on volar flexion to the dorsal side (Fig. 6).

While these axes for the dorsovolar and for the radioulnar excursions of the radiocarpal joint are grossly related to the long and short diameters of the articular surface of the radius, these movements are in reality more complicated. First of all, it has been shown already by Langer[6] that the radial abduction is accompanied by volar flexion and ulnar abduction by dorsiflexion of the proximal row. Secondly, in the lateral motion it is the scaphoid which undergoes the greatest shift, moving radially in ulnar abduction as much as 1 cm. (Fick,[5] III, pg. 371).

3. Capsule and ligaments

It is of clinical interest to observe that the capsule of the radiohumeral joint is rather roomy, having its narrowest point in the middle where the lunate articulates with the radius.

a) Longitudinal ligaments are the ulnar and the radial collaterals. The ulnar arises from the styloid of the ulna and fastens to the triangular cartilage, reaching the pisiform to combine with the pisometacarpal ligament (Fig. 7). It becomes tight on radial abduction. The radial collateral arises from the styloid of the radius and becomes interwoven with the capsular apparatus, from which it is difficult to separate. It reaches the scaphoid and, crossing the intracarpal articulation, passes on to the trapezium and the first metacarpal. It becomes tight on ulnar abduction (Fig. 7).

b) The transverse and oblique ligaments of the wrist

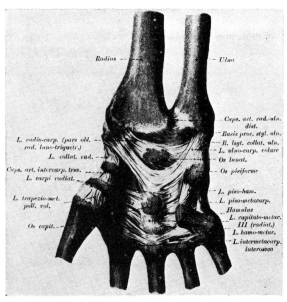

FIG. 7. Volar view of the wrist showing the ligamentous structures. (Fick)

form a great number of heavy strands on both volar and dorsal surfaces. They are: on the volar side, the volar arcuate and the volar radiocarpal and ulno-

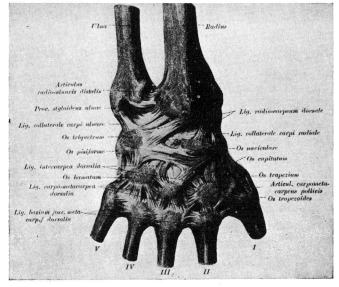

FIG. 8. Dorsal view of the wrist showing the ligamentous structures. (Fick)

carpal (Fig. 7), and on the dorsal side, the dorsal radiocarpal and the dorsal arcuate ligaments (Fig. 8).

All of these form a strong tiepiece reinforcing the various carpal articulations and maintaining the carpal arch. The two oblique volar ligaments, namely, the volar ulnocarpal and the volar radiocarpal, converge toward the midline at the lunate and the head of the capitate (Fig. 7). They form a heavy arcuate tiepiece across the carpus, the ligament of Henle.

The dorsal ligaments are less symmetrically arranged. There is the above mentioned oblique one from the radius to the triquetrum, the dorsal radiocarpal ligament, and a transverse ligament from the scaphoid to the triquetrum, the ligamentum carpi dorsale transversum (Fig. 8). The ligaments from the radius to the triquetrum, both volar and dorsal, are the ones which secure the wrist to the forearm in pro- and supination. Supination tightens the volar and pronation the dorsal ligament.

B. THE INTRACARPAL ARTICULATION

Between the proximal and distal carpal rows there is a compound joint, the intracarpal articulation. Here, both articulating bodies have a convex and a concave portion. In the ulnar half the convexity is formed by the hamate and the capitate, and the concavity by the adjacent surface of the triquetrum, the lunate, and a portion of the scaphoid. In the radial half, the convexity is formed by the rest of the scaphoid, and the concavity by the lesser and greated multangulum (Fig. 2). Specifically, the distal portion of the scaphoid articulates with the greater and lesser multangulum, the body of the scaphoid and the lunate with the capitate, and the triquetrum with the hamate bone.

In radial abduction, the capitate glides ulnarly along the socket formed by the bones of the proximal row, and vice versa, in ulnar deviation, it glides radially. Ulnar deviation causes a gap between scaphoid and greater multangulum; radial deviation closes the joint surfaces tightly. In dorsiflexion of the hand, the distal row moves volarly, and in volar flexion, dorsally.

Except for the capsular septa to the individual joints separating the various carpal bones, the capsule bridges the intracarpal articulation. It is more relaxed dorsally and more tight volarly. There are volar and dorsal ligaments reinforcing the intracarpal articulation. The dorsal ones have mostly a stellate form, spreading fanshape from the hamate, the capitate, the trapezoid and trapezium upward to the first row and downward to the bases of the metacarpals (Fig. 8). The volar ligaments are more transversely directed, going from the capitate to the multangula radially and the hamate ulnarly (Fig. 7). In this articulation, one can speak of a fixed transverse axis only in pure dorsovolar movement. If accompanied by radial abduction with the scaphoid fixed, the axis goes radioulnarly from the head of the capitate to the hamate; in midposition, the lunate is fixed and the axis goes from the trapezium through the capitate to the middle of the hamate.

The optimum excursion range is from radiodorsal to volarulnar, an arrangement which fits in admirably with the distribution and the pairing of the muscles about the wrist.

C. THE COMBINED MOVEMENTS OF BOTH CARPAL JOINTS
(FIG. 9)

From the diagram representing the ranges of the combined radiocarpal and intracarpal articulations (Braune and Fischer[1]), one sees that the combined excursion range in dorsovolar direction is 168°. The midposition of the joint is not a straight 180° extension because, as mentioned above, the lower end of the radius is slightly bent forward. This gives an advantage to volar flexion at the expense of dorsal extension. It can further be noted from the curve representing the excursion field of the *radiocarpal articulation*, that the excursion range in this joint is principally in volar direction (Fig. 9), which makes it the principal joint for the volar flexion and har-

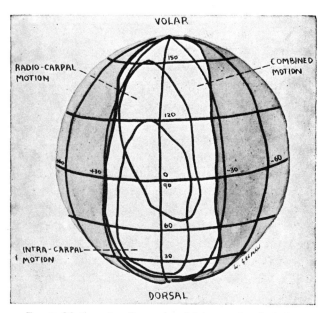

FIG. 9. Motion of radiocarpal and intracarpal articulations.

monizes with the fact that the ligamentous apparatus of the radiocarpal articulation is rather lax on the dorsal side in contrast to the tighter volar ligaments.

In the *intracarpal articulation,* on the other hand, the excursion field extends in the opposite direction. It is wider dorsally and radially; this articulation contributes more to the dorsiflexion and the radial abduction of the hand.

When these two excursion fields are *combined* (Fig. 9), one notices first that the ab- and adductory ranges are greatest in midposition of the hand as a whole. Also, the combined flexion-extension range is greatest in midposition and rapidly diminishes with ab- or adduction.

In the specimen, the greatest range of dorso- and volar flexion at or about midposition would amount to about 170°, of which 70° is due to the radiocarpal and about 100° to the intracarpal joint. The ab- and adductory range has also its maximum at midposition, amounting to 125°. These values, taken from the ligamentous specimen, are much greater than in the living, where muscular checks considerably restrict the ranges of motion. In the living these ranges show great individual differences and they are usually greater in women and in loosely knit individuals. Volar flexion, on the whole, is carried out to a greater degree in the radiocarpal and to a lesser degree in the intracarpal articulations. The latter, again, contributes more to dorsiflexion. Ulnar abduction takes place principally in the radiocarpal joint, and radial abduction in the intracarpal joint. The combined volar excursion field in the living is 84°, the dorsal 64°, a total of 148°, as against 170° in the specimen. The combined radial abduction is 30°, ulnar from 30-50°, a total of 80°, against a maximum of 125° in the specimen.

D. THE CARPOMETACARPAL ARTICULATION

1. The carpometacarpal joint of the thumb

If one considers the greater multangulum as a rudimentary metacarpal, this joint would range with the metacarpophalangeal joints of the fingers which it approaches in function more than the carpometacarpal articulations of the other rays.

It is a saddle joint which has two principal axes: a dorsovolar for ab- and adduction, and a radioulnar for flexion and extension; in addition, there is rotation about a longitudinal axis, three degrees of freedom of motion in all.

The plane of ab- and adduction does not strictly coincide with the plane of the hand, as it does in ab- and adduction of the fingers where the axis for the movement is strictly dorsovolar. In the thumb, this movement occurs about an axis which forms an angle with the ab- and adductory axis of the fingers of about 25° opening dorsally. The range of motion about this axis is 35-40°. Adduction is checked by contact with the index finger and abduction by the tension of the adductor pollicis. Flexion and extension or opposition and reposition occur about a transverse axis, which is also oblique to the transverse axis of the metacarpophalangeal flexion-extension movement of the fingers, forming with the axis of the latter an angle of about 25° opening ulnarly. The range of motion in this plane is 45-60°. The length rotatory motion is limited since an independent length rotation is not suitably provided for by this saddle joint. DeBois-Raymond[4] gives it a pronatory range of 30°.

Opposition of the thumb is combined with pronation, and the plane of the nail changes in full opposition from the frontal to the sagittal, i.e., 90° or more. Since the axis for extension and flexion is oblique, complete opposition of the thumb is a combination of movements. This opposition starts with abduction, then follows flexion, and is finally finished with adduction, all in the carpometacarpal joint. It is a circumductory movement in which the thumb describes the surface of a cone with a wide base. According to Bunnell,[2] the thumb can be extended to behind the plane of the palm; it can be abducted to 90° with the radial border of the hand. Forward flexion is associated with pronatory rotation after the first third of the movement, and in maximum opposition the plane of the nail is almost parallel again with that of the palm, which means a rotation of close to 180°.

2. The carpometacarpal articulation of the fingers

The second metacarpal articulates principally with the trapezoid and, with smaller facets also with the trapezium, the capitate, and the third metacarpal. The third metacarpal articulates principally with the base of the capitate and has smaller facets contacting the bases of the second and fourth metacarpals; the fourth articulates with the opposing edges of the capitate and hamate and with the adjacent bases of the third and fifth, and the fifth articulates with the hamate and the base of the fourth metacarpal (Fig. 2). All articulations are heavily supported by strong transverse and weaker longitudinal ligamentous tracts, both volarly and dorsally (Figs. 7, 8), which greatly restrict mobility in these joints.

There is practically no motion and, owing to the intricate articular relationship with neighboring joints and the heavy ligamentous reinforcements, isolated dislocation is rare, but strain of the carpometacarpal arch from relaxation of these ligaments is not uncommon.

E. THE METACARPOPHALANGEAL ARTICULATIONS

1. The metacarpophalangeal joint of the thumb

This joint allows volar flexion and dorsal extension with a slight degree of ab- and adduction, though the latter motion is much more restricted than that of the fingers. It carries two sesamoids on the volar side, a larger radial and a smaller ulnar, both increasing the arm of leverage for the short flexor of the thumb. The arrangement of the ligaments is the same as in the metacarpophalangeal joints of the fingers except that the arcuar bundles of the collaterals go directly to the respective sesamoids, which, in addition, are united by a strong intersesamoid ligament.

On the whole, the excursions in this articulation are much more restricted than in the fingers. The thumb loses in this articulation some of what it gains by the free mobility in the carpometacarpal joint. According to Fick,[5] the flexion-extension range is between 50-70°, and it is the dorsiflexion which is mainly restricted although there are many people whose thumb is hyperextensible in the metacarpophalangeal joint.

2. The metacarpophalangeal joints of the fingers

They have their convex body in the metacarpal head which articulates with the shallow concavity of the basal phalanges. The curve of the heads is sharper dorsoventrally than that of the sockets, but in the radioulnar direction, convex and concave portions of the joint show more congruity. The heads appear "pinched" in lateral direction, becoming narrower in front, in contrast to the thumb where the joint surfaces are broader and more quadratic. Sometimes the volar portion of the joint surface of the head is separated from the dorsal by a cartilage-covered transverse ridge. This is of clinical interest because it may be one of the reasons for the phenomenon of the snapping finger. In dorsoventral direction the angular value of the head is, according to Fick,[5]

180°. He concedes to the basal phalanx, which has a much greater radius, an angular value of only about 20°, although most observers place it much higher. This discrepancy of the radii leaves the joint capsule rather slack. The atmospheric pressure is of some consequence and on pulling the finger a cracking noise is sometimes heard as the two articular surfaces separate.

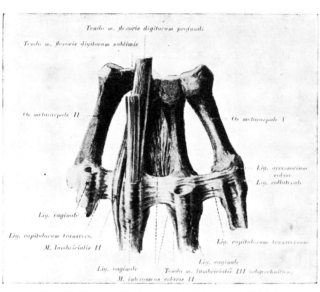

FIG. 10. Fibrocartilaginous vaginal ligament. (Fick)

This joint permits three degrees of freedom of motion: dorsal and volar flexion in the sagittal, ab- and adduction in the frontal, and pro- and supination in the transverse plane. Actually, only two movements are under voluntary control, namely, flexion-extension and ab- and adduction. Pro- and supination accompany flexion and extension as instantaneous (momentary) motion.

Ligamentous reinforcements and checks:

a) The most powerful reinforcement of the joint capsule is the volar fibrocartilaginous plate. This plate also forms the dorsal side of the fibrocartilaginous vaginal ligament which encloses the flexor tendons. The proximal end of the plate continues laterally as the intermetacarpal ligament (Fig. 10). This plate restricts dorsiflexion but its primary function is to form part of the gliding apparatus for the flexor tendons.

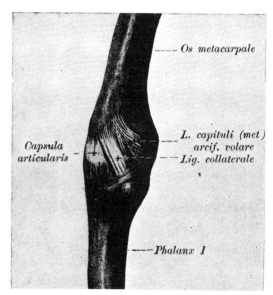

FIG. 11. Collateral ligaments of the metacarpophalangeal joint. (Fick)

b) The sides of the joint are occupied by the collateral ligaments (Fig. 11). They converge in their oblique course volarly and distally. It is particularly the tension of these collaterals which restricts motion. They become tight in flexion as they follow the path of the forward gliding basal phalanges. They are relaxed in extension.

The excursion range of the dorsovolar movement in this joint is 110-120°; the phalanx can be brought to right angle with the metacarpal in flexion and it can be extended 20° beyond the straight angle. This range varies greatly in different individuals and even in the different fingers of the same hand. There is also a certain interdependence of the four fingers in extension movement due to intertendinous bridges between the extensor tendons. Independent flexion and extension of the individual fingers can be acquired by training, as in a pianist or a violinist, and the old idea of cutting the anastomosing tendinous bands between the extensor tendons has little to recommend it.

The ranges of lateral motion differ with the individual fingers. All side motions become free as the fingers go into extension, while in flexion they are checked by the tension of the lateral ligaments. The freest is the movement of the idex finger, with a total of 60° ab- and adduction. Then comes the little finger with 50° of mostly abduction, then the middle and the ring fingers with about 45°.

In addition to the checking effect of the collateral ligaments, one must also consider the tendinous sleeve which forms part of the extensory apparatus of the fingers. In flexion, this sleeve moves distally over the joint and so inhibits side motion, while in extension it moves proximally, leaving the joint free for ab- and adduction.

So far as pro- and supination of the fingers is concerned, it has already been mentioned that it is not under voluntary control. As the fingers go from flexion into extension, they rotate axially from a pronatory to a supinatory position, and vice versa, but this movement is entirely passive, "trackbound," so to speak. It is closely connected with the spreading of the fingers in extension, and with their cupping and crowding together when they go into flexion.

F. THE INTERPHALANGEAL JOINTS

They are true hinge joints, having only one degree of freedom of motion. The convex body is formed by the heads of the phalanges, the basal and midphalanges, respectively; the concave bodies are furnished by the bases of the middle and endphalanx. The interphalangeal joint of the thumb shows no essential difference.

The angular values of the convex bodies of the basal and middle phalanges is 180°, that of the respective sockets is 90°. However, in the endphalangeal joint the difference between the angular values is usually less, which explains why the range of motion is restricted to 60-70°.

In full flexion, the tension of the dorsal capsular structures acts as a check. Lateral motion does not occur, being effectively checked by the lateral ligaments of this joint. The ligamentous apparatus is constructed on similar principles as in the metacarpophalangeal joint (Fig. 11). We find here the same rectangular fibrocartilaginous plate (ligamentum accessorium volare) and the same collaterals, except that their course is less oblique and they are less tight than in the metacarpophalangeal joint. The flexion range in the midphalangeal joint averages 100-110°, and in the endphalangeal joint, 60-70°. Bunnell[2] places the excursion of the midphalangeal joints at 110-130° and that of the endphalangeal joints at 45-90°, against the flexion-extension range of the metacarpophalangeal joint of 90-100°. In many loose-jointed individuals a considerable degree of hyperextension is possible in either joint, and especially in the thumb and little finger.

II. THE DYNAMICS OF THE WRIST AND FINGERS

The muscle pattern for the hand and fingers is essentially identical with that of the foot and toes; but, whereas the function of the foot has undergone a

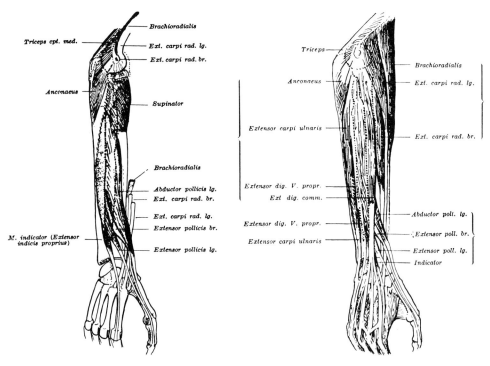

FIG. 12. Dorsal muscles of the forearm. (Strasser)

FIG. 13. Dorsal muscles of the forearm. (Strasser)

philogenetic recession in the sense that stabilization has taken precedence over mobility, in the hand and fingers the case is reversed: its function has become more refined and mobility has become the principal requirement.

A. REVIEW OF HAND AND FINGER MUSCLES

With the exception of the pronator teres and the quadratus, the brachialis and the supinator, all muscles of the forearm cross both wrist joints and many of them the metacarpophalangeal and finger joints as well. The fact that they develop rotatory moments about all these joints makes for a very complicated dynamic situation.

There are two main groups: the volar, which comes from the internal condyle of the humerus and the volar aspect of the forearm; and the dorsal, which arises from the external condyle and the dorsal side of the forearm.

1. The dorsal group

This group, again, can be divided into three layers: a superficial, a deep, and an intermediate one.

a) *The superficial layer.* This layer consists of the supinator brevis, the long and short extensors carpi radialis, the extensor carpi ulnaris, and the

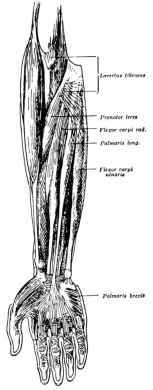

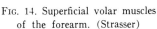

FIG. 14. Superficial volar muscles of the forearm. (Strasser)

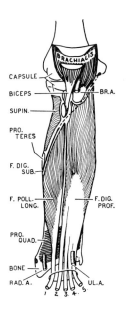

FIG. 15. Deep volar muscles of the forearm. (Grant)

brachioradialis. Only the extensor carpi radialis longior and brevior are purely wrist muscles, inserting into the bases of the second and third metacarpals, respectively (Figs. 12, 13).

b) The *deep layers* are the oblique muscles, or the so-called "outcroppers." They comprise the extensor pollicis longus and brevis, the abductor pollicis longus, and the extensor indicis. Arising from the dorsal surface of the ulna, radius, and interosseous membrane, they run distally in oblique direction, crossing over the tendons of the extensors of the wrist to gain the thumb and index finger (Figs. 12, 13).

c) The *intermediate layer* consists of the common extensor and the extensor digiti minimi. It arises from the lateral humeral condyle and the radial collateral ligament of the elbow and runs between the superficial and the deep layers to the fingers, respectively (Figs. 12, 13).

2. The volar group

This has two layers, a superficial and a deep one.

a) The muscles of the superficial layer arise from the internal humeral epicondyle as a common muscle mass. The pronator is the most radial, then follows the flexor carpi radialis running to the bases of the second and third metacarpals, then the palmaris longus which ends in the palmar aponeurosis, and, finally, the flexor carpi ulnaris which goes to the pisiform and the fifth metacarpal base (Fig. 14).

b) The deep layer is composed of the finger flexors. The sublimis arises from

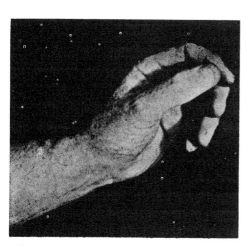

FIG. 16. Position of rest of fingers and wrist. (Bunnell)

the internal epicondyle of the humerus, the coronoid process of the ulna, and from the volar surface of the radius and ends in four divergent tendons at the bases of the middle phalanx of the second to the fifth fingers. The deep common flexor arises from the ulna and the interosseous membrane, crossing the wrist with the deep flexor underneath the volar carpal ligament to reach, after perforating the tendons of the superficial flexors, the endphalanx of the second to the fifth fingers; finally, the long flexor of the thumb, the deepest of the long forearm muscles, comes from the volar surface of the radius and goes to the base of the endphalanx of the thumb (Fig. 15).

B. THE ACTION OF THE MUSCLES OF THE WRIST

The neutral or rest position of the wrist, i.e., the one in which all muscles are under equal tension, is 3° ulnar abduction and 12° dorsiflexion. Bunnell[2] considers the rest position of the hand slightly different, namely, 20° dorsiflexion and 10° ulnar abduction. The neutral position in which all muscles are evenly balanced is this position of the wrist with the forearm midway between pro- and supination, the fingers slightly flexed in all joints, and the thumb in partial opposition (Fig. 16).

1. The volar movement

This movement is produced by the flexor carpi ulnaris, the flexor carpi radialis, and the long finger flexors. The shortening distance of these muscles is as follows: flexor carpi ulnaris, 3.9 cm.; flexor carpi radialis, 3.8 cm.; flexor digitorum profundus, 4.2 cm.; and flexor sublimis, 4.5 cm. (Fick[5]).

2. The dorsal extension

This is carried ont by the extensors of the fingers, the extensor pollicis longus, the two radial extensores carpi, the ulnar extensor carpi, and the extensor pollicis longus. Their contractile distance for this range is as follows: extensor digitorum communis, 4 cm.; extensor carpi radialis longior, 3.4 cm.; extensor carpi radialis brevior, 4 cm.; extensor carpi ulnaris, 2.1 cm.; extensor pollicis longus, 2.5 cm.; and extensor indicis, 3.8 cm.

3. The ulnar abduction

This movement is enacted by the flexor and extensor carpi ulnaris and extensor digiti minimi. In this direction the shortening of the muscles is as follows: extensor carpi ulnaris, 2 cm.; flexor carpi ulnaris, 1.4 cm.

4. The radial abduction

This is produced by the extensor pollicis longus, the extensor pollicis brevis, the abductor pollicis longus, the extensors carpi radialis longior and brevior, the flexor carpi radialis, and the extensor indicis proprius. The contraction distances in this direction are as follows: extensor carpi radialis longior, 3.6 cm.; extensor carpi radialis brevior, 1.4 cm.; extensor pollicis longus, 1.4 cm.; abductor pollicis longus, 2.1 cm.; flexor carpi radials, 0.3 cm.; and extensor indicis, 0.7 cm. The above figures relating to the distance of contractile shortenings are those of DeBesser.[3]

C. THE FUNCTIONAL PATTERN OF WRIST MOTION

The description of the part which the different muscles play in the movement of the wrist about its two axes must be supplemented by an account of the synergistic combinations in which they perform their usual functions. For this purpose the relation of the axes of the joint to the direction of the different muscles must be taken into consideration. As the diagram (Fig. 17) shows, all muscles have some obliquity to the planes of motion, i.e., no muscle has its axis strictly perpendicular to one or the other of the two joint axes. There is no pure extensor or flexor or a pure ab- and adductor, but because of their oblique course all muscles develop secondary components. Therefore, if a motion in either sagittal or frontal plane is to be accomplished by action of a muscle, its secondary component must be neutralized by its antagonist.

The diagram shows that there is a definite antagonistic pairing of the muscles and one can distinguish three principal pairs.

The extensor carpi radialis longior and brevior are paired with the flexor carpi ulnaris and the two long flexors of the fingers: the former are extensors and radial abductors, the latter are flexors and ulnar abductors. The excursion plane of these muscles runs from the radiodorsal to the ulnovolar. This cor-

responds to the natural flexion-extension movement of the wrist which is
directed from radial extension to ulnar flexion.

The second pair is the extensor digitorum communis and the extensor indicis proprius against the flexor pollicis longus and the flexor carpi radialis. The
former extends the wrist and slightly abducts it ulnarly; the latter flexes the
wrist and abducts it radially. The excursion plane of this pair of muscles comes

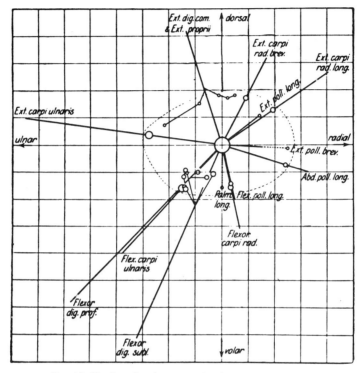

FIG. 17. The functional pattern of wrist motion. (Braus)

close to, but does not precisely reach, the sagittal plane. It deviates slightly
dorsoulnarly to ventroradially.

The third pair is composed of the extensor carpi ulnaris against the extensor
pollicis brevis and the abductor pollicis longus. Here, also, movement occurs
almost but not quite in the frontal plane because ulnar abduction is associated
with slight extension and radial abduction with slight flexion.

D. THE MUSCLE ACTION OF THE FINGERS

The long finger flexors as well as the extensors are effective upon all three
finger articulations in cooperation with the intrinsic muscles of the hand.

The flexor tendons of the fingers, the deep as well as the superficial, emerge
from under the transverse volar ligament, enclosed in the ulnar bursa, and
then diverge in the palm toward the individual fingers. In the free interval
between the ulnar and the digital sheath, the deep flexor takes on the lumbricals

on the radial side. They, then, continue their course united within the volar digital sheath until the superficial tendon splits apart to let the deep flexor reach its insertion at the endphalanx, while the superficial tendon inserts its two slips into the midphalanx of the finger (Fig. 18).

The interossei and lumbricals coming from the palm reach the extensor tendons in oblique direction and leave a triangular field, which is occupied by the triangular ligament lying directly in the line and transmitting the pull of these intrinsic muscles. As they reach the extensor tendon, the latter splits to

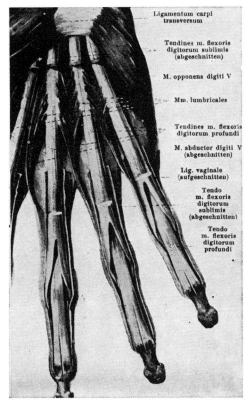

Ligamentum carpi
transversum

Tendines m. flexoris
digitorum sublimis
(abgeschnitten)

M. opponens digiti V

Mm. lumbricales

Tendines m. flexoris
digitorum profundi

M. abductor digiti V
(abgeschnitten)

Lig. vaginale
(aufgeschnitten)

Tendo
m. flexoris
digitorum
sublimis
(abgeschnitten)

Tendo
m. flexoris
digitorum
profundi

FIG. 18. The volar hand and finger muscles.
(Spalteholz)

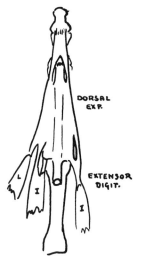

DORSAL
EXP.

EXTENSOR
DIGIT.

L

I

I

FIG. 19. Intrinsic muscles streaming into extensory apparatus.
L = Lumbrical
I = Interosseous

allow its midportion to fasten into the midphalanx while the lateral slips with the tendons of the interossei, which on the radial side are joined by the lumbricals, unite again to reach the endphalanx (Fig. 19).

Between these two systems, the flexory and extensory, are placed the intrinsic muscles which fill the intermetacarpal space in the palm. There are seven of these, four occupying the dorsal and three the volar side of the interosseous space. Taking the middle finger as the midline, the volar interossei (one for the index, one for the ring finger, and one for the fifth finger) bring the fingers toward this line. The dorsal interossei (two for the middle finger, one for the index, and one for the ring finger) move the fingers away from the middle finger and spread the fingers. Thus the interossei control the voluntary side motion of the fingers in the metacarpophalangeal joints, though passive spreading of the fingers occurs to some extent with extension and passive crowding with flexion of the fingers (Fig. 20).

The function of the intrinsic muscles is by no means restricted to ab- and adduction in the metacarpophalangeal joint. They also have a strong flexory effect on this joint and, due to their attachment to the extensor apparatus, they have a strong extensory effect upon the mid- and endphalangeal joints.

The intrinsic muscles act, so to speak, as moderators between the long ex-

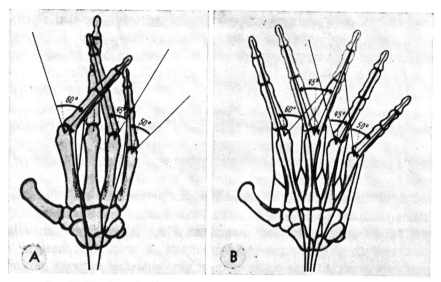

Fig. 20. The ab- and adductory action of the intrinsic muscles on the fingers.
(Lanz & Wachsmuth)

tensors and the long flexors of the fingers. These muscles are not perfect antagonists and they do not neutralize each other completely. Their respective rotatory moments are not equal and opposite to any of the articulations of the fingers; for instance, the rotatory moment of extensors in the metacarpophalangeal joint is greater than that of the volar flexors. Furthermore, when the hand is strongly flexed in the wrist, the closure of the fingers becomes impossible because the finger flexors are already contracted almost to their limit, and therefore they become actively insufficient. Similarly, when the fingers are flexed in the interphalangeal joints it is not possible to flex fully the metacarpophalangeal joints without straightening the interphalangeal joints at the same time. When the wrist is maximum extended, the extensors can extend the metacarpophalangeal joint but it would be difficult for them to extend the interphalangeal joints because they also become actively insufficient.

It is, therefore, the province of the intrinsic muscles to establish the equilibrium and to come to the aid of the long finger muscles whenever these threaten to become insufficient. Accordingly, these intrinsic muscles, by flexing the metacarpophalangeal joint, assist the long flexors to hold the equilibrium against the powerful action of the long extensors; and, when they extend the interphalangeal joints, they help the long extensors which would not otherwise be able to hold the equilibrium against the more powerful flexory moment of the long finger flexors.

Another effect of the lumbricales which is often overlooked results from their anatomic relation to the deep flexor tendons. When the lumbricales contract to extend the fingers in the interphalangeal joint, they necessarily exert a pull upon these tendons from which they arise, and thereby they relax the distal end of the long flexors so that they become unable to resist the action of the lumbricales as they extend the interphalangeal joint (Recklinghausen[8]). The cooperation of the intrinsic muscle is most essential for the carrying out of the finer movements of the fingers.

The shortening distances of the long finger muscles have been determined by Bunnell[2] as follows: the flexor profundus, 35.6 mm.; the sublimis, 35.2 mm.; the extensor communis and indicis proprius, 16 mm.

E. THE ACTION OF THE MUSCLES OF THE THUMB

The midposition in the carpometacarpal joint of the thumb is between maximum opposition and maximum extension. There are four special muscles forming the thenar. The adductor arises with a transverse and an oblique portion from the volar surface of the third metacarpal. The opponens comes from the transverse ligament and the greater multangulum and goes to the body and head of the first metacarpal. The short flexor of the thumb arises with a superficial head from the transverse ligament and with a deep head from the multangula, the capitate and the second metacarpal, and runs in two portions to the radial and ulnar sesamoid and to the basal phalanx of the thumb. The short abductor comes from the carpal ligament and the tuberculum of the scaphoid and goes to the basis of the basal phalanx of the thumb (Fig. 21).

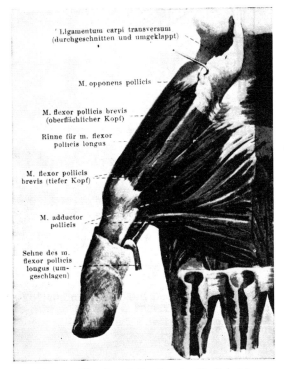

FIG. 21. Muscles of the thumb. (Spalteholz)

Adduction of the thumb is performed by the adductor and the radial head of the first dorsal interosseus; abduction is carried out by the abductor pollicis longus, extensor brevis, and the abductor pollicis brevis; opposition is produced by the opponens which, being attached to the radial side of the carpus, is together with the flexor brevis the principal muscle producing a pronatory movement of the thumb. Extension is carried out by the long and short extensors of the thumb.

From extension and reposition to flexion and opposition the sequence of muscle action is as follows: abduction and extension, flexion, opposition and adduction. In full opposition the thumb is able to sweep the bases of all fingers when they are extended in the metacarpophalangeal joints.

III. THE COMPUTATION OF WORK OF THE MUSCLES OF THE WRIST AND FINGERS

A. THE WORKING CAPACITY OF THE MUSCLES OF THE WRIST

In basing the working capacity on the product of the absolute muscle power and the contraction length (visible work), we find that of the flexors of the wrist to be 4.613 kgm. against only 1.945 kgm. of the extensors, and that of the ulnar abductors 0.6 kgm. against 0.739 kgm. for the radial abductors. This tallies with the cross-section areas of the muscles, that of the volar flexors being 33.4 cm.^{2} against only 16.72 cm.^{2} of the dorsal extensors; the total cross-section area of the ulnar abductors is 10.3 cm.^{2}, and that of the radial 11.12 cm.^{2}

To estimate the actual strength of these muscles which they display under isometric condition, i.e., when the muscle is not allowed to shorten, these values of the respective cross-section areas are essential. But the strength of the muscle, as explained before, rapidly decreases with increasing contraction. To judge their practical efficiency when the muscles perform visible motion, the contraction distance becomes a factor, and one has to think in terms of work (force times distance) rather than of absolute muscle power.

B. THE RELAXATION ANGLE OF THE WRIST MUSCLES

We have on other occasions stressed the fact that when the wrist is in a muscular midposition, i.e., 12° dorsiflexion and 3° ulnar abduction, all muscles are under equal tension. Under Sherrington's law of reciprocal innervation, the antagonist gradually releases its tension as that of the synergist increases. In the case of flexion of the wrist, the tension of the antagonistic extensors disappears only after 1/12 of the flexion range has passed. This 1/12 is called the field of common tension.

Now since the muscles, synergist as well as antagonist, are under equal tension in midposition, the question arises as to what degree a muscle contracts before its antagonist relaxes completely to assume its natural length. Recklinghausen[8] investigated this problem and found as follows:

The flexor carpi radialis relaxes fully at 8° dorsiflexion; the extensor carpi radialis requires even 24° volar flexion before it becomes completely relaxed.

Finger extensors also require a certain flexion position before their innervational tone fully disappears. It follows that while the most suitable position for a stable wrist is the midposition, where all muscles are under equal tension, the most stable position for the fingers is slight flexion. The combined position of fingers and wrist at this point is the so-called functional position of the entire hand, both of wrist and fingers (Fig. 16).

C. THE HAND AS AN INSTRUMENT OF GRIP

Fist closure is either hook-like or vise-like. In the latter case, a uniform pressure is exerted on the object held, while in the former, pressure is exerted

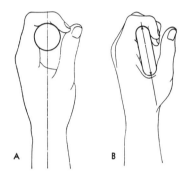

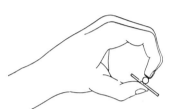

FIG. 23. Small object held between fingertips. (Recklinghausen)

FIG. 22. *A*, hook-like grip. *B*, vise-like grip. (Recklinghausen)

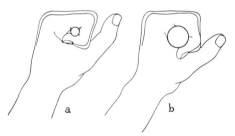

FIG. 24. The clamping fist. (Recklinghausen)
a. smaller, b. larger object

against the phalanges only, and not against the palm (Recklinghausen[8]) (Fig. 22).

In another case the object held in the hand may be small so that when the fingers close upon it their tips touch only the tip of the thumb (Recklinghausen[8]) (Fig. 23). In both instances the flexors profundus and sublimis furnish the gripping force .The intrinsic muscles do participate as they are not only flexors of the metacarpophalangeal joints but also extensors of the interphalangeal joints.

1. The clamping fist

Because the ratio of moments of the finger flexors in the three joints is seven (metacarpophalangeal) to five (midphalangeal) to one (endphalangeal joint), it is obvious that the endphalangeal joint is not engaged so much in closing the fist as the other two (Fig. 24). In this type of closure the object is pressed by all articulations, though with different strength. The thumb and the thumb web take part in the pressure.

2. The hook fist

In the hook-shape fist, as for instance when hanging from a horizontal bar, pressure is confined to the phalanges, the resistance is restricted to the inter-phalangeal articulations; and thumb and thumb web remain free (Fig. 22A).

3. The egg-shape fist closure (Fingertips touching)

A small object is held firmly, for instance, a pen. This is also carried out with considerable force, and the strength of both flexors and extensors is necessary to stabilize the tips of the fingers.

4. Opening the fingers

In opening the fingers, no particular resistance is to be overcome; however, flexing the wrist aids by virtue of the passive insufficiency of the extensor muscles in the extension of the fingers, and this is made use of automatically in certain pathological conditions.

It has been pointed out that the so-called position of function of the wrist is essential for the best display of flexory power of the fingers. In fact, this position puts the finger flexors somewhat beyond their natural length and takes up a good deal of the slack which is produced by the flexion of the three finger joints. At the same time, the tension which the finger flexors display in flexing the interphalangeal joints is also antagonistic to the efforts of the wrist extensors to hold the hand in a functional position. Therefore, a certain amount of flexor strength is bound up or absorbed by this stabilizing mechanism. According to the computation of Recklinghausen,[8] the strength of the finger flexors can be utilized only to about 71% of their total strength. There is no danger, however, that fist closure will suffer to any degree so long as the finger flexors and the extensors of the wrist are intact. On the other hand, any weakness of the wrist extensors, as, for instance, in musculospiral paralysis or in poliomyelitis, greatly interferes with the strength of the finger flexors and with their ability to close the fist forcefully and to grip objects firmly.

IV. SUMMARY

1. The radiocarpal articulation forms its convex body by the proximal carpal row consisting of scaphoid, lunate, and triquetrum, while the concave portion consists of the articular surface of the radius and the triangular cartilage. The joint has an anteroposterior and a side-to-side curve of which the anteroposterior is the sharper with an angular value of 67° for the radius and 113° for the carpal row, and the side curve is the flatter with an angular value of 75° for the radius and 109° for the carpal row.

2. The joint has two degrees of freedom of motion, one in dorsovolar and one in radioulnar direction. Midposition of this joint alone is slight volar flexion and ulnar abduction. The axis for dorsovolar movement corresponds approximately to the greatest diameter of the articular surface of the radius; that for radioulnar movement is perpendicular to it.

3. The movement of the joint is gliding: in radial abduction the proximal carpal row glides ulnarly; in ulnar abduction, radially; in dorsiflexion, volarly; in volar flexion, dorsally. Radial abduction is accompanied by volar flexion, ulnar abduction by dorsiflexion of the proximal carpal row.

4. Ligamentous reinforcements are the ulnar and radial collaterals, the former checking radial, the latter ulnar abduction, and the transverse and oblique ligaments of the wrist, both volar and dorsal. They consist on the volar side of the volar arcuate and the volar radiocarpal, and on the dorsal side of the dorsal radiocarpal and dorsal arcuate ligaments. All of these form strong tiepieces reinforcing the various carpal articulations and maintaining the carpal arch. The ligaments from radius to triquetrum, both volar and dorsal, secure the wrist in pro- and supination.

5. The intracarpal articulation is between proximal and distal carpal rows. Both joint constitutuents have a convex and a concave half. In the ulnar half the convexity is formed by the hamate and capitate, the concavity by the triquetrum, lunate, and a portion of the scaphoid. In the radial half the convexity is formed by the rest of the scaphoid and the concavity by the greater and lesser multangulum. The movement here is also gliding. In radial abduction the capitate glides ulnarly, in ulnar abduction, radially. In dorsiflexion the distal row moves volarly, in volar flexion, dorsally. Volar and dorsal intracarpal ligaments reinforce the intracarpal articulation. Only in straight dorsovolar movement is there a fixed transverse axis; if accompanied by radial abduction, the scaphoid fixed, the axis goes radioulnarly from capitate to hamate; in midposition, the lunate fixed, from trapezium through capitate to hamate. The optimum excursion is from radiodorsal to ulnovolar.

6. In combined movements of both joints, ab- and adductory ranges are greatest in midposition of the hand as a whole, which is 12° dorsiflexion and 3° ulnar abduction. In the radiocarpal joint, the excursion range is principally volar and ulnar; in the intracarpal joint, dorsal and radial. The combined volar excursion in the living is 84°; the dorsal, 64°; a total of 148°. The combined radial abduction is 30°; the ulnar, 30-50°; a total of 80°.

7. The carpometacarpal articulation of the thumb is a saddle joint with a dorsovolar axis for ab- and adduction and a radioulnar for flexion and extension; there is also rotation about a length axis. Ab- and adduction are in a plane inclined volarly 25°. The range is given variously from 40-60°. The flexion and extension plane also diverges from that of the fingers by an angle of 25° opening ulnarly; the range is 45-60° Opposition of the thumb is combined with pronation of over 90°, and in maximum opposition, according to Bunnell, of 180°.

8. The carpometacarpal joints of the fingers have practically no motion, although they are heavily fortified by strong ligaments to sustain the carpo-metacarpal arch.

9. The metacarpophalangeal articulation of the thumb permits volar and dorsiflexion and a slight degree of ab- and adduction. It carries two sesamoids to increase the leverage of the short flexor of the thumb. The flexion-extension range is 50-70°.

10. In the metacarpophalangeal joints of the fingers the heads have a sharp dorsoventral curve and appear pinched laterally. A transverse ridge sometimes separates the dorsal from the volar part of the joint surface, which may be one of the causes of the snapping finger. The angular value of the metacarpal head in volar direction is 180°. The joint has three degrees of freedom of motion: dorsovolar, lateral, and length rotatory; only the first two are under voluntary control. The strongest reinforcement of the capsule is the volar fibrocarti-laginous plate; it restricts dorsiflexion. Laterally, the joint is reinforced by the collateral ligaments, which become tight in flexion and relaxed in extension.

11. The dorsovolar excursion range of the metacarpophalangeal joint is 110-120°. The intertendinous bridges between the extensor tendons interfere with the independent movements of the fingers, but independence can be acquired by training (pianist, violinist). A tendinous sleeve forming part of the extensor apparatus moves distally in flexion, inhibiting side motion; moving proximally in extenson, it leaves the joint free for ab- and adduction. The ranges for lateral motion are 60° for the index finger, 50° for the little finger, and 45° for middle and ring finger. Pro- and supination are not under voluntary control but the fingers go automatically in pronation on volar, and in supination on dorsiflexion.

12. The interphalangeal joints are pure hinge joints; the convex bodies of the basal and middle phalanges have an angular value of 180°, their respective sockets, 90°. The ranges for the midphalangeal joints are 100-110°, for the endphalangeal joints, 60-70°. Flexion is checked by the dorsal capsule, lateral motion by the lateral ligaments which are arranged similarly as in the meta-carpophalangeal joints.

13. The dorsal groups of the muscles of wrist and fingers have three layers: The superficial consists of the supinator brevis, the long and short extensors carpi radialis, the extensor carpi ulnaris, and the brachioradialis. The deep layer is formed by the outcroppers, the extensor pollicis longus and brevis, the abductor pollicis, and the extensor indicis. The intermediate layer consists of the common extensor and the extensor digiti minimi.

14. The volar group has two layers: the superficial one is composed of the pronator teres, the flexor carpi radialis, the palmaris longus, and the flexor carpi ulnaris; the deep layer consists of the common finger flexors, deep and superficial, and the long flexor of the thumb.

15. The neutral position of the wrist, of 12° dorsiflexion and 3° ulnar abduction, is also the one in which all muscles are under equal tension. Volar

flexion is produced by the flexor carpi ulnaris, flexor carpi radialis, and the finger flexors; they have shortening distances of from 4.5 cm. to 3.8 cm. Dorsal extension is carried out by the extensors of the fingers, the two radial extensors of the wrist, the extensor carpi ulnaris, and the extensor pollicis longus. The shortening distance of these muscles is between 2.1 cm. and 4 cm. Ulnar abductors are the flexor and extensor carpi ulnaris (shortening distance, 1.4 cm. to 2 cm.); and radial abduction is performed by the extensors pollicis longus and brevis, extensors carpi radialis longior and brevior and flexor carpi radialis. Their contraction length varies from 3.6 cm. for the long radial extensor of the wrist to 0.3 cm. for the flexor carpi radialis.

16. There is a definite functional pattern of wrist motion based on the pairing of muscles with their direct antagonists:

a) The extensors carpi radialis longior and brevior are paired with the flexor carpi ulnaris and the two long flexors of the fingers for movement which runs from radiodorsal to ulnovolar. This is the natural flexion-extension plane.

b) The extensor digitorum communis and the extensor indicis proprius are paired with the flexor pollicis longus and flexor carpi radialis. This excursion plane is not quite sagittal but deviates slightly in dorsovolar to ventroradial direction.

c) The extensor carpi ulnaris is paired with the extensor pollicis brevis and the abductor pollicis longus; this movement is radioulnar, almost precisely in the frontal plane.

17. Between the flexory and extensory systems of the fingers the intrinsic muscles are placed as the third link of a three-cornered balance. They act as moderators between long extensors and long flexors, because these latter muscles are not the perfect antagonists which could neutralize their movements in the finger joints. The rotatory moment of the extensors in respect to the metacarpophalangeal joints is greater than that of the flexors. The rotatory moment of the finger flexors in respect to the interphalangeal joint is greater than that of the extensors. This discrepancy is equalized by the interossei which assist in flexion of the metacarpophalangeal and in extension of the interphalangeal joints. Besides, the interossei and the lumbricales take care of the lateral movement in the metacarpophalangeal joints when the fingers are extended and when the lateral ligaments as well as the dorsal extensor sleeve no longer restrain lateral motion.

18. Adduction of the thumb is carried out by the adductor and the radial head of the first dorsal interosseus; abduction by the abductors pollicis longus and brevis and the extensor brevis; opposition by the opponens and flexor brevis. From extension and reposition to flexion and opposition, the sequence is: abduction, extension, flexion, opposition, and adduction.

19. The working capacity of the flexors of the wrist is 4.613 kgm. against 1.945 kgm. of the extensors; 0.6 kgm. for ulnar and 0.739 kgm. for radial abduction.

20. The antagonistic extensors of the wrist completely relax after 1/12 of

the flexion range has been passed. Expressed in degrees, the flexor carpi radialis relaxes at 8° dorsiflexion, the extensor carpi radialis at 24° volar flexion.

21. Fist closure can be vise-like or hook-like. In the clamping fist, the object is pressed uniformly in all articulations; in the hooked fist, as in hanging from a bar, pressure is confined to the phalanges, the thumb web remains free; in the egg-shape fist closure, a small object such as a pen is held between finger tips, but the position requires considerable force of flexors and extensors to stabilize the finger joints in a semiflexed position.

Flexion of the wrist facilitates extension of the fingers and opening the fist. The functional position of the wrist puts the finger flexors somewhat beyond the natural length. Flexing the interphalangeal joints places the extensors of the wrist under abnormal tension. A certain amount of flexor strength as well as extensory power is absorbed by the mechanism of wrist stabilization. According to Recklinghausen, only 71% of actual strength of the finger flexors is available for flexion of the finger joints. On the other hand, weakness of the wrist extensors, by failing to maintain the position of function, greatly interferes with the strength of the finger flexors and the ability to carry out a forceful closure of the fist.

BIBLIOGRAPHY

1. BRAUNE, W., and FISCHER, O.: Das Handgelenk. *Abh. d. Kgl. Sächs. Ges. d. Wiss.*, 1887.
2. BUNNELL, S.: *Surgery of the Hand*, 2d ed., Philadelphia, J. B. Lippincott, 1948.
3. DEBESSER, E.: L'action mecanique des muscles des doigts et du poignet. *Diss. Lausanne*, 1899.
4. DEBOIS, RAYMOND R.: Über die Oppositionsbewegung. *Arch. anat. Physiol., Physiol. Abt.*, 1896.
5. FICK, R.: *Handbuch der Anatomie und Mechanik der Gelenke*, Vol. I. Jena, G. Fischer, 1904; Vol. III. Jena, G. Fischer, 1911.
6. LANGER, C.: Mittellage der Gelenke. *Ztschr. Ges. d. Ärzte*, Wien, 1856.
7. LANZ, T., and WACHSMUTH, W.: *Praktische Anatomie.* Berlin, J. Springer, 1935.
8. RECKLINGHAUSEN, V. H.: *Gliedermechanik und Lähmungsprothesen*, Vols. I and II. Berlin. J. Springer, 1920.
9. RAUBER-KOPSCH: *Lehrbuch und Atlas der Anatomie des Menschen.* Fr. Kopsch, Leipzig. Georg Thieme, 1922.

Lecture XXXI

THE PATHOMECHANICS OF PARALYSIS
OF HAND AND FINGERS

I. REVIEW OF NORMAL MECHANICS OF WRIST
AND FINGER FUNCTION

As THE shoulder joint with its three degrees of freedom of motion controls the orbit of the movements of the extremity as a whole, and as the elbow joint by its ability to lengthen and shorten the arm dominates its reach, so the hand and fingers placed at the periphery of the limb establish, so to speak, contact with the outside world by their ability to grasp and hold objects. They represent, in a sense, the business end of the entire shoulder-arm complex. To perform this function requires an extraordinary amount of mobility involving seventeen degrees of freedom of motion, of which no less than five degrees are under control of the fingers. The degree of disfunction of the hand and fingers produced by paralysis, and the possibilities of repair must be evaluated on functional rather than on purely anatomic grounds.

A. THE WRIST

1. Extension

To give the wrist the stability which furnishes the optimum condition for the grip, both extensors and flexors of wrist and fingers are necessary. Being the basis for the proper use of the fingers, wrist extension is more a static than a dynamic function, and stability precedes mobility in importance.

2. Flexion

Here, the dynamic function takes precedence over the static as the wrist movement follows the momentum generated in the more proximal joints. It must be kept in mind, however, that active stabilization of the wrist in extension, which is so important for the function of the fingers, calls for flexors as well as for extensors. The extensors are able to provide stability only with the antagonistic support of wrist or finger flexors.

Whenever the stability of the wrist cannot be procured by muscle action, either natural or supplied by muscle transference, it may become necessary to sacrifice motion in the wrist entirely by arthrodesis. In this case, the flexors of the wrist become available for supplanting the extensors of the fingers and thumb.

3. Ab- and adduction

Movements occur in any position of the flexion-extension range, but they have their optimum in the above-mentioned position of function. Inability to ab- or adduct the hand actively is not a serious handicap so long as the hand is held in midposition and is not in contracture, as, for instance, the ulnar deviation in the arthritic hand.

4. Closing of the fist

In the closing of the fist the flexors of the fingers and the extensors of the wrist are in close synergistic relation.

a) Forceful gripping of a large object depends largely on the functional position of the wrist. The special task of the intrinsic muscles is in this situation to accentuate the grip in the metacarpophalangeal joints (Fig. 1) (Recklinghausen[7]).

b) Finger tips against the thumb for gripping smaller objects (Fig. 2). The muscles involved are again the flexors of the fingers and the intrinsic muscles.

Fig. 2. Fingertip grip for small objects. (Recklinghausen)

Fig. 1. Forceful grip. (Recklinghausen)

This time, the latter must overcome the long flexors in the interphalangeal articulation and accentuate the action of the finger extensors in these joints. This gives the fingers a sort of pill roller hand position, for which the functional position of the wrist is advantageous but not indispensable. The special effort of the intrinsics easily leads to fatigue (writer's cramp).

5. Opening of the fist

It would seem to require only the extensory function of the common and proper extensors of the fingers and of the wrist to accomplish the opening of the fist. However, flexion of the wrist, by causing a passive insufficiency of the extensors, facilitates the opening of the fist; this maneuver is used frequently in paralysis of the finger extensors (spastic hand).

6. Opposition of the thumb

It involves the thenar muscles and the adductor, with particular cooperation of the short abductor. The latter assists the opponens. The short flexes the metacarpophalangeal joint and, indirectly, extends the interphalangeal joint of the thumb. Paralysis of the short flexor interferes with the extension in the interphalangeal joint necessary to meet the tips of the fingers.

II. THE PATHOKINETICS OF PARALYTIC DISABILITIES

A. THE WRIST

1. Paralysis of the extensors (Recklinghausen[7]) (Fig. 3)

It has been explained above why paralysis of the extensors has a weakening effect upon the flexing ability of the fingers. The principal wrist extensors are the extensor carpi radialis longior and the extensor carpi ulnaris. One of these gives sufficient extension for ordinary use of the hand, except that the hand is drawn into an ab- or adductory position for lack of antagonistic check. The extensor carpi radialis brevior assists in extension, but alone it is unable to neutralize the ulnar adduction of the extensor carpi ulnaris.

2. Paralysis of the flexors (Fig. 4)

Many disturbances arise from the paralysis of the wrist flexors. Aside from the fact that the stabilizing effect of the flexors upon the extension of the wrist

Fig. 3. Paralysis of wrist extensors. (Recklinghausen)

Fig. 4. Paralysis of wrist flexors. (Recklinghausen)

is missing and must be substituted by the long flexors of the fingers, paralysis of the flexor carpi ulnaris is especially troublesome because there is no other muscle to abduct the hand ulnarly in flexed position. Duchenne[2] mentions the case of the violinist with paralyzed flexor carpi ulnaris of the left hand who could not reach the high notes on the stringboard because of inability to adduct the flexed wrist ulnarly.

B. THE FINGERS

1. Paralysis of the finger extensors (Recklinghausen[7]) (Fig. 3)

Paralysis of the finger extensors does not interfere seriously with the extension of the two interphalangeal joints, since this since function is performed by the interossei and lumbricales. On the other hand, a contracture of these intrinsic muscles leads to hyperextension of these joints as well as flexion contracture of the metacarpophalangeal joints, as exemplified in the arthritic "pill roller" hand (Fig. 5).

2. Paralysis of the finger flexors (Fig. 6)

A certain tension of the opposing muscles is necessary for the forceful extension of the finger joints. In the absence of the long flexors of the fingers, there is no such antagonistic control and voluntary extension by the interossei becomes excessive. The tension of the flexors, profundus, and sublimis no longer opposing the interossei, the latter hyperextend the interphalangeal joints.

Furthermore, the extensors of the fingers which have a considerable rotation moment in respect to the metacarpophalangeal joint, find themselves opposed only by the intrinsic muscles, so that the equilibrium is overthrown in favor of hyperextension of the metacarpophalangeal joint. This leads to the formation of a clawhand though the deformity does not assume the degree it does when the interossei are also paralyzed.

3. Paralysis of the interossei (Fig. 7)

Loss of interossei deprives the fingers of the ability to extend the middle and terminal phalanges, and it deprives the metacarpophalangeal joints of

FIG. 6. Paralysis of finger flexors. (Recklinghausen)

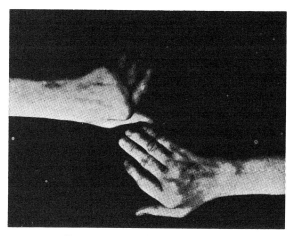

FIG. 5. Pill-roller hand. (Steindler)

FIG. 7. Paralysis of interossei. Claw hand. (Recklinghausen)

sufficient flexory strength to prevent their being overbalanced by the finger extensors. The result is the well-known clawhand. So far as the lateral motion of the fingers is concerned, loss of interossei does not entirely abolish it, because this motion can be carried out by the extensor indicis proprius and the extensor digiti minimi when the fingers are extended in the metacarpophalangeal joint. This is particularly true for the fifth finger, which has its own proper extensor and abductor.

According to Duchenne,[2] the force displayed by the interossei in extending the interphalangeal joints is less than that in ab- and adducting the fingers in the metacarpophalangeal joints.

4. Paralysis of the thenar muscles

A) AB- AND ADDUCTION

The long abductor of the thumb abducts the first metacarpal and at the same time abducts the wrist joint, but the muscle is not only an abductor, as the name implies, but also has a supinatory effect on the first metacarpal.

The short extensor abducts the first metacarpal, extends the metacarpophalangeal joint of the thumb, and, at the same time, the interphalangeal joint goes into flexion. It also supinates the first metacarpal. To a lesser extent, abduction of the first metacarpal is carried out by the abductor brevis and the radial head of the flexor brevis, likewise, has an abductory effect.

When the long abductor and the short extensors are lost, the first metacarpal goes into adduction approaching the second metacarpal (Duchenne[2]) (Fig. 8). If only one of these muscles is paralyzed, the abductory loss is not great, the remaining muscles substituting for the lost one. In case of paralysis of the abductor longus, the short extensor abducts the thumb but, at the same time, the first metacarpal goes into slight extension. The loss of the short extensor is somewhat more disturbing because the thumb loses extension as well as some of the abduction.

The principal adductor of the thumb is the adductor pollicis together with the ulnar head of the flexor brevis. If the adductor is paralyzed, adduction of the thumb is still possible to some degree by the extensor pollicis longus. Loss of adduction restricts approachment of the thumb to the fingers in all positions, especially in the plane of the palm where only the lateral head of the first interosseous has a slight adductory action. Paralysis of the adductor produces an early dorsal shrinkage of the first interosseous space.

B) EXTENSION AND FLEXION PARALYSIS

Principal extensors are the long and short extensors of the thumb and, to a degree, the long abductor. The long extensor extends the metacarpophalangeal and interphalangeal joints.

When the long extensor is paralyzed, the first metacarpal drops forward and ulnarly. The opponens and the short flexors are the direct antagonists and bring the thumb forward and into opposition with the metacarpophalangeal joint flexed.

The principal flexors are the long and short flexors of the thumb. The former flexes the endphalanx, the basal phalanx remaining in midposition. The latter flexes the metacarpophalangeal joint and the endphalanx remains extended. They are not the only flexors but they are joined by the short abductor and the opponens.

C) OPPOSITION AND REPOSITION

These are more composite movements in which axial rotation of the first metacarpal participates.

The sequence of the circumductory movement of the thumb, beginning with reposition and extension to full opposition is as follows: (1) extension by the long and short extensors; (2) abduction (radially) by the abductors longus and brevis; (3) then flexion combined with pronation by the long and short flexors and the abductor brevis; (4) then opposition by the opponens and the flexor brevis; (5) and, finally, adduction by the adductor.

The three muscles of the thenar which oppose and flex the first metacarpal are the opponens, the short abductor, and the superficial head of the short flexor. They place the thumb in a sagittal or oblique sagittal plane, but, in order that the thumb may reach the other fingers, it must further rotate and approach the medial border of the hand. This requires the assistance of the adductors.

At this point the metacarpal has described almost two-fifths of a circle and the plane of the thumb nail is inclined to the plane of the finger nails at 135°.

The loss of the short abductor, opponens, and superficial head of the flexor

Fig. 8. The abnormal position of the thumb following paralysis of the short extensor and long abductor. (After Duchenne)

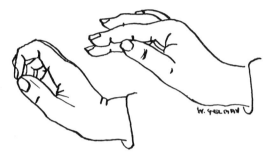

Fig. 9. Paralysis of thenar muscles. (After Duchenne)

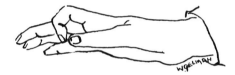

Fig. 10. Paralysis of all thenar muscles except adductor. (After Duchenne)

brevis still permits opposition of the thumb to the fingers, but the first metacarpal is not sufficiently abducted for the thumb to meet the tips of the extended fingers. They have to flex in the interphalangeal joints for that purpose. If the fingers are extended, the tip of the thumb recedes to about the proximal interphalangeal joints of the fingers (Fig. 9).

The short abductor is an important muscle because it permits the thumb to be opposed to the flexed index finger so as to hold a small object such as a pen between them. If the short flexor is lost, the thumb cannot be opposed against the two last fingers (Duchenne[2]).

The opposition movement terminates with adduction. Even if all other muscles of the thenar eminence are lost, the adductor permits holding objects between the proximal phalanx of the thumb and the palm of the hand. The patient succeeds by strong contraction to press the thumb against the lateral border of the second metacarpal, and by extension of the proximal phalanx of the thumb he can meet the index finger (Fig. 10).

From a practical point of view, it means that in thenar palsy (for instance, in poliomyelitis) a substitution of the opposition movement must take into consideration the full axial rotation as well as the approachment of the first metacarpal to the palm.

III. THE RECONSTRUCTION OPERATIONS IN PARALYSIS OF THE WRIST AND FINGERS

The kinetic situation must be evaluated from the following angles: position, stability, adequacy of strength of donor muscle, direction of pull, angle of application, and excursion length.

A. THE PARALYSIS OF THE EXTENSORS OF THE WRIST AND FINGERS

1. Tendon transplantation

Inasmuch as the transplanted flexor muscles must supply both the extension of the wrist and that of the fingers, the total excursion length serving for wrist and fingers must be considered in relation to contraction distance of the recipients.

A) EXTENSORS (RECIPIENTS)

For the extensor proprius indicis and extensor communis the average for all fingers is given as 55 mm to 35 mm (Bunnel[1]), or 40 mm (Strasser[8]). For the extensor carpi radialis brevior it is 37 mm (Bunnell), 40 mm (Strasser). For the extensor carpi radial is longior, 37 mm (Bunnell), 34 mm (Strasser). For the extensor pollicis longus (wrist alone), 28 mm (Bunnell); all joints, 58 mm (Strasser).

B) FLEXORS (DONOR MUSCLES)

For the flexor digitorum sublimis, full flexion of all joints, Fick[3] estimates the contraction length at 80 mm; Bunnell, at 53 mm. For the flexor profundus, full flexion of all joints: Bunnell, 53 mm; Fick, 85 mm. For the flexor carpi ulnaris: Fick, 39 mm; Bunnell, 33 mm. For the flexor carpi radialis: Fick, 38 mm; Bunnell, 40 mm.

The figures reported by Strasser[8] and Fick[3] are identical with those of E. Weber.[9]

The following operative plans are in use:

1) The flexor carpi ulnaris to the common extensor of the fingers; the pronator teres to the extensor carpi radialis longior; the flexor carpi radialis to the abductor longus; and the palmaris longus to the extensor pollicis longus.

Assuming that the direction of the transplanted muscles can be preserved, the weakest point in this plan, both from the point of muscle strength and contraction length, is the use of the palmaris longus for the extensor pollicis longus.

2) The flexor carpi radialis to the extensor pollicis longus and the abductor

pollicis longus; the flexor carpi ulnaris to the common extensors; and palmaris longus to the extensor pollicis brevis.

3) The flexor carpi radialis to the extensor pollicis longus; the flexor carpi ulnaris to the extensor communis; and the palmaris longus to the extensor brevis and the abductor longus pollicis.

Bunnell[1] states that all these schemes as well as a number of others work well. This may be expected because both in point of strength and excursion range the donor muscles meet the requirements of the recipients.

2. The arthrodesis of the wrist combined with tendon transplantation. Kinetic analysis

Situations arise where the extensors are paralyzed and there are not enough donor muscles available from the flexor side. The minimum number of muscles to provide for extension of the wrist and fingers is three. If only two strong flexors are available, the wrist can be ankylosed, thus relieving the donor muscles of extending and stabilizing the wrist joint. In this case, the most common procedure is to use the flexor carpi ulnaris for the common extensors and the flexor carpi radialis for the abductor and short and long extensors of the thumb, provided these muscles are available; if the flexor carpi radialis is paralyzed, the pronator teres and palmaris longus must be used as substitutes. The necessity of stabilizing the wrist firmly calls for a more radical type of arthrodesis than is used in spastic paralysis, preferably one in which a graft is used. An advantage is that the excursion range for the extension of the fingers and thumb remains adequate since the substituting muscles (flexor carpi radialis and ulnaris) can spend all their power of contraction upon the finger joints and no strength is used up in extending the wrist.

On the whole, the situation for the reconstruction is favorable in paralysis of the extensors, whether it be due to a peripheral nerve lesion or to infantile paralysis, because the tendon exchange rests on sound kinetic considerations. Transplantation alone gave us 80% and the combination of arthrodesis with transplantation, 77% satisfactory results.

B. THENAR PALSY. MECHANICS OF RECONSTRUCTION

If one considers that the thumb is moved by four short muscles (abductor brevis, flexor brevis, opponens and adductor) and by four long muscles (extensor pollicis longus, extensor brevis, abductor longus, and flexor longus), it is clear that in thenar palsy not all short muscles can be substituted individually, nor can all long muscles be used for donors.

The missing components of the circumductory movements of the thumb in thenar palsy are flexion, opposition, pronation, and adduction. In peripheral paralysis, due to injury to the median nerve, the adductor and half of the short flexor are preserved. In anterior poliomyelitis, they often are not. To carry out the entire circular movement which the thumb and its metacarpal describe over the surface of a cone, the tendon of the donor muscle must produce abduction in substitution of the abductor brevis, then flexion and pronation in substitution of the flexor brevis and opponens, and, finally adduction in substitution of the adductor to maintain the thumb firmly in the position of opposing the fingers.

The tendon of a substituting muscle must, therefore, have an oblique course, being attached to the base of the basal phalanx or the distal end of the metacarpal in such a way to swing it in a circle around the greater multangulum.

The author's operation accomplishes this by using the lateral half of the flexor pollicis longus, which is fastened to the ulnar side of the base of the basal phalanx (Fig. 11). It has the advantage of simplicity, but it also has the disadvantage that the transposed tendon forms a very acute angle with the axis of the thumb; therefore, its rotatory component about the greater multangulum is sometimes too small to carry out circumductory movement in a wide enough circle, and the opposition obtained is sometimes not sufficient. Another point which requires attention is that the long flexor is given a double anchorage by the operation; half of it is left attached to the distal phalanx, the other half is fastened to the ulnar side of the basal phalanx.

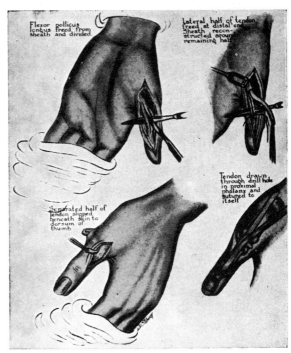

FIG. 11. Author's method of flexor plasty of thé thumb. (Steindler)

Inasmuch as this feature of double anchorage is used in other operations, a short discussion of the kinetic situation is in place. Assuming that both attachments act on their point of application with the same force, their rotatory effect is different since it is inversely proportional to the masses to be moved. It can readily be seen that in this case the endphalanx will go into easy flexion before the more massive metacarpal unit follows; by the time flexion of the endphalanx is complete, the muscle has lost so much of its contractile power that not enough is left to swing the metacarpal around. This difficulty can be easily avoided by placing the endphalanx in flexion at and after the operation, a point that has been neglected by many surgeons.

The operation of Bunnell avoids these difficulties. He uses the flexor carpi ulnaris or the flexor sublimis of the fourth finger as donor; the tendon of these muscles is elongated by a tendon graft taken from the palmaris longus and it

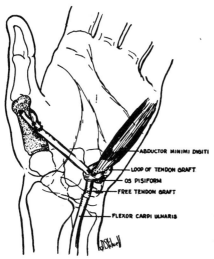

FIG. 12. Bunnell's method for restoration of opposition of the thumb. (Steindler)

is passed through a loop at the pisiform prepared from another piece of the palmaris tendon. The end of the donor tendon emerging from this loop then traverses the base of the palm to become attached through drill holes into the base of the basal phalanx. This method is particularly useful where the long flexor of the thumb is not available and it has the undoubted advantage of furnishing a more forceful leverage for the circumduction (Fig. 12). In a modification of this author, a tendon graft is anchored to the basal phalanx of the thumb and to the neck of the fifth metacarpal, and is activated by the palmaris longus which is hooked into the tendon and pulls it proximally in T-fashion (Fig. 13).

For opponens weakness, Goldner and Irwin[4] advise the transplantation of the flexor digitorum sublimis of the fourth finger into a dorso-ulnar site of the proximal phalanx of the thumb, provided all long muscles are in good condition; if the extensor pollicis longus is weak, the extensor carpi radialis is sutured to the extensor pollicis, in addition.

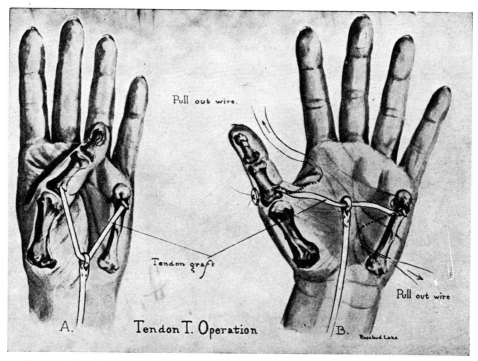

FIG. 13. Bunnell's modified method for restoration of opposition of the thumb. (Bunnell)

From the kinetic point of view, all these procedures are rational and their selection should be based on the ground of technical difficulties. We have found, for instance, that the simple plasty of the flexor pollicis longus with transposition of the outer half dorso-ulnarly to the base of the basal phalanx of the thumb is entirely satisfactory for producing thumb opposition, provided the muscle is in normal condition and that one observes the kinetic requirement of placing the endphalanx in flexion, as we have explained above.

The situation becomes aggravated by contracture deformities which follow poliomyelitis. Irwin and Eyler[5] call attention to the fact that when the thenar muscles are paralyzed the extensor pollicis longus acts as adductor to effect the pinch between the thumb and the index finger to hold objects (Fig. 14). This results in weakening of the palmar portion of the multangulometacarpal

FIG. 14. Complete paralysis of all thenar muscles. Extensor pollicis longus acting as adductor. (Redrawn from Irwin & Eyler)

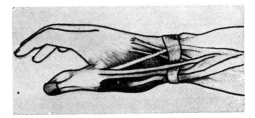

FIG. 15. Irwin and Eyler's transposition of the extensor pollicis longus. (Irwin & Eyler)

joint with anterior subluxation. Their treatment consists in restoring opposition by re-routing the extensor pollicis longus to a more radial position (Fig. 15), which then cooperates with the abductor pollicis longus. This eliminates the necessity of arthrodesing the carpometacarpal joint of the thumb.

If, in addition to the subluxation of the first carpometacarpal joint due to the action of the extensor pollicis longus, there is also a displacement of the tendon into the thumb web, then the radial side of the capsule of the carpocarpal joint of the thumb becomes stretched and an ulnar deviation results. This requires a carpometacarpal arthrodesis in 20-30° flexion, in addition to the re-routing of the extensor pollicis longus. The authors also find that a flexion contracture of the interphalangeal joint of the thumb requires metacarpophalangeal arthrodesis in hyperextension.

C. PARALYSIS OF INTEROSSEI. MECHANICS OF SUBSTITUTION

Lexer[6] used strips from the medial and lateral sides of the flexor digitorum sublimis to the extensor apparatus of the fingers. However, this operation produced adhesions which in the end destroyed the mechanical effect of the transplants (Fig. 16).

Bunnell[1] solved the problem in a more physiological way. He uses the tendons of the sublimis which are detached at their insertion and then passed through the lumbrical canals to be attached to the lateral bands of the extensor apparatus of the fingers (Fig. 17). Thus, the sublimis, which normally flexes the middle phalanx, becomes an extensor of the joint, i.e., its own antagonist. The only counterbalance against it is provided by the long flexor.

Here is a situation in which the superficial finger flexor is sacrificed for the purpose of substituting the missing interossei. The question may arise whether the loss of the flexors interferes essentially with the gripping ability of the

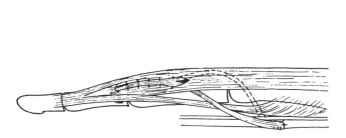

FIG. 16. Lexer's method of substitution for paralysis of interossei. (Lexer)

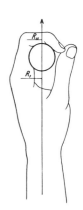

FIG. 18. Fist closed over rod. (Recklinghausen)

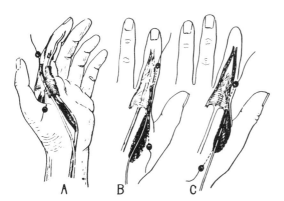

FIG. 17. Bunnell's method of substitution for paralysis of interossei. (Bunnell)

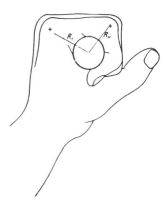

FIG. 19. Strong grip using all finger joints. (Recklinghausen)

fingers. The rotation moments of the flexors on the metacarpophalangeal joints, the midphalangeal, and the endphalangeal joints have a ratio of 7:5:1 (Recklinghausen[7]). This ratio excludes the interossei and the lumbricales. The superficial flexor being more powerful in respect to the metacarpophalangeal and midphalangeal joints, it would seem that the loss of this muscle does greatly interfere with the strength of the grip. However, if the intrinsic muscles are included, one finds that added flexor strength in the metacarpophalangeal joint greatly increases the efficiency of the grip.

The situation is somewhat different for the interphalangeal joints. If the fist is closed over a ring or rod, the external resistance depends on the diameter of the enclosed object. It becomes more evenly distributed as the angles of the two articulations, metacarpophalangeal and proximal interphalangeal, approach each other (Recklinghausen[7]) (Fig. 18). The interossei and lumbricales add their share in the flexory moment in the metacarpophalangeal joint. In the interphalangeal joint, however, they are antagonists of the flexors. The more flexory strength is demanded in the metacarpophalangeal joint, i.e., the heavier the intrinsic muscles are engaged in flexing this joint, the lesser becomes the rotation moment in the midphalangeal joint, and, in the absence of the superficial flexor, a critical misproportion may result. These patients may do fairly well in such performances as hanging on a bar, but they may do poorly in occupations requiring strong grip of all finger joints, such as swinging an axe (Fig. 19).

IV. SUMMARY

1. Both extensors and flexors are necessary to give stability to the wrist. On the whole, the effect of the extensors on the wrist is more static; that of the flexors, more dynamic. If stability of the wrist cannot be provided sufficiently by muscle action, motion in the wrist may have to be sacrificed by arthrodesis to secure a stable position of function which is indispensable for the proper action of the fingers.

2. Inability to abduct or adduct the hand does not entail a severe handicap, provided there is no contractural disalignment.

3. Forceful gripping of larger objects depends mainly on the functional position of the wrist; it is the special task of the intrinsic muscles in this situation to accentuate the gripping power in the metacarpophalangeal joints against the extensory action of the extensors in these joints.

4. In gripping smaller objects by the tips of the fingers, the intrinsic muscles must overcome the flexory effect of the long finger flexors in order to extend the fingers and to give them the pill roller hand position necessary for this purpose.

5. Making the extensors of the fingers passively insufficient by strongly flexing the wrist facilitates the opening of the fist.

6. Opposition of the thumb requires the thenar muscles including the adductor and the short abductor. Paralysis of the short flexor interferes with the extension in the interphalangeal joint, necessary to meet the tips of the fingers.

7. Paralysis of the extensors of the wrist weakens the flexory power of the fingers. One of the radial extensors is sufficient for the extension of the wrist, but the hand is drawn into radial abduction without the neutralizing effect of the extensor carpi ulnaris.

8. Paralysis of the flexors of the wrist interferes with its stabilization. Loss of the flexor carpi ulnaris deprives the flexed hand of ulnar abduction.

9. Paralysis of the finger extensors does not interfere seriously with the extension of the interphalangeal joint, which is taken care of by the intrinsic

muscles. But a contracture of these muscles unopposed by the extensors leads to a pill roller hand deformity.

10. Paralysis of the finger flexors leaves the extensors unopposed and results in a clawhand deformity.

11. This deformity is more marked in paralysis of the intrinsic muscles, such as follows ulnar nerve paralysis.

12. Paralysis of the long abductor and short extensor of the thumb causes the first metacarpal to go in adduction; in paralysis of the adductor, adduction is still possible to some degree by the extensor pollicis longus, but the approachment of the thumb to the second metacarpal is impaired. When the extensor longus is paralyzed, the first metacarpal drops forward and ulnarly. When the short abductor, opponens, and superficial head of the flexor brevis are lost, the thumb can no longer meet the extended fingers, but its tip remains opposite the proximal interphalangeal joint. If the short flexor alone is lost, the thumb cannot be opposed against the last two fingers.

13. In reconstructive operations involving tendon transference, position, stability, adequacy of donor strength, direction of pull, angle of application and excursion length must be considered.

14. The most common operative plans of tendon transplanation for paralysis of the extensors of wrist and fingers are:

a) Flexor carpi ulnaris to common extensor of fingers; pronator teres to extensor carpi radialis longior; flexor carpi radialis to abductor longus; and palmaris longus to extensor pollicis longus.

b) Flexor carpi radialis to extensor pollicis longus and abductor pollicis longus; flexor carpi ulnaris to common extensor of fingers; and palmaris longus to extensor pollicis brevis.

All these schemes should work out well because the donor muscles are adequate in strength and excursion range.

15. If the donor material is insufficient for both the stabilization of the wrist and the extension of the fingers, the wrist can be ankylosed and the flexor carpi ulnaris used for the common finger extensor, and the flexor carpi radialis for the abductor and short and long extensors of the thumb. The arthrodesis permits the substituting muscles to spend all their contractile strength upon fingers and thumb.

16. In thenar palsy, the donor muscle must produce flexion, pronation and opposition, for which purpose it must have an oblique course suitable to swing the thumb in a circle around the greater multangulum.

17. The author's operation of using the lateral half of the long flexor tendon of the thumb and attaching it to the ulnar side of the base of the basal phalanx meets this requirement, although, due to the small angle of application, the rotatory moment in respect to the carpometacarpal joint is sometimes inadequate. It is necessary to place the endphalanx in flexion; otherwise, the contracting flexor longus spends itself in flexing the endphalanx and has not enough tension left for a forceful circumductory movement.

18. Bunnell's operations are very useful in cases where the flexor pollicis longus is paralyzed or weak. The circumduction is accomplished by using the flexor carpi ulnaris or the sublimis of the fourth finger. The tendon is led through a loop at the pisiform, elongated by a graft from the palmaris longus tendon, and fastened through drill holes into the base of the basal phalanx of the thumb. In a modification by the same surgeon, a tendon graft is anchored to the basal phalanx of the thumb and to the neck of the fifth metacarpal, and the palmaris longus is hooked into it, its pull on the graft bringing the thumb into opposition. Others (Goldner and Irwin) transplant the flexor sublimis of the fourth finger to a dorsoulnar site of the proximal phalanx of the thumb. All these methods seem rational from the kinetic point of view.

Contracture deformities following poliomyelitis complicate the situation. In paralysis of the thenar muscles the extensor pollicis longus acts as an adductor and may produce an anterior subluxation of the multangulo-metacarpal joint. This may require re-routing of the extensor pollicis longus more radially, or, if there is a displacement of the tendon into the web of the thumb, an additional arthrodesis of the carpometacarpal joint may become necessary (Irwin and Eyler).

19. Substitution of the paralyzed interossei is carried out after Bunnell by detaching the tendons of the sublimis from their insertion and passing them through the lumbrical canals to be attached to the lateral bands of the extensor apparatus of the fingers, thus transforming the muscle into an extensor of the interphalangeal joints.

20. The sacrifice of the sublimis has some effect upon the flexory efficiency in the interphalangeal joint and a critical misproportion may arise in these joints, which have only the deep flexor left. This deficiency may become manifest in grips which require strong flexion power in all finger joints, as, for instance, the swinging of an axe.

BIBLIOGRAPHY

1. BUNNELL, S.: *Surgery of the Hand.* Philadelphia, J. B. Lippincott, 1948.
2. DUCHENNE, G. B.: *Physiology of Motion.* Transl. E. B. Kaplan. Philadelphia, J. B. Lippincott, 1949.
3. FICK, R.: *Handbuch der Anatomie und Mechanik d. Gelenke, III.* Jena, G. Fischer, 1911.
4. GOLDNER, J. L., and IRWIN, C. E.: Paralytic thumb deformation. *J. Bone & Joint Surg., XXXII A:*3, 627, July 1950.
5. IRWIN, C. E., and EYLER, D. L.: Surgical rehabilitation of hand and forearm disabled by poliomyelitis. *J. Bone & Joint Surg., XXXIII A:*4, 825, Oct. 1951.
6. LEXER, E.: Über die Ulnaris Ersatzoperation. *Handbuch d. ärztl. Erfahrung im Weltkriege, Vol. II.* Leipzig, John Ambrose Barth, 1931.
7. RECKLINGHAUSEN, N. v.: *Gliedermechanik und Lähmungsprothesen,* Vol. I. Berlin, J. Springer, 1920.
8. STRASSER, H.: *Lehrbuch der Muskel- und Gelenkmechanik, IV.* Berlin, J. Springer, 1917.
9. WEBER, E.: Über das Verhalten der Vorderarmmuskeln an den Hand- und Fingergelenken. *Würzburg Verh.,* 1880.

Lecture XXXII

THE ARM AS A WHOLE

I. ORIENTATION

THE distinction between the open and the closed kinetic chain to which reference has been made before is as essential in the analysis of the complex movements of the upper extremity as it is in the less complicated kinetic situations of the lower. The freedom of motion which, under the conditions of the open chain, is accorded to the peripheral portion of the limb develops a high degree of terminal acceleration. The motor impulses reach the terminal joint through one or more proximal articulations, and each of these contributes its share to the final motor effect. The absence of any peripheral restraint makes it possible to accumulate acceleration from joint to joint and to give the motor performance the desired effect. An example is the throwing of a ball.

There are other situations in which there is also no significant external resistance but in which the muscles of the proximal joints are used for the stabilization of certain positions and attitudes necessary to give the terminal motor action the desired accuracy and precision. Acceleration is suppressed by posture. This type of motion is represented by far the greater majority of the everyday motor performances of the upper extremities; or, the objective is to develop both speed and accuracy, which involves both postural and dynamic effort: for example, playing tennis or golf.

All these situations mentioned above have in common that no external resistance opposes the action of the muscles. The only difference between them is that the movements designed for speed have a greater ratio of free movement and a lesser intrinsic restraint by stabilization. Those which are designed for accuracy have a greater ratio of "lost" motion, i.e., stabilization, and a lesser for the development of acceleration.

Speaking of external resistance, it should not be overlooked that the masses of the moved portions themselves resist motion in the sense that any acceleration imparted to a body in rotatory motion is inversely proportional to its mass $(a = \dfrac{R}{I})$. A muscle acting both on its proximal and peripheral end with equal tension, therefore, imparts accelerations to the peripheral and proximal lever arm commensurate inversely with the weight $(W = m.a.)$ of the respective body parts.

In the case of moving the peripheral end of the arm freely, the mass of the hand, for instance, is infinitely smaller in comparison with the rest of the body, which represents the mass of the proximal lever arm.

If all of the other joints of the body are stabilized, then this latter mass represents a solid unit. The result is that flexion of the wrist brings forth a visible movement of the hand but no appreciable movement of the rest of the body because the acceleration imparted to it is comparatively extremely small.

In contrast, there are situations in which the peripheral portion of the joint is absolutely fixed by large or absolute external resistance: for instance, a rigid post or bar. In this case the mass of the peripheral end forms a unit with the fixed external object, which may be anchored firmly to the ground. It is this external resistance which represents a mass infinitely greater than that of the whole body. Since the peripheral end is held immovable the muscle acts with its proximal end only, imparting to the rest of the body a visible acceleration. When one pulls himself upon a horizontal bar against one's body weight, the forearm is immobilized and the upper arm and body become the movable parts of the system.

Or, it may be that both the central and the peripheral lever arms of the lever system are being resisted by masses so great that the muscle force applied against them is entirely inadequate to produce visible motion at either end of the lever arm. For instance, trying to lift the back end of an automobile, neither it nor the ground against which one strains gives and no visible motion is produced. Here is an instance of an absolutely closed kinetic chain. All muscle action is isometric; no shortening occurs. When we speak of a closed kinetic chain in the ordinary sense, however, we do not mean that visible motion is suppressed by external resistance at both ends. We must consider those situations rather in which a peripheral resistance is large enough to produce an inversion of the muscle action from the usual peripheral to a proximal motor effect, as, for instance, in chinning against a horizontal bar.

In performing a certain motor act which engages a series of joints, these may be set in motion at the same time or concomitantly, or in sequence or consecutively. It lies in the nature of things that where great external resistances must be overcome, as, for instance, in heavy pushing or lifting the muscles of all joints are liable to work simultaneously. On the other hand, where a certain terminal speed is the objective, the muscles of the various joints are more likely to act consecutively, that is, in certain sequence, as is the case in throwing a ball or hitting it with a tennis racquet. In motor events representing the open kinetic chain, the movements in the different joints should be expected to occur consecutively, whereas in the so-called closed kinetic chain, where the effort is directed against an external resistance, the action of the joints is more likely to be concomitant. It follows that wherever rigid stabilization of all articulations is the primary object, the joints of the upper extremity are likely to act simultaneously.

II. COMBINED MOVEMENTS

A. SHOULDER AND ELBOW

When the elbow is flexed actively by the brachialis, whether free or resisted, a passive movement in the shoulder accompanies it. If the flexion movement is free, the arm moves backward in the shoulder because the common center of gravity of the system falls behind the center of the joint (Fig. 1). When both joints are being moved together by the biceps, and the peripheral end is free (open kinetic chain), then flexion of the elbow is combined with forward flexion of the arm in the shoulder joint (Fig. 2). On the other hand, when the forearm is fixed by an external resistance (closed kinetic chain), the flexion of the elbow is associated with forward movement of the shoulder blade and backward movement of the arm (Baeyer[1]) (Fig. 3).

B. ELBOW AND WRIST

With the hand freely movable and relaxed, the forearm assumes in flexion of the elbow a position of pronation, and in extension of the elbow a supinatory position. If the hand is fixed by an external resistance so that pro- and

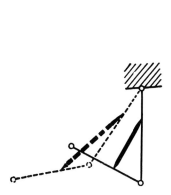 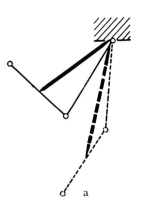 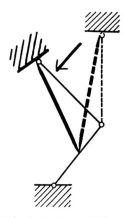

FIG. 1. Flexed arm following center of gravity.

FIG. 2. Flexion of biceps. Open kinetic chain.

FIG. 3. Flexion of biceps. Closed kinetic chain.

supination of the radius is locked, then the ulna must rotate around the radius in inverse direction, which movement is carried out in the shoulder. Flexion of the elbow is then associated with inward rotation and extension of the elbow with outward rotation in the glenohumeral joint.

C. SHOULDER, ELBOW AND WRIST

In the open kinetic chain, the translatory movement transmitted from the shoulder to the hand causes correlated movements in all three joints.

In translatory direction, the action of the biarticular muscles is as follows: In direction of proximation (toward the shoulder), the long head of the biceps flexes the shoulder and the elbow and the flexors of the forearm flex the elbow, wrist and fingers until the fingers are touching the shoulder joint. In direction of distention (away from the shoulder), the long head of the triceps adducts the humerus and extends the elbow; the extensors of the forearm extend wrist and fingers, giving the extremity the maximum length.

In the forward swing with the elbow extended, the triceps neutralizes the flexory action of the brachialis and biceps in the elbow joint, while in the shoulder joint, the long head takes part in the forward swing of the arm, neutralizing the extensory action of the triceps. In the backward swing, again, the triceps neutralizes the flexory action of the brachialis and biceps, keeping the elbow in extension.

One notices that in for- and backward swing there is the same counter-current shift (Fig. 4) of the biarticular muscles which occurs in the for- and backward swing of the lower limb; and, in translatory lengthening and short-

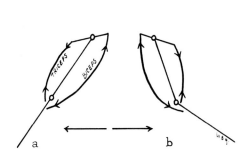

FIG. 4. Countercurrent swing of biarticular muscles. *a*. backward swing. *b*. forward swing.

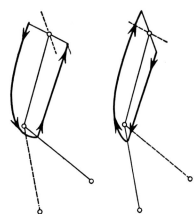

FIG. 5. Concurrent shift in lower extremity.

ening of the arm, the same concurrent shift of the biarticular muscles is observed as in the lower extremity (Fig. 5).

III. THE ANALYSIS OF SOME OF THE ARM MOVEMENTS UNDER OPEN KINETIC CHAIN CONDITIONS. NO EXTERNAL RESISTANCE

A. BALL THROWING

The object is to give the missile the greatest possible speed as it is released by the hand. For this purpose there must be a summation of accelerations which are developed by the movement of the various articulations increasing from the proximal to the distal ones. This requires an extremely well-planned sequence of events. One motion must follow the preceding one while the latter is still increasing in speed, i.e., before the motion of the next proximal joint has become completed. It is obvious that momentum cannot be accumulated and transmitted from the proximal to the distal joint and from there to the missile unless the action of the joints sets in in sequence at optimum time, that is, the time when the proximal joint has developed its peak of velocity and acceleration. Conforming with these postulates, the pitcher uses a combination of movements of the lower extremity, trunk, and upper extremity, all culminating in producing the greatest possible rotatory momentum. First, he winds himself up by an axial rotation of the lower extremities and the trunk, a movement

FIG. 6. The wind up and delivery of a pitcher.

developed systematically from below upward. The upper extremity, held in the meantime in dorsiflexion and elevation in the shoulder, flexion in the elbow and extension in the wrist, is ready to go into forward flexion of the shoulder and extension of the elbow the instant the rotatory momentum from the trunk reaches the shoulder. The shoulder is then brought forcibly forward; the elbow is extended, then the wrist is extended and the missile is delivered (Fig. 6).

The sequence in which these movements occur from joint to joint is of greatest importance. Precipitate simultaneous action in all joints would interfere with the proper transmission of acceleration; again, retardation of the sequence would make the effect of the preceding joint useless because its

FIG. 7. The discus throw. (Bresnahan & Tuttle)

acceleration has already decreased. It may be difficult to convince one of the consecutive character of these motions because of their rapidity which deceives the eye; but a slow motion picture will reveal the truth. Throwing a baseball is a particularly instructive case in which the objective of the locomotor combination is the development of speed and in which there is no ex-

Fig. 8. The shot-put. (Bresnahan & Tuttle)

ternal resistance. As we shall see later, the mechanical constellation is entirely different where the external resistance is considerable (closed kinetic chain).

B. DISCUS THROWING (FIG. 7)

The accumulation of momentum requires a proper stance, a preliminary full circle swing of the arm, then a pivoting on the left foot first, then on the right, and finally the delivery and release of the missile. As Bresnahan and Tuttle[2] describe it, the initial stance is with the feet approximately at right angles to the direction of the throw, the weight evenly distributed between the two feet, trunk erect, and knees variously bent. Then comes the preliminary arm swing from front to back three times around to establish proper balance and to develop momentum. With the last swing the trunk turns to the left by pivoting on the left foot. Then, maintaining momentum, the athlete shifts his weight to the right, making a half turn on the right, and then with vigorous forward thrust of the right leg an "explosive" whip of the arm is carried out and with a powerful snap of the wrist the missile is delivered. Here are manifestly three sources of consecutively accumulated momentum: the preliminary circling swing, the rotation of the trunk on the alternatingly pivoting feet, and the terminal forcible extension of the elbow and flexion of the wrist as the disc is released.

C. SHOT PUTTING (FIG. 8)

The 12-pound shot constitutes some external resistance but the situation of an open kinetic chain still applies because here also speed is the object and the motor act is a dynamic performance. As in the disc throwing, momentum is accumulated in a series as follows: preliminary left leg forward swing, body shift to the right with the weight on the right foot, right knee moderately bent, forceful extension of the right knee, then backward shift to the left, forceful extension of the arm in shoulder and elbow, delivery with flexion movement of the wrist.

D. THE ROLE OF THE ARMS IN MAINTAINING BODY EQUILIBRIUM

1. Standing on toes (Fig. 9)

There is a corresponding change in the position of the common center of gravity with every change of position of the arms. For this reason they are useful as equilibrators. When one rises on his toes and the body threatens to fall backward, throwing the arms forward will regain the equilibrium. In fact, all oscillations of the line of gravity can be controlled by the movement of the arms for- or backward or sideways to the effect that the line of gravity can be made to fall into the supporting surface of the ball of the foot. This equilibrium can be accentuated by more forceful movement of the arms or by increasing the mass of the arm when balancing a heavy rod, as does the tight-rope walker.

2. The role of the arms in running and jumping

A similar balancing mechanism is carried out by the arms in running; essentially, the play of the upper extremity is the same as in the gait, namely, contralateral to the leg, except that during the run the swing of the arms is more forceful. They assist both in propulsion and in restraint (Fig. 10).

In the jump, as soon as the feet have left the ground, no more changes are possible so far as the body as a whole is concerned; that is, the body, having abandoned contact with the floor, can no longer acquire additional acceleration through the reaction from the floor but it must move forward in the air with the velocity acquired while still in ground contact. If the swing of the arms is to increase the total of the living force, the action must occur simultaneously with the actions of the body while the feet are still on the ground.

To clear the hurdle in the high jump, the feet are drawn up as high as possible on the trunk. In the broad jump, the feet must be extended forward as far as possible so that in landing the body pivots over the landing feet. Here, again, the upward swing of the extended arms serves to produce the

FIG. 9. Maintaining equilibrium standing on toes.

FIG. 10. Role of arms in running. (Bresnahan & Tuttle)

restraint, i.e., a backward acceleration which neutralizes the forward acceleration produced at the takeoff. As the jumper lands, he flexes the hip and knees in order to absorb the forward momentum.

After the jumper's feet have left the ground, he can do comparatively little to increase his forward acceleration. He can, however, by drawing up his extremities, change the configuration of the body as a whole from a cylindrical to a more global form, diminishing thereby the inertia resistance and preserving acceleration during the flight $(a = \frac{R}{I})$.

IV. MOVEMENT OF THE UPPER EXTREMITY IN A CLOSED KINETIC CHAIN

There are situations in which the upper extremity moves against an external resistance large enough to interfere with the free peripheral motion of the extremity or even to arrest visible motion altogether. This resistance forms then with the extremity a closed kinetic chain. When one lifts a heavy weight or moves a very heavy object, it furnishes the external resistance which closes the kinetic chain. The acceleration given to these objects by the contracting muscles is inversely commensurate to their masses. The mass of the external resistance may be so great that no visible motion can be produced at the peripheral end of the lever system. This situation is represented when one chins himself on a horizontal bar. The peripheral resistance is absolute since the bar is fixed. The central resistance consists of the rest of the body. In rowing, the resistance of the water against the oars may be considered the external resistance; the forward bending body becomes

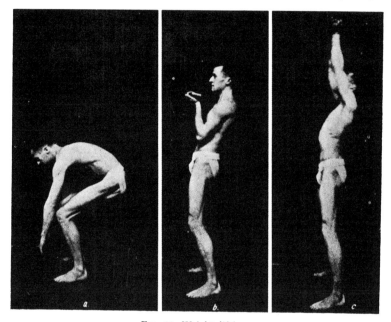

FIG. 11. Weight lifting.

a proximal synergist to the elbow flexors and the backward bending body to the extensors of the elbow against the external resistance of the water.

On the whole, the characteristic feature of the closed kinetic chain is the display of force, whereas the creation of speed or acceleration is the central effort of the open kinetic chain.

A. WEIGHT LIFTING (FIG. 11)

As long as the weight to be lifted has its weight center close to the line of gravity of the body, the rotatory moment this weight produces in the different articulations is comparatively small. Consequently, the weight lifter is particularly anxious to choose his stance so that the weight is as close as possible to his line of gravity. Then only the translatory stress has to be overcome.

This explains the type of form used in weight lifting. The lifter puts himself first in a preparatory position in which the joints of the lower extremity are flexed, giving the extensors the necessary tension for the next act. Now, by simultaneously extending all of the joints of the lower extremity, he is able to lift the heavy iron bar straight vertically upward, holding the weight with arm extended about the level of the thighs. Now the work of the upper extremity sets in. It starts with flexion of the elbow while the trunk is thrown backward, always holding the forearm close to the line of gravity to minimize the rotatory effect of the load. Gradually, the weight rises until it passes above the shoulders to reach the chin. The problem is now to clear the chin. He cannot do so by passing the weight forward without increasing its rotatory effect. Therefore, the head is strongly dorsiflexed to allow the weight to pass. Now comes the last act of raising the weight above the head. This, of course, requires the extension of the flexed elbow and the forward elevation of the arm in the shoulder joint, the most difficult stage in the whole procedure. The method used for this is an follows: with the weight in place below or at the chin, the legs are again flexed in hip and knee; then, by a sudden forceful and simultaneous extension of knees and hip, a considerable upward acceleration is developed. This is of great advantage to the following extension of the elbow and elevation of the arms. It only must be seen to that the direction of this lifting is close to the line of gravity; therefore, the chin had to be gotten out of the way by extension of the neck. Then the combined action in elbow and shoulder assisted by the upward acceleration generated by the lower extremities permits lifting of the weight until the arms are perpendicular and the elbow completely extended.

B. CHINNING ON THE HORIZONTAL BAR

The situation is similar to weight lifting, except this time it is not the bar which is lifted by the arms but it is the body. The hands holding the bar are the fixed points against which the upper arm is moved against the forearm and the body against the upper arm. Here, also, the point is to hold the body close to the arm. It starts with flexion of the elbow until the body is pushed up against the bar. When this is reached, the grip must be changed so that the elbows can be carried backward; the chin is now made to clear the bar by thrusting the head forward over it and, with an ultimate simultaneous extensory effort in elbow and shoulder, the body is lifted until the elbow joint is in full extension.

C. BOXING

Boxing is a peculiar combination of speed and force. The upper extremity is poised to meet an external resistance and, at the same time, it is so handled as to meet the resistance with the greatest possible momentum. In contrast with throwing, for instance, where the main objective is to impart to the extremity the greatest possible speed, in boxing the principal aim is force;

and yet it is essential for developing the desired punch to give it the necessary acceleration. This is done mainly by the play of the lower limbs and the movement of the trunk especially the shoulders. The upper extremity is held comparatively rigid and delivers the impact accumulated by the rest of the body. Here, the element of stabilization of all joints of the upper extremity is in the foreground to the effect that all momentum developed by the body may be transmitted without loss and the blow may lack nothing in force.

One can readily see that it is difficult to speak of purely open or purely closed kinetic chains in the ordinary daily locomotor performances, and even in sport events this distinction cannot be carried through. Most of the athletic feats are combinations in which speed and force participate at various ratios. So far as the upper extremity is concerned, some situations appear definitely of the dynamic type in contrast to others which are more of what might be called the static type; they are represented respectively by the open and by the closed kinetic chain. The aptitude for sports emphasizing more the open or the closed chain type may be reflected in the constitutional makeup of the athletic trainee. But, on the whole, it will be found difficult to develop an athlete for speed without developing his strength at the same time, and vice versa.

V. SUMMARY

1. Compound movements of the upper extremity are best classified on the basis of whether they are arranged in an open or in a closed kinetic chain, i.e., without or with interfering external resistance. An example of the first is throwing a ball; of the second, chinning oneself on a horizontal bar.

2. In the process of for- and backward swing on one hand and of lengthening and shortening the extremity on the other, the biarticular muscles behave exactly as do those of the lower extremity: They move in a concurrent shift against their antagonist in shortening and lengthening, and in countercurrent shift in for- and backward swing; the joints involved are shoulder, elbow and wrist.

3. With the hand free, the forearm assumes a pronatory position in flexion, a supinatory in extension. With the hand fixed, the ulna moves around the radius in opposite sense, the center of motion being the glenohumeral joint.

4. Ball throwing is an example of an open kinetic chain. To accumulate acceleration, the joints move in a certain sequence, as follows: the wind-up, an axial rotation of the trunk, transmitted to it through the lower extremities; the shoulder goes from elevation and dorsiflexion into forward flexion, then the elbow from flexion to extension, the wrist from extension to flexion, delivering the missile. All movements are so timed that the next joint sets in when the momentum of the preceding one has reached its peak, and neither before nor after it.

5. Disc throwing is a similar performance except that momentum is ac-

cumulated first by circling the arm, then by a trunk rotation brought about by change of pivoting from the left foot to the right.

There are three sources of accumulating momentum: the preliminary circling of the arm, the rotation of the trunk on the alternatingly pivoting feet, and the terminal whip action consisting in forcible extension of the elbow and flexion of the wrist.

6. The external resistance of the 12-pound shot is greater that that in the preceding sports, but the objective is here also of dynamic nature, i.e., development of speed. This is accomplished by a rotatory movement of the trunk followed in sequence by extension of the arm in shoulder and elbow, and the delivery of the missle by a forceful flexion movement of the wrist.

7. In standing on the toes and in running and jumping, the upper extremities act as stabilizers to maintain body equilibrium. Their action is directed proximally toward the body. They control the position of the common center of gravity. Throwing the arms forward prevents falling backward, and vice versa the backward thrown extremity saves the body from falling forward. This is particularly effective in the jump, where the forceful extension of the lower extremities imparts to the body a forward and upward acceleration, further increased by violent upward jerking of the arms. The result is loss of body equilibrium. Then, in landing, it is regained by the upward swing of the extended arms, acting as restraint.

8. Only if the external resistance is such that visible motion of the peripheral part is grossly restrained or impeded, can one speak of a closed kinetic chain.

9. Weight lifting is performed under conditions of essentially increased peripheral resistance and although visible motion is still accomplished, it requires special positions and attitudes of the body to overcome the gravitational rotation moment produced by the weight, which otherwise could not be negotiated. Specifically, the backward bending of the body to bring the weight to chest height and the throwing back of the head to clear the chin, are maneuvers designed to minimize the rotational effect of gravity.

10. Chinning oneself on a horizontal bar represents a truly closed kinetic chain because the external obstacle represented by the fixed bar cannot be overcome. Instead, all visible motion is proximal, i.e., the body is moved against the immobilized arm and forearm.

11. Boxing is a combination of force and speed. The principal aim is force. The upper extremity is poised to meet an external obstacle with the greatest possible momentum. It is held comparatively rigid. Yet, in order to develop the desired punch, a certain acceleration must be acquired. This is done by the play of the lower limbs and the movements of trunk and shoulders. The stabilization of the joints of the upper extremity has the effect that all momentum accumulated by the movements of legs and body is transmitted without loss of force.

12. It can be seen that it is difficult to speak of purely open or closed kinetic

chains in ordinary locomotor performances or even in sport events. Most of these are combinations in which speed and force are developed at ratios varying with the specific motor act.

BIBLIOGRAPHY

1. BAEYER, H. v.: Die Synapsis in der allgemeinen Gliedermechanik. *Report 2d. Intern. Orthop Congress,* London, 1933.
2. BRESNAHAN, Z. T., and TUTTLE, W. W.: *Track and Field Athletics.* St. Louis, C. V. Mosby, 1937.

Lecture XXXIII

SPRAINS OF THE JOINTS OF THE UPPER EXTREMITY

I. SPRAINS OF THE SHOULDER JOINT

A. CAPSULAR AND LIGAMENTOUS STRAIN OF THE GLENOHUMERAL JOINT

A DISTINCTION must be made between the simple capsular sprain or distention and actual tears of the musculotendinous cuff which reinforces the capsular apparatus. The resistance of the capsule to tear is considerably less at its inferior and lower portions than it is in its upper and outer portions where it is heavily reinforced by tendons and ligaments. According to the experiments of Fessler,[2] separation of the capsule from either head or the glenoid required 200 kg. for the upper outer, and only 60-100 kg. for the inner and lower portion of the capsule. The latter broke under a leverage of 15.4 kg. with a lever arm of 17 cm., that is, under a moment of 262 kgcm. It is safe to assume that minute tears occur infinitely more often than gross anatomical lesions.

Mechanics of glenohumeral strain

At the anterior aspect the three portions of the glenohumeral ligament as well as the coracohumeral restrict the outward rotation of the hanging arm.

The glenohumeral ligament has three portions, an upper, a middle and an inferior one, and it constitutes the reinforcement of the anterior capsule. All portions restrict outward rotation of the arm.

The coracohumeral ligament restricts the axial rotation of the humerus. It sends a ligamentous bridge over the bicipital groove. This bridge consists of firm bundles running from the greater to the lesser tubercles of the humerus which are intimately interwoven with the coracohumeral ligament. A tear of this ligament on forced rotation affects the security of the biceps tendon.

Abduction of the arm in the frontal plane is always associated with outward rotation. As the arm is moved in this plane from the hanging down to the upright perpendicular position, the humerus rotates about its length axis about 90° outward. Should this rotation be inhibited so that the humerus retains its inward rotatory position, then the head may be levered out of the glenoid, or the greater tuberosity may be shorn off by the acromion, or a tear may occur in the portion of the capsule which is interwoven with the tendon of the supraspinatus.

B. LIGAMENTOUS STRAIN OF THE CORACO- AND ACROMIOCLAVICULAR JOINTS

The trapezoid portion of the coracoclavicular ligament restricts movement of the scapula forward. A blow or a force pushing the shoulder forward and inward may break this ligament. The conoid portion restricts backward movement of the scapula; a forcible back thrust will rupture this ligament. The acromioclavicular ligament restricts the displacement of the clavicle backward. It requires the break of both the coracoclavicular and the acromioclavicular ligaments to produce a substantial overriding of the clavicle on the spine of the scapula.

C. THE AVULSIONS

Inasmuch as the traumatizing force is directed against the tendon or the ligament or the capsular apparatus proper, and the avulsion of small pieces of cortical bone is secondary to it, the avulsions should be considered in connection with the soft tissue strains rather than classified as fractures.

The most common of these avulsions is that of the greater tuberosity at the

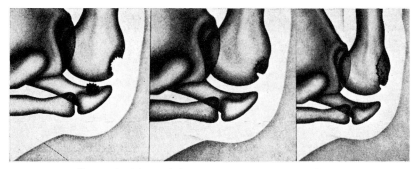

FIG. 1. Avulsions of the greater tuberosity. (Steindler)

insertion of the supraspinatus tendon. The mechanogenesis of this tear fracture is the same as that of the tear of the supraspinatus tendon. In contrast, the contusions of the tuberosity are produced by pressure on the humeral head and are caused by direct blow or by fall. Consequently, displacement of the fragment may be looked for in avulsions as they are still under the influence of their respective tendons, whereas the shear fractures, caused by contusion, tend to keep from being displaced so long as they are not disturbed by passive movement (Fig. 1).

D. THE PATHOMECHANICS OF MUSCLE RUPTURES

1. The biceps

Acute ruptures are rare, and they occur by sudden and forceful movement. Chronic rupture is also a mechanical event in the last analysis, but it involves an already degenerated tendon which is almost ready to break until a last, and often insignificant effort completely separates the tendon.

2. The deltoid

This rupture is very rare and usually occurs indirectly by forceful contraction of the muscle against a rigid obstacle.

II. SPRAINS OF THE ELBOW JOINT

Although the humero-ulnar portion of the elbow joint is generally considered to be a true hinge joint, its constituents do not show perfect congruity. In extreme extension the entire medial aspect of the olecranon shows a gap, and in extreme flexion the lateral border shows a dehiscence (Fick[3]). There is, in fact, no absolute fit of the articular surfaces in any position, the best adaptation being at an angle of 120°, which is about the midposition. A lateral movement of the ulna against the trochlea was first demonstrated by Langer,[5] and it amounts, according to Hultkranz,[4] to 1½-2 mm. This small shift has no effect upon the actual motion, but it merely shows that the joint is not precisely "trackbound" (Zwangsläufig).

The radiohumeral portion which has two degrees of freedom of motion, namely, flexion-extension and pro- and supination, is loose, secured by ligaments and muscles only, and has no natural skeletal checks as has the ulnohumeral portion where in full extension the olecranon abuts against the humerus. There is also a certain length shift of the forearm bones. Extension causes a shift of the radius distally, due to the spiral-shape curving of the capitellum (Fig. 2), and it is associated with a slight ab- and adductory

Fig. 2. Distal shift of radius on extension. (Fick)

shift in the elbow joint. In radial abduction the ulna shifts distally, and in ulnar abduction, the radius (Fick[3]). Finally, there are considerable differences in the cubital angle. Not only does the total cubital angle between the axes of humerus and ulna vary between 154-180°, but there are also variations in the relative ratio between the humeral and the ulnar portion of this angle. All these factors contribute to make the joint vulnerable to stresses of all kinds.

The thinnest portion of the elbow capsule is posteriorly around the fossa olecrani, and the strongest reinforcements are provided by the ulnar and radial collateral ligaments. Fessler's[2] experiments show for the medial collateral a traction resistance of 100-120 kg.; for the lateral, 80-160 kg.; and for the interosseous membrane, a stress resistance of 40-80 kg.

The whole humero-ulnar joint resisted longitudinal traction up to 85-230 kg., while the radiohumeral joint had a resistance against displacement of only 15-40 kg. These figures taken from the ligamentous specimens do not give, however, the true resistance in vivo where the muscles contribute so much to the stabilization of the joints. But they present the relative values of the ligamentous reinforcement and convey an idea of what may be expected when

the muscles are relaxed or paralyzed and external forces of the above
magnitude are applied.

A. THE COLLATERAL LIGAMENTS

1. Internal collateral ligament

The internal collateral ligament has three portions. The strongest bundle
goes from the internal epicondyle to the internal border of the ulna. It slides

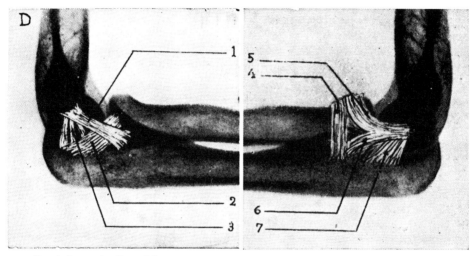

FIG. 3. Internal collateral ligament.
(H.-Ros Codorniu)
1 internal collateral ligament
2 ligament of Cooper
3 posterior bundle of internal collateral ligament

FIG. 4. External collateral ligaments of the elbow.
(H.-Ros Codorniu)
4 annular ligament
5 anterior bundle of the lateral collateral ligament
6 medial bundle of the lateral collateral ligament
7 posterior bundle of the lateral collateral ligament

over the synovial membrane which covers the medial surface of the trochlea,
and it is reinforced by Cooper's ligament running forward from the olecranon
to the coronoid and posteriorly by some strands from the trochlea to the
olecranon fossa. A fiberous bundle encircles the ulnar nerve. It comes under
strain particularly by forces which abduct the forearm and tend to increase the
carrying angle. When this angle is increased naturally or from traumatic
causes, tension stresses develop in this ligament (Ros-Codorniu[7]) (Fig. 3).

2. External collateral ligament

The external collateral also has three portions: an anterior, a medial and a
posterior, and, in addition, includes the annular ligament of the radius. It
has a double function: it secures the lateral aspect of the elbow joint and
maintains the radius in proper relation to the capitellum (Ros-Codorniu[7])

Stresses in the direction of adduction of the forearm cause strain of this ligament (Fig. 4).

The annular portion, arising from the sigmoid fossa of the ulna, covers head and neck of the radius, keeping the latter in alignment with the ulna. A downward pull on the radius brings first under tension the three bundles of the radial collateral, then transmits it to the annular ligament, and it may cause the latter to slip off the head and neck and may even come to lie in the radiohumeral joint.

B. THE ANTERIOR LIGAMENTS

The anterior capsule is reinforced by two ligamentous strands: a triangular one from the trochlea to the ulnar and anterior border of the coronoid process, and a lateral bundle from above the trochlea to the radial border of this process. Both ligaments come under strain by forced extension of the elbow. The posterior capsule is so well covered by the anconeus running from the lateral epicondyle to the posterior border of the ulna, and the humeral and ulnar heads of the supinator brevis, that no further ligamentous reinforcements are necessary. The movements producing strain of these ligaments are, therefore, abduction for the internal collateral, adduction and traction for the external collateral, and hyperextension for the anterior ligamentous structures.

C. THE TENNIS ELBOW

The mechanical explanation of this injury is a forced pro- or supinatory motion with the elbow extended, for instance, in the backhand stroke in tennis, whence the name. Such violent movements may lead to an actual tear of some of the fibers at the origin of the wrist extensors, especially of the extensor carpi radialis brevior, or they may secondarily involve a neighboring bursa, as suggested by Osgood[6] and more recently confirmed by other investigators; or, it may be an actual impingement of a torn capsule or a synovial fringe in the radiohumeral joint. The cases differ accordingly in the local clinical manifestations, although the mechanogenesis is the same in either instance.

III. SPRAIN OF THE WRIST

A complex mechanical unit such as the wrist with its numerous articulations being entrusted with kinetic performances of the greatest variability, naturally needs ample ligamentous reinforcement to protect it. This complex system of reinforcement can be divided into volar, dorsal, and lateral groups: Volarly, there is the anterior radioulnar ligament, the radiocarpal and the ulnocarpal ligament (Fick[3]) (Fig. 5). The dorsal ligaments are the dorsal radioulnar ligament, the dorsal radiocarpal and a transverse midcarpal ligament. The lateral ligaments are the ulnar and the radial collateral carpal ligaments (Figs. 5, 6)

Under certain circumstances, all of these can become strained: the radial collateral by forced ulnar abduction; the ulnar collateral by forced radial abduction; dorsally, the radiocarpal and the transverse midcarpal ligament by forced flexion; and, volarly, the radiocarpal and the ulnocarpal ligaments by forced extension (Fick[3]).

A word should be said about the volar and dorsal radioulnar reinforcing ligaments. Pro- and supination, when tried in the cadaver, appear to be solely

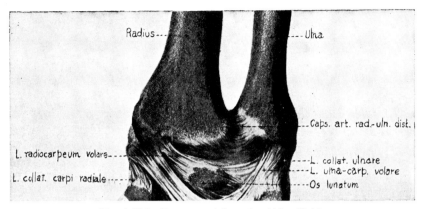

FIG. 5. Volar ligaments of the wrist. (After Fick)

a motion of the radius around the head of the ulna, the latter being the fixed point, and the common axis of the upper and lower radioulnar articulation strikes the head of the ulna.

In vivo, this motion is somewhat different. While the radius circles around the ulna, the latter also circles around the radius in inverse direction. That is,

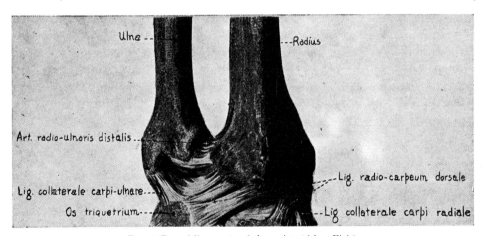

FIG. 6. Dorsal ligaments of the wrist. (After Fick)

as the radius carries out a pronatory movement, the head of the ulna moves against it in supinatory direction, and vice versa. Since the ulna cannot rotate lengthwise in the elbow joint, this would be difficult to understand. But when one observes the humerus one will find that with pronation of the wrist the humerus rotates outward, thereby imparting a supinatory twist to the ulna, while in supination it rotates inward causing the ulna to pronate (Boppe[1]).

This has some bearing upon the strain of the reinforcing ligaments of the radioulnar articulations: the posterior is tight in pronation; the anterior tightens in supination, stabilizing the wrist in this position, as does also the volar, ulnocarpal ligament.

There is another factor which favors the development of sprains of the wrist. The mechanics of finger motion require the tendons to crowd volarly into the volar carpal canal with the result that on leaving the canal the tendons change their direction and diverge to reach the fingers. The carpal canal is formed by the lateral pillars, the pisiform and the uncinate on the ulnar, and the scaphoid tubercle and the ridge of the greater multangulum on the radial side. The maintenance of this hollow trough against the lateral strain of the muscles which it encloses then falls upon a strong ligamentous tiepiece, the volar retinaculum or volar transverse ligament. Only the distal portion of the ligament between the ridge of the trapezium and the hook of the hamate is tight. The proximal half depends upon the flexor carpi ulnaris to tighten it (Steindler[8]).

Hyperextension will strain or even tear the volar carpal ligament. Points of tenderness develop then at the pisiform and the greater multangulum. The volar carpal ligament also preserves the carpometacarpal arch. It may become relaxed after sickness or general disability when its supporting structures lose tone. Its palmar concavity flattens out so that the hook of the unciform is no longer directed inward but looks forward or even outward.

General sprains and distortions of the joint capsule and its reinforcements as a whole are all brought about by forced movements against strong resistance. The resistance to tear of the joint capsule of the wrist, is according to Fessler[2], 184 kg.

V. THE CALCULATION OF WORK PERFORMED

The trapeziometacarpal joint of the thumb is reinforced by strong volar and dorsal ligaments. These ligaments restrict abduction and opposition; forced movements in this direction are liable to produce sprain. In general, one will find that the reinforcement of the capsule is ample on the ulnar and volar side of the joint but less so on the lateral radial side where the extensors and the abductors protect the joint. Hence signs of sprain develop usually on the dorsal and radial side underneath these tendons.

V. SPRAIN OF THE METACARPOPHALANGEAL JOINTS OF THE THUMB AND FINGERS

Sprains of these joints are frequent and are usually the result of pull, while contusion of these joints is caused by fall or blow. According to Fessler,[2] the tear resistance of the metacarpophalangeal joint of the thumb is 65-100 kg. and that of the second to fourth fingers 75-83 kg. We recall that the capsule is reinforced by the two collaterals and the arcuar volar ligament (Fig. 7). The latter tightens in dorsiflexion; in volar flexion the collaterals resist lateral motion (Fig. 8). There is comparatively little reinforcement on the dorsal aspect of the joint except the expansion of the extensor apparatus of the fingers.

The strain produced by excessive dorsiflexion, therefore, must overcome the strong volar ligament and especially the volar fibrocartilaginous plate which serves as a gliding track for the flexor tendons. Dorsiflexion must be extreme before a true sprain or tear or even a dorsal dislocation can occur (Fick[3]).

The situation is similar in the case of the metacarpophalangeal joint of the thumb. Hyperextension is checked by the marginal crest of the metacarpal head. But acting as a fulcrum it can produce strain and tears of the volar

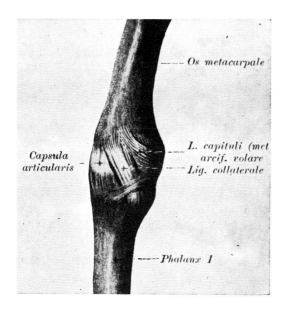

FIG. 7. The collateral metacarpophalangeal ligaments.
(Fick)

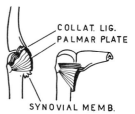

FIG. 8. Flexion of metacarpophalangeal joints tightens collateral ligaments. (Grant)

reinforcements of the capsule. If the capsule tears, its proximal portion may become impinged between head and socket which, in the case of dislocation, is a considerable obstacle to reduction. Habitual hyperextension of the thumb is commonly observed in people with relaxed anterior ligaments.

VI. SUMMARY

1. Outward rotation of the hanging arm is restricted by the three portions of the glenohumeral as well as the coracohumeral ligament. The latter sends fibrous bundles over the bicipital groove; a tear of these bundles affects the security of the biceps tendon.

2. Abduction of the arm is associated with outward rotation. If this is inhibited, the head may be levered out of the glenoid in abduction or a tear of the supraspinatus tendon may result.

3. The acromioclavicular articulation is secured by the coracoclavicular ligament which resists both for- and backward displacement of the scapula

against the clavicle and by the acromioclavicular ligaments reinforcing the joint. Rupture of both of these ligaments is necessary to permit the clavicle to override the spine of the scapula.

4. The most common avulsion is that of a portion of the greater tuberosity at the insertion of the supraspinatus tendon. In contrast to contusion, displacement is common in avulsions because of muscle pull.

5. Acute ruptures of the biceps by sudden contraction are rare. Chronic ruptures occur usually on the basis of previous degenerative changes of the tendon.

6. The elbow joint although almost a true hinge shows gaps between olecranon process and humerus, the closest fit being at 120° flexion. There is an insignificant lateral ulnar shift on motion and also a length shift of the forearm bones on flexion and extension. Furthermore, a decreased cubital angle (cubitus valgus) predisposes the joint to strain.

7. The strongest reinforcement of the joint is provided by the radial and ulnar collateral ligaments; both have three portions: an anterior, a posterior and a middle portion. The ulnar collateral is strained by abduction, the radial by adduction. Besides, the ulnar has a fibrous bundle encircling the ulnar nerve, which becomes involved when this bundle is exposed to strain. The radial collateral secures the lateral side of the elbow joint and maintains the radius in proper relation to the capitellum by means of the orbicular ligament with which it is interwoven. Downward pull on the radius stretches first the radial collateral, then the orbicular ligament, which latter may slip off the radial head and neck.

8. The anterior capsule is reinforced by a triangular ligament from the trochlea to the ulnar, and a lateral from the trochlea to the radial border of the coronoid process. It becomes tight in extension.

9. The mechanics of the tennis elbow is strain on forced pro- and supination of the extended joint, causing either a tear of the extensor carpi radialis brevior tendon or irritating a bursa underlying this tendon, or causing an impingement of a torn capsule or of a fat fringe in the radiohumeral joint.

10. The wrist is protected laterally by the collateral radiocarpal and ulnocarpal ligaments, volarly by the anterior radioulnar, radiocarpal and ulnocarpal, and dorsally by the dorsal radiocarpal, dorsal radioulnar and a dorsal midcarpal ligament. All of these become strained in certain positions: the ulnar collateral by radial, the radial collateral by ulnar abduction; the dorsal radiocarpal and transverse midcarpal by forced flexion, and the volar radiocarpal and ulnocarpal by forced extension.

11. The fact that in vivo not only the radius circles the ulna in pro- and supination but that the ulna circles the radius in opposite sense has a bearing upon the strain of the ligaments of the radioulnar articulation. The anterior becomes tight in supination, the posterior in pronation.

12. The volar retinaculum or volar transverse ligament acts as a strong tiepiece maintaining the carpal arch. Hyperextension will strain or tear the

volar carpal ligament. When it relaxes the palmar concavity of the carpal groove flattens out under strain.

13. The carpometacarpal joint of the thumb is susceptible to sprain on its ulnar and volar side, less so dorsally and radially where it is protected by the extensors and abductors of the thumb.

14. The metacarpophalangeal joints of thumb and fingers sprain frequently. The capsule is reinforced by the volar arcuate and the collateral ligaments. The latter become tight in flexion, preventing side motion in the metacarpophalangeal joints; the former tighten on extension. Tear of the capsule of the thumb often causes impingement of its proximal portion, which in case of an accompanying subluxation or dislocation becomes an obstacle to reduction.

BIBLIOGRAPHY

1. BOPPE, M.: *Advant bras physiologie, Ombredanne et Mathieu.* Traité de Chirurgie orthopedique, Vol. III, Paris, 1937.
2. FESSLER, J.: Festigkeit der menschlichen Gelenke mit besonderer Berücksichtigung des Bandapparates. *Ztschr. f. Chir.,* 1906.
3. FICK, R.: *Spezielle Gelenk- und Muskelmechanik III.* Jena, G. Fischer, 1911.
4. HULTKRANZ, J. L.: *Das Ellbogengelenk und seine Mechanik.* Jena, 1897.
5. LANGER, C.: Die Bewegungen der Gliedmassen, insbesondere der Arme. *Wr. Med. Wchnschr.,* 1859.
6. OSGOOD, R. G.: Radiohumeral bursitis. *Arch. Surg., 4:*402, 1922.
7. ROS CODORNIU, A. H.: *El codo, sus Fracturas y Luxaciones.* Madrid, Ed. Cirugia del aparato locomotor, 1945.
8. STEINDLER, A.: *Traumatic Deformities of the Upper Extremity.* Springfield, Illinois, Charles C Thomas, Publisher, 1946.

Lecture XXXIV

THE PATHOMECHANICS OF SPRAINS
OF THE LOWER EXTREMITY

I. SPRAINS OF THE HIP JOINT

A. CAPSULAR AND LIGAMENTOUS SPRAINS

THE tension resistance of the hip joint with its reinforcing ligaments is very high. Fessler[3] found it to be 240-650 kg., with an average of 380 kg., aside from the resistance of the muscular apparatus.

The weaker portions of the capsule are situated, according to to Fick,[4] anteriorly under the iliopsoas, medially near the incisura acetabuli, and pos-teriorly at the incisura ischi-adica, roughly corresponding to the three rays of the cart-ilaginous junction uniting the three pelvic bones. The strongest portion of the cap-sule is where it is reinforced by the iliofemoral ligament with its two portions, the superior and the anterior (Fig. 1). Between it and the pubofemoral ligaments, run-ning from the os pubis to the femur, there is a thinned-out area. Sprains of the anterior portion of the iliofemoral lig-ament are caused by forcible extension of the hip, while the superior portion of the ligament tightens also in in-ward rotation and adduction.

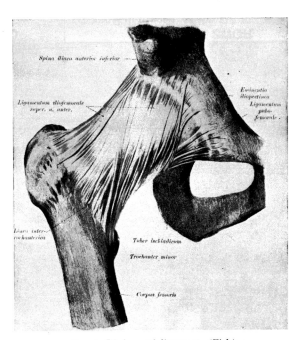

FIG. 1. Iliofemoral ligament. (Fick)

The pubofemoral ligament (Fig. 1) is strained by abduction, the ischio-femoral (Fig. 2) by inward rotation of the femur either in flexed or in extended position of the joint.

The two remaining ligaments are the *ligamentum teres* and *the orbicular* ligament. The teres (Fig. 3), though of little mechanical significance, is of considerable strength and can support a stress of 15-57 kg. according to Brause.[2] It always tears in traumatic dislocations. The orbicular ligament (Fig. 4), being closely interwoven with the longitudinal ligaments, follows their movement. It constricts the capsule, securing the rotatory movements of the

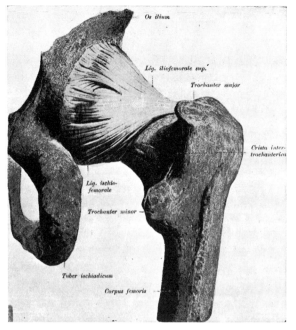

FIG. 2. Ischiofemoral ligament. (Fick)

head which it encircles. It is strained by abduction but does not necessarily tear in dislocation as the head may leave the socket above this ligament, either in front or in back, through the above-mentioned weaker areas of the capsular apparatus.

B. MUSCLE SPRAINS

Muscle sprains occur by direct blow but more often through excessive ultraphysiological movement, usually against external resistance. Violent sprain may be accompanied by avulsions.

Excessive *abduction* causes sprain of the adductors: for instance, in doing the so-called split, one limb straight forward, the other backward, both thighs touching the ground. In the forward leg pectineus, adductor brevis, gracilis and longus are sprained and may tear, in the backward leg the adductor magnus.

Violent *backward extension* causes sprain of the iliopsoas: for instance, when a runner trying to come to a sudden stop throws his body forcefully backward and causes a passive stretching of the muscle; or it may be produced actively by forceful flexion of the hip as in kicking a football. Severe iliopsoas strain is often accompanied by avulsion of the lesser trochanter. Strain of the rectus femoris with avulsion of the anterior inferior spine of the os ilii may occur in running. Such an avulsion is often seen in skiing. The mechanism is as follows: in the skijump the body is inclined forward to become perpendicular to the downward slope. If the latter becomes steeper and the forward inclination of the body does not follow it promptly, the center of gravity of the trunk falls behind the hip

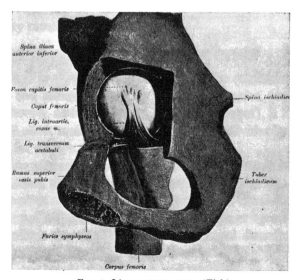

FIG. 3. Ligamentum teres. (Fick)

joint axis and the skier would fall backward. To avoid this he jerks the body violently forward as he is ready to land, which is done by forceful contraction of both the iliopsoas and the rectus femoris. If the moment of the backward inclined body is too great for the strength of the rectus femoris muscle, then its violent contraction leads to tear and avulsion of the anterior inferior spine (v. Saar[5]).

Another sprain which occurs by *forceful flexion of the hip with extended knee* is that of the hamstrings with occasional avulsion of the apophysis of the tuber ossis ischii.

Traumatic exostosis or myositis ossificans develops after strain of the adductor muscles. It is typical in horseback riding and mostly due to the inexperience of the novice who overstrains the adductors which secure him in the saddle. The result is

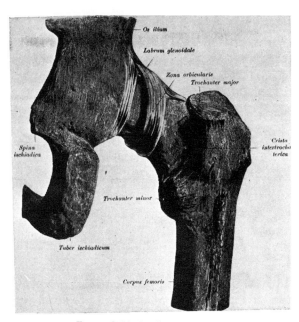

FIG. 4. Orbicular ligament. (Fick)

the *"rider's sprain."* The anatomical bases are the contusions and hematomas in these muscles which accompany small tears. The trouble is ushered in by rather vague pains causing the person to walk with legs apart to save the adductors (v. Saar[5]).

II. PATHOMECHANICS OF THE SPRAINS OF THE KNEE JOINT

A. SPRAIN OF THE COLLATERALS

The collaterals are checkers of the lateral and rotatory movements of the knee joint but their static function goes beyond this as they sustain the knee under weight bearing and prevent distention under traction. The traction resistance of the knee joint is given by Fessler[3] at an average of 315 kg.

1. The internal (tibial) collateral

This ligament is more exposed to sprain, first due to the natural valgity of the knee and then because of its intrinsic connection with the capsular apparatus. According to Smillie,[7] the mechanism of the partial sprain is most commonly a minor rotatory strain with the knee flexed and less often an extension strain. The complete sprain or avulsion occurs by a heavy blow against the outside of the extended leg, heavy enough to break through the protective action of the sartorius. The mechanism is similar to that producing injury of the medial meniscus, except that there is no locking. Complete rupture, which is caused by violent abduction strain, is seldom an isolated injury but most often associated with lesions of the medial meniscus or a fracture of the external tibial condyle.

2. The external collateral

Lesions of this ligament are not so common since it is better protected and separated from the joint by the popliteus tendon. The mechanism of forced adduction is more often caused by direct than by indirect violence. Lateral instability in the sense of genu varum is the result of complete rupture of the ligament.

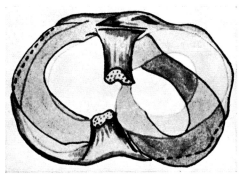

FIG. 5. Motion of the menisci with inward rotation of the tibia with knee in full extension. (Schaer)

B. THE MENISCUS SPRAIN

To appreciate what happens to the menisci when an external force is applied to them, one must recall their movement during normal activity of the knee. The position of the menisci changes constantly and is characteristic for each phase of motion. In extension both menisci fill almost completely the space between the joint surfaces. In flexion without active rotation, aside from the track-bound rotation connected with the flexion-extension mechanism, the menisci wander gradually backward and therefore can no longer be palpated in front of the joint. The backward motion can go so far in extreme flexion that the posterior border of the menisci overhangs the posterior borders of the tibial joint surfaces (Schaer[6]).

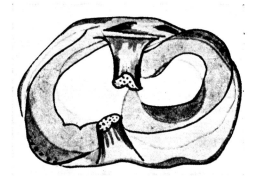

FIG. 6. Motion of the menisci with outward rotation of the tibia with knee in full extension. (Schaer)

In flexion, rotation becomes more free. If such rotation is carried out by the leg the menisci move with the femur. In inward rotation of the leg the lateral meniscus moves backward and somewhat inward while the medial moves forward into about the same position it occupies when the knee is in full extension (Fig. 5).

In outward rotation of the leg the position is reversed. The internal meniscus now moves backward and the external moves forward (Fig. 6). The forward displaced external meniscus can be palpated in front at the lateral aspect of the joint.

1. The medial meniscus

The usual mechanism of meniscal strain is flexion and rotation. With few exceptions the force is indirect. An old explanation is that the excessive outward rotation of the leg while the knee is flexed pulls the internal meniscus (which follows the rotation of the femur as we have remarked above) farther into the joint space and then in rapid extension the meniscus is caught between the internal condyles of tibia and femur. Ordinarily, it pulls back promptly into position, but if it is wedged in between the result is that under the torsion effect it splits lengthwise and becomes displaced into the interior of the joint. This explanation also holds for other than longitudinal tears: for instance, the ruptures of the anterior and posterior horns which result in free tabs; it also applies to the transverse tears. Other mechanisms are the squatting with outward rotated legs; when it is then followed by a sudden extension of the knee, the same impingement process sets in. But this is not the only mechanism. It seems that simple ab- and adductory forces in the sense of genu valgum and varum may lead to meniscus injuries. Also, extreme hyperextension may produce a compression fracture of the anterior portion of the menisci.

In contrast to the frequent meniscal injuries by indirect forces, those caused by direct force are rare and the forces applied must be considerable. Schaer[6] could produce in the cadaver minimal tears by direct blow only when the knee was in extreme flexion and inward rotation. The point is that injury to the menisci is only possible when flexion is superimposed upon internal or external rotation (Smillie[7]) and that while the collateral ligaments tear by ab- and adduction strain, the principal factor in meniscal injury is rotatory torsion compression.

2. The lateral meniscus

The less common mechanism underlying injuries of the lateral meniscus is flexion and inward rotation of the tibia. The lateral meniscus, being more mobile and not closely interwoven with its collateral ligament as is the medial meniscus, does not sustain the backward pull and, therefore, is not subject to become loosened from its coronary ligament. Here, the grinding rotatory motion produces not so much longitudinal splits as transverse or so-called parrot-beak tears (Smillie[7]).

C. SPRAINS OF THE CRUCIATES

According to the observations of Brantigan and Voshell,[1] the anterior cruciate controls forward gliding of the tibia on the femur and in association with the lateral ligaments also lateral mobility. Together with the posterior cruciate and the collaterals it also controls hyperflexion and hyperextension.

The posterior controls backward gliding of the tibia on the femur. The brothers Weber[8] already found that on cutting both cruciates the security of the knee was not affected in extension, the joint being stabilized by the collaterals, but that in flexion for- and backward movement of the tibia was possible and they concluded that the cruciates stabilize the knee in all positions of flexion. Together, they also check extreme flexion, an observation corroborated by R. Fick.[4] Consequently, a sprain or rupture of both of them may occur by hyperflexion, a sprain of the anterior by hyperextension, and a rupture of both by posterior or anterior dislocation of the knee.

So far as rotation is concerned, it is principally under control of the collaterals and only when these are torn is further rotation checked by the cruciates. A violent rotation, therefore, must rupture the collaterals before the checking action of the cruciates comes into play. The anterior cruciates may also be sprained by abduction but only after rupture of the medial collateral ligament.

D. SPRAINS OF TENDONS AND THEIR APONEUROSES

1. Sprain of the extensor apparatus of the knee

A) RUPTURE OF THE QUADRICEPS

It occurs by direct injury, fall of a heavy weight upon the knee which strikes above the patella, or a heavy load pinning the knee against the ground. It also may be caused by a violent muscle contraction against a strong external resistance. Partial ruptures which leave a portion of the quadriceps extensor still attached to the knee cap are not uncommon. In athletes a few fibers of a portion of the muscle may tear also after a violent kick or when the quadriceps is contracted and receives a hard blow. The part injured is usually the rectus but frequently rupture of the medialis, the intermedius or the lateralis occurs in similar manner. Avulsion of the quadriceps from the patella is observed also in older people from a stumble or when missing a step going downstairs when the quadriceps is suddenly extended from a flexed position in order to maintain equilibrium. In acute cases of partial rupture any extension attempt is suppressed by pain.

B) AVULSION OF THE PATELLAR LIGAMENT

The mechanism is sudden forceful contraction of the quadriceps against heavy external resistance. The tear is usually complete so that active extension of the knee is entirely lost, in contrast to the quadriceps rupture where some of the muscle fibers still remain attached. In the avulsion of the patellar ligament from the patella, the tear involves also the extensions on either side. A tear of the extensory apparatus usually accompanies fracture of the patella. In this case also the extensory power is entirely lost. The proximal fragment of the patella is pulled up by the rectus.

2. The avulsion of the tibial tubercle (Osgood-Schlatter's disease)

This is seen in young people before the fusion of this apophysis with the epiphysis of the tibia has taken place. The tibial tubercle is a beak-shape forward and downward extension of the upper tibial epiphysis in which separate small centers of ossification appear before it fuses with the epiphysis and the latter

with the shaft, which is about the age of 18 in boys and somewhat earlier in girls. Up to that time the epiphyseal plate is an area of lesser resistance. A strain of this line may be brought about by sudden flexion of the knee, the pull of the patellar tendon transmitting itself to the epiphyseal plate. The inner hamstrings which stream into the tibial tuberosity in front prevent displacement, but repeated flexion movement of the knee constantly sustains the strain.

In the adult the tibial tubercle is firmly united and resists avulsion. It may be brought about, however, by a very violent maneuver of passive flexion such as occurs in poorly advised attempts to overcome stiffness of the knee. Usually only a part of the patellar tendon goes with the avulsed fragment.

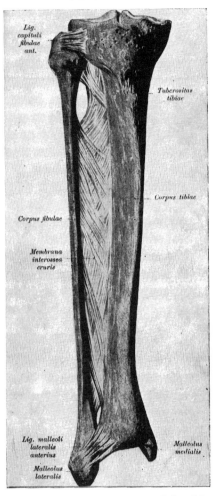

Fig. 7. Ligamentous structures of the tibia and fibula, anterior view. (Fick)

III. THE PATHOMECHANICS OF SPRAINS OF FOOT AND ANKLE

Next to the hand and the wrist, the mechanics of foot and ankle are more complicated and intricate than those of any other joints of the body. Adding to this the fact that the static demands on this structure are exceedingly great and that the process of locomotion, and especially that of propulsion to which these articulations are subservient, is a highly complex procedure, one must expect that a great variety of disturbances and deficiences will make their appearance under appropriate circumstances. So far as these develop from merely static causes, that is, under weight-bearing stress alone, they have been dealt with under the heading of static disabilities (Lecture XXIII). Only the purely traumatic events involving foot and ankle are to be considered at this time. A great multitude of structures is affected and there is a close connection between the mechanism of the injury and the particular structure which is injured.

A. SPRAINS OF THE LOWER TIBIOFIBULAR ARTICULATION

Three ligamentous structures reinforce this joint: the anterior tibiofibular ligament which runs obliquely from tibia to fibula in front of the articulation and checks for- and backward displacement of the fibula; the posterior tibiofibular ligament, also running downward obliquely from tibia to fibula at the posterior aspect of the joint, checks for- and backward as well as lateral displacement; and the intermediate ligament which runs in the same direction and checks the lateral separation of the fibula from the tibia (Fick)[4] (Fig. 7). When these ligaments are sprained or torn, the spring-like action of the malleolar mortise, which in dorsiflexion accommodates the broader anterior half of the body of the astragalus, is lost. The mortise no longer hugs the astragalus tightly.

B. THE SPRAINS OF THE ANKLE JOINT

This joint has very powerful reinforcing ligaments which secure it from all sides. The combined resistance of these to tension stress is considerable. Fessler[3] found it to amount to an average of 248 kg.

The lateral ligaments

Both tibial and fibular collateral ligaments have three principal portions, each. The fibular resists supination, the tibial pronation, with their midportions. At the same time, the anterior and posterior portions have their own checking effect on the flexion and extension range.

1. The tibial collateral ligament (Fig. 8)

a) The anterior portion has the longest fibers. They come from the inner malleolus and run forward over the talus and the scaphoid. They check the plantar flexion of the foot as well as pronation of the astragalus.

b) The middle or tibiocalcaneal portion runs from the internal malleolus to the sustentaculum. It is the principal checker of pronation and resists the raising of the lateral border of the foot.

c) The posterior tibiofibular ligament has the shortest

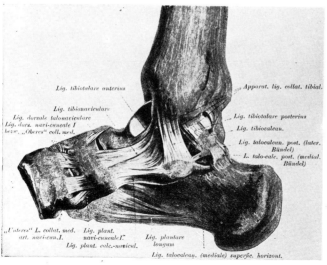

Fig. 8. Medial ligaments of foot and ankle. (Fick)

fibers; it runs from the posterior surface of the internal malleolus to behind the posterior facet of the articulation between astragalus and os calcis. It checks dorsiflexion of the foot as well as pronation.

One will understand, therefore, that a rupture of this ligament does more than produce pathological pronation. It also permits an unusual degree of plantar and dorsiflexion in the ankle joint. When under strain, trigger points of tenderness develop not only below but also behind and in front of the malleolus, with respective flexion or extension attempts at the ankle joint; repair of all three portions is therefore necessary. The mechanism of sprain is forced pronation.

The position of the ankle at the moment of sprain is important; whether it is in plantar flexion such as in going downstairs, or in dorsiflexion such as in climbing up a hill; depending upon it, either the anterior or the posterior strand comes under tension, together with the pronatory strain.

2. The fibular collateral ligament (Fig. 9) (Fick[4])

This ligament also consists of three major portions:

a) The ligamentum fibulotalare anterius runs from the anterior border of the outer malleolus horizontally or slightly upward over the sinus tarsi to the head of the astragalus. It checks supination and also plantar flexion of the ankle joint.

b) The ligamentum fibulocalcaneum runs from the tip of the fibula downward

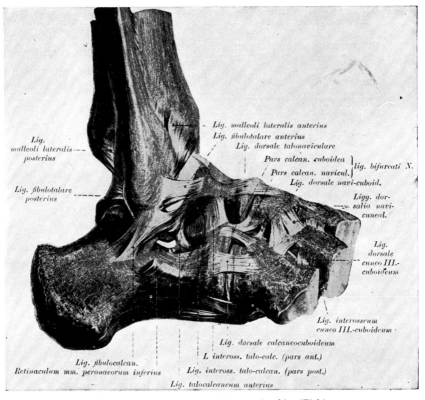

FIG. 9. Lateral ligaments of foot and ankle. (Fick)

and backward underneath the peroneal sheath to the outer side of the os calcis. It checks supination and also dorsiflexion of the ankle joint.

c) The ligamentum fibulotalare posterius is the strongest of the three. It runs from the posterior aspect of the external malleolus, in slightly downward direction, to the posterior margin of the astragalar body (Stieda's process). This ligament is a checker of supination as well as of dorsiflexion of the ankle joint.

Sprain of this ligament occurs by inversion of the foot from a stumble or a misstep or indirectly by violent outward rotation of the tibia, as in golf for instance. Here also the position of the ankle joint determines whether a part or all of the ligaments come under stress. The development of painful areas and the requirements of repair parallel those of the internal collateral.

C. THE SPRAINS OF THE SUBASTRAGALAR JOINT

Here are also three principal ligaments which secure this articulation.

1. The anterior talocalcaneal ligament (Fig. 9) extends from the lower surface of the talus to the upper of the os calcis at the posterior wall of the sinus tarsi. It checks the for- and backward gliding of the astragalus on the os calcis and is also a checker of supination.

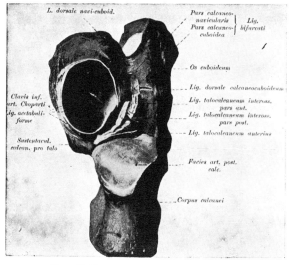

FIG. 10. Interosseous ligaments. (Fick)

2. The talocalcaneal interosseous ligament (Fig. 10), anterior portion, also runs from the underside of the talus to the upper surface of the os calcis; it checks supination and lateral displacement of talus.

3. The posterior portion of the ligamentum talocalcaneum interosseum (Fig. 10), lying behind the former, runs from the os trigonum or Stieda's process of the talus to the upper and medial surface of the os calcis. This ligament checks forward and lateral movement of the talus.

The prime function of these ligaments is to secure the relationship between astragalus and os calcis, especially so far as for- and backward gliding is concerned. In their pro- and supinatory checking effect they are greatly assisted by the respective portions of the tibial and fibular collateral ligaments. The interosseous ligaments lie so deep that an isolated sprain hardly ever occurs, but they take part in strain which involves the collaterals.

D. SPRAINS OF CHOPART'S JOINT

1. The talonavicular portion (Fig. 11)

a) The fibrocartilaginous cup which forms the lower and medial portions of the astragaloscaphoid articulation is heavily reinforced by the calcaneonavicular ligament running from the sustentaculum to the scaphoid. It prevents the head of the astragalus from gliding downward into the space between the scaphoid and sustentaculum. Acute sprains occur by violent abductory motion of the forefoot or directly by heavy blow. In the mechanics of the pes planus where it is under chronic static stress, it is of particular importance (Fig. 10).

b) The plantar or lateral calcaneoscaphoid ligament is often called the lower key to Chopart's joint (Fig. 8). From the anterior edge of the posterior articular facet of the os calcis it runs to the lower and lateral edge of the scaphoid, forming a firm plantar tiepiece between os calcis and scaphoid.

c) Dorsally, a talonavicular ligament

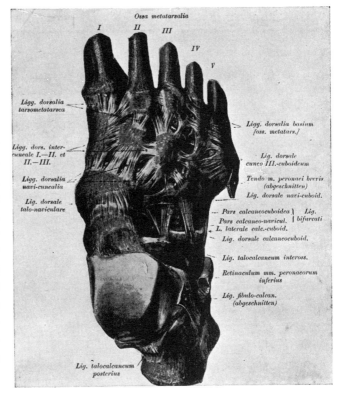

FIG. 11. The dorsal ligaments of the foot. (Fick)

running from the neck of the astragalus to the dorsum of the scaphoid checks plantar flexion of the forefoot in this joint (Fig. 8).

d) There is also the above-mentioned anterior portion of the internal deltoid ligament which reaches from the tibia over the talus to the scaphoid (Fig. 8).

e) Finally, a bifurcated ligament runs dorsally from the anterior inner edge of the os calcis across to the scaphoid while the other shank runs forward to the cuboid. The former is also called the dorsal key to Chopart's joint (Fig. 11). It prevents the scaphoid from being pried away from the edge of the os calcis and it also presses it to the astragalus (Fig. 11).

All ligaments secure the head of the astragalus in its relation to the scaphoid. As a forceful abduction causes a sprain of the medial, so a forceful extension of the forefoot sprains the volar and a forceful flexion the dorsal ligamentous system.

2. The calcaneocuboid portion

This junction is secured by three ligaments:

a) A calcaneocuboideum dorsale, which is really the above-mentioned sagittal strand of the bifurcate ligament, prevents dorsal displacement of the cuboid (Fig. 11).

b) A lateral calcaneocuboid ligament checks displacement of the cuboid laterally (Fig. 11).

c) The more massive ligamentum plantare is a long ligament starting at the tuber of the os calcis. The superficial layer goes forward as far as the metatarsal of the second to the fifth ray; the deeper layer is the ligamentum plantare breve which also runs from the posterior process of the os calcis and ends at the cuboid (Fig. 8).

All three are powerful tiepieces of the longitudinal arch, securing the close apposition of the cuboid to the os calcis.

Traumatic sprains of these ligaments are frequent. Forceful adduction of the forefoot sprains the lateral system, producing characteristic tenderness at the site of the calcaneocuboid junction. If the supinated forefoot is caught and flexed volarly with force, the dorsal system becomes sprained and develops a tell-tale trigger point dorsally. The sprain of the long plantar ligament, which is seen most often in the static flatfoot, produces a characteristic tenderness deep at the sole of the foot lateral to the tender area along the medial margin of the plantar fascia.

E. THE SPRAIN OF THE SCAPHOCUBOID JUNCTION

This junction, accessible in the middle of the dorsum of the foot, is subject to sprain which manifests itself by a tender point laterally to the navicular bone. Three small ligaments secure the articulation, namely, a deep and a superficial dorsal scaphocuboid ligament and the interosseous. All prevent separation of the scaphoid from the cuboid; they become sprained by direct blow or fall on the feet (Fig. 11).

F. THE SPRAIN OF THE SCAPHOCUNEIFORM JUNCTION

This is caused by forced abduction and dorsiflexion of the forefoot. The painful area is immediately in front of the scaphoid at the dorsum but it can be followed around the medial border to the plantar surface. The joint is secured by three distinct ligaments: a dorsal, a medial, and a plantar scaphocuneiform (Figs. 8 and 11). They secure the three cuneiforms against the scaphoid preventing abnormal motion. Acute sprain of this joint is caused by heavy blow or by a fall on the feet when the shock is received by the forefoot, the ankle being held in equinus position. A violent compression of the forefoot under a heavy load (wheels of a car) may sprain or rupture the ligaments securing both Chopart's joint and the scaphocuneiform junction. This is important to remember because these ligaments do not heal readily unless weight bearing is eliminated and the foot is immobilized.

G. THE SPRAINS OF LISFRANC'S TARSOMETATARSAL JOINT

This is a row of articulations comprising the junction of the three medial meta-carpals with the three cuneiforms and the two lateral with the cuboid. They form a broken line with recesses for the basis of the longer second metacarpal. They are secured by dorsal, plantar, and interosseous ligaments (Figs. 11 and 12), and they sustain the arch of the metatarsal bases which is comparable to the arch formed by the bases of the metacarpals in the hand. While the plantar and the interosseous ligaments check dorsiflexion, the bases are held together by a trans-

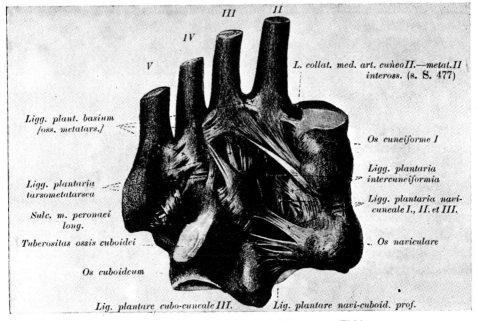

FIG. 12. The plantar tarsometatarsal ligaments. (Fick)

verse intermetatarsal band. The tiepiece of the arch is the peroneus longus. Except for heavy blows and crushes which may destroy the articulations as well as their ligaments, sprain is produced by indirect force. Relaxation or sprain of these ligaments produces a transverse area of pain at the level of Lisfranc's joint. A periosteal reaction of the first metatarsotarsal joint is not uncommonly seen in people with rather high longitudinal arch. It appears in the form of a dorsal exostosis at this articulation.

IV. SUMMARY

1. Sprains of the anterior portion of the iliofemoral ligament is caused by forcible extension of the hip, sprain of the superior portion by adduction and inward rotation. The pubofemoral comes under stress in abduction, the ischio-femoral in inward rotation.

2. The ligamentum teres tears in traumatic dislocation, the orbicular is strained by abduction.

3. Excessive abduction strains the adductors; violent backward extension of the hip strains the iliopsoas and is often accompanied by avulsion of the lesser trochanter. Rectus femoris strain is associated occasionally with avulsion of the anterior inferior spine. This often occurs in skijumping. Forceful flexion of the hip with the knee extended puts a strain on the hamstrings and may cause avulsion of the tuber ossis ischii. "Rider's sprain" of the adductors may result in hematomas and myositis ossificans.

4. The internal (tibial) collateral is more exposed to sprain because of the natural valgity of the knee and because it is intimately connected with the capsular apparatus. The mechanism is similar to that producing injury to the medial meniscus but there is no locking. Complete rupture by violent abduction is usually associated with meniscus lesions.

5. Sprains of the external (fibular) collateral ligament are less common, usually caused by direct violence.

6. Meniscal sprains require rotation and flexion. The mechanism of medial meniscus injury is indirect, by outward rotation, flexion and abduction. It is pulled farther into the joint space and caught between the condyles of femur and tibia. Splitting lengthwise, the detached portion is displaced into the interior of the joint. The mechanism for anterior and posterior tears as well as the separation of the coronoid ligament is similar. However, simple abductory forces may produce a meniscus lesion and extreme hyperextension may cause a compression fracture of the anterior portion. Minimal tears could be obtained in the cadaver by direct blow. In general, however, injury to the menisci is only possible when flexion is superimposed upon rotation of the tibia, while the collaterals tear by ab- or adduction strain.

7. Injuries to the lateral meniscus are less common because of its greater mobility. Grinding rotatory motion produces more transverse and "parrot beak-like" tears than longitudinal.

8. Sprain or rupture of both cruciates may be caused by hyperflexion, by posterior or anterior dislocation of the knee; a sprain of the anterior cruciate may be produced by hyperextension. Rotation is primarily under the control of the collaterals, which must rupture first before the checking action of the cruciates comes into play.

9. Rupture of the quadriceps is more often a direct injury than caused by violent contraction of the muscle. Partial ruptures, leaving portions of the muscle still attached to the knee cap are not uncommon. The part injured is usually the rectus femoris. In older people, an avulsion of the quadriceps from the patella may follow a stumble.

10. Avulsion of the patellar tendon results from forcible contraction of the quadriceps and involves the extensor apparatus on either side so that active extension of the knee is completely lost.

11. Avulsion of the tibial tubercle (Osgood-Schlatter's disease), seen in young people before the fusion of the tubercle with the tibial epiphysis has taken place, is brought about by sudden flexion of the knee transmitted through the patellar tendon to the epiphyseal plate.

12. The lower tibiofibular articulation is reinforced by the anterior, posterior and intermediate tibiofibular ligaments. When these are strained the spring-like action of the malleolar mortise is lost and the mortise no longer hugs tightly the broader anterior half of the astragalus as the foot goes in dorsiflexion.

13. The ankle joint is reinforced by the powerful tibial and fibular collateral ligaments, each of which has an anterior, a middle and a posterior portion. The anterior portion of the tibial collateral checks plantar flexion; the middle portion, pronation; the posterior, dorsiflexion and pronation. Rupture of this ligament not only produces pathological pronation but also permits unusual degrees of plantar and dorsiflexion. The mechanism of the sprain is forced pronation.

The anterior portion of the fibular collateral checks supination and plantar flexion; the midportion, supination and dorsiflexion; and the posterior portion, also supination and dorsiflexion. Sprain of this ligament is produced by forced supination and its effect on plantar and dorsiflexion is similar to that of the tibial collateral.

14. The subastragalar joint also has three principal reinforcing ligaments: the anterior, the posterior, and the interosseous talocalcaneal. Their prime function is to secure the relationship between os calcis and talus, checking especially for- and backward gliding. They also check pro- and supination in conjunction with the tibial and fibular collaterals.

15. The talonavicular portion of Chopart's joint is formed by a fibrocartilaginous cup, the cup-shape ligament, heavily reinforced by the calcaneonavicular ligament running from the sustentaculum to the scaphoid. It prevents forward and downward gliding of the head of the astragalus. Acute sprains occur by violent abduction of the forefeet. A plantar or lateral calcaneoscaphoid ligament, the so-called lower key to Chopart's joint, forms a tiepiece between os calcis and scaphoid. Plantar flexion is checked by a dorsal talonavicular ligament. A bifurcated ligament from the os calcis across to the scaphoid prevents the latter from pulling away from the os calcis. All ligaments secure the head of the astragalus to the scaphoid.

16. The calcaneocuboid portion of Chopart's joint is secured by a dorsal, a plantar and a lateral calcaneocuboid ligament. All three form powerful tiepieces for the longitudinal arch and secure close apposition of the cuboid to the os calcis. Forceful adduction of the forefoot sprains the lateral; forceful volar flexion, the dorsal; and forceful dorsiflexion, the plantar ligaments. Sprain of the long plantar ligament is often observed in the static flatfoot.

17. The scaphoid is tied to the cuboid by three small ligaments, all preventing separation between these two bones. A sprain results in local tenderness laterally to the scaphoid.

18. A sprain of the scaphocuneiform junction is caused by forced abduction and dorsiflexion produced by heavy blow or fall on the feet or by violent compression, for instance, from the wheels of a car.

19. A sprain of Lisfranc's tarsometatarsal joint weakens the arch formed by the metatarsal bases, which are held together by the transverse metatarsal bands. Relaxation or sprain produces a transverse area of pain over Lisfranc's joint.

A periosteal reaction in form of a dorsal exostosis is often seen in people with a high longitudinal arch.

BIBLIOGRAPHY

1. BRANTIGAN, O. C., and VOSHELL, A. F.: The mechanism of the ligaments and menisci of the knee joint. *J. Bone & Joint Surg., XXIII:*44, 1941.
2. BRAUSE, W.: *Über die Funktion des Ligamentum teres am menschlichen Hüftgelenk.* Leipzig, Bose Programme, 1875.
3. FESSLER, J.: *Die Festigkeit der menschlichen Gelenke mit besonderer Berücksichtigung des Bandapparates.* Munich, 1893.
4. FICK, R.: *Lehrbuch der speziellen Gelenk- und Muskelmechanik.* Jena, G. Fischer, 1911.
5. v. SAAR, G.: *Die Sportsverletzungen.* Stuttgart, F. Enke, 1914.
6. SCHAER, H.: *Der Meniscusschaden.* Leipzig, G. Thieme, 1938.
7. SMILLIE, I. S.: *Injuries to the Kneejoint.* Baltimore, Williams and Wilkins, 1946.
8. WEBER, W. and E.: *Mechanik der menschlichen Gewerkzeuge.* Göttingen, Dieterischsche Buchhandlung, 1836; and Berlin, J. Springer, 1894.

Lecture XXXV

THE PATHOMECHANICS OF THE MORE COMMON FRACTURES OF THE UPPER EXTREMITY

I. INTRODUCTION

A. EXTERNAL FORCE AND INTRINSIC RESISTANCE

IN A SPECIFIC traumatic incident which produces a discontinuity of bone it is impossible to subject to a mechanical analysis all the different external forces and their combinations which are responsible for the failure of bone. All one may state is that a force or a combination of forces have resulted in overcoming the stress resistance. On the other hand, the intrinsic resistance of the bone to external violence and the condition under which it holds its equilibrium against gravitational and muscular stresses is, at least for normal bone, a matter of record based on extensive studies of its internal structure and the latter's relation to gravitational or muscular stresses.

To a considerable extent, therefore, disalignments can be analyzed on static and dynamic grounds. The unknown quantity in the whole situation, then, is not so much the intrinsic anatomical makeup of the injured area as is the external force which produces the injury. In a general way, the intrinsic equilibrium of bone against external forces has been discussed previously. It is reflected by the arrangement of the trabecular system of the bones on one hand, and by gravitational stresses and rotatory components of the controlling muscles on the other. Once a discontinuity of the bone by external violence has occurred, the subsequent displacement of the fragment can be anticipated and the mechanical forces to bring about realignment can be planned and devised accordingly.

This is in contrast to the pathological fractures; here we have nothing to go on in determining the resistance to the external forces because the unit stress resistance of the bone becomes as much of an unknown quantity as is the external force. For this reason, we believe that an analysis from the purely mechanical standpoint is too hopeless a proposition to venture into it.

B. THE MECHANICAL ESSENTIALS OF ALIGNMENT

It can be stipulated in a general way that alignment requires: a) restoration of length and b) positioning of the lower controllable fragment to meet the position of the upper uncontrollable one.

The former point requires traction to overcome the axial pull of the muscles; the latter the placing of the distal fragment in a position corresponding strictly to that of the upper in all three planes of reference, frontal, sagittal and transverse; in other words, a position exactly adjusted to the pathological muscle equilibrium in which the proximal fragment finds itself after being separated from the distal portion of the bone.

There is no question that these principal postulates apply to all fractures. Broad contact of all fracture planes can only be achieved by overcoming the overriding produced by longitudinal muscle pull, and by adjusting the position of the distal, manageable fragment to that of the proximal one by proper positioning of the former.

Fundamental as these requirements are, it will be observed that they cover the process of adaptation of fragments only and not the maintenance of this adaptation during the process of consolidation.

Two more postulates must, therefore, be added: The first is the reformation of the trabecular systems which have been interrupted by the fracture. It will be recalled that the principal trajectory systems are pressure sustaining. The reason for this is obvious. Aside from the fact that in the weight-bearing bones gravitational stresses act most powerfully upon the bone the muscles surrounding the bone develop, by virtue of their acute angle of application, longitudinal or stabilizing components greatly in excess of their rotatory. As has been explained in the discussion on Wolff's law, a certain amount of axial pressure is favorable to bone formation, as can be seen in the stimulating influence of weight bearing. This is also borne out clinically in the early consolidation which occurs in the so-called compression arthrodesis. On the other hand, distraction of the fragments due to excessive traction retards or even prevents consolidation. Quite apart from actual interposition of tissue, we know that bone tolerates pressure much more than tension, which is often followed by absorption.

The other postulate bears on the permanency of the alignment and has to do with the permanent reorganization of the trajectories. The soft callus remains pliable and yields to the pull of the muscles. The rotatory component of the muscle tries to bend the soft callus and develop pressure stresses on the concavity and tension stresses on the convexity of the bend. Tension stresses lead to absorption of bone, which undergoes deformation. There are, therefore, two elements which may interfere secondarily with alignment and consolidation: the delay in restoration of axial pressure trabeculae because they are crushed by the fracture; and late rotatory deformities due to a persisting muscular disequilibrium.

From the practical point of view, those fractures in which the condition for permanent alignment can be fulfilled are stable fractures; those in which these conditions are lacking, either because of primary injury to the trabecular system or due to secondary loss of musclar equilibrium, are unstable. This distinction has been made strikingly clear by Evans[5] for the subtrochanteric fractures of the femur.

It is not to be inferred that under favorable conditions, principally circulatory, malaligned fractures do not consolidate. On the contrary, deformities are only too commonly seen after fractures of the femoral shaft or in trochanteric or supracondylar fractures. Where circulatory conditions are precarious as in the femoral neck, or where restoration of the stress trabeculae becomes difficult as in the intertrochanteric fracture, or where as in the lower end of the radius or the femoral shaft it becomes difficult to control the musclar balance, special measures are necessary to prevent malalignment.

It would be impossible in this and in the following lecture to venture into a kinetic analysis of all fractures. We shall confine ourselves to present the pathomechanics of only a few of the more common situations as they occur in the long bones of the extremities.

II. COMMON FRACTURES OF THE SHOULDER JOINT

A. FRACTURE OF THE HUMERAL NECK

1. Length rotatory equilibrium (transverse plane)

The proximal fragment is in neutral position, an equilibrium being established between the external rotators (supra- and infraspinatus and teres minor with a cross section area of 24.2 cm.^{2}) and the inward rotators (subscapularis with a cross-section area of 21 cm.^{2}). The distal fragment, on the other hand, is in strong inward rotation under the influence of the powerful pectoralis major and latissimus, which are no longer opposed by the outward rotators.

2. Ab- and adductory equilibrium (frontal plane)

The proximal fragment is held in abduction by the supraspinatus. The distal is adducted by the pectoralis and latissimus being opposed only by the mid-portion of the deltoid. (Cross-section area of the abductors is 22 cm.^{2}, of the adductors, 31.6 cm.^{2}) An axial component of the deltoid pulls the lower fragment upward and causes it to override the upper medially (Fig. 1).

B. FRACTURE OF THE SHAFT (Steindler[13])

1. Below the pectoralis major and above deltoid insertion

A) Length Rotatory Equilibrium (Transverse Plane)

The upper fragment is held in neutral position by the equilibrium established between the inward rotating pectoralis major, latissimus and teres major and the outward rotating muscles of the rotator cuff, with the inward rotators slightly prevailing.

The lower fragment has neither outward nor inward rotators except for the mutually neutralized anterior and posterior portions of the deltoid.

B) The Ab- and Adductory Equilibrium (Frontal Plane)

In this plane the normal equilibrium is completely overthrown. The adductors acting on the upper fragment hold it strongly adducted. The distal fragment is abducted and overrides the proximal laterally (Fig. 2).

2. Fractures of the humeral shaft below the deltoid insertion

A) LENGTH ROTATORY EQUILIBRIUM (TRANSVERSE PLANE)

There is no rotatory disalignment of the upper fragment because equilibrium between outward and inward rotators is fairly well established. The lower fragment, however, is apt to fall into inward rotation against the upper under the effect of gravity alone.

B) AB- AND ADDUCTORY EQUILIBRIUM (FRONTAL PLANE)

Here the deltoid definitely prevails against the adductors forcing the upper fragment into abduction. The lower fragment being in adduction overrides the

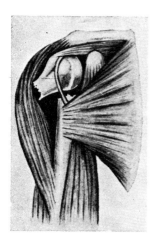

FIG. 1. Fracture of shaft of humerus above insertion of pectoralis major. (Watson-Jones)

FIG. 2. Fracture of shaft of humerus below the pectoralis major insertion and above the deltoid insertion. (Watson-Jones)

FIG. 3. Fracture of shaft of humerus below the deltoid insertion. (Watson-Jones)

upper medially but the amount of overriding is moderate and less than in the case of fracture through the neck (Fig. 3).

The upward pull of the lower fragment is the result of the pull of the deltoid as well as of the biceps, triceps and coracobrachialis. It is, consequently, greater wherever the fracture line is above the deltoid insertion. It can easily be overcome by traction when the fracture is below the deltoid tubercle, which is the ideal site for the "hanging cast." So far as length rotatory and abductory disalignment is concerned, the most critical situation is the high fracture through or below the neck. Here the upper fragment is in strong abduction and outward rotation and it is necessary to place the lower in a corresponding position after the overriding is overcome.

C. DISLOCATION AND FRACTURE DISLOCATION
OF THE SHOULDER

One reason for the comparatively high frequency of shoulder dislocation is the gross incongruency between the humeral head and the glenoid fossa; there is only a small area of contact between the two articular surfaces. Another reason is the slackness and the redundancy of the capsular sac, which lacks reinforcing ligaments in its anterior inferior portion. Likewise, the muscular protection of the joint, which is so abundant at the superior, anterior and posterior aspect, is entirely missing in its inferior portion where the great vessels and nerves descend into the axilla.

When the arm is abducted, an outward rotation accompanies abduction to facilitate the clearance of the greater tuberosity under the acromial edge.

If abduction occurs too suddenly and forcefully, the capsule may tear and the head dislocate through the rent in the anterior-inferior portion; the outward

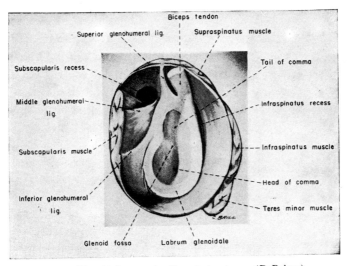

FIG. 4. Glenoid fossa and adjacent structures. (DePalma)

rotators may or may not be able to follow the head as it leaves the glenoid cavity. If they do not, they tear away from the greater tuberosity or they avulse it, separating it from the head. In either event, they come to lie like a shroud over the glenoid, preventing the head from reentering the glenoid cavity. This is the primary reason for the early irreducibility of the dislocation. Later, the dense scar forms a new capsular sac for the head. The position of the arm indicates the type of dislocation. In the subglenoid the arm is lengthened and only slightly abducted. In the subcoracoid dislocation the arm is in springy abduction and in outward rotation and the axis of the humerus points to the deltopectoral groove.

The recurrent or habitual dislocation occurs in abduction and outward rotation. It was formerly believed that a rupture of the capsule was necessary (Seidel[12]), although Malgaigne[9] already stated that incomplete dislocation without capsular rents was possible. It is now recognized that the essential capsular injury is the detachment of the base of the labrum from the glenoid rim anteriorly while the capsule and the glenohumeral ligament remain attached to the labrum and the defect lies between these structures and the neck of the scapula (Watson-Jones[14]).

Inasmuch as there still seems to be some controversy in regard to the relationship of the labrum to the joint capsule, it should be pointed out that, according to Fick[6] (I, pp. 169), the joint capsule is attached normally to the bone beyond the border of the labrum leaving the latter free, very much as

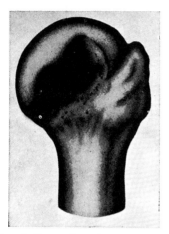

the menisci of the knee, so that the border of the labrum can be pulled away from the glenoid forming a synovial recess, as has been described and demonstrated recently by De Palma[4] (Fig. 4). At any rate, the capsule is stripped off the neck of the scapula and becomes redundant enough so that the head can leave the glenoid fossa without leaving the capsular sac. The deep groove which the sharp anterior edge of the glenoid digs into the posterior aspect of the head is the mechanical result of pressure upon the dislocated head (Fig. 5).

FIG. 5. Excavation of posterior surface of humeral head. (Steindler)

The *fracture dislocation of the humeral head* is initiated by a subglenoid or subcoracoid dislocation. The pathomechanical process is that the dislocation forces the head into abduction to lie below the glenoid or the coracoid process. Then, as the abducting force continues, the head becomes impinged and fractures. Being no longer contained in the socket, the free head turns around, sometimes as much as 180°, so that its articular surface points to the shaft and its fracture line to the axilla. Separated from the outward rotators by the forward dislocation, the head is now controlled only by the subscapularis, which gives it strong inward rotation.

It is obvious that only as long as the head can be made to re-enter into the joint and assume normal relation to the glenoid cavity is a controlling position of the lower fragment of any avail. This means a combination of traction in strong abduction, reducing the head by manual pressure at the same time. It is, therefore, not surprising that many cases defy purely mechanical methods of reduction and become operative problems.

III. COMMON FRACTURES AT THE ELBOW

A. THE SUPRACONDYLAR FRACTURE (ROS CODORNIU[10])

This is the most frequent of all fractures of the elbow, though it is eminently a fracture of the growing age.

1. The mechanism

The mechanism has been studied and described by Kocher,[8] who distinguished two types: one with posterior displacement of the distal fragment (Fig. 6a) and one with a forward displacement (Fig. 6b). The mechanogenesis is essentially different in these two types: The one with backward displacement is a distinct hyperextension fracture from indirect violence. A fall

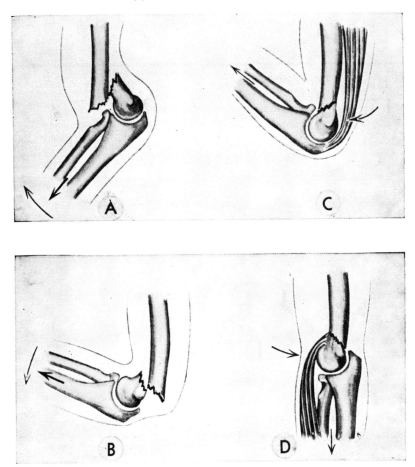

FIG. 6. Supracondylar fracture of humerus. *A*. Posterior displacement fracture. B. Anterior displacement fracture. *C*. Reduction of *A*. *D*. Reduction of *B*. (Watson-Jones)

upon the palm of the hand with the forearm extended is the most common cause, though not the only one. Occasionally, it is caused by direct shear when the patient falls upon the slightly flexed elbow and also by a shearing force when the lower humerus is hit while the forearm is braced against a solid resistance.

2. The fracture line

In the usual extension fracture the line runs obliquely from anterior distally to posterior proximally and the forearm is displaced backward. Anteriorly, the periosteum is broken: otherwise, the posterior displacement could not take place. Posteriorly, however, the periosteum usually strips off some distance without breaking, which is important for the reduction. Flexion of the forearm is not all that is required. It is also of great advantage to have an intact posterior periosteal hinge to maintain the fragment in position. Ros Codorniu[10]

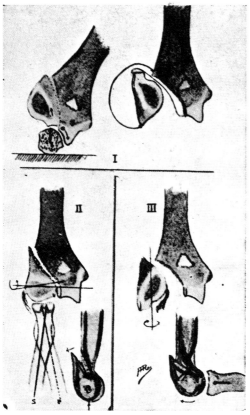

FIG. 7. Fractures of external condyle of the humerus. *I*. Mechanics of the injury and rotation of the fragment about the sagittal axis. *II*. Rotation of the fragment about a horizontal axis. *III*. Rotation of the fragment about a longitudinal axis. (H.-Ros Codorniu)

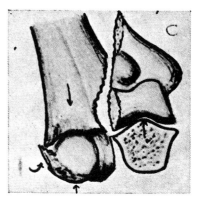

FIG. 8. Fracture of the internal condyle. (H-Ros Codorniu)

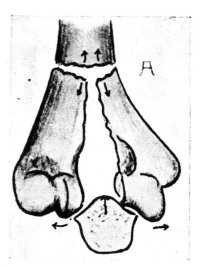

FIG. 9. Bicondylo-trochlear fracture. (H.-Ros Codorniu)

believes that many failures are due to interposition of periosteum as well as of portions of the brachialis. The mechanism of reduction maneuvers resolves itself, therefore, in 1) traction in the axis of the limb which opens the fracture line and allows the lower fragment to impact itself on the upper, eventually pressing the lower fragment forward for this purpose; and 2) graduated flexion of the elbow (Fig. 6c).

In the rare supracondylar fracture with forward displacement the mechanism as well as the fracture line are reversed. The latter goes from backward distally to forward proximally. Consequently, the fragment tends to slide forward and upward. The mechanism of reduction is traction and extension of the elbow (Fig. 6d).

B. THE DIACONDYLAR FRACTURE

Here the fracture line runs through and not above the condyles of the humerus and the lower fragment carries the trochlea as well as the capitellum.

The small fragment is displaced backward as in the supracondylar fracture and the same mechanism of reduction is applied. *Epiphyseal separation* of the growing years occurs as a direct fracture by fall upon the flexed elbow or, more frequently, indirectly by fall upon the hand with the elbow extended. In the former case the displacement is forward; in the latter, backward as in the supracondylar fracture.

C. SPLIT FRACTURES OF THE HUMERAL EPIPHYSIS

Ros Codorniu[10] distinguishes three subgroups: the separation of the external condyle, that of the internal condyle, and the bicondylar Y or T fracture in which both condyles are pried apart.

1. Fracture of the external condyle (Fig. 7)

A direct upward and outward thrust caused by a blow against the dorsal aspect of the flexed ulna or a fall on the elbow with the arm in abduction, or the upward thrust by the head of the radius against the capitellum when the arm is extended, may pry off the external condyle of the humerus. Aside from being displaced upward, the lower fragment is rotated about its anteroposterior axis by the pull of the muscles coming from the external epicondyle. It is quite obvious that, except for a simple condylar fracture without much displacement, manipulation reduction will be difficult. Rotation of the fragment, especially, is not amenable to mechanical measures.

2. Fracture of the internal condyle

The oblique fractures of the internal humeral condyle (Fig. 8) have a similar mechanism to that of the external; the fracture first goes through the trochlea and extends upward, displacing the condyle upward and inward. The elbow joint goes into adduction, decreasing the carrying angle (cubitus varus), which is the opposite to what happens in fractures of the external condyle where an increase of the angle and a cubitus valgus results.

3. The bicondylo-trochlear fractures (T, V, and Y fractures)

Almost always the traumatizing force strikes directly the dorsal aspect of the flexed elbow. The ulna is the separating factor (Fig. 9). The force is transmitted to the trochlea by the crest of the olecranon fossa of the ulna which fractures the trochlea, separating it from the capitellum, and which, penetrating into the humeral metaphysis, pries the two condyles apart. This results in a T, Y- or V-shape fracture with two lower and one upper fragments: reduction by manipulation is practically impossible. Skeletal traction through the olecranon is required for the adaptation of the fragments.

4. Cubitus valgus and varus

These are frequent residual deformities in the wake of condylar fractures. Due to the displacement of the distal fragment, the transverse axis of the elbow joint is deflected, slanting inward and downward in the case of cubitus valgus and outward and downward in cubitus varus. As the elbow is flexed and extended, humerus and forearm are no longer in the same plane as they are normally when the elbow axis strictly bisects the angle between humerus and ulna. As long as this deflection is strictly in the frontal plane, i.e., due to upward displacement of the condyles, the range of motion is not impaired and the disability is almost entirely cosmetic except for the effect the cubitus valgus deformity may have on the ulnar nerve. However, if the fragments rotate in the transverse plane, i.e., displaced forward or backward, then the projecting fragment blocks motion and flexion is incomplete.

D. MONTEGGIA'S FRACTURE

1. The extension or anterior type

This lesion combines a fracture of the upper third of the ulna with a dislocation of the radius in the radiohumeral joint. The general opinion is that it occurs by direct violence through a blow against the forearm, although it would be difficult to visualize that at this level the violence should not result in a fracture rather than a dislocation of the radius. At least for the so-called extension type, in which the ulna is fractured with anterior convexity and the radial head is dislocated forward, the explanation of Evans[5] is much more plausible. This surgeon considers it a pronation injury caused by a fall upon the pronated arm with the body swinging around it to increase the pronotory stress (Fig. 10). Fracture of the ulna occurs first in pronation, and as the ulna fractures its forward angulation acts as a fulcrum which forces the radius to dislocate forward. Sometimes the latter also fractures near its upper end (Fig. 11). It will be recalled that when the radius is fixed as in a fall upon the hand, pro- and supination occur by rotation of the ulna around the radius which would involve a length-rotatory movement of the humerus in the shoulder joint when the elbow is straight at the moment of fall; or, if slightly bent, a supinatory movement (against the radius) occurs in which the humerus describes the surface of a cone with common base at the elbow joint (Fick[6]). When reduction is attempted the conditions are reversed; the forearm is the movable, the body the fixed element. Traction effaces the buckling, and supination of the radius by producing a relative pronation of the ulna corrects the length-rotatory disalignment of the latter; manual pressure upon the head of the radius under traction reduces the dislocation.

Evans[5] proved this type of mechanogenesis experimentally when he produced in the cadaver an anterior Monteggia fracture by forced pronation. He, futhermore, accomplished reduction of the deformity by means of his closed

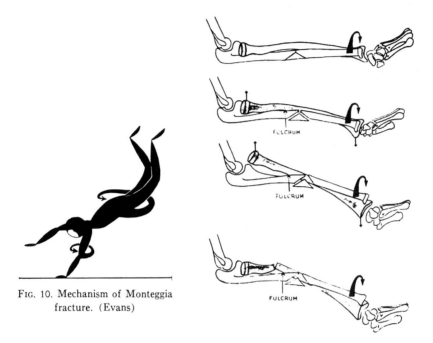

FIG. 10. Mechanism of Monteggia fracture. (Evans)

FIG. 11. Pattern of injury following forced pronation of the forearm. Fracture of the ulna followed by dislocation of the head of the radius, or fracture of the shaft of the radius, depending upon the degree of the pronatory force. (Evans)

method in 9 out of 11 cases. The maneuver consisted in full supination and, when necessary, manual pressure upon the head of the radius.

2. The flexion or posterior type of Monteggia's fracture

The less frequent flexion type (10-15°, according to Watson-Jones[14]) seems to be produced by direct violence in form of a blow against the forearm. Watson-Jones states that all that is necessary to secure alignment is traction, which locks the fragments of the ulna and which allows the backward dislocated head of the radius to slip back into the socket. Operative reduction is seldom necessary.

IV. COMMON FRACTURES AT THE WRIST

A. PATHOMECHANICS OF COLLES' FRACTURE

The usual mechanism is a fall upon the hand with the elbow in extension. If the wrist is in extension, a typical Colles' fracture occurs. The fall drives the hand backward and radially so that the normal anterior curving of the lower end of the radius is lost and the lower fragment is tilted backward and is impacted into the shaft on the radial side. This oblique fracture line is thus directed from volar distally to dorsal proximally and the site of the fracture is anywhere from 12-25 mm above the styloid of the radius, though in the so-called "higher" Colles' fracture it may be as far as 3-4 cm above it. The displacement represents the well known silver fork or bayonet-shape deformity. The principal mechanism of reduction is adaptation of the fragments by traction and release of the radial impaction if such is present by ulnar abduction and flexion. Watson-Jones[14] objects to disimpaction of the fragments by increasing the backward displacement, as is practiced by some. Traction is accomplished by pulling separately on thumb and fingers (Boehler[1]), with countertraction applied to the arm. The logical position after reduction is pronation of the distal fragments (Bulfaro[2]).

If the fall occurs with the wrist in flexion, a so-called reverse Colles' fracture is produced in which the fracture line rises in opposite direction, namely, from dorsal distal to volar proximal. The distal fragment then may remain in contact with the proximal, being simply bent forward, or it may be entirely displaced forward. If the hand is in radial abduction at the fall, the result is fracture of the radial styloid and the shock is transmitted especially to the scaphoid (Bulfaro[2]).

B. EPIPHYSEAL SEPARATION

The displacement of the lower radial epiphysis most common in children and in adolescents has the same mechanogenesis, namely, a fall upon the extended wrist. As in all epiphyseal separations, the fracture line is really in the metaphysis, of which a marginal fragment remains attached to the lower fragment (Watson-Jones[14]). Of particular interest is the fate of the lower radioulnar articulation. The radial part of it is displaced together with the lower fragment. The triangular cartilage follows the displacement of the radius to which it is closely attached, and the head of the ulna separates from the radial fragment and becomes deflected volarly.

C. FRACTURE OF THE SCAPHOID

A similar mechanism is that of fracture of the carpal scaphoid. Bunnell[3] states that in dorsiflection of the hand the proximal half of the bone is covered by the radius while the distal half projects beyond it. If, in a fall on the palm, the wrist is forced into dorsiflexion and at the same time is radially abducted, then the impact of the fall is received by the tuberosity of the scaphoid which is the distal portion, while the pointed styloid of the radius presses against the midportion of the bone. According to Schnek,[11] the usual cause is a fall upon the more or less dorsiflexed wrist with radial abduction.

The fracture of the tuberosity of the scaphoid itself is infrequent, and either a tear fracture of the radial intercarpal ligament, the fan-shaped ligament running from the tuberosity to the radial border of the trapezium, or, more usually, a sheer fracture with a clear fracture line between tuberosity and body of the scaphoid.

The fracture of the body through the middle of the bone is a true bending fracture starting at the concave distal surface which faces the capitate and then continuing toward the convex side of the bone.

V. SUMMARY

1. In analyzing the mechanics of fractures, the unknown quantity is the direction and combination of the external forces rather than the inherent resistance of the bone, the details of which have been ascertained to a certain extent by extensive investigations of the physical properties of bone. Consequently, given a certain fracturing force or a combination of such forces, their effect can be to a large extent anticipated so far as the site and nature of the fracture are concerned. This is not possible in pathological fractures where the physical properties of the affected bone are unknown.

2. Essential requirements for alignment and reduction are traction to produce length and positioning to align the controllable distal fragment to the uncontrollable proximal one. Further requirements for the permanency of alignment and reduction are: first, undisturbed formation and reorganization of the callus by close apposition and avoidance of diastasis, and, secondly, avoidance of deforming muscle pull by maintaining the distal fragment in positions corresponding to the pathological equilibrium of the proximal fragment until full reorganization of the callus has taken place.

3. In fractures of the humeral neck the proximal fragment is in neutral rotation but it is strongly abducted by the supraspinatus. The distal fragment is strongly inward rotated as well as adducted by the pectoralis and latissimus.

4. When the shaft is fractured below the insertion of the pectoralis and above that of the deltoid, the upper fragment is in neutral rotation and strongly adducted; the lower fragment, not carrying any rotators, is also in neutral rotation but it is abducted and overrides the proximal laterally.

5. In fracture of the humerus below the deltoid insertion the upper fragment is in neutral rotation but strongly abducted. The lower fragment is inward rotated and in adduction, overriding the upper medially.

The upward pull of the lower fragment is due to pull of the deltoid as well as of the biceps, triceps and coracobrachialis. It is least when the fracture line is below the deltoid insertion, which is the ideal site for a hanging cast.

6. Dislocation of the humerus is produced by forcible abduction, the head leaving the socket through a rent in the anteroinferior portion of the capsule.

The outward rotators tearing away from the greater tuberosity cover the glenoid and prevent reduction. In the habitual dislocation the essential capsular injury is the detachment of the labrum from the glenoid rim anteriorly, while capsule and glenohumeral ligament remain attached to the labrum so that the defect lies between these structures and the neck of the scapula. The joint capsule is attached to the bone beyond the border of the labrum. The head leaves the glenoid without leaving the capsular sac, which is stripped off from the scapular neck.

7. The fracture dislocation of the humeral head starts with dislocation by abduction, then the impinged head fractures. The head is controlled only by the subscapularis and turns around sometimes 180°, its fracture surface facing the glenoids. As long as the head cannot be brought back into the socket, no position of the lower fragment has any control over the reduction of the fracture.

8. The most frequent fracture of the elbow is the supracondylar; the hyperextension type with backward displacement of the lower fragment is more common. The fracture line goes from anterodistal to posteroproximal. Anteriorly, the periosteum is broken, posteriorly, there is a periosteal hinge; this latter is important for the maintenance of the reduction. In the rare supracondylar fracture with forward displacement, the fracture line is reversed. Reduction is accomplished by traction and extension.

9. In dicondylar fractures the line is through the condyles with posterior displacement of the fragments, and the same reduction maneuver of traction in flexion is used as in the supracondylar fracture. The same applies to the epiphyseal separation seen in young patients.

10. Fractures of the external condyle are caused by fall on the extended elbow or by blow on the dorsum of the flexed ulna. The fragment is rotated forward by the pull of the extensor muscles. Rotation of the fragment cannot be corrected by manipulation.

11. Fracture of the internal condyle has a similar mechanism. The fracture goes through the trochlea and extends upward, separating the internal condyle. A cubitus varus often results from this fracture, while a cubitus valgus often follows fracture of the external condyle.

12. In bicondylar T, V or Y fractures the force strikes the dorsal aspect of the flexed elbow, the crest of the olecranon fossa of the ulna prying the condyles apart. Reduction by manipulation is difficult if not impossible.

13. Cubitus valgus or varus, as residual deformities, cause little functional impairment as long as the elbow axis is deflected in the frontal plane only. If there is a rotatory displacement in the transverse plane, however, the projecting fragment blocks flexion.

14. The anterior type of Monteggia's fracture, in which the ulna is fractured with anterior convexity and the radius is dislocated forward, is not caused by direct violence but is, according to Evans, a pronation injury produced by fall on the pronated arm with the body swinging around to increase the pronatory

stress. Evans was able to reduce this fracture by traction, full supination, and manual pressure on the head of the radius.

The flexion or posterior type of this fracture is produced by a blow against the forearm and requires only traction for its reduction according to Watson-Jones.

15. The mechanism of Colles' fracture is a fall on the extended elbow. Reduction is accomplished by traction, releasing the radial impaction by ulnar abduction and flexion. The fracture line is from volar distal to dorsal proximal. In the reversed Colles' fracture with forward displacement of the distal fragment the fracture line runs in opposite direction. Separation of the lower radioulnar articulation with rupture of the triangular cartilage and volar displacement of the head of the ulna is a frequent complication.

16. The mechanism of the fracture of the scaphoid is a fall on the palm when the hand is radially abducted. The styloid of the radius impinges against the bone and fractures it, usually at the midportion. Fracture of the tuberosity of the scaphoid is either a tear fracture of the radial intercarpal ligament or, more usually, a shear fracture separating the tuberosity from the body of the scaphoid.

BIBLIOGRAPHY

1. BOEHLER, L.: *Technik der Knochenbruchbehandlung, 3d Ed.* Wien, W. Maudrich, 1932.
2. BULFARO, J. A. H.: *Fracturas del extremo inferior del radio.* Tesis de doctorado Univers. Nac. Buenos Aires, 1946.
3. BUNNELL, S.: *Surgery of the Hand, 2d Ed.* Philadelphia, J. B. Lippincott, 1948.
4. DePALMA, A. F.: *Surgery of the Shoulder.* Philadelphia, J. B. Lippincott, 1950.
5. EVANS, E. M.: Pronation injuries of the forearm with special reference to the anterior Monteggia fracture. *J. Bone & Joint Surg., XXXI B:*4, 578, Nov. 1949.
6. FICK, R.: *Lehrbuch der Speziellen Gelenk- und Muskelmechanik.* Jena, G. Fischer, 1911.
7. FICK, R.: *Spezielle Muskel- und Gelenkmechanik. III.* Jena, G. Fischer, 1911.
8. KOCHER, TH.: *Les Fracture de l'humerus et du Femur.* 1904.
9. MALGAIGNE: *Traité de Fractures et des Luxations.* Paris, Balliere, 1855.
10. ROS CODORNIU, A. H.: *El codo, sus fracturas y luxaciones. Cirugia del aparato locomotor, Vol. II.* Madrid, 1945.
11. SCHNEK, F.: Die Verletzungen der Handwurzel. *Ergebn. d. Chir. u. Orthop. Vol. 23,* Berlin, J. Springer, 1930.
12. SEIDEL, H.: Die habituelle Schulterluxation. *Ergebn. d. Chir. u. Orthop., 10:*1012, 1918.
13. STEINDLER, A.: *Traumatic Deformities of the Upper Extremity.* Springfield, Illinois, Charles C Thomas, Publisher, 1946.
14. WATSON-JONES, R.: *Fractures and Joint Injuries.* Edinburgh, E. and S. Livingstone, 1943.

Lecture XXXVI

THE PATHOMECHANICS OF THE MORE COMMON
FRACTURES OF THE LOWER EXTREMITY

I. THE HIP JOINT

A. THE FRACTURE OF THE NECK OF THE FEMUR

The earlier classification of this injury was strictly anatomical. Ashley
Cooper (quoted by Putti[16]) first distinguished between the extra- and intra-
capsular types. German authors speak of the medial and lateral neck fractures,
and Delbet divides the fractures in subcapital, cervical, and cervicotrochanteric.
All these distinctions are purely anatomical, having some bearing upon the
circulatory situation, but they do not take into account important mechanical
factors which are at least of equal influence in determining the stability of the
reduction and the prospect of consolidation.

A step closer to the mechanical concept is Henning Waldenström's division
into ab- and adduction fractures because it involves the separation of pressure

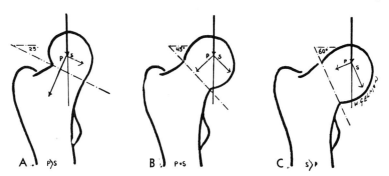

Fig. 1. Relation of pressure and shear to fracture plane.

and shear. Boehler[1] pointed out that the position of the fragments in the abduction fracture is much more favorable to impaction and consolidation.

At any rate, this distinction between pressure and shear is the basis of the
important contribution of Pauwels.[14] According to this author, the fracture of
the femoral neck is divided into three distinct classes. In the first the fracture
surface is exposed essentially to pressure: Delbet's abduction fracture (Fig.
1a); in the second it is subject to shear as well as to pressure (Fig. 1b). Both
components are equal as the inclination of the fracture plane reaches 45°. In
the third group the superincumbent load develops shear and bending stresses
with tension at the lateral and pressure at the medial contour of the fracture site
(Fig. 1c). According to Putti,[16] all three of these classifications have their

610

clinical importance. The purely topographical one indicates that the prospect of healing is the better the farther distal the fracture; the second morphological one implies that fractures in abduction heal much better than those in adduction; the third mechanical one bases the prospect of consolidation upon the direction of the fracture line.

The mechanism varies. It is often a direct blow by a fall, often an indirect stress, a tripping over a carpet, stumbling on an uneven floor. The violence required to produce this fracture is also variable. Sometimes a trifling incident suffices, since the stress resistance of the femur changes with constitution and age.

Inasmuch as the direction of the traumatizing force is unpredictable, one can draw conclusions as to the site of the fracture only from the knowledge of the structure of the femur and what the critical points are in respect to stress resistance.

For those rare fractures which occur from a strictly vertical force, such as a fall upon the feet or knees, we can, in general, apply the principle of the femur being an excentrically loaded column. Under this concept the shearing stress should be maximal near the head and the bending stress maximal near the trochanteric region.

Evans, Lissner and Pederson,[4] using the stress coat method, produced under tension stresses cracks which run perpendicular to the direction of the stress. They found that the diameter of the neck as well as the angle of inclination influenced the degree of bending stress, and that its maximum was at the middle of the bone in the sagittal and probably closer to the trochanteric region in the frontal plane.

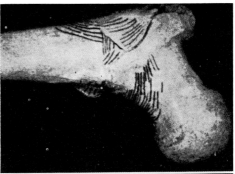
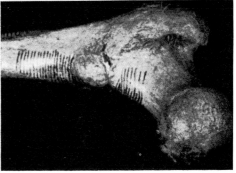

Fig. 2. Abduction loading. Tension stress at medial contour of neck with spreading to trochanteric area and medial contour of shaft. (Evans, Pedersen & Lissner)

This coincides with the mathematical computation of stress distribution which has been discussed in a former lecture. Evans, Lissner and Pederson[5] also found that a dynamic load of 15.8 inch lbs. of energy will produce the same deformation as 400-715 lbs. statistically applied, which means that the living force developed by a block of 7.9 lbs. weight falling 2 inches could produce this deformation. The stress coat patterns resulting from the dynamic loading were almost entirely restricted to the proximal third of the femur.

Studying the regional variations of the physical properties of the femur, Evans and Lebow[6] demonstrated that the highest ultimate tensile strength of bone and the highest module of elasticity to the tension, together with the greatest elongation under tension (1.27%) is found at the middle third of the femoral shaft while, in contrast, the proximal third of the shaft has the lowest tensile strength and the lowest module of elasticity, and is also the softest. One is, therefore, justified in concluding that the upper one-third of the femur is the critical area for bending stresses, static or dynamic, and hence the most liable to sustain bending fractures.

1. The site of the fracture

Evans, Lissner, and Pederson[5] showed that under a static load the first cracks in the stress coat test appeared at the lateral contour of head and neck, repre-

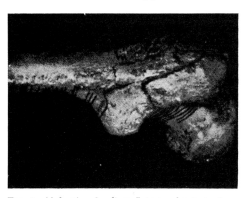

senting the line of the lateral neck fracture. Such a fracture could be produced at this site by a vertical static load of 1280 lbs.

In "abduction" loading the load was applied to the greater trochanter while the femur rested on the head and medial condyle. This produced tension stresses at the medial contour of the neck and the stress coat creases appeared first at its medial aspect, spreading under increasing load down to the medial contour of the shaft

FIG. 3. Abduction loading. Intertrochanteric fracture produced by abduction loading. (Evans, Pedersen & Lissner)

(Fig. 2); in this arrangement of loading an intertrochanteric fracture was produced (Fig. 3). At the same time, a load applied to the greater trochanter may cause a subcapital fracture which would indicate that the critical area in lateral loading where the load or fall strikes the greater trochanter extends all the way along the medial border of the neck to the medial contour of the shaft: Thus, if the force is directed vertically against the greater trochanter, of if in a fall the force is directed along the axis of the neck, a transverse subcapital shear may well result in an abduction fracture.

Torsion produces stress cracks in the shaft in form of spirals (Fig. 4). They appeared in the experiment of the above named investigators[5] under 116-448 in. lbs. of torque. In our experiments on dry bone we obtained torsion failure under a stress of 1135-1670 inch lbs. (Fig. 5). If the fracture occurs by a simple twist without any other external force, it is likely to be a pathological fracture.

The separation of the upper epiphysis of the femoral head is not strictly a traumatic event, although episodes of acute slipping of an already weakened epiphyseal plate occur after traumatism. In this condition the displacement of the head differs from that of the subcapital fracture. It is displaced backward and inward of the shaft, and it is, therefore, flexed and abducted and inward rotated in relation to the shaft which is poised in opposite direction to that of the head.

2. The forces prevailing in ab- and adduction fractures

A) ABDUCTION FRACTURE

The abduction fracture line is nearly horizontal; the head is placed squarely on the fracture line. There is no more shear. All force is pressure; the weight of

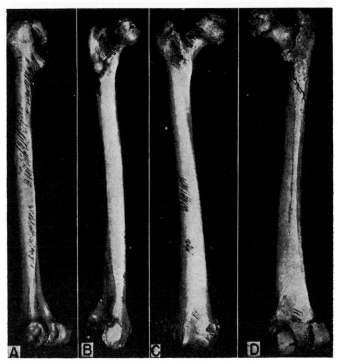

FIG. 4. Torsion stresses producing spiral fracture of shaft. *A, B,* and *C* show stresscoat pattern following torsion stresses. *D* shows spiral fracture of shaft produced by torsion. (Evans, Pedersen & Lissner)

the body impacts the fragment and immobilizes it. The shearing component is proportional to the sine of the angle of the fracture line with the horizontal, therefore, in this case it is minimal or zero; the pressure component being commensurate to the cosine of this angle, all force is, therefore, pressure. Shear being absent, all abduction fractures unite under the pressure force which is favoring rather than hindering union.

B) ADDUCTION FRACTURE

In the adduction fracture union also depends on the obliquity of the fracture angle, and it is most unfavorable when the latter is vertical in which case all force of the superincumbent weight is shear (Pauwels[14]). Watson-Jones[20] states that if the fracture angle is less than 30° the fracture will unite soundly without any treatment. From 30-50° angle the shearing stress must be met by internal fixation. From 50-90°, even internal fixation often fails.

c) THE ROTATION OF THE PROXIMAL FRAGMENT

1) In the *subcapital adduction* fracture which is intraarticular, no muscular forces act upon the proximal fragment. The latter is in abduction and some internal rotation in relation to the distal fragment.

2) In the *lateral cervical* fracture the fragment is also held in abduction, inward rotation and slight flexion.

3) The greatest rotatory discrepancy exists in the *inter- and pertrochanteric* fractures. If the iliopsoas remains attached to the proximal fragment, as is the

FIG. 5. Spiral fracture of femur produced by torsion stress at 1,135 inch lbs.

case in the subtrochanteric fracture, then the muscle flexes the upper fragment strongly while the gluteus medius attached to the greater trochanter abducts it. Hence the reduction and the subsequent immobilization calls not only for abduction and inward rotation but also for considerable flexion of the distal fragment.

In general, the mechanism of reduction follows naturally from the position of the proximal fragment to which the distal must become aligned; traction overcomes the upward displacement of the lower fragment which is due to the pull of the longitudinal muscles of the thigh; abduction and inward rotation bring the fracture line into full contact.

B. THE INTERTROCHANTERIC FRACTURE

Not only the site of the fracture and the relative portion of the fragment is a decisive factor for consolidation, but also the amount of disorganization which the fracture has caused in the internal architecture of bone. In all weight-bearing long bones the degree of destruction which has occurred in the weight or pressure sustaining trabecular systems profoundly affects the subsequent course. We have intimated before that the restoration of these systems is indispensable for the maintenance of the reduction and the safety and permanency of proper consolidation.

Comminution of the principal trabecular system not only makes reduction difficult but it also enhances subsequent deformation produced by weight pressure or muscle pull. This is true for all joints, not only for the weight-bearing ones. The secondary deformation one sees in a comminuted Colles' fracture is a case in point.

Destruction of the principal weight-sustaining system occurs with particular frequency in the trochanteric fracture where the weight-bearing lamellar trajectories running along the medial contour of the shaft are often crushed beyond the possibility of spontaneous repair.

From this mechanical point of view, E. M. Evans[3] distinguishes between stable (72%) and unstable (28%) fractures, the stability depending on the condition of the cortical bone on the medial side of the neck and shaft after reduction. This trajectory system forms a weight-sustaining buttress. If it is not displaced, the fracture is stable; if it is crushed and the overlapping fracture lines cannot be reduced, it is unstable and a coxa vara develops subsequently even though consolidation may occur. The fracture line usually runs from proximal lateral to distal medial. If the obliquity is reversed so that the line runs from distal lateral to proximal medial, the subsequent deformity is likely to be a medial displacement of the shaft with destruction of the medial cortical buttress (Evans[3]). It will be observed that the failure starts on the lateral convex contour under the tension element of the bending force, and it proceeds then medially to destroy the medial trajectory under the compression element of the bending stress.

C. MECHANICS OF RESTORATION OF PELVIC SUPPORT LOST IN UNUNITED FRACTURES

In the numerous situations in which the normal pelvic support is eliminated by ununited fractures of the neck, by dislocation or subluxation of the femoral head the mechanical problem presents itself of how the effect of the weight-sustaining trajectories of the shaft can be transmitted directly to the acetabular roof. This can be accomplished by osteotomies. They either by-pass the disabled proximal end of the femur and align the femoral shaft directly with the axis of weight bearing going through the center of the shaft; or, else, they realign the fracture line into a more horizontal position which eliminates the shearing stress and places the upper fragment more in a straight line with the shaft; in either case, the effect is that the anatomical axis of the femoral shaft comes close to becoming the mechanical axis.

A number of operative procedures such as the osteotomies of Lorenz,[9] Kirmisson,[7] Schanz,[17] McMurray,[10] and Leadbetter[8] have these objectives in view. Leadbetter's method deals with the problem of converting the bending and shearing elment to which the fragments are exposed into pure axial pressure stresses by an osteotomy which gives the fracture line a horizontal instead of a vertical direction. A similar method is Luck's procedure of horizontalization of the fracture line. The osteotomy plane being horizontal, all weight stress becomes pressure, and the bending and shearing component is eliminated.

In other cases of ununited fracture of the neck of the femur where anatomic reconstruction of the fractured neck is not possible, it may be advisable to direct the shaft of the femur so as to lie in line immediately underneath the fractured head. This is the principle of the osteotomies of Kirmisson, Lorenz, and McMurray in which the shaft is placed underneath the fractured head so that its axis is aligned with the axis of the head resting in the acetabulum.

A third principle is transmitting the weight stresses through the femur directly to the pelvic ring by angulation osteotomy such as the "low" osteotomy of Schanz.[17] The angulation of the shaft in respect to the frontal and in cases of dislocation of the head also in the sagittal plane must be carefully calculated.

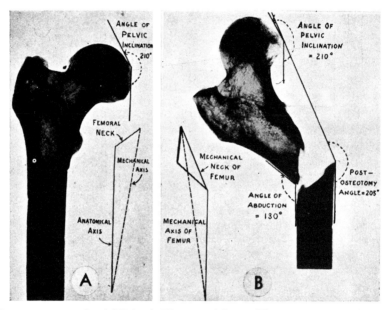

FIG. 6. Postosteotomy angle of Milch. *A*. The normal femur. The angle of the neck may be determined, for clinical purposes, by the anatomical axis and a tangent drawn from the head to the upper end of the shaft. The interposition of the neck displaces the anatomical axis laterally, and gives rise to axial divergence between it and the mechanical axis, which lies in the axis of the limb. (Reproduced from *The Journal of Bone and Joint Surgery*, XXIII, 592, July 1941.[2]) *B*. The ostosteotomized femur (Schanz type). The postosteotomy angle is determined "by the anatomical axis and a tangent drawn from the head to the upper end of the shaft." As a consequence of abduction, the axial relationship has been reversed so that the mechanical axis has shifted laterally, or the anatomical axis has shifted medially. (Reproduced from *The Journal of Bone and Joint Surgery*, XXIII, 592, July 1941.[2])

To this end, Milch[12] determines what he calls the postosteotomy angle from the anatomical axis of the femur and a tangent drawn from the head to the lower end of the shaft (mechanical axis). When the shaft is abducted after the osteotomy the axial relationship is reversed so that the mechanical axis (i.e., head to center of knee joint) shifts laterally while the anatomical axis shifts medially (Fig. 6). In this pelvic support osteotomy the upper fragment is to be closely aligned to the side of the acetabulum. The shaft must be abducted to an angle which will result in its being strictly perpendicular to a line drawn through the acetabular rim so that the weight-bearing line falls fully into the axis of the shaft.

II. THE KNEE JOINT

A. THE SUPRACONDYLAR FRACTURE

It takes extreme and usually indirect violence to produce a fracture of the supracondylar region. Only one muscle remains attached to the distal fragment, namely, the gastrocnemius, and it forces it into position of flexion. The difficulty of controlling this short fragment and to bring it into apposition with the femoral shaft is emphasized by Watson-Jones[20] who states that neither im-

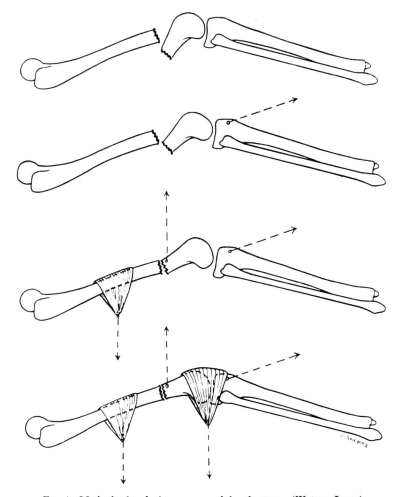

FIG. 7. Method of reducing supracondylar fracture. (Watson-Jones)

mobilization in a straight splint with skin traction nor the use of the double inclined splint with the knee in 45° flexion and skeletal traction from the tibia is adequate for alignment of the fragments. The flexed position and tibial skeletal traction in line with the femur must be combined with an upward skeletal traction of the fragment itself and counterbalanced by traction applied to the thigh (Fig. 7). Only in this way the powerful flexory effect of the gastrocnemius is overcome.

Quite in contrast to the supracondyler fracture, we find that in separation of the lower epiphysis of the femur the lower fragment is rotated forward due to the action of the quadriceps. The fragment is under the control of the extensors which, acting through the patella force the fragment forward while the gastrocnemius with most of its fibers originating above the line of separation has no control over it.

Posteriorly, the separation line follows closely the epiphyseal plate, while anteriorly, a piece of the anterior cortex of the femoral metaphysis often goes with the fragment. In this injury reduction requires traction followed by acute flexion of the knee.

B.   THE PATHOMECHANICS OF FRACTURES OF THE TIBIAL PLATEAU

Structurally, the wide expanse of the tibial plateau projecting beyond the narrower shaft represents a region of weaker resistance. The overhang makes it susceptible to crush and shear; the cancellous structure offers less mechanical resistance to pressure stresses than the heavy cortex of the tibial shaft.

A blow transmitted through the femur, as in fall upon the feet, is only inadequately absorbed by the menisci; and the natural valgity of the knee exposes the outer condyle of the tibia especially to crushing and shearing forces. In severe abduction of the knee it is the medial collateral ligament which is the first to be subjected to strain and when it breaks the jamming together of the external condyles of the femur and tibia develops a considerable force which has a high pressure as well as shearing component. It is the external condyle of the tibia which gives way and the force of the shearing component can be recognized by the amount of lateral displacement of the tibial fragment.

One will understand why similar fractures of the medial condyle of the tibia are comparatively rare. They also require an external force, usually acting indirectly and producing a marked degree of adduction. They are, as a rule, associated with tear of the lateral collateral ligament and, not infrequently with injuries to the external popliteal nerve.

The small fragment is difficult to control. Mechanical pressure with or without a vise can bring the fragments together if there is not too much displacement or comminution. Any gross displacement, however, would indicate injury ot the lateral meniscus and would make its removal advisable. The knee joint tolerates irregularity of contour very poorly and since a hairline adaptation cannot be accomplished except in the mildest cases, most observers are inclined to operative measures which should include the repair of the ruptured collateral ligament.

III. FOOT AND ANKLE

A. MALLEOLAR FRACTURE

All types of external forces are engaged in the production of this fracture: pressure, bend, shear and torsion; but they usually act in combination: bending being commonly associated with torsion, abduction with external and adduction with internal rotation. According to whether one or the other of these mechanisms prevails, one may subdivide the cases into external rotation abduction, and internal rotation adduction fractures which involve bending and torsion, and in true vertical compression fractures caused by translatory forces.

1. External rotation fractures (Watson-Jones[20])

The force is usually indirect, a violent abduction of the foot with outward rotation. The first effect is a sprain of the central and, according to the position of the ankle joint, whether in plantar or dorsiflexion, also a sprain of the anterior or posterior portions of the internal collateral ligament.

Watson-Jones[20] distinguishes several degrees of this type:

a) If the deltoid ligament holds, a fracture of the lateral malleolus develops

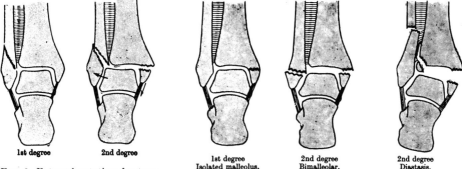

1st degree   2nd degree

FIG. 8. External rotation fracture of the ankle. (Watson-Jones)

1st degree   2nd degree   2nd degree
Isolated malleolus.   Bimalleolar.   Diastasis.

FIG. 9. Abduction fractures of the ankle. (Watson-Jones)

with a spiral fracture line running from lateral proximal to medial distal (Fig. 8).

b) An avulsion of the medial malleolus is associated with the spiral rotation fracture of the fibula (Fig. 8).

2. Abduction fracture

a) The internal malleolus suffers a straight fracture at its base (Fig. 9).

b) A transverse fracture of the fibula at various levels is associated with the avulsion fracture of the internal malleolus (Fig. 9). The astragalus is displaced outward.

c) Avulsion of the internal malleolus and bending fracture of the fibula is associated with a shear fracture at the lateral edge of the tibia; the astragalus is displaced laterally (Dupuytren's fracture) (Fig. 9).

3. Compression fracture

a) A vertical force with the ankle joint in calcaneus position produces a fracture of the anterior lip of the tibia. The astragalus is displaced forward (Fig. 10A).

b) A vertical force with the ankle joint in equinus position causes shear fracture of the posterior lip of the tibia. The astragalus is displaced backward (Fig. 10B).

Manual reduction is easy in simple malleolar fracture without astragalar displacement; in the group of lateral displacement of the body of the astragalus, reduction becomes more difficult because of the uncertainty of controlling the tibial fragment.

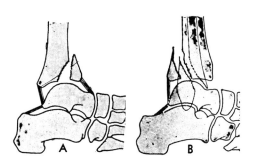

FIG. 10. Compression fractures of the ankle. *A.* shearing off of anterior lip. *B.* shearing off of posterior lip. (Watson-Jones)

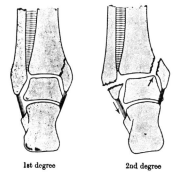

1st degree   2nd degree

FIG. 11. Adduction internal rotation fractures of the ankle. (Watson-Jones)

In the group where the posterior lip of the tibia has been shorn off the astragalus moves backward, being free to do so because of the rupture of the anterior portion of the deltoid ligament.

Reduction in equinus position would seem to place the posterior fragment in alignment; however, in case of rupture of the posterior capsule and the posterior shank of the external collateral, the fragment remains uncontrollable by conservative maneuver.

Where both malleoli are fractured the astragalus is displaced laterally and, impinging against the lateral edge of the tibia, it pries off a fragment (Dupuytren's fracture). The ligamentum tibiofibulare is ruptured and the small fragment being uncontrollable, a satisfactory reduction cannot be obtained by manipulation.

4. The adduction internal rotation fractures (Watson-Jones[20])

The mechanism of the injury is reversed; the foot goes into supination and inward rotation and the tibia is in relative outward rotation. Here, also, the force is usually indirect; a fall or a misstep. Previous strain of the lateral collateral ligament with relaxation or rupture predisposes to the injury.

a) In milder degrees there is a shear fracture of the inner malleolus without displacement (Fig. 11).

b) As the fibular collateral ligament is ruptured or avulsed with a fragment of the fibula, the astragalus becomes displaced medially. Both types can be reduced manually (Fig. 11).

5. The combination of bimalleolar fracture with fracture of the anterior or posterior lip of the tibia

The reduction of this trimalleolar fracture depends upon the integrity of the anterior or posterior attachments of the fragments. When the posterior lip of the tibia is shorn off and the posterior tibiotalar ligaments are intact, apposition can be accomplished by strong plantar flexion; if the anterior lip is shorn off, then, by dorsiflexion. This does not apply to the severe comminutions with many fragments. Here, manipulative reduction is usually futile.

B. THE PATHOMECHANICS OF THE FRACTURE OF THE OS CALCIS

The many clinical variations of this fracture have led most authors to adopt a voluminous classification; mechanically, they all represent the effect of three types of physical forces, singly or in combination: tension (or avulsion), compression, and shear.

A better understanding of the operation of these forces can be acquired by studying the normal architecture of this bone.

The spongious substance is arranged in three trabecular systems. One system of longitudinal trabeculæ runs from the mechanical center of the bone near the upper surface forward and backward and transmits pressure received from above. A second, arcuar system crosses the first at right angles, from backward to forward, resisting the tension stresses of the horizontal component of the oblique gravitational stresses. A third system in the posterior process of the bone resists tension or traction by the heel cord (Ojeda[13]) (Fig. 12).

1. The avulsion fracture of the os calcis

The so-called "duck bill" fracture of the posterior process, the first description of which is ascribed to Petit,[15] is an avulsion fracture produced by the tension of the tendo achillis. According to Soubeyran and Rives,[18] there are two areas at the heel separated by a crest below·which is the insertion of the tendo achillis, while the upper portion is not subject to traction. Therefore, according to these authors, only if the fracture line is below the crest is it an avulsion fracture; if it is above, they believe that it is the impingement of the posterior border of the astragalus which pries off the fragment. This may well explain cases of this type of fracture without lesion of the heel cord (Watson-Jones[20]) (Fig. 13).

2. The compression fracture of the os calcis

If we consider the os calcis as a supported beam, the fracturing forces are really bending and shear. But, because of the impaction of the astragalus into the crushed os calcis they are commonly designated as compression fractures. They are usually caused by the impact of a fall from above or a violent shock from below. Malgaigne[11] first called attention to the role that the fall of the body plays in producing this fracture.

A) Fall Upon the Feet

This is by far the most frequent mechanism. The os calcis becomes crushed, the talus being driven into it.

The extent of the fracture depends upon the size of the living force which strikes the os calcis, i.e., the height of the fall and the weight of the body, and on how much of this force is absorbed by the ground. It depends on the position of the foot in striking the ground (Ojeda[13]), and it also depends upon the age and the constitution of the patient and upon whether the fall was unexpected or whether some muscular defense was installed to absorb the shock.

1) So far as the living force (mv^2) is concerned, Ehalt[2] found that normally the height necessary is 1-2 m., and that it hardly occurs by falls of less than

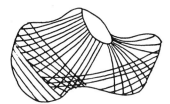

Fig. 12. Trabecular system of os calcis. (Ojeda)

Fig. 13. Duckbill fracture of os calcis. (Watson-Jones)

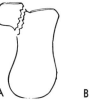

Fig. 14. Isolated fracture of os calcis. A. sustentaculum. B. outer wall of body. (Watson-Jones)

0.5 m., though cases of fall of only 20 cm. resulting in fracture of the os calcis are reported.

2) Soft ground lessens the shock and in this case a fracture of the astragalus is more likely to develop.

3) The fractures are rare in children and adolescents.

4) So far as the suddenness of the fall and the unpreparedness of the patient are concerned, Ojeda[13] points out that flexion of the hip and knee and plantar flexion of the ankle may act as considerable shock absorbers.

5) An important factor is the position of the foot at the moment it strikes the ground. If it is at right angle, the shock develops a reaction both at the heel and the ball. Soubeyran and Rives[18] estimate that two-thirds of the shock concentrates at the heel and one-third at the ball. If the foot is in extreme equinus position so that the trabecular system of the os calcis is pointed almost vertically, either the neck of the astragalus breaks (Tanton[19]) or its posterior process. If the foot is in supination, a fracture of the sustentaculum may occur (Watson-Jones[20]) (Fig. 14), or a longitudinal shear fracture of the outer wall is produced.

B) Fracture by Impact from Below

This mechanism has become of interest through experiences in naval warfare when sudden explosions hurl the deck upward, or in terrestial warfare by explosion of land mines. Fractures of the os calcis in longitudinal or oblique direction have been observed.

The assumption is that the upward directed force of the rising deck and the downward directed weight constitute a couple of forces which are neither co-planar nor colinear; the result is a shearing effect producing oblique or even transverse fractures of the os calcis (Ojeda[13]) (Fig. 15).

From the practical point of view, the all-important question so far as the fracture disability is concerned, is whether or not the fracture involves the subastragalar joint and, if so, to what extent.

1) Fractures not involving the joint (Watson-Jones[20]). The vertical fracture of the posterior process causes little disability; in the "duck bill" fractures, if unreduced, disability is due to a projecting spur since complete avulsion of the tendo achillis is not usual. Also, the simple fracture of the sustentaculum, which is rare as an isolated fracture, does not leave much disability.

2) Fracture involving the joint. Diminution or reversal of the tuber-joint angle follows crush fractures of the body of the os calcis and impaction of the

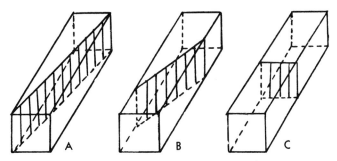

Fig. 15. Shearing effect on the os calcis, producing oblique or transverse fractures. *A* and *B*. oblique type fractures. *C*. transverse type fractures. (Ojeda)

talus. The outer wall of the os calcis buckles; the outer half of the posterior articular facet becomes displaced. In severe joint injury, both posterior and anterior joint surfaces are crushed. Degenerative arthritis of the subastragalar joint is almost sure to follow.

IV. SUMMARY

1. The important contribution of Pauwels to the mechanical concept of the fracture of the neck of the femur is his distinction between pressure and shearing stresses. Accordingly, he differentiates between fracture lines exposed principally to pressure stresses (Delbet's abduction fracture); fractures more or less equally exposed to pressure and shear, i.e., those with inclination of the fracture plane up to 45°; and fracture lines exposed essentially to bending and shear (adduction fractures).

2. The areas of stress were established by Evans, Lissner, Pederson and Lebow through the stress coat method. It was established that the maximum of bending stress on vertical loading was at the middle of the femur in the sagittal, and nearer the trochanteric region in the frontal plane. As to the physical properties of the femur investigated by this method, it was found

that the highest ultimate tensile strength was at the middle of the shaft, the lowest at the proximal end. Hence the upper third of the femur is the critical area for bending stresses and the most liable to sustain bending fractures.

3. Under a static load the first cracks in the stress coat appear at the lateral contour of the neck. Fracture could be obtained at this site by a static load of 1280 lbs.

4. On the other hand, when the load was applied perpendicular to the axis of the femur the first stress coat creases appeared on the medial side of the neck and an intertrochanteric or a transverse subcapital abduction fracture could be produced.

5. Torsion cracks appeared under 116 to 448 inch lb. torque; in our experiments torsion failure occurred under a stress of 1135 to 1670 inch lb.

6. The displacement in separation of the epiphysis is different from the adduction fracture; the head is displaced backward and inward of the shaft.

7. In abduction fracture the fracture line is nearly horizontal; all forces acting on it are pressure forces; there is little or no shear.

8. In the adduction fracture union depends on the obliquity of the fracture angle which determines the degree of shear. Fractures with angles less than 30° will unite, from 30-50° they must be treated by internal fixation, from 50-90° they fail even under operative treatment, according to Watson-Jones.

9. The proximal fragment is in abduction and inward rotation in the subcapital adduction fracture, in abduction, inward rotation and flexion in the lateral cervical fracture. In the intra- and pertrochanteric fracture the proximal fragment is strongly flexed and abducted.

10. The course of the intertrochanteric fracture depends upon the degree of destruction of the weight- or pressure-bearing trabecular systems. Comminution of these systems enhances subsequent deformity. On this basis, Evans distinguishes between stable (72%) and unstable (28%) fractures. The fracture line runs usually from proximal lateral to distal medial. If the line is reversed the subsequent deformity is likely to be a medial displacement of the shaft. Failure usually starts on the lateral convex contour under the tension effect of the bending force.

11. When union fails, the question arises of restoring the pelvic support independent of the discontinuity of the fractured neck. This is done by osteotomies which shift the shaft directly under the head, thereby eliminating largely the angle of inclination and its bending effect. This is the objective of the methods of Luck, Lorenz, Kirmisson, McMurray and Leadbetter. By realigning the shaft directly with the head the shearing stress of the vertical fracture line is eliminated and replaced by pressure stress of the head upon the shaft. The angulation osteotomies of Schanz and the bifurcation of Lorenz aim at the direct transmission of the weight stress from the pelvis to the shaft. The degree of abduction given to the shaft in these angulation osteotomies can be carefully calculated, according to Milch, from the anatomical axis of the

femur and a tangent from the head to the lower end of the shaft (mechanical axis). When the shaft is abducted, the angle between these two lines becomes reversed, and abduction should be carried to the point where the anatomical axis is vertical to a line drawn through the acetabular rim.

12. In the supracondylar fracture, the gastrocnemius attached to the lower fragment rotates it backward into a position of flexion. Reposition requires not only skeletal tibial traction and flexion of the knee in a double inclined frame, but also upward traction of the lower fragment itself to overcome the powerful flexory effect of the gastrocnemius. In contrast, in separation of the lower femoral epiphysis the line is below the insertion of the gastrocnemius and the fragment is rotated forward and is under the effect of the quadriceps. Reduction requires traction followed by acute flexion of the knee.

13. The wide expanse of the tibial plateau and its cancellous structure make it susceptible to crush and shear fractures. The natural valgity of the knee exposes especially the outer condyle of the tibia to injury which can be brought about directly by blow or fall or by severe abduction of the knee, straining or breaking the internal collateral. Fracture of the medial condyle of the tibia requires usually an indirect external force. It causes marked adduction and is associated with tear of the fibular collateral ligament and injury to the external popliteal nerve. Any gross displacement of the fractured plateau indicates injury to the meniscus and makes its removal advisable. Operative procedures should include the repair of the collateral ligament.

14. In *external rotation* fracture of the ankle the force is a violent abduction and outward rotation of the foot. Its first effect is sprain of the central and, according to the position of the foot, also of the anterior or posterior portions of the internal collateral ligament. If this ligament holds a fracture of the lateral malleolus develops with spiral fracture line. An avulsion of the medial malleolus is also associated with spiral rotation fracture of the fibula.

15. In the *abduction* fracture, either the medial malleolus is merely fractured at its base; or it is associated with a transverse fracture of the fibula and the astragalus is displaced outward; or there is also associated a shear fracture of the lateral edge of the tibia with lateral displacement of the astragalus (Dupuytren's fracture).

16. A compression from a vertical force produces a fracture of the anterior lip of the tibia if the foot is in calcaneus, and a fracture of the posterior lip if the foot is in equinus position. In the first case the astragalus moves forward, in the second case, backward. If the posterior capsule is ruptured as well as the posterior shank of the external collateral, the fragment of the posterior lip remains uncontrollable. Also, in Dupuytren's fracture the small fragment from the outer edge of the tibia cannot be controlled by manipulation.

17. The mechanism of the adduction-internal rotation fractures is reversed, the foot going into supination and adduction while the tibia remains in relative outward rotation. Milder cases present a shear fracture of the inner

malleolus; in more severe cases the fibular collateral ligament is ruptured or a fragment of the fibula is avulsed and the astragalus is displaced medially. Both cases yield to manual reduction.

18. Bimalleolar fractures associated with shearing off of the anterior or posterior ledge of the tibia depend for reduction upon the attachments of the fragment being intact. In this case manipulative apposition can be accomplished, placing the foot in equinus for posterior and in calcaneus position for the anterior lip fractures.

19. There are many types of fracture of the os calcis included in the clinical classification but from the mechanical point of view they all represent three types of forces: avulsion, compression, and shear. The "duck bill" fracture of the posterior process is an avulsion type if the fracture line is below the crest which divides the area of the tendo Achillis insertion from the free area above. If the fracture line is above, it is believed to be an impingement against the posterior border of the astragalus which shears the fragment off the os calcis.

20. The usual type is a compression fracture, the fracturing forces are bending and shear. The most frequent mechanism is fall upon the feet. It was found that a fall of not less than 1-2 m. height was necessary to produce the fracture, but the condition of the ground, the suddenness of the fall are determining factors.

21. The position of the foot in striking the ground is important. If it is at right angle two thirds of the shock is concentrated at the heel and one third at the ball. In extreme equinus with the foot pointing almost vertically downward, either the neck of the astragalus or its posterior process breaks. If in supination, a fracture of the sustentaculum or a longitudinal shear fracture of the outer wall may result.

22. Impacts from below such as deck explosions may produce fractures of the os calcis in longitudinal or oblique direction.

23. The practical point is whether or not the fracture involves the subastragalar joint and, if so, to what extent. Fractures not involving the joint are the "duck bill" fractures; little disability is caused by simple vertical fractures and isolated fractures of the sustentaculum. Fractures involving the joint severely are the crush fractures resulting in diminution or even reversal of the tuber-joint angle and the impactions of the talus. In these cases the outer wall of the os calcis also buckles and the outer half of the posterior facet of the subastragalar joint becomes displaced. In all these cases a degenerative arthritis of the subastragalar joint is almost certain to follow.

BIBLIOGRAPHY

1. BOEHLER, L.: Behandlung der Schenkelhalsbrüche. *J. Internat. de Chir., 1:6*, 693, 1936.
2. EHALT, W.: Die Verwertung des Tubergelenkwinkels in der Beurteilung von Verletzungen und Erkrankungen des Fersenbeines. *Monatsschr. f. Unfallheilk., 40:25*, 1922.
3. EVANS, E. MARVIN: Trochanteric fractures. *J. Bone & Joint Surg., 33B:2*, May 1951.
4. EVANS, F. G., LISSNER, H. R., and PEDERSON, H. E.: Deformation studies of the femur under vertical loading. *Anat. Rec., 101:225*, 1948.

5. EVANS, F. G., LISSNER, H. R., and PEDERSON, H. E.: The role of tensile strain in the mechanism of femoral fractures. *J. Bone & Joint Surg., 33A:2*, 485, April 1951.

6. EVANS, F. G., and LEBOW, M.: Regional differences of some of the physical properties of the human femur. *J. Appl. Physiol., III:9*, March 1951.

7. KIRMISSON, E.: De l'osteotomie soustrochanterienne dans la luxation de la hanche. *Revue d'Orth. V:137*, 1894.

8. LEADBETTER, G. W.: Cervical axial osteotomy of the femur. *J. Bone & Joint Surg., 24:4*, 713, Oct. 1944.

9. LORENZ, A.: Behandlung der irreponiblen Hüftverrenkung und Schenkelhalsfrakturen. *Wien. klin. Wchnschr., 32:997*, 1919.

10. McMURRAY, T. P.: Ununited fracture of the neck of the femur. *J. Bone & Joint Surg., 18:319*, 1936.

11. MALGAIGNE, J. F.: Mémoirs sur les fracture par ecrasement du calcaneum. *J. de Chir., I:63*, 1843.

12. MILCH, H.: The postosteotomy angle. *J. Bone & Joint Surg., 26A:2*, 394, April 1944.

13. OJEDA, J. D. R.: Fracturas del calcaneo. Madrid, *Publicaciones del Instituto Nacional de Prevision*, 1946.

14. PAUWELS, F.: *Der Schenkelhalsbruch: ein mechanisches Problem*. Stuttgart, F. Enke, 1935.

15. PETIT, J. L.: Fracture del calcaneum par action musculaire. *Mem. de l'acad. des Sciences*, Paris, 1722.

16. PUTTI, V.: *Cura Operatoria Delle Fratture del Collo del Femore*. Bologna, L. Cappelli, 1940.

17. SCHANZ, A.: Behandlung der veralteten angeborenen Hüftverrenkung. *Münch. Med. Wchnschr., LXIX:930*, 1922.

18. SOUBEYRAN, P., and RIVES, A.: Fracture de calcaneum. *Rev. de Chir., 47:429*, 1913.

19. TANTON, J.: Fracture de calcaneum. *Nouveau Traité de Chirugie, Le Dentu et Delbet. II:40*, 1915.

20. WATSON-JONES, R.: *Fractures and Joint Injuries*. Edinburgh, E. and S. Livingstone, 1943.

Part IV

THE GAIT

Lecture XXXVII

ON THE MECHANICS OF THE GAIT

I. HISTORICAL

THE pioneer work on the mechanical analysis of the human gait was carried out by the brothers Weber[14] who presented their classical treatise on the subject in 1836. They observed and measured the inclinations of the trunk, the alternations of swinging and standing phases, the relations between step duration and step length; they established the rhythm of alternation which appears in walking and running, setting up separate patterns for each, and they initiated investigations into muscular effort involved in propulsion and restraint. While their contributions were observational rather than calculative, they nevertheless have laid the foundation to the scientific concept of the mechanics of the human gait as it is accepted today.

Further progress on these lines was made by Carlet[3] in Marey's laboratory who used kymographic recordings of the length and duration of the step, the swinging and standing time, and the inclination of the body. Vierordt[13] registered space and time relationship of the phases of the gait more indirectly by footprint methods and by tracing the movements of the different parts of the body with colored fluids projected to the floor.

Marey[7] was the first to use photography for the analysis of the movements of the body for which purpose certain reference points of the body were marked by light stripes on the dark-clad individual, similar as, later, Muybridge and, following him, Londe[6] applied their snapshot picture method of recording movements (1882).

Once photography was established in the service of recording the phases of the gait, the way was open for the classical studies of W. Braune and O. Fischer[2] in 1895 whose ingenious experimental methods and brilliant mathematical calculations today dominate the study of the human gait. Starting with the morphological description of the gait and the visualization of the path of progress of the different landmarks, these investigators proceeded to the calculation of velocities and accelerations to arrive, finally, at the computation of the forces which are operative in this complicated locomotor performance. It was a gigantic undertaking to determine the moving forces through the indirect way of first establishing the pathways, then the velocities, the accelerations, and, finally, the muscular effort involved in the human gait.

The next problem was to determine the part the individual muscles play in human locomotion. In the direction of the myokinetics of the gait outstanding work was done by R. Scherb.[9] He established the pattern of the sequence of muscle action of the lower extremities in the gait. His earlier studies were carried out by purely palpatory methods on the treadmill, to be elaborated later by the introduction of electromyography. To these investigations of the mechanics of the gait significant contributions were made by American research in the last three decades. One should mention in this connection the work of Schwartz and Vaeth,[11] especially of the former who since 1932 studied the problem and constantly improved the methods. His first experiments with his basograph enabled him to make exact records of the distribution and gravitational reaction from the floor both under normal and pathological conditions, using subsequently a pneumographic method (Schwartz and Heath,[10] Schwartz, Heath and Wright[12]) and the electrobasograph for recording. Schwartz reported in 1947 observations of quantitative pressure changes at the heel and the metatarsal heads for which purpose this author used as measuring instrument carbon discs with isometric sensitiveness, with which he produced pressure curve diagrams both of the normal foot as well as the foot in equinus, calcaneus and varus deformity. One should appreciate this direct measuring of ground pressure in contrast with the circumstantial and laborious way by which Braune and Fischer[2] arrived at the determination of pressure forces.

In the field of electromyographic investigations of the function of individual muscles engaged in the gait the most outstanding work was contributed by Inman.[5] His contributions on the function of the hip muscles, both qualitative and quantitative, are especially valuable for the study of the gait, particularly since they in many respects substantiate the early work of Scherb.

We cannot leave this somewhat sketchy account of how our knowledge of the human gait was accumulated without mentioning the name of Hans v. Baeyer. While his work is not confined specifically to the human gait, it is elementary for the understanding of coordinate muscle function in general (Baeyer, H. v., Die Synapsis in der allgemeinen Gliedermechanik, Verhandlungen, II. Internat. Orthop. Kongr. 1932).

Duchenne's dictum that isolated muscle action is not a natural process is coordinated by Baeyer's most illuminating work on the synapses in general. It was he who emphasized and studied muscle function in relation to external resistance as links of what he called a closed kinetic chain. In contrast to the current purely anatomic concept, he focused attention on a bipolar proximal and peripheral effect of muscle action, which became of particular importance in the analysis of gait and posture.

II. THE DEVELOPMENT OF THE HUMAN GAIT FROM THE QUADRUPEDAL TO THE ALTERNATING BIPEDAL TYPE

In the end the gait represents a translatory progression of the body as a whole; since all motion of the body consists of rotatory movement of one part against another, it is clear that this translatory end effect must be the result of a series of elementary rotatory motions. How this translatory end effect is brought about by a combination of rotatory movements has already been explained on a former occasion.

Generally speaking, the human gait is a rhythmic play in which by a system of alternating swing and support the lower extremities propel the body forward, balancing upon them the pelvis and the superimposed trunk.

A. THE QUADRUPEDAL GAIT

If we follow the philogenetic development of the human gait from the quadrupedal stage, we note that in the quadruped the center of gravity lies within the floor area bounded by the four supporting extremities. Locomotion is carried out in a pattern of alternating lengthening and shortening the extremities by which the hind legs, in extending, push the body forward, while the front legs, in shortening, pull it forward: a sort of push and pull proposition (Fig. 1).

B. THE BIPEDALISM

The abandonment of the front legs as means of support and the achievement of the bipedal gait was conditioned upon elevating the center of gravity so as to lie over the supporting area of the feet. The function of locomotion being

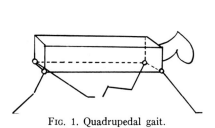

FIG. 1. Quadrupedal gait.

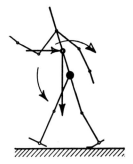

FIG. 2. Alternating bi-pedalism.

taken over by the lower extremities, that of the upper became restricted to assist in balancing the trunk over the pelvis.

In order to secure a forward propulsion it is necessary to place the extremity to the ground in an oblique downward direction so that the reaction from the floor can develop both a horizontal and an upward or vertical acceleration. Thus, as the extremity is lengthened by the extensors, the center of gravity is moved upward and forward by the vertical and horizontal components. Bipedal locomotion, therefore, is a combination of rhythmic forward propulsion and elevation of the body.

We have arrived at this point at what might be called the saltatory type of bipedalism. It is the locomotion of the frog or the kangaroo which, by suddenly extending the obliquely and backward displaced lower extremity, give themselves an upward and a forward acceleration. However, at the moment when the animal leaves the ground it has lost its balance and on landing it would fall forward. It is for this reason that before landing the animal places its jumping hind limbs in a forward diagonal direction. Meeting the ground in this position they now develop a backward horizontal component and act as springs or buffers to neutralize the momentum which has developed during the jump. This is the principle of restraint which becomes operative when the gait reaches the bipedal stage.

C. THE ALTERNATING BIPEDALISM

This is the pattern of the human gait. Only one extremity is used at a time, either as propelling or restraining force. The propulsion is carried out by the leg being placed on the ground in a backward diagonal direction and in slight flexion. The restraint is produced at the same time by the other leg, the heel of which is placed on the ground in forward diagonal direction at the moment when the propelling other leg is preparing to leave the ground (Fig. 2).

The details of propulsion and restraint are as follows: At the end of the swinging period the limb is placed on the ground in forward diagonal direction, heel first, and the knee is flexed and the ankle dorsiflexed. Muscle contraction restrains the forward and downward acceleration until a position is reached where the center of gravity of the forward moving body comes to lie squarely over the supporting foot. This is the period of restraint. Then, by a sudden extension in hip, knee and ankle, the body is given a forward and upward acceleration. This is the period of propulsion.

The human gait may, therefore, be described as an alternating play between the two extremities, one in touch with the ground producing in sequence restraint and propulsion, while the other swinging freely carries with it the forward momentum of the body. At the completion of the swing the leg again touches the ground, first restraining the body in its forward path and then again assuming the role of propulsion. The moment the center of gravity of the forward moving body has passed beyond the supporting base the balance is lost and remains lost during the entire propulsion phase of the standing leg. It is only regained as the heel of the other leg touches the ground and the leg exercises a restraining function. It is quite proper, therefore, to designate the human gait as a constant play between loss and recovery of the equilibrium. Naturally, the restraint which recovers the equilibrium can be effective only within certain limits. This ability varies with the type and speed of locomotion. There is for every type and speed a limit beyond which recovery of equilibrium by restraint is no longer possible. As long as the person walks with average and uniform velocity, the necessary restraining effort is comparatively small. It is different when forward progression is to be checked suddenly, that is, if one is to come to an abrupt stop, especially in rapid walking or in running. Then a considerable restraining effort must be developed by the leg as it touches the ground, and one observes in addition that at the end of the run the body and arms are thrown backward to add to the moment of restraint.

III. THE MORPHOLOGICAL DESCRIPTION OF THE GAIT

The brothers Weber[14] introduced the basic concept that the body resting upon the pelvis is carried forward by the momentum created by the action of the lower limbs and that it maintains during the gait its relationship to the pelvis by rhythmic movements. Obviously, both the propelling and the restraining impulses created by the lower extremities must be imparted to the trunk through the hip joint, to which trunk and pelvis respond by rotatory movements in all planes; sagittal, frontal and transverse. They move with the rise and fall of the hip joint as the lower limbs are shortened or lengthened; they respond to impulses from the lower extremity by rotating backward and forward in the horizontal and right and left in the frontal plane.

All this makes the pattern of the gait rather complicated. The best way to present it is by starting with a simple morphological description of the human gait as it appears in walking or running. On the basis of this general pattern of the gait the path curves which the body and its parts describe during the gait can be plotted. From these curves the velocity and acceleration can then be computed, which finally leads to calculating the muscular forces operative in the gait.

A. THE WALK

In general, the two extremities alternate in their periods of support and swing. The supporting phase begins with the moment the heel is set on the ground, and it ends with the "take-off," that is, the moment when the big toe leaves the ground after the deploy of the foot. The swinging phase begins as the leg leaves the ground to swing forward, and it ends the moment the heel again touches the ground. The entire period in which one limb covers these two phases is called the double step.

In walking these two periods are not equal in duration. The supporting period lasts longer than the swinging period. The faster the walk the less is the time difference between the two periods, and the more they approach each other in duration. But it never occurs in walking that the time of support is shorter than the time of swing. Herein lies a cardinal difference between the mechanism of walking and that of running. In the latter, the time of support is never as long as the time of the swing.

1. Description of the phase of support (Fig. 3)

From the moment the heel of the forward placed leg touches the ground until the hip joint with the forward movement of the body comes to lie vertically over the ankle joint, the limb acts as restraint. From the moment when the hip joint stands vertically over the ankle joint to the moment when the foot finally leaves the ground with the take-off, the limb assumes an increasingly backward direction and acts as propulsion.

While the limb is in contact with the ground, that is, during the supporting phase, it changes its shape twice: it has a maximum and a minimum length. As the heel touches the ground the limb is flexed in the knee and dorsiflexed in the ankle. The limb now rotates around the heel while the trunk and pelvis are balanced at the proximal end of the extremity. As the hip joint gradually overtakes the heel the limb assumes a backward diagonal direction, thereby giving the body a forward acceleration. Extension of the limb is carried out first in the knee joint alone, but soon there is added to it the plantar flexion of the foot, and the increase of the angle between foot and leg adds considerably to the length of the extremity. The body reaches its maximum height at the middle of the supporting phase. Then, by raising the heel the foot deploys itself forward from the ground-similar to the deploy of a wheel. It is a rocking motion which has the advantage that it is being carried out without friction. This deploy also increases the step length by the measure of the foot.

2. The description of the swing

The forward swinging leg shares with the trunk the forward momentum imparted to it by the propulsion of the other leg. It has also an independent

FIG. 3. Phase of support.
(Braus)

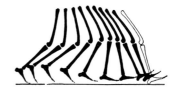

FIG. 4. Phase of swing. (Braus)

forward pendulum movement since it has left the ground in a backward diagonal position. During the period of swing the leg also changes its shape twice, that is, it has a maximum and a minimum length just as the standing leg has during the period of support. As it swings forward it shortens by the flexion of the knee and dorsiflexion of the ankle joint (Fig. 4), reaching its maximal shortening in perpendicular position. The brothers Weber[14] measured the amount of shortening and found it to be almost one-ninth of the entire length of the extremity.

As the leg continues to swing forward beyond the perpendicular it gradually lengthens, though not to full length because a certain amount of flexion must be left for the restraint as the heel touches the ground.

The brothers Weber[14] believed that the swing is automatic, that is, that it occurs under the laws of a physical pendulum, and that it therefore does not require any active muscular effort. To this, Fischer[4] took exception on the ground that some active extension in the knee must be necessary to give the leg the position in which the heel is set to the ground. Otherwise the leg, if only under the effect of the pendulum law, would lag behind, a point which is fully corroborated in the technique of artificial limb making where an accelerator is necessary for adequate extension of the knee.

3. The description of the double support

In the ordinary walk the two periods of swing and support are not so arranged that one ends when the other begins. This is the case only when one walks with the greatest possible velocity and cadence, that is, when the walk borders on the run. In all other situations the two periods are not strictly alternating but they are rather syncopated in the sense that the standing period overlaps the swinging period at both ends; before the beginning and after the ending of each swing there is a period in which both limbs are on the ground (Fig. 5). This is called the period of double support. When one walks

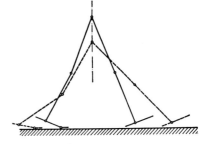

FIG. 5. The walk: diagram of phases 'of support (——) and swing (⌒). (Weber) Note periods of double support between single support and swing periods.

FIG. 6. Relation of hip level to speed of gait.

slowly each of these periods is about one-half the duration of the period of the swing.

The experiments of Braune and Fischer[2] show that of the 31 phases in which they divided the double step the duration of the single step occupied 13 phases, or 0.498 second; the duration of the double support was 0.077 second, or approximately five of the 13 phases; the duration of the double support plus the single support occupied 18 phases, or 0.57 second, while the swing alone occupied only 0.4 second. When one walks with a slower cadence, say 100 steps per minute, then this period of double support becomes somewhat longer, being up to two-tenths of the single step, or 0.1 second.

4. The difference between slow and fast walk

There are several points of difference. In fast walking the trunk is more inclined and the periods of double support are shorter. The step length is greater and the cadence, i.e., the number of steps per minute, is higher. A basic difference between the slow and fast walk is the height at which the hip joints move forward. The higher the hips are above the floor the slower is the walk, and the lower the hip joints are the faster it is (Fig. 6). Also, the higher the femoral head stands above the floor the lesser is the oblique slant at which the lower extremities are set on the ground and, consequently, the shorter is the step; on the other hand, the lower the head is the greater is the inclination of the limbs and the greater is the step length. For the man of average height, Fischer[4] found the angle of the oblique extremity with the vertical, or the angle of spread to be from $44\frac{1}{2}°$ to $52\frac{1}{2}°$.

Also, the greater the obliquity of the backward leg the greater is its forward acceleration during the swing. If the femoral heads are carried high the length

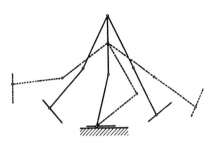

FIG. 7. Relation of hip level to degree of shortening of leg.

difference between shortening and lengthening become less, and, therefore, less force is developed both for restraint and propulsion (Fig. 7). Consequently, one of the differences between fast and slow walking is the amount of the ultimate shortening and lengthening of the leg when it is in contact with the floor.

So far as the swinging time is concerned there is little difference between that for a small arc in the slow gait and that of the larger arc in the fast walk. But since the angular values of the arcs differ the angular velocity of the larger arc becomes commensurately greater than that of the smaller arc. There is a certain correlation between cadence and step length; for the individual of a certain height a definite cadence is as natural as a certain length of the step.

5. The oscillations of the trunk in walking

A) VERTICAL

The vertical oscillations of the body are due to the vertical components of the forces of propulsion and restraint. The Weber brothers[14] found these up and down movements to be 32 mm., varying little with the size of the step in ordinary walking. The upward movement is imparted to the body by the vertical upward component which the supporting leg develops as it goes into extension. The downward movement of the body, or the drop of the body, which follows occurs under the effect of gravity, but this fall must also be restrained by the extensors of the supporting extremity. These elevations in vertical direction attain their maximum and minimum during the single step.

B) LATERAL

The lateral or side to side movement of the trunk attains its maxima and minima during the double step. These lateral oscillations in the frontal plane are imparted to the pelvis also by an upward directed component which the supporting leg develops as it extends.

C) ROTATION

The rotation of the trunk in the horizontal plane is likewise produced by the obliquity of the supporting limbs; the backward oblique leg imparts to the pelvis a forward and the forward oblique leg imparts to the leg a backward rotation. These rotations of the pelvis in the transverse plane are partly the effort of the horizontal component which develop in the phases of restraint and propulsion during the supporting period, and partly they are produced by the forward momentum of the swinging leg.

B. THE RUN

We have seen that the velocity in walking is limited first by the step length which is determined by the possible span between the legs while both are on the ground, and then also by the time necessary for the leg to complete its swing. According to the Weber brothers,[14] this time cannot be less than half of the swinging time of a physical pendulum which has the length and weight of the leg.

On the other hand, the run is a type of locomotion in which there is no theoretical limit of velocity on such mechanical grounds. The principal object

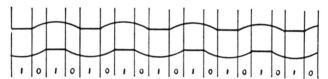

Fig. 8. The run: diagram of phases. (Weber) Note absence of period of double support (▭) and appearance of phase of double float (⬎) instead.

of running is to increase the speed of progression. This is accomplished by causing the body to float in the air with each step for a short period of time. During this time the legs hang onto the trunk, flying with it, as it were, and the step is thereby made larger than the span of the extremities would otherwise provide.

This is called the period of double float. During it neither leg is in touch with the ground. Thus, in contrast to the walk where the standing periods overlap each other in the double support phase, we have here a syncopation or overlapping of the swinging periods (Fig. 8). In other words, the swinging of one leg begins while the other is still in the air. Thus, comparing the two periods of support and swing, we find that in walking the supporting time and in running the swinging time is the longer. Measurements of the Webers[14] show that on the average the steps made in running are twice as large as those in walking and that the ratio of cadence of walk and run is as 2:3. In the ordinary run the vertical movements of the trunk, that is, the up and down movements, are rather less than they are in the walk, e.g., 20-30 mm.

Another difference is also that in running the extremities are placed on the ground in a more flexed position than in the walk. Then, when they are extended for propulsion they can develop a much greater forward acceleration. While in walking the body rises highest when vertically in line with the supporting leg; in running, on the contrary, it is lowest at the moment when the hip joint is vertical over the heel and then rises gradually with the extension of the extremity.

The swinging time can be increased still more by letting the swinging leg not only complete its whole swing but even allow it to swing back for a moment before the limb is set to the ground. The muscular effort here is less because the degree of shortening and lengthening is less than it is in the ordinary run, and the hips are held higher. But the step length is considerably increased though the velocity may be diminished (Fig. 9).

There is really no strict borderline between walk and run. As one walks with increased cadence the time of double support gets shorter and finally disappears at a cadence of about 180 per minute. The Weber brothers,[14] examining this point of transition, found it at a step length of 82 cm. to have a step duration of 0.32 second or a cadence of about 180 per minute.

It is difficult to speak of an upper velocity limit in running except from the viewpoint of the expendable muscular force. What makes the difference in

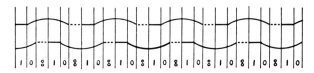

FIG. 9. The sprint: diagram of phases. (Weber) Note the prolonged swinging time, the periods of double float before and after periods of single support. This indicates the period in which the body moves over the forward extended leg, before the real support (——) is assumed by the leg.

velocity is the force of propulsion and the length of the step it produces. In one instance the femoral head is carried low which makes for more shortening and more forceful subsequent extension of the limbs. This increases the forward acceleration and at the same time carries the body farther through the air; there is more forward and less upward acceleration produced. In another instance, as in the overland sprint, the hips are held high; the acceleration produced by the propelling leg is less because there is not so much effort spent in extension, the direction of the propelling leg being less oblique; there is more upward acceleration than in the first case.

Still another difference between walking and running is that the lack of double support in the latter makes it more difficult for the runner to change direction. The faster one walks the more difficult becomes this change, and if there is some difficulty in the fast walk there is much more in running. This explains why it is easier to drill troops in walking when all men keep steps of equal length and duration. If one wants to drill troops in double quick time the duration of the step or cadence must be chosen to fit the swing of the fastest cadence of the men who have the longest legs. On the other hand, the length of the step should be chosen so as to approach as much as possible the maximum step length of the man with the shortest legs. Therefore, also, in running or in double quick walk it is better to let the men proceed in single line so that every one can adjust to his own natural step length and cadence.

IV. THE GRAPHIC PRESENTATION OF THE GAIT (FIG. 10)

The work of Braune and Fischer[2] has been mentioned in the introductory remarks to this lecture as the most outstanding contribution to the study of the gait. By very elaborate experiments and painstaking calculations these two investigators were able to present the first scientifically exact analysis of human locomotion.

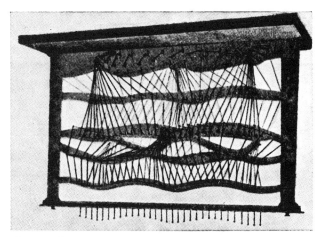

For observation they used the photographic method aided by a system of Geisler tubes. These were attached to certain points of the dark-clad body of their model, namely the head, the shoulder joints, the hip joints, the wrists, the knees, and the feet, eleven in all. By means of a battery of four cameras placed in suitable positions it was pos-

FIG. 10. The graphic presentation of the gait. (Braune & Fischer)

sible to trace the routes of these tubes and to determine the path curves of the different parts in all three planes.

For presentation of the path curves a point 90 cm. above the floor was taken as the origin of a three-coordinate system. From this point a straight line in sagittal direction served as the X axis, a straight line in the frontal plane as a Y axis and a straight vertical line as the Z axis of this coordinate system.

The plane containing the X and Y axes or the XY plane is a horizontal plane upon which would be projected the horizontal movements of the body viewed from above. Similarly, the YZ plane would give the projections of these movements in the frontal, and the XZ plane would give the projection in the sagittal plane.

Braune and Fischer[2] found that the sequence and the type of movements of the different parts of the body were, under normal conditions, essentially the same in all individuals. All would carry out similar oscillations and torsions of the body. There were only quantitative but no qualitative differences making up the individual pattern of the gait.

A. THE PATH CURVES

1. Head, hip and shoulders (Fig. 11)

The path curves described in the sagittal, frontal and transverse planes are almost entirely regular sine curves. The vertical sine curve, that is, the one projected in the sagittal plane (X and Z coordinates), has a wave length equal to the single step (Fig. 10). The horizontal sine curve, that is, the one projected upon the transverse plane (X and Y coordinates), is twice as long, e.g., it has its maximum and minimum in a double step. The frontal sine curve, that is, the one projected on the frontal plane (YZ coordinates) also has a wave length equal to the double step.

The movements of the shoulder and hip are carried out in opposite directions; as one goes up the other on the same side goes down in the frontal plane; as one goes forward the other on the same side goes backward in the horizontal plane; the horizontal back- and forward oscillations of hip and shoulder have about five times the amplitude of the vertical up and down oscillations.

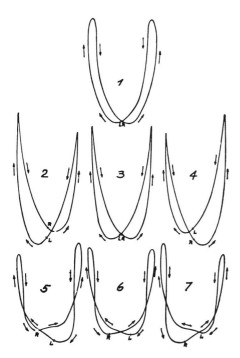

The vertical oscillations have their minimum (lowest point) as the heel is placed forward and another minimum at the end of the standing phase. They have their maxima at the height of the swing and in the middle of the standing period when the hip is vertically over the foot. Thus the vertical oscillations up and down have two maxima and two minima for each double step.

FIG. 11. Frontal projection of path curves (Y Z plane). On each side complete sine curve with single step (up and down); and one sine curve with double step (right and left). (Braune & Fischer) 1, Vortex. 2, Center of left shoulder joint. 3, Mid-point of shoulder line (bihumeral axis). 4, Center of right shoulder joint. 5, Center of left hip joint. 6, Mid-point of hip line (bi-coxal axis). 7, Center of right hip joint.

The horizontal oscillations of the pelvis have their forward maximum as the swinging leg is set to the ground. Then the rotation is reversed and the pelvis rotates backward while on the other side it rotates forward. It reaches its backward maximum as the other heel is placed to the ground. Then, as the standing limb goes into the phase of propulsion, the pelvis again moves forward. It, therefore, finds itself in the same position (forward rotation maximum) at the end of a double step.

The sideways oscillations of the pelvis in the frontal plane likewise have their maximum in the standing period when the hip joint comes to lie vertically over the foot; and, again, it finds itself in the same position at the end of the double step. The frontal and horizontal movements of the shoulder are the same except that they run in opposite directions to those of the pelvis.

2. The trunk movements

In describing the movements of the trunk three points are to be considered, namely, the movements of the shoulder line, the movements of the hip line, and the movements of the trunk line.

A) THE TRUNK LINE MOVEMENTS

If we call the hip line one which unites the center of both hip joints and the shoulder line one which unites both shoulder joints, then the trunk line can be defined as a line drawn from the middle of the shoulder line to the midpoint of the hip line.

This line is used to indicate the movements of the trunk during the gait. Since the trunk line is about five times longer than either hip and shoulder line, one may expect that the oscillations will be proportionately larger. This trunk line carries out three oscillations during the gait.

Shortly after putting the right leg to the ground the trunk line is deflected to the right, and after the left leg is put down it is deflected to the left. The total amplitude at the middle of the shoulder line averages only 1.5 cm., but it varies considerably with certain individuals.

B) THE HIP LINE MOVEMENTS

1) The frontal plane. The vertical maximum is reached with the middle of the swinging period and the minimum shortly after the swinging leg is set to the ground. At the moment the heel touches the ground the hip line starts rising again and reaches a new maximum about 0.1 second later. Then it reaches a minimum toward the end of the standing period. The maximum of the right hip joint which occurs during the period of support corresponds to the minimum of the left hip joint occurring during the swinging phase.

2) The horizontal plane. These horizontal oscillations have a greater amplitude than the ones in the frontal plane; according to Braune and Fischer,[2] about five times as great. The forward movement of the hip line corresponds with the forward swing. As soon as the leg is set to the ground, heel first, the curve reverses itself; the side of the hip line moves backward as the other side moves forward.

C) THE SHOULDER LINE MOVEMENTS

1) The frontal plane. These movements are reciprocal with those of the hip line. In the frontal plane the line has its maximum with the minimum of the hip line, and vice versa. The amplitude in this plane is about three-fifths of the amplitude of the hip line.

2) The horizontal plane. The forward movement of the shoulder line corresponds to the backward movement of the hip line on the same side, and vice versa. The amplitude of the shoulder line motion in the horizontal plane is about two-fifths of that of the hip line in the same plane.

B. THE PATH CURVES OF THE COMMON
CENTER OF GRAVITY (FIG. 12)

The common center of gravity represents the total mass of the body and, therefore, the total resistance which the body offers to movement. Before beginning with the description of its pathways a few fundamental facts should be recalled.

First of all, the body depends for locomotion not only upon the rotatory action of its own muscles but also upon external forces such as gravity. That is to say, were the body floating in space with no external force applying to it, no amount of muscular effort would be able to move the center of gravity even a fraction of an inch. In other words, without an external force, in this case

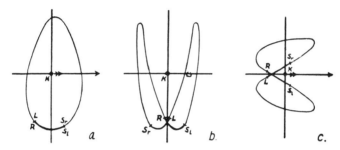

FIG. 12. Path curves of the center of gravity of the body in walking. *a.* Projection in the sagittal plane, showing vertical movement up and down and backward and forward movements oscillating about the plane of progression. *b.* Frontal plane; vertical movements and movements right and left. *c.* Transverse plane; for- and backward movement and movement right and left.

   R   right support
   L   left support
  Sr right swing
  Sl left swing

gravity, locomotion of the body as a whole would be impossible. The effect of the muscles is then to be judged by the effect they have on gravitational stresses, augmenting or decreasing or neutralizing them as the case may be. Even where body parts are moved against each other, such as in flexion of the hip, the inertia plays the role of an external force commensurate with the mass of the body part which is to be moved ($I = mg^2$); under the terrestrial condition the mass of the body is given by its weight over the gravitational acceleration ($m = P/g$). This reduction of the partial masses of the body into a common center of gravity simplifies the situation to a point where the effect of the moving forces upon the mechanism of the gait as a whole can be calculated.

In the last analysis, progression is the result of a counterpressure or reaction produced by the extremities contacting the floor. Due to the obliquity with which the feet are placed on the floor, these reactions so produced are oblique to all three orientation planes: sagittal, frontal, and transverse. Consequently, they can be resolved into vertical and horizontal components, giving rise to movements in these three planes. All three components change with the degree of obliquity of the standing leg. If the extremity is vertical under the hip joint for instance, its downward pressure produced by extension of the leg will give a reaction which has practically only a vertical component while the horizontal component which produces propulsion or restraint will be minimal. On the other hand, if the limb is placed maximally oblique, say 45° or more, then it will develop a horizontal component which is equal or even exceeds the vertical.

The knowledge of the movements of the pathways and their graphic presentation then leads us to recognize and calculate the velocity and acceleration and in the end to the determination of the forces which are active in the different periods and phases of the gait. We shall start, therefore, with the presentation of the path curves the common center of gravity describes during the gait.

How the situation of the common center of gravity was established experimentally has been taken up in a previous lecture. Braune and Fischer[2] ascertained it from the integration of the partial masses of the different body parts on the principle that the common center of gravity of two masses will lie in a line connecting the two partial centers of gravity and at the distance from the partial centers which is inversely proportionate to the masses of the two parts. The center of gravity of a man 170 cm. tall lies about 90 cm. above the floor.

1. The path curve of the common center of gravity in the sagittal plane (Fig. 12a)

In these up and down movements the center of gravity describes a sine curve twice during the double step, the length of the curve being equal to the single step. It attains its maximum at about the middle of the swing and in the standing phase shortly after the foot has been placed obliquely on the ground. There are, therefore, two maxima and two minima in the double step.

2. Frontal plane

In the frontal plane the lateral movements of the common center of gravity are as follows: at the beginning of the supporting period the center of gravity is always on the side of the supporting leg. The moment of transition of the center of gravity to the other side always coincides with the putting down of the other leg to the floor. The movement has one minimum and one maximum with each double step (Fig. 12b).

3. Transverse plane

In the transverse plane the center of gravity carries out forward and backward movements. As in the frontal plane, the complete curve here covers the double step. In this plane the center of gravity moves forward commensurately with the movements of the pelvis, that is, the maximum backward movement occurs at the moment of the take-off, while the maximum forward point is reached as the heel is set to the ground; thus the maximum forward coincides with the maximum backward movement of the other, swinging leg and vice versa (Fig. 12c).

C. THE VELOCITIES AND ACCELERATIONS OF THE COMMON CENTER OF GRAVITY

The method of Braune and Fischer[2] is to determine the velocity by first recording the path curves. Then these curves are brought into relationship to time simply by having their progression marked on a paper containing time lines.

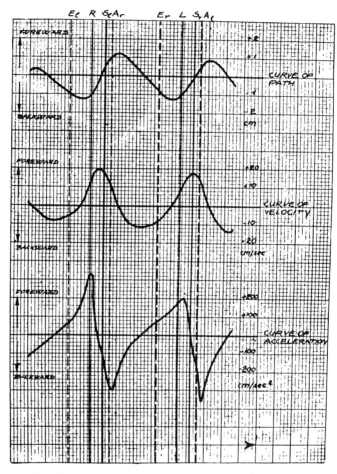

Fig. 13. Diagram of path curves, velocity curves and acceleration curves of center of gravity in the walking direction (for- and backward). Transverse plane. (Braune & Fischer)

R and *L* the setting down of right or left heel.

Sr and *Sl* beginning of right and left swing.

Ar and *Al* beginning of right and left support (sole).

Er and *El* end of right and left support (sole).

To determine the velocity by means of the path curves at a given moment, one only has to draw a tangent to the curve at this point and construct a right-angle triangle the horizontal cathete of which has the length of 1 cm. at the time grid. Then the length of the vertical cathete will indicate the number of the velocity units.[1]

[1] The velocity is proportionate to the trigonometric tangent of the angle which is formed between the tangent at a particular point of the curve and the abscissa.

In the plot used by Braune and Fischer[2] each millimeter corresponds to a time interval of 1/100 of a second. The ordinate represents the unit of 1 cm. (cm./sec.) velocity.

As we compare the path curve and the velocity curve we note as follows (Fig. 13).

The velocity curve always crosses the abscissa (the base line) at the moment when the path curve shows its positive or negative maximum. At each maximum the tangent drawn on the curve is zero because the angle is zero, the tangent being parallel to the abscissa. Conversely, one notices that the velocity is maximal whenever the path curve crosses the base line (or abscissa), that is, when the deviation of the path from the midline is at minimum.

This means: velocity is always minimal when the path curve reverses itself, that is, at the point where the swing has reached its greatest deflection. The velocity is greatest when the swinging excursion is least, that is, when the swing crosses the neutral position or base line.

To determine the acceleration of the center of gravity one begins with the velocity curve and then relates it again to time ($a = v/t$) (Fig. 13). Just as the velocity is measured by the trigonometric tangent of the angle between a point of the path curve and the abscissa, so also the acceleration is determined by the trigonometric tangent of the angle for any special point of the velocity curve.[2]

The acceleration unit is indicated by cm./sec.[2].

In studying the acceleration curves we again notice that the curve crosses the abscissal base line at the moment when the velocity has reached its maximum on the positive or negative side of the curve, and when at the same time the path curves have reached their minimum, that is, when the path crosses the base line. Consequently, all maxima of acceleration coincide with the minima of velocity and the maxima of excursion. All minima of acceleration coincide with the maxima of velocity and the minima of excursion.

D. THE INTERPRETATION OF PATH, VELOCITY AND ACCELERATION OF THE COMMON CENTER OF GRAVITY

It is now possible to correlate the values of excursion, velocity and acceleration of the movements of the center of gravity during the gait.

1. In the progression plane. Forward and backward oscillations (Fig. 14)

During a single step the center of gravity oscillates backward and forward in relation to a frontal plane which moves in a uniform manner in the direction of the gait, the so-called progression plane. The amplitude of these back- and forward oscillations is, according to Braune and Fischer,[2] about 12 mm.

Shortly before the heel of the swinging leg is placed on the ground and while

---

[2] Again in the table used by Braune and Fischer there is a reduction of scale in that the unit of velocity is not related to seconds but 1/10 of a second. So, one unit of time represents 10 units of volocity ($v = \dfrac{s}{t}$), or 100 units of acceleration ($a = \dfrac{s}{t^2}$).

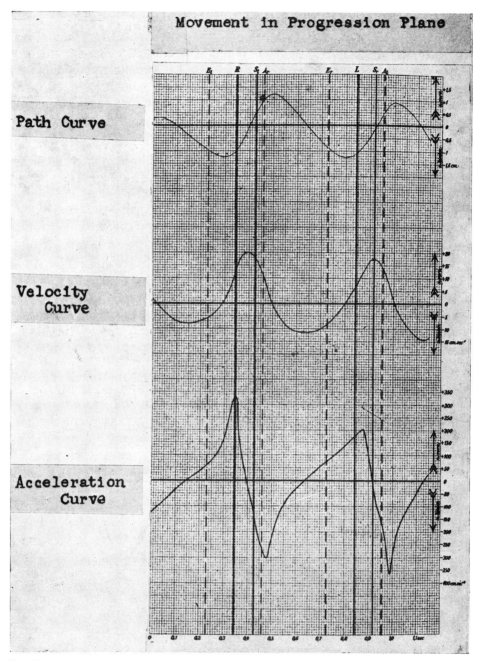

FIG. 14.

R and L the setting down of right or left heel.
Sr and Sl beginning of right and left swing.
Ar and Al beginning of right and left support (sole).
Er and El end of right and left support (sole).

the other, standing leg is already in deploy, the center of gravity is farthest behind this frontal progression plane. It has, therefore, at this moment zero velocity. It now changes its direction in respect to the frontal plane, approach-

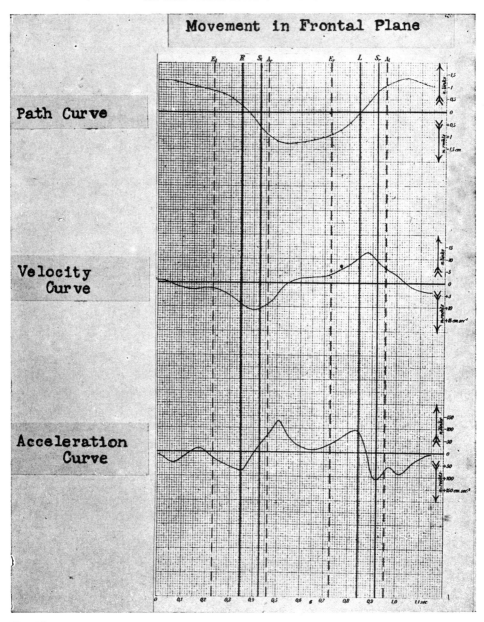

FIG. 15

R and L   the setting down of right or left heel.
Sr and Sl   beginning of right and left swing.
Ar and Al   beginning of right and left support (sole).
Er and El   end of right and left support (sole).

ing it from behind with increasing velocity. It passes then through the plane with maximum velocity (and therefore with minimum acceleration, as shown above) and reaches its greatest forward distance at the moment when the sole of the standing leg is in contact with the floor. It then can reverse itself before the take-off and again reach its farthest backward point when the heel of the

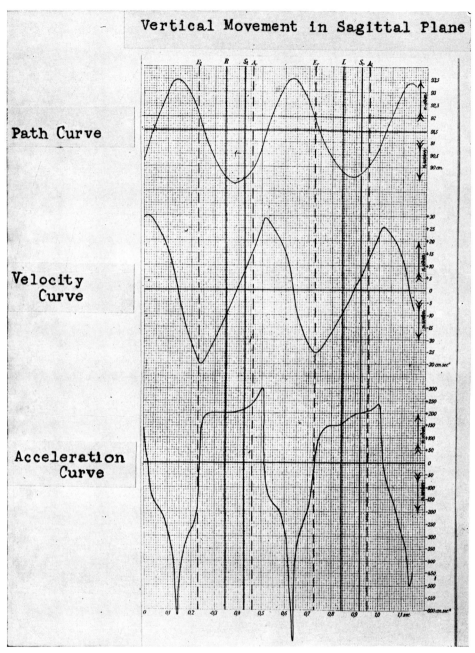

FIG. 16.

R and L   the setting down of right or left heel.
Sr and Sl   beginning of right and left swing.
Ar and Al   beginning of right and left support (sole).
Er and El   end of right and left support (sole).

other leg is down. The total curve from forward to backward occupies a single step. The cycle from forward to forward maximum occupies a double step. It will be observed that as the center of gravity reaches its maximum distance behind or in front of the progression plane its velocity is zero; therefore, it moves only with the general forward velocity of the gait. Its forward pendulum

Absolute Value of the Total Velocity of the Respective Movements of the Common Center of Gravity in cm sec^{-1}. (Three Experiments)				Absolute Value of the Total Acceleration of the Movement of the Common Center of Gravity in cm sec^{-2}. (Three Experiments)			
1	29,2	20,2	25,7	1	188	220	215
2	28,3	22,2	19,8	2	172	330	234
3	22,8	26,2	14,8	3	221	166	310
4	14,2	24,2	17,5	4	341	175	350
5	10,3	19,3	25,1	5	452	228	234
6	21,1	12,3	30,8	6	289	430	145
7	30,3	17,2	29,6	7	182	346	216
8	26,4	25,2	24,9	8	227	225	305
9	19,2	26,6	23,6	9	283	218	426
10	18,1	19,8	27,7	10	393	290	358
11	22,4	16,8	27,9	11	216	303	256
12	22,5	18,5	28,8	12	245	243	484
13	21,1	19,0	37,0	13	364	240	194
14	25,5	19,0	26,7	14	392	315	376
15	27,5	23,8	16,3	15	204	403	306
16	23,0	27,6	9,0	16	203	259	381
17	16,0	19,8	16,1	17	391	267	373
18	15,3	13,3	26,7	18	355	286	286
19	22,9	11,0	29,4	19	228	420	147
20	26,9	18,6	23,5	20	86	253	270
21	22,4	23,7	20,3	21	182	135	260
22	17,4	21,9	22,1	22	219	201	334
23	15,9	16,7	26,1	23	251	238	283
24	18,7	14,6	22,7	24	275	238	223
25	20,4	16,0	21,6	25	280	195	258
26	19,3	16,5	24,6	26	283	229	188
27	22,2	18,1	23,4	27	394	355	120
28	25,4	28,3	16,7	28	191	439	404
29	22,4	28,2	15,5	29	215	225	361
30	16,8	19,9	21,4	30	337	260	161
31	16,2	12,4	24,8	31	445	284	109

FIG. 17. (From O. Fischer)   FIG. 18. (From O. Fischer)

movement adds to the average velocity of progression while its backward movement diminishes it.

2. In the frontal plane. Sideways oscillations (Fig. 15)

The center of gravity attains its greatest distance from the midline shortly after the support of the standing leg is shifted to the whole sole. At this point there is an inversion of motion: the velocity is zero, the acceleration is maximal. Then follows the period of double support and the velocity increases as the center of gravity again approaches the midline. The duration of the oscillation in this plane is the double step.

3. In the sagittal plane. The up and down movement of the center of gravity (Fig. 16)

If one lays a horizontal plane through the resting center of gravity of the body, this plane is situated in a man of 170 cm. height about 90 cm. above the

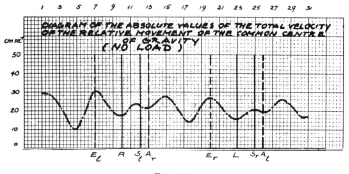

FIG. 19.

floor. In forward progression the center of gravity oscillates about this plane.

It passes through this plane with a maximum velocity in downward direction as the standing leg deploys the ball of the foot. It then moves below this plane with decreasing velocity and reaches the lowest level in the middle of the period of double support with zero velocity. Then the center of gravity reverses its curve, reaching the horizontal plane of reference from below with increasing velocity and zero acceleration. Then it rises above the plane and reaches its highest point again as the standing foot is full on the ground.

4. The compound velocities and accelerations of the center of gravity during the gait

As the problem has been made easier by integrating the partial centers of gravity into a common center of gravity, and by then investigating the path curves, velocities and accelerations in all three perpendicular planes of reference, so it can be further simplified by integrating the different movements into a single compound movement of the center of gravity and then consider the compound velocities and accelerations.

It is understood that the path curves so obtained are not flat but are compound curves which have three dimensions. The compound velocity then represents the resultant of three components in perpendicular planes. These total velocities were calculated by Braune and Fischer[2] (Fig. 17, 19) for all 31 subdivisions of the double step. Similarly, the accelerations were computed from these compounds in three planes (Fig. 18). Finally, from the values of the total acceleration, the values of the floor reactions or floor pressure were computed (Fig. 20).

V. THE CALCULATION OF WORK PERFORMED
IN WALKING

It can be seen from Braune and Fischer's computation that the values of the normal or perpendicular reactions of the floor fluctuate in a man weighing about 60 kg. from a maximum of 76 kg. to a minimum of 31 kg. (Fig. 20).

The diagram (Fig. 21) shows (O. Fischer[4]) that at the moment when the standing foot begins its deploy from the floor the downward pressure nearly equals the weight of the body. From then on it rises rapidly up to the moment

Values of the Normal Reaction of the Floor in Regard to the Pressure
Normally Exerted on the Floor by our Feet, in Kg.

No.	First Experiment	No.	First Experiment
1	67,08	17	35,37
2	50,20	18	37,46
3	46,26	19	45,36
4	38,36	20	57,20
5	31,66	21	66,90
6	41,53	22	67,67
7	48,53	23	68,09
8	70,49	24	69,83
9	70,67	25	70,73
10	70,73	26	71,32
11	70,97	27	72,94
12	71,50	28	54,39
13	72,88	29	47,33
14	76,29	30	38,66
15	50,52	31	32,14
16	47,33		

FIG. 20. (O. Fischer)

when the other foot is on the ground. It then falls in the second half of the supporting phase to below half the body weight. But the pressure stays at that level only momentarily and again increases rapidly, reaching body weight again toward the end of the standing period.

The external forces which act upon the body during the gait are proportional to the acceleration of the common center of gravity which represents the total mass of the body $(F = am)$. If the acceleration of the center of gravity is greater or lesser than the gravitational acceleration of 981 cm./sec., then an external force must be operative which either augments or diminishes the gravitational acceleraton. The acceleration (Y) which this external force (K) develops can be determined by the equation of $K: Y$ as Y to 981. The forces at play which so change the gravitational acceleration in the gait are the muscles. The normal gravitational force is represented by the weight of the body and the variation of this gravitational acceleration indicates the action of these external muscular forces.

In order to make the calculation of the visible work performed in the gait

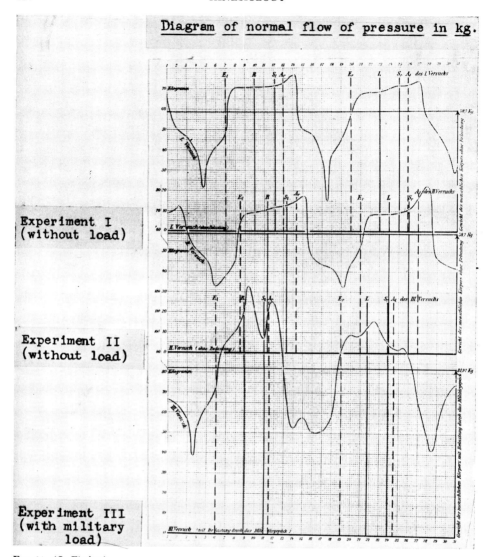

Diagram of normal flow of pressure in kg.

Experiment I
(without load)

Experiment II
(without load)

Experiment III
(with military
load)

Fig. 21. (O. Fischer)

R and L　the setting down of right or left heel.
Sr and Sl　beginning of right and left swing.
Ar and Al　beginning of right and left support (sole).
Er and El　end of right and left support (sole).

still simpler, the lesser oscillations of the body are neglected and we calculate only the work performed in progression, in elevation, and in the swing. But the visible work alone will not tell the whole story since a great deal of energy is absorbed in the restraint. One restraining component, for instance, is the one which controls or slows down the fall of the elevated body; another restrains the pendulum movement of the swing.

The most important is the sagittal component of the force of restraint. In forward propulsion it receives the reaction from the floor. To reestablish the equilibrium restraint must retard or reverse the forward acceleration of the body. Fortunately, investigations of Marey[7] and of Braune and Fischer[2] have shown that there is a definite ratio between the visible propulsion and the invisible restraint. We know that the energy which is expended in the restraint amounts to about 52% of the visible work of forward and upward propulsion. If it takes a certain amount of force of the propelling leg to give the center of gravity a certain amount of forward acceleration, then the forward extended leg in the next moment produces a restraint which amounts to 52% of the energy expended in propulsion. The same is true in the elevation or lowering of the body during the gait. The extension of the extremity gives the center of gravity an upward acceleration and raises the body to a certain height; it then restrains the falling body likewise by the extensor muscles, which effort produces no visible motion. Therefore, if we calculate the work of visible motion in the gait we must always add to it 52% to include approximately the amount of energy expended in nonvisible work of restraint.

Amar's[1] computation of the work in walking is as follows: He divides the entire work into three components:

T_1, the work of moving the body up and down.

T_2, the work of moving the body forward.

T_3, the work of swinging the leg forward in pendulum movement.

T_1: If a man's weight (W) is 65 kg. and with each step he rises a distance of 4 cm., then the work including the 52% calculated for the restraint is $65 \times .004 \times \dfrac{152}{100} = 3.952$ kgm. per step.

T_2: This represents the horizontal component of the reactions from the floor while T_1 is the vertical. This horizontal component also has a propelling and a restraining factor. To evaluate this horizontal propelling force one must know the velocity with which the body is kept in motion because the living energy for this motion equals $\frac{1}{2} mv^2$, where m, the mass, is the weight or 65 kg. over the gravitational acceleration and v is the average velocity which in slow walking amounts to 0.6 m. per step. Thus T_2 would be $\frac{1}{2} \times \dfrac{65}{9.81} (0.6)^2 \times \dfrac{152}{100} = 1.812$ kgm. per step.

T_3 is the energy of the pendulum movement of the swinging leg. Its value is $\frac{1}{2} \omega^2$ (angular velocity) times I (moment of inertia for the whole leg) times the factor $\dfrac{152}{100}$ for the restraint. ω is, on the average, 126° or, expressed in radians (linear value), $\omega = \dfrac{126\pi}{180}$ or $\dfrac{42\pi}{60}$. T_3 is then $\frac{1}{2} \times \dfrac{42^2\pi^2}{60^2} \times 0.146$ (the inertia of the whole leg) $\times \dfrac{152}{100}$, or 0.281 kgm. per step.

Adding now T_1 to T_2 and $T_3 = 3.952 + 1.812 + 0.281$, we have a total of 6.045 kgm. of work performed per step in horizontal walking. Assuming now the step length to be 0.778 m., we arrive at the value of $\dfrac{6.045}{0.778} = 7.712$ kgm.

(Amar[1]) per meter of walk. This would give a total of 7.712 kgm. of work per kilometer for walking at a moderate speed of 5 km. or a little over 3 miles per hour and, consequently, for each kg. of body weight the individual would expend $\frac{7.712}{65}$, or 0.119 kgm. of work per meter of progress.

When walking with a load (Fig. 22) the general character of the step is the same except that the step length is decreased and the period of support, especially that of double support, is prolonged. The foot is placed flat on the ground and the contraction of the calf muscle is more intense. The vertical oscillations are decreased both due to the reduction of the step length and also because the

FIG. 22 Walking with a load.

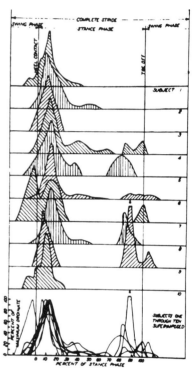

FIG. 23. Gluteus maximus summary curves. Ten subjects at 95 steps/min. level walking. (Inman)

knees are slightly flexed in order to lower the center of gravity and to give the body greater stability. However, when the load is carried on the head, the center of gravity rises and the body is less steady. The load should therefore be carried low, hanging from the shoulders.

VI. SPECIAL MUSCLE DYNAMICS OF THE GAIT

In regard to the action of individual muscles in the gait, especially the sequence and time of their activity, we refer to the electromyographic studies of Inman[5] and to the myokinetic investigations of Schwartz[10, 11, 12] and Scherb.[9]

A. THE HIP MUSCLES

1. The extensors

According to Scherb,[9] the gluteus maximus starts its action at the beginning of the double support and extends it over the single support phase. Inman[5] likewise places the beginning for the gluteus maximus at the early stance shortly after the heel strikes the ground. In ascending there appears an added intensity at midstance point. In descending the contraction begins at the end of the swing. The intensity of contraction varies with the cadence. Inman states that the gluteus maximus acts definitely to prevent forward dropping of the pelvis and also to hold the pelvis back from rotating during the propulsion period of the other leg (Fig. 23).

2. The abductors

Scherb[9] places the action of the gluteus medius synchronously with the maximus, that is, from the beginning of the double to the end of the single support, acting in the frontal plane to stabilize the pelvis. According to Inman,[5] also, the maximum activity occurs as soon as the stance is assumed. We notice particularly in the early stage of the stance that the forward leg exerts a backward oblique thrust which not only elevates the pelvis on this side, giving it an adductory position, but also rotates it backward. Both motions are restrained by the abductors.

3. The adductors

According to Scherb,[9] the action of these muscles starts at the beginning of the single support and ends at the beginning of the subsequent double support. Inman,[5] however, places the peak of contraction at the last moment of the stance phase. It is a biphasic contraction, the lower peak occupying the earlier period of the stance phase. Since during the stance period the hip goes from forward flexion (heel contact) to extension (take-off), one must expect that the secondary flexory component (in extension) and the extensory component (which accompanies hip flexion) of this adductor group comes into play.

In addition, during the period of double support the backward extremity gives a forward and upward thrust to the pelvis and, at the same time, rotates the hip outward. The forward extremity thrusts the pelvis into backward rotation, that is, the hip joint rotates inward.

B. THE KNEE MUSCLES

1. The extensors

The quadriceps begins its action, according to Scherb,[9] in the double support phase and carries it into half of the swinging period.

Inman[5] also finds that it has an early stance phase, i.e., in the period of double support. The forward swing of the leg as it leaves the ground is carried out by the hip flexors, including the rectus. In faster speeds the natural swinging time of the limb is not adequate so that the leg would lag behind and, therefore, the quadriceps action will be needed to bring the leg forward in time. At low speed of walking the swinging velocity is small, but even here Fischer[4] believes that the same action of the rectus femoris is necessary.

2. The flexors

Scherb[9] finds that the biceps, gracilis and semimembranosus action starts at the beginning of the double support and lasts almost through the entire standing phase in the period following the double support.

However, Inman[5] finds that the primary action of the hamstrings is to restrain the forward swing of the leg, particularly when the swing is accelerated (Fig. 24). The peak of this activity is toward the end of the swinging phase. Hamstrings and rectus femoris maintain in the forward and backward swing the relations of a countercurrent movement against each other, mutually checking their action upon the knee joint. This is still increased in rapid walking when the forward inclined trunk adds to the tension of the hamstrings.

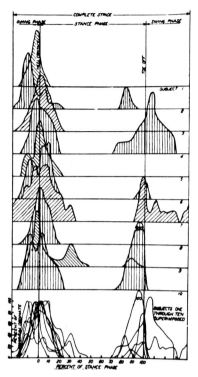

Fig. 24. Hamstring Group summary curves. Ten subjects at 95 steps/min. level walking. (Inman)

C. THE ANKLE MUSCLES

1. The dorsiflexors

Tibialis anticus and extensors of the toes are both active in the swinging phase which is explained by the necessity of dorsiflexion of the foot during this period. But the maximum activity is displayed during the standing period very shortly after the heel strikes and, especially, as the foot gradually goes into plantar flexion. In other words, the tibialis anticus is active both in the standing and in the swinging phase. The explanation is simple. During the swinging phase the extensors hold up the foot in dorsiflexion to clear the ground, as only the weight of the foot has to be overcome. In the standing period, however, the foot goes gradually from dorsi- into plantar flexion with a measured and directed control carried out by the dorsiflexors. We shall point out later, in the discussion of the pathomechanics of the gait, that the paralysis of the extensors greatly interferes with the take-off because of the lack of antagonistic restraint of the flexor muscles. It is the same situation as the weakness of the grip when the extensors of the fingers are paralyzed.

2. The plantar flexors (calf group)

Gastrocnemius and soleus are active during the entire stance phase, the tension increasing in the latter half from the moment when the hip is vertical over the ankle and when the period of propulsion begins with the deploy of the foot. But even before the actual deploy the tension of the calf muscles is necessary to prevent dorsiflexion of the foot when the sole is flat on the ground. As the center of gravity moves gradually forward the tension of the calf muscles increases to find its maximum in the latter half of the supporting period, shortly before the take-off. It is the action of the gastrosoleus which lifts the heel as the deploy begins, while the common flexors of the toes contract a little later to dig the toes into the ground preparatory to the take-off.

VII. SUMMARY

1. From the pioneer work of the brothers Weber to the latest myokinetic contribution of Inman there has been a steady progress in the study of the human gait in which the names of Marey, Amar, Braune and Fischer, and Scherb and Schwartz occupy a prominent place.

2. The human gait is an alternating bipedalism. In contrast to the saltatory type of bipedalism, propulsion and restraint during the standing, and for- and backward swing during the swinging phase alternate between the two lower extremities. Restraint is produced by the extremity being placed in forward diagonal, and propulsion by it being placed in backward diagonal direction.

3. Muscular effort restrains the forward acceleration which the body was given by the preceding propulsory action, thereby regaining the equilibrium which had been lost by the propulsion. Gait can, therefore, be described as an alternating loss and recovery of body equilibrium.

4. The effort of restraint is comparatively small in slow and uniform walking but it increases in fast walking and in running, and when the direction of the walk changes or when a sudden stop is made.

5. The supporting phase begins with the moment the heel is set on the ground and it ends with the take-off. From the point when the forward placed leg touches the ground to the point where the hip lies vertically over the ankle joint the limb acts as restraint. From then to the end of the take-off it acts as propulsion. While on the ground the leg changes it shape twice, i.e., it has its minimum at the beginning and its maximum in the middle of the phase of support.

6. Besides its independent pendulum movement, the swinging leg shares with the trunk the forward momentum imparted to it by the propulsion of the other leg. It also changes its shape during the swing to a minimum length in perpendicular and a maximum length in forward oblique position. The total shortening amounts to one ninth of the entire extremity length. The swing is not entirely automatic but must be accelerated by action of the quadriceps.

7. The double support is the period between swing and single support in which both feet are on the ground. Its duration depends on the velocity of

walking, being shorter the faster the walk, and it disappears when the walk borders on the run. The duration of the single step in moderate walking is, according to Braune and Fischer, 0.49 second, the double support period, 0.077 second, the double plus single support, 0.57 second, and the swing, 0.4 second.

8. Differences between slow and fast walk are as follows: In the fast walk the trunk is more inclined, the step length is greater, and the cadence is higher. The higher the hip joint the slower the walk; the less the obliquity the shorter is the step and the less the forward acceleration during the swing; and the less force is developed during stance both for restraint and propulsion. A certain correlation exists between natural cadence and step length.

9. In walking the trunk carries out oscillations in three planes: The vertical oscillations up and down have their maximum and minimum during a single step and amount to 32 mm.; the lateral or side-to-side movement has a maximum and a minimum during the double step; and the rotations of the trunk in horizontal plane forward and backward also have their maximum and minimum during a double step.

10. The object in running is to increase the speed of progression. This is accomplished by the body floating in the air with each step, thereby making the step length longer. The period when the legs are in the air is the period of the double float. In contrast to the walk, the overlapping periods are not periods of support but periods of swing; while in the walk the supporting period is the longer, in the run it is the period of swing. The ratio of cadence in walk and run is 2:3; the vertical oscillations are less than in the walk: only 20-30 mm.

11. Other differences between walk and run are: In the run the extremities are placed to the ground in more oblique position and are more flexed; consequently, they develop greater forward acceleration. In walking the body is highest when vertically over the leg; in running, it is the lowest. The swinging time in the run can be increased by allowing the leg to swing back after completion of the forward swing, before it is set on the ground, which considerably increases the step length. The borderline between walk and run at a step length of 82 cm. has been found to be a step duration of 0.32 second, or a cadence of 180 steps per minute. Another difference is that the lack of double support makes it difficult for the runner to change direction or to come to a sudden stop. Troups are drilled better for double quick walking if the cadence is chosen according to the men with the longest, and step length according to the ones with the shortest legs

12. Braune and Fischer used for their graphic presentation of the gait the center of gravity as the origin of a three-coordinate system, with the X axis in sagittal, the Y axis in frontal, and the Z axis in vertical direction. They found that sequence and type of movements of the different parts of the body in the gait were very constant and showed only quantitative individual differences.

13. The path curves described by head, hip and shoulders are almost regular sine curves. The vertical up and down sine curve has a wave length equal to the single, the horizontal curve (backward and forward movement) as well as the

frontal (sideways) curve has a wave length equal to the double step. Movements of shoulder and hip are in opposite direction. The vertical oscillations have their minima at the moment of the forward placing of the heel and the end of the standing phase, their maxima at the height of the swing and the middle of the standing phase. The horizontal oscillations have their forward maxima with the setting down of the heel, their backward maxima with the setting down of the other heel. The frontal oscillations have their maxima when the hip of the standing leg is vertical over the foot.

14. The trunk inclines in the frontal plane to the side of the standing leg as the foot is put on the ground, with a total amplitude at the middle of the shoulder line of 1.5 cm. In the sagittal plane it is maximally inclined forward just before setting down of either leg. These oscillations have the length of a single step and amount to 2.5 cm. at the middle of the hip line.

15. The vertical maximum up and down of the hip line is reached at the middle of the swinging period and the minimum after the swinging leg is set down; then it starts rising again reaching a new maximum at the middle and a new minimum at the end of the standing period. The horizontal oscillation has its forward maximum as the foot is set down; then swing reverses itself. The movements of the shoulder line are reciprocal with those of the hip line.

16. The common center of gravity represents the total resistance the body offers to motion. The body depends for locomotion upon external forces such as inertia and gravity. If it were floating in space with no external force applying to it, no amount of muscular effort would be able to move the center of gravity. The reduction of the partial masses of the body into a common center of gravity greatly simplifies the calculation of the moving forces operative in the gait.

17. In the sagittal plane the up and down movements of the center of gravity describe a sine curve twice during the double step. The maxima are at the middle of the swing and shortly after the foot is placed to the ground. In the frontal plane the movement has one minimum and one maximum during the double step. The center of gravity is on the side of the supporting leg; the transition to the other side coincides with the setting down of the other foot. In the transverse plane the curve also has the duration of the double step, the center moving commensurately with the pelvis; the forward maximum is at the take-off, the backward maximum as the heel is set down.

18. The velocity curve always crosses the base line, i.e., it is at the minimum, when the path curve shows its positive or negative maximum and the velocity is at maximum when the path curve crosses the base line, i.e., is at its minimum. This means that velocity is minimal when the swing is farthest out and on the point of reversing itself, and at maximum when the swing is at the neutral base line.

19. The acceleration line crosses the base line (is minimal) when the velocity has reached its maximum and when at the same time the path curve has reached its minimum. Conversely, the acceleration is at maximum when the velocity is

at minimum, i.e., at the terminal point of the swing, and then the path curve is at maximum, i.e., farthest out from the base line.

20. Correlating all values for excursion, velocity and acceleration for the center of gravity we find:

a) The center of gravity oscillates during a single step back and forth of a frontal plane, the so-called progression plane, which moves uniformly in direction of the gait. This excursion amounts to 12 mm. The center is farthest behind this plane shortly before the swinging leg is placed on the ground; it is farthest forward when the sole of the standing leg is in full contact with the floor. The total curve from forward to backward occupies a single, the whole cycle from forward to forward a double step.

b) In side oscillations in the frontal plane the center of gravity is farthest from the midline after the support of the standing leg is shifted to the whole sole. The duration of this oscillation is the double step.

c) In the up and down movement in the sagittal plane the center of gravity is highest when the standing foot is full on the ground; it reaches the horizontal base plane (90 cm. above the floor in a man of 170 cm. height) with maximal velocity as the standing leg deploys and attains its lowest level with zero velocity in the middle of the period of double support.

21. Compounding the three dimensions of the path velocity and acceleration curves the total velocities and total accelerations and finally the total floor pressures have been calculated by Braune and Fischer for each of their 31 subdivisions of the gait. Their computations showed that the floor pressure in a man weighing 60 kg. fluctuated in these different phases from 76 kg. maximum to a minimum of 31 kg. These fluctuations of increase and decrease of the gravitational stress are produced by active muscle work. At the moment the standing foot begins to deploy the downward pressure nearly equals the body weight. It then falls in the second half of the supporting phase below half of the body weight. It then rises rapidly to its maximum at the end of the standing period.

22. It is much simpler to calculate the work performed in walking as a whole without consideration of the pressure fluctuations. One has only to eliminate secondary movements and calculate only the effort spent in up and down movement of the body, in its horizontal progression and the force expended in the swing. To these three components of visible motion one must then add a factor which represents the restraint; for instance that developed by the forward planted heel or the resistance which curbs the downward fall of the body in the up and down movements. Braune and Fischer found that for each type of visible motion there is a coefficient of restraint which amounts to 52% of the effort expended in the respective visible movements.

23. On this basis Amar calculated the work of walking for these three essential components (T_1, T_2, T_3).

T_1 is the work of moving the body up and down: If a man's weight is 65 kgm. and the distance of elevation 40 mm. ($= 0.04$ m.) this work would, including 52% for restraint, amount to $65 \times 0.04 \times \dfrac{152}{100}$, or 3.952 kgm. per step.

T_2 represents the forward progression. On the basis of 65 kgm. weight, calculating the living energy developed during the step, $\frac{1}{2}$ m. v^2, the average velocity amounts to 0.6 m. per step and again the coefficient of restraint is $\dfrac{152}{100}$.

Hence the equation would be $\frac{1}{2} \times \dfrac{65}{9.81} \times 0.6^2 \times \dfrac{152}{100}$, or 1.812 kgm. per step.

T_3 represents the work of the swinging leg. The work is calculated as $\frac{1}{2}\, \omega^2\, I$, where ω is the swinging angle of an average 126° and the inertia of the whole leg $= 0.146$. Expressing 126° in linear values: $\dfrac{126}{180}\pi$ or $\dfrac{42}{60}\pi$, the equation would be $\frac{1}{2} \times \dfrac{42^2\pi^2}{60^2} \times 0.146 \times \dfrac{152}{100}$, or 0.281 kgm. per step.

For all three movements the work adds up to 6.045 kgm. per step, or for a step length of 0.778 m., $\dfrac{6.045}{0.778} = 7.712$ kgm. per meter, or $\dfrac{7.712}{65} = 0.119$ kgm. of work per kg. and meter of progress.

24. So far as the work of the individual muscles is concerned, we owe a good share of our knowledge to the work of Scherb and Inman.

25. Of the hip muscles the gluteus maximus begins action at the beginning of the double support or at the early stance and extends it over the single support phase. It prevents the pelvis from dropping forward during the propulsion period of the other leg. The gluteus medius acts synchronously with the maximus as soon as stance is assumed. The adductors start action at the beginning of single support, with the peak of contraction at the end of the stance phase.

26. Of the knee muscles the quadriceps begins' to contract at the period of double support and continues its action into the swinging period. The flexors of the knee start contracting at the beginning of the double support and carry on during the entire standing phase. According to Inman, their primary function is to restrain the forward swing.

27. Tibialis anticus and extensors of the toes are both active in the swinging phase. Their maximum activity, however, is during the standing period shortly after the heel strikes. In this period their function is to regulate and to restrain the action of the calf muscles as the foot goes gradually from dorsal to plantar flexion.

28. Gastrocnemius and soleus are active during the entire stance phase. Even before actual deploy the tension of the calf muscles prevents dorsiflexion of the foot when the sole is flat on the ground. The maximum tension of the calf muscles is shortly before the take-off. The common flexors contract to dig the toes into the ground preparatory to the take-off.

BIBLIOGRAPHY

1. AMAR, J.: *The Human Motor.* New York, E. P. Dillon & Co., 1920.
2. BRAUNE, W., and FISCHER, O.: Der Gang des Menschen. *Abh. d. Mathem. Phys. Klasse d. Kgl. Sächs. Ges. d. Wiss.,* Leipzig, S. Hirzl, 1895.
3. CARLET: Essai experimental sur la locomotion humaine, étude de la marche. *Ann. de Sciences. Nat. Sect. Zool.,* XV, 1872.
4. FISCHER, O.: *Der Gang des Menschen, Part III.* Leipzig, B. G. Teubner, 1900.
5. INMAN, V. T.: Pattern of muscle activity in the lower extremity. *Report of the Nat'l Res. Council on Studies of Human Locomotion, I:1,* Subpart 3.
6. LONDE: La photochromographie appliquée aux études médicales. *Internat. mediz. Photogr. Monatschr., II:9,* 1891.
7. MAREY: Sur la reproduction par la photographie de divers phases de vol des oiseaux. *Compt. rendu, 94:683,* March 9, 1882.
8. SCHERB, G.: *Kinetisch diagnostische Analyse von Gehstörungen.* Stuttgart, F. Enke, 1952.
9. SCHERB, R.: Mitteilungen zur Myokinesiographie. *Ztschr. f. Orthop. Chir.,* 48 and 49.
10. SCHWARTZ, R. P., and HEATH, A. L.: The pneumographic method of recording gait. *J. Bone & Joint Surg., 14:783,* Oct. 1932.
11. SCHWARTZ, R. P., and VAETH, W.: A method for making graphic records of normal and pathological gait. *J.A.M.A., 90,* Jan. 14, 1928.
12. SCHWARTZ, R. P., HEATH, A. L., and WRIGHT, J. W.: Kinetics of the human gait. *J. Bone & Joint Surg., XVI:343,* 1934.
13. VIERORDT: *Über das Gehen des Menschen im gesunden und kranken Zustande.* Tübingen, 1881.
14. WEBER, W. and E.: *Mechanik des menschlichen Gehwerkzeuge.* Göttingen, Dieterich, 1836.

Lecture XXXVIII

THE PATHOMECHANICS OF THE GAIT

INTRODUCTION

IT IS said that the human gait is almost as distinctive of the individual as is his face, in spite of the fact that there are no fundamental differences in the pattern of the normal gait. The distinctive feature can, therefore, consist only in slight quantitative variations in cadence or oscillation. Even in a definitely pathological gait one cannot say that there is a complete overthrow of the normal pattern of walking. Due to these quantitative differences there is a large borderline of gaits still within physiological limits which differ from the average merely by a peculiar accentuation or suppression of one or the other of the phases of the gait. These may make the gait appear ridiculous or even absurd, but they are no new inventions. Even the grotesque characters with their capricious exaggerations which are presented on the stage for their comic effect are made up simply of the same old pattern in somewhat distorted arrangement; it is not a new gait, but merely a caricature of the old.

Not even the cripple who has good reasons behind his locomotor deficiency is able to create new mechanics of walking. Fundamentally, the design of walking follows the old inborn pattern. It is the same score even though it may be badly out of tune.

So, before going from the normal to a manifestly pathological gait it is well to scrutinize certain types of cases whose gait is merely a variant of the normal gait. One group of transitional or borderline cases is characterized by exaggeration or suppression of the normal oscillations and by changes in step length or step duration; another by dynamic deficiencies still within physiological limits.

I. THE BORDERLINE GAIT

A. PHYSIOLOGICAL EXAGGERATION OF THE PERIODS OF GAIT

1. Exaggeration of oscillations

A) THE SAILOR GAIT

At the moment the limb is set to the ground the oblique upward thrust of the extremity is not adequately met by the gluteals and by the tensor to hold the pelvis in firm grip. As a result, the pelvis drops for a moment on the unsupported side as the other limb leaves the ground with the takeoff. This makes for a noticeable increase of the lateral as well as the horizontal oscillations of the pelvis. The shoulders also assume a more pronounced swing because, as the pelvis drops on the unsupported side, the trunk must swing to the supported side to retain balance. In addition, the thrust which the pelvis receives by the setting down of the leg is directed not only upward but backward as well. This produces an increase in the pelvic inclination and in the lumbar lordosis. The so-called swinging gait in which these pelvic oscillations are moderately increased is, to a certain degree, a female sex characteristic.

B) THE PROCESSIONAL GAIT

This is an exaggeration of the swing in the sense that its duration is increased. The swinging limb is allowed not only its full pendulum amplitude, but is also permitted to reverse itself for a short space before the heel is set to the ground. This has a dual effect: it increases the time of swing and, due to the lesser degree of obliquity of the forward leg, it produces less restraint. The result is a greatly decreased cadence and a greatly increased period of support. For this reason it is especially adaptable to processions and funerals. It is an uneconomical gait so far as the relation of effort to visible motion is concerned.

C) THE GOOSE STEP

This is the height of economic waste in walking. Similarly as in the processional gait, the swing is prolonged over the full amplitude by a phase of reversal. Then, to make matters worse, the whole foot is slammed to the ground with scant obliquity and the individual is deprived of the best part of the deploy.

2. Exaggeration of restraint

An example of this type is the so-called "mincing" gait. Here, the swing is abnormally restrained in order to reduce the step length without increasing or even with decreasing the cadence. It is an attempt to walk like a person with much shorter legs. It is very uneconomical and it is designed to convey the appearance of daintiness. But the comical aspect of it is that the individual walks with the step length of a smaller individual and the cadence of a larger one; because of this discrepancy, the whole performance is not convincing.

B. BORDERLINE GAITS WITH DYNAMIC DEFICIENCIES

1. The incoordinate gait

This gait, which has so many real pathological corollaries, can also be found under physiological conditions as a temporary aberration. The main object of such a gait is to enlarge the supporting area in the frontal plane in order to secure lateral balance of the body. For this purpose the legs are spread. The individual strides wide. In a manner, this eases the reciprocal effort of the abductors and adductors on the contralateral sides. One sees this type of gait in people who are in a weakened condition, for instance after prolonged illness, or in older people who have become less foot-sure, especially when walking on uneven ground, and in some other individuals in whom coordination has become temporarily affected for more or less obvious reasons.

2. The slouch or malposture gait

Malposture interferes with the whole pattern of the gait. Every phase of it is slowed down and retarded. As much as possible is left to gravitational forces and as little to active muscular effort. The amplitude of the swing is decreased and because no effort is made for the leg to catch up with the swinging angle of the thigh it lags behind and, in consequence, the limb is set to the ground with the knee in flexion. Because the reaction from the floor is diminished there is less back and upward thrust coming from the forward leg, and the pelvis and trunk oscillations are diminished. The step length is shortened and the cadence is less. Since the weakness of the leg action imparts only weak accelerations to the body, they are easily controlled and little muscle effort is required for restraint. Just as little is expended in propulsion. The whole impression of this type of gait is one of studied languish.

3. The gait of fatigue

In many respects it is similar to the former. The individual keeps the body bent forward, the knees and hips are flexed, the steps are short, and the cadence is diminished. The gait is slow and shuffling and the period of double support is prolonged with the lowering of the cadence. The principal point is, however, that the flexion of hip and knees lowers the center of gravity. Its excursions are less and the stability of the body is increased. The lack of muscular reserve which is due to fatigue causes the individual to become more circumspect in managing his muscular effort and the periodic displacements of the center of gravity are reduced to a minimum. We find this gait in heavy workers who return from their task or in a football players coming from the field.

II. THE PATHOLOGICAL GAIT

The great variety of definitely pathological gaits can be classified under the following headings:

a. Asymmetry of the gait caused by inequality of the legs without restriction of motion.

b. Asymmetry or abnormalities of the gait caused by contracture or ankylosis of the joints of the lower extremities.

c. Abnormalities of the gait caused by static instability of the supporting joints.

d. Abnormalities of the gait due to intolerance of weight bearing. The antalgic gait.

e. Abnormalities of the gait caused by dynamic deficiencies, paralytic gait, flaccid or spastic.

A. INEQUALITY OF THE LOWER LIMBS. THE SHORT-LEG GAIT

A moderate shortening up to $1\frac{1}{2}$ inches in the adult can be equalized by dropping the pelvis on the affected side. When the shortening exceeds this amount, the patient prefers to equalize the length by an equinus position of the ankle. When the shortening is excessive, beyond 3 to 4 inches, then the patient may shorten the other leg by flexing the knee. These three expedients, therefore, suggest the degree of shortening, whether slight, medium or excessive.

The short leg walk shows changes in all three cardinal planes:

1. The most conspicuous feature is in the *frontal plane* (Fig. 1). The dip or the drop of the pelvis begins at the moment of support with the planting of the heel and continues through the entire period of support. The dynamic effect of the dip is a more violent upward and backward thrust of the pelvis. Consequently, the frontal plane oscillations of the pelvis have a greater amplitude on the affected side. They are very fatiguing and the individual easily prefers a compensatory equinus position of the shorter leg. He is much more comfortable with an extension of the heel which, however, should compensate the full amount of shortening only in the case that the affected limb has full control over the movement in the hip, knee and ankle. During the standing period, when the pelvis is dropped, the trunk hangs over on the affected side, and this is especially noticeable when the individual walks fast or runs. In excessive shortening, for instance, in congenital defect of the femur, the patient resorts to the third expedient, namely, to shorten the other leg by flexion in knee and hip, or, rather, he applies this in combination with an equinus position of the ankle on the affected side. Obviously, during the swinging period the limb must swing through in position of flexion; the shortening drop increases to a maximum in the middle of the swing of the sound leg. But we notice that in the standing period of the sound leg, while the short leg swings through, the flexion is also maintained in the joints of the longer leg for the reason that the extension in all joints of the longer leg would entail an upward lift of the center of gravity and this requires a considerable muscular effort. Consequently, if the shortening in the adult reaches four or five inches or more, the patient is better served with a patten, or, if the shortening is so great that a patten shoe would be too heavy, he would be better off with a prosthesis.

2. In the *horizontal plane* the backward thrust of the pelvis which the forward placed short leg produces also is greater than that of the sound side. We recall that as the pelvis rotates backward under the backward thrust the hip goes in adduction and inward rotation. It is up to the inward rotators and adductors of the other leg, therefore, to check these impulses and, at the same time, to prevent the sound leg from going into too much abduction and outward rotation.

3. In the *sagittal plane* (Fig. 2) the efforts of the patient to lengthen the shorter or to shorten the longer leg are rewarded in the first place by the fact that extensive up and down movement of the center of gravity is avoided. In the previous lecture we have seen that of all the muscular requirements placed on the gait, that of elevating the center of gravity and that of checking its gravitational drop are the most exacting. Both in walking and running, these movements are held within narrow limits. The same principle holds also for the short leg limp. The surest way to avoid early fatigue and exhaustion is to

equalize the leg length by one or the other of the expedients mentioned above. For this reason, also, the question of operative equalization of the legs becomes important whenever feasible, not for the cosmetic effect alone but also because it may save undue exertion of the leg muscles and restore the general alignment of the trunk.

B. THE CONTRACTURE LIMP

Two elements enter into the composition of the disability. First, the contracted limb is too short for any phase of the gait where propulsion or restraint depends on periodic lengthening. Secondly, the position of contracture may be incompatible with the particular position which either swing or restraint or propulsion requires in order to become effective. For instance, in equinus the

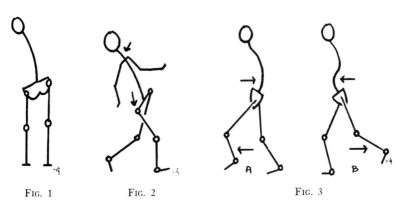

FIG. 1   FIG. 2   FIG. 3

FIG. 1. Short leg gait. Dip of pelvis at moment of support.
FIG. 2. Short leg limp with pelvis on affected side dropping.
FIG. 3. Flexion ankylosis of one hip, showing excurison of lumbar spine.
A, Forward swing. B, Backward swing.

foot is in poor position to receive the impulse of restraint; or, a flexed hip and knee are incompatible with the requirements of propulsion. From the mechanical point of view, contracture and ankylosis can be considered together as it makes little difference for the gait whether the restriction of motion is intra- or extraarticular.

1. Ankylosis or contracture of the hip joint (Fig. 3)

It is remarkable how well people can walk with one hip ankylosed. The other hip acts as the center of motion for all oscillations in the frontal and horizontal planes. This greatly increases the amplitude of the pelvis on the affected side during the period of swing. In the sagittal plane, the for- and backward swing of the leg is taken up partly by the lumbar spine and partly by the other hip. Inasmuch as the mobility of the system now depends upon the other hip joint and the lumbar spine, it is an added burden to these structures and it must be considered under what condition this substitution can be carried out with the least strain on either lumbar spine or the other hip.

What is the best position of the ankylosed hip? Since the position of the ankylosis must be a compromise between position of function and the strain placed on the other hip and on the lumbar spine, the best position of an ankylosed hip is slight flexion and neutral rotation. The question now arises, is it proper in cases of a shortened limb to ankylose it in position of abduction? This was in former years the favored position for a shortened extremity.

From the kinetic point of view this is not sound reasoning. The abducted position entails an obliquity of the pelvis which must be neutralized by the contralateral shift of the trunk and which places considerable strain on the contralateral abductors and sacrospinalis. Lengthening of the limb by extension shoe seems preferable. We have never seen much benefit coming from giving the ankylosed hip an abductory position.

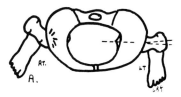

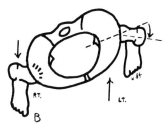

FIG. 4. Ankylosis of one hip and lumbar spine showing backward thrust of whole body on sound hip as ankylosed hip swings forward.

FIG. 5. Ankylosis of hip. *A*, Ankylosis of right hip in external rotation. *B*, As right hip is carried forward, the pelvis rotates counterclockwise and the sound hip goes into internal rotation.

The other point is what is the desirable degree of flexion? It is erroneous to believe that the lumbar spine can always take on the burden of substituting for flexion and extension of the hip. If the lumbar spine is ankylotic (Fig. 4), for- and backward movement can still be carried out by transferring the center of motion to the opposite hip joint in such a way that the forward swing of the ankylosed limb will have to be absorbed by hyperextension in the sound hip joint. But it can readily be understood that the range which this joint furnishes in this direction is small compared with the substitutionary range the lumbar spine can provide for. People can walk with one ankylosed hip and a stiff lumbar spine; but the gait is asymmetrical and particularly the swinging period of the affected side is short and the swing has a small amplitude.

Particularly difficult is it for the patient to manage an ankylosed hip if it is at the same time in outward or inward rotation. For straight for- and backward flexion the lumbar spine movements become compensatory; but they have no or little effect on the length rotation. For compensation of the length rotatory deformity the center of motion is the sound hip joint. It is in this joint where the ankylosed hip must be carried far forward in order to adjust the knee joint to the plane of progression (Fig. 5). This entails a considerable effort of the rotators of the sound hip, both external and internal, to direct and secure the sound leg in inward rotation. While the lumbar spine adjusts the movement in the sagittal plane, the rotation of the pelvis in the horizontal plane which keeps the knees aligned to the plane of progression is entirely a function of the contralateral hip joint. It is for this reason that in osteoarthritis of the hip the outward rotation is the most disturbing element and requires correction even if the position in the frontal and sagittal planes is suitable.

When the affected leg is on the ground and the free leg swings forward, the ball of the foot on the ankylosed side pivots on the floor while the swinging sound leg carries the pelvis with it. The strain which this outward rotation of the foot places upon the ankle and knee joints is a major factor of the disability.

Adduction contracture or ankylosis, if combined with outward rotation, adds to the precariousness of the situation. In the swinging period the pelvis on the affected side must be raised, for which purpose the trunk sways to the opposite side. So far as the affected ankylotic leg is concerned, the difficulties are really greater in the swinging than in the standing period, except for the static strain which the adduction and outward rotation places on the knee and ankle joints.

Bilateral ankylosis of the hip, even if the position is otherwise favorable, eliminates walking in the normal pattern and substitutes a different mechanism. The principal factor of it is that the center of motion is transferred to the ball or the heel of the foot. Each swing means a pivoting upon the ball or heel of the other leg. The momentum imparted to the feet comes from the trunk and the upper extremities. These patients truly walk from their arms and trunk and they move their arms forcefully forward to impart a rotatory momentum to the trunk to be transmitted through the lower extremities to the feet. The gait, as must be expected, is extremely laborious and the question of mobilizing at least one of the hip joints becomes very urgent.

2. Ankylosis and contracture of the knee (Fig. 6)

In flexion contracture it is the standing period which is more affected than is the swing. For one thing, it becomes more difficult to dorsiflex the ankle sufficiently in order to place the heel on the ground. This interferes with the restraint. The flexion of the knee diminishes the forward obliquity of the limb and, consequently, the upward and backward impact from the floor is weakened. This, therefore, sets a limit to the velocity of the gait, and as the speed of walking is increased the short leg limp becomes more marked because most of the restraint which is commensurate with this new rate of speed falls now upon the sound limb. The swinging period is also shorter because of the shortening of the leg. Otherwise, there is little interference with the gait as long as the flexion is moderate, that is, up to 30°. Beyond that point there is the typical short-leg gait.

It is somewhat different in the extension contracture or ankylosis (Fig. 7). Here, the limb is definitely too long, and since it cannot shorten itself while swinging through it must either be circumducted by abduction in the hip joint or the homolateral pelvis must be raised up by placing the standing hip in abduction.

The functional position of the ankylosed knee is 30° flexion and neutral rotation. This position serves best the static requirements of the gait. The ex-

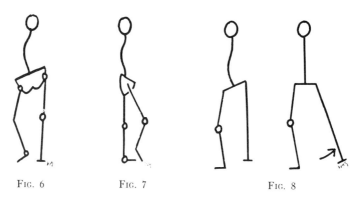

Fig. 6   Fig. 7   Fig. 8

Fig. 6. Standing with one knee ankylosed in flexion.
Fig. 7. In swinging forward of leg with knee ankylosed in extension the pelvis is elevated on the involved side, or the involved hip is abducted and circumducted, the pelvis remaining level.
Fig. 8. Standing with one foot fixed in equinus.

tension ankylosis, on the other hand, is a serious obstacle in the gait and such a position must be avoided in favor of 30° flexion whenever it becomes necessary to arthrodese the knee.

3. The contractures and ankylosis of the ankle and foot

We assume that the deformity, irrespective of its underlying cause, is stationary and that the foot does not change its configuration under weight bearing. The situation must be viewed from two aspects: First, how does the deformity interfere directly with restraint and deploy and takeoff? Secondly, what is the remote effect of the deformity on the proximal joints of the extremity so far as the mechanism of the gait is concerned?

A) THE ANKLE JOINT

If the ankle joint is fixed in equinus by contracture or ankylosis, the immediate effect is the loss of deploy in the standing period. Because the toes touch the ground first instead of the heel, the restraint received through the toes and transmitted upward is necessarily weak and the limit of speed is accordingly lowered. On the other hand, the propulsion is less compromised because the foot is already in the equinus position and ready for the takeoff through the big toe (Fig. 8).

In the rigid equinus position the ankle becomes an angular lever and rotation occurs about the ball of the foot. This gives the leg a powerful backward thrust during the entire standing period and it stabilizes the knee. In the swing phase, however, the relative lengthening of the leg by the equinus position of the foot makes floor clearance more difficult; the patient has to adopt the so-called steppage gait by raising the knee (Fig. 9).

We have stated that in the equinus the restraint is more affected than the propulsion and that this limits somewhat the speed of progression. The main point is that the burden of the restraining impact falls on the ball of the foot causing the anterior arch to become relaxed. Because of lack of proper deploy, the standing period is somewhat shortened.

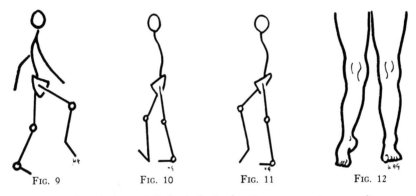

FIG. 9.   FIG. 10.   FIG. 11.   FIG. 12.

FIG. 9. In swinging period of gait the fixed equinus causes a steppage gait.
FIG. 10. Standing period of gait with fixed calcaneous.
FIG. 11. Standing period of gait with fixed equino-varus.
FIG. 12. Fixed equino-varus with compensatory genu valgum.

The opposite situation arises in the fixed calcaneus. Here, the restraint is in full force, being received by the heel. But the propulsion is impaired because of the lack of plantar flexion of the foot. The result is that the period of support is also shortened while the swinging period is normal. This discrepancy becomes more marked as the gait is accelerated (Fig. 10).

B) THE SUBASTRAGALAR AND MIDTARSAL JOINTS

The deficiency in the gait caused by fixed contracture or ankylosis in the subastragalar joint and in the midtarsal joint affects more the propulsion than it does the restraint. In the fixed varus or valgus contracture or in ankylosis in these positions, one does not expect any change in either the period of support or that of the swing if the foot is rigid and there is good weight-bearing tolerance. What is changed, however, is the dynamic effect of the propulsion because in neither instance is a true deploy or takeoff possible. This defect becomes noticeable in fast walking or running (Fig. 11).

There are, however, secondary static effects of the deformity which involve the gait. In the inveterate clubfoot there is often a relative increase of the external rotation of the tibia due to the adduction of the foot; furthermore, the fact that the contact surface of the foot does not lie under, but medial to, the tibial length axis causes the reaction from the floor to produce a powerful horizontal component. This component leads to a deviation of the knee joint in the sense of a genu valgum in the frontal and at the same time to a deformation in the sagittal plane in the sense of a genu recurvatum (Fig. 12).

The condition for the gait is somewhat more favorable for the rigid pes valgus because some semblance of a takeoff can be accomplished by strongly pronating the inner border of the foot and the great toe.

C. THE GAIT IN STATIC DISABILITIES OF THE HIP JOINT

Here, the change in the gait is brought about primarily by the altered relation of the pelvis to the lower extremities. The fault lies essentially in the inability of the hip muscles to immobilize the rim of the pelvis during the standing

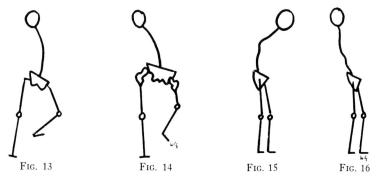

Fig. 13.   Fig. 14   Fig. 15   Fig. 16

FIG. 13. Positive Trendelenburg on affected side in coxa vara.
FIG. 14. Congenital dislocation of the hip with positive Trendelenburg.
FIG. 15. Stance in dorsal tuberculosis. Trunk flexed forward.
FIG. 16. Stance in lumbar tuberculosis. Trunk thrown backward

period so that the latter gives under the weight pressure and the swinging side drops instead of rising. (Trendelenburg sign) (Fig. 13). The mechanical factor behind this insufficiency is the relaxation of the pelvitrochanteric muscles and the shortening of the tensor fasciae. As a result of the latter, the pelvis assumes an increased inclination and a lumbar lordosis. The anatomic reason for this deficiency is the elevation of the greater trochanter, which in all positions of the hip stands now above the Roser-Nelaton line. In the coxa vara this deficiency expresses itself in the waddling gait and the increased lumbar lordosis. Because of the restricted mobility of the hip joint the forward swing of the limb occurs in marked outward rotation. In the congenital dislocation of the hip there is added to it the factor of instability because the dislocated limb telescopes under weight bearing as the head rises, and the lordosis becomes rhythmically accentuated in the period of support.

Besides the insufficiency of the abductors, the weakness of the extensors of the hip causes the body to sway backward as the heel is planted to the ground. The trunk in inclined to the dislocated side in the period of support and there is a corresponding oscillation of the shoulder line (Fig. 14).

In the slow walk the periods of support and swing are of normal duration and the gait is not particularly handicapped. With the increase of speed, however, both lateral and sagittal oscillations of the trunk are very much exaggerated and the effort of balancing the body becomes so exhausting that fatigue sets in early.

D. THE ANTALGIC GAIT

If any part of the extremities or of the spinal column becomes intolerant to weight bearing because of pain, a characteristic antalgic gait develops. If the seat of such intolerance is in the spinal column, or if it involves parts of the pelvis or the lower extremities on both sides, the gait remains symmetrical.

1. The gait in weight-bearing intolerance of the spine

The most striking example of this type of gait we find in spinal tuberculosis. The patient avoids jarring and concussion of the spinal column. The gait is guarded, slow, and it is particularly the restraint which is suppressed. The steps are short; both swinging and supporting periods are reduced. The position of the trunk relative to the pelvis depends upon the site of the lesion. In cervical tuberculosis the neck is bent forward. The gait is particularly restrain because of the high mobility of the cervical spine and its readiness to respond freely to concussion and jar.

In the dorsal spine the collapse of the vertebral bodies forces the trunk in forward flexion. The gait resembles the spondylarthritic except that it is more guarded (Fig. 15). In lumbar tuberculosis the spine is hyperextended and the trunk is thrown backwards. The posture is similar to that of gluteus maximus paralysis but the back- and forward oscillations are suppressed (Fig. 16).

2. The gait in sacrolumbalgia

A) MEDIAN SACROLUMBAR STRAIN

In the great majority of cases the trigger point at the lumbosacral junction appears or becomes accentuated on backward bending. Yet the sacrospinalis muscles are contracted bilaterally; this is necessary to stabilize the forward flexed position of relief which is assumed in order to release pressure upon the strained ligaments or impingements of the spinous processes. The gait is slow, the steps are shortened so as to minimize the normal backward and forward oscillations during the standing period (Fig. 17).

B) LATERAL SACROLUMBAR STRAIN

If, as happens so often, there is a unilateral involvement of the soft structures, particularly of the insertion area of the sacrospinalis in the sacral triangle or of the quadratus lumborum on one side, then the stance is asymmetrical. The body inclines to the side of the lesion so that any strain on the injured tissue is avoided (Fig. 18). In the gait the body is bent forward and the lateral inclination is accentuated.

c) THE SCIATIC GAIT

The first one to describe the antalgic gait in lumbosacralgia associated with sciatica was Gussenbauer (quoted by K. Cramer[1]). Ehret (quoted by Cramer[1]) showed in 1899 that the sciatic nerve relaxes 5-6 cm. From the latter observer also comes the suggestion to immobilize leg and body in the antalgic position (the "as is" position of Putti[5]).

Whether the sciatic radiation is due to the presence of a herniated disc pressing upon a nerve root or to the impingement of the nerve in the inter-

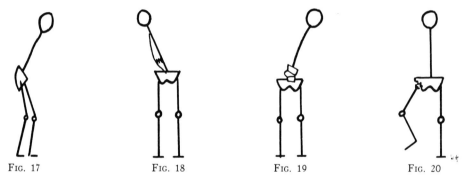

FIG. 17.   FIG. 18.   FIG. 19.   FIG. 20.

FIG. 17. Stance in mid sacrolumbar strain—symmetrical.
FIG. 18. Asymmetrical stance in unilateral sacrolumbar strain—tilt to side of lexion.
FIG. 19. Antalgic position in herniated disc—trunk to unaffected side.
FIG. 20. Antalgic position in tuberculosis of hip in flexion, abduction and external rotation.

vertebral foramen or to any other cause, the posture is most likely asymmetrical and the trunk is usually inclined to the sound side and bent forward. The explanation lies in the weight-bearing intolerance of the affected side. The supporting period is, therefore, shortened on this side and so is the opposite swinging period. If impingement of the nerve root in the intervertebral foramen is the cause it is also believed that the position improves the patency of the foramen (Fig. 19). It has been shown that contralateral tilt causes some tension in the sciatic roots. Whether this has any effect in freeing the root from disc pressure is not established. A striking feature in some cases is the change of position, the so-called scoliosis sciatica alternans. Inclination to the affected side, to all appearances, relaxes the roots somewhat but at the same time increases the pressure stress on the disc. This may be the explanation for the change in position.

The gait is, under all conditions, very guarded, and the steps are shortened. The cadence is low and there is a definite limp on the affected side due to the marked reduced time of support.

3. The coxalgic gait

The principal factor in the coxalgic gait is not so much the position of the affected limb as is the duration of its weight-bearing period. This period being shortened, a definite coxalgic limp develops. In regard to the position of the hip there are several factors to be considered.

In the first place, the inflamed hip joint in the earlier stages goes into a mid-position of flexion, abduction and outward rotation. Then, also, as Inman[2] pointed out, the center of gravity being thrown toward the affected side by the tilt of the body, the tension of the abductors is lessened. The coxalgic gait in the true sense of the word is one which is assumed because tolerance both to weight-bearing stress and to pressure stress by muscle tension is diminished by the inflammatory condition. In a way, it is a contracture gait because the hip is held in a protective position, which insures the least tension and least irritation of the sensitive inflamed synovial membrane. As stated above, this position for the hip joint is slight flexion, abduction and outward rotation. The knee is then placed in a compensatory degree of flexion and the ankle joint is in plantar flexion (Fig. 20). Then, in walking the toes and not the heel are placed first on the ground, the shock of restraint is lessened, the period of support is shortened. While the other limb swings forward, the horizontal rotation of the pelvis is carried out by pivoting on the ball of the affected leg. The takeoff is weak and as the affected limb swings forward the pelvis makes a wide circle around the sound hip in order to bring the outward rotated limb into the progression plane.

Seen in the frontal plane, the sound standing limb has to clear the affected swinging one by strongly elevating the diseased side of the pelvis. In the sagittal plane, the forward and backward oscillations of the trunk are restricted.

In contrast, the so-called psoas gait, that is, one caused by a psoas abscess or by psoitis, is much freer. Only the extension is limited in the hip joint. The lumbar spine is lordodic, hip and knee are in flexion, the body is inclined forward and the ankle is in equinus. But it is not in abduction and, therefore, no special effort is necessary by the other leg to clear the affected limb off the ground; and, since there is no outward rotation, the horizontal oscillations of the pelvis are more symmetrical.

4. The gonalgic gait

The inflamed or irritated knee assumes a midposition of about 25° flexion and of inward rotation. The ankle is held in equinus in order to lessen the jar of the restraint and also to give the limb proper length. Since the hip is free there is no difficulty with the horizontal or frontal rotation of the pelvis during both the standing and the swinging periods. But the gait is markedly asymmetrical because of the restriction of the supporting period on the affected side. Faster walking, especially, produces a decided limp.

5. The podalgic gait

More often than not the antalgic asymmetry of the gait is due to functional disabilities of the foot: a corn, a callus, a strain of the longitudinal or transverse arch, a sprain of the ligaments of the ankle, and even a chronic inflammatory condition of the ankle joint such as tuberculosis may still be compatible with weight bearing. The position in which the foot is held throws considerable light on the nature of the pathological condition. It will be held in equinus in the case of a painful heel spur, whereas the equinus position is avoided in favor of a calcaneus if there is a tender callus or a corn or any degree of relaxation of the anterior arch of the foot. The strain of the longitudinal arch causes the patient to walk on the lateral border of the foot. The type of gait then corresponds to the respective contracture gait. The more painful the condition, the shorter is the supporting period and all the more noticeable is the limp.

If the ankle joint is affected it is held in equinus to reduce the shock of

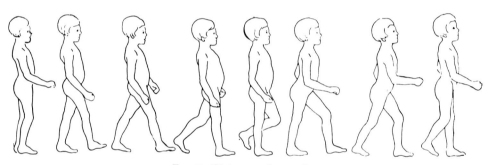

FIG. 21. Gluteus maximus gait.

restraint. In this case the period of support is reduced to the point where the patient hobbles on the affected side.

There is a characteristic difference between the antalgic limp due to local tenderness and the limp caused by circulatory deficiencies such as endarteritis or thromboangiitis obliterans. Here the limp is also marked; it does not appear, however, with the first step but only after a variable distance of walking, and the foot is held in normal position.

E. THE PATHOMECHANICS OF THE GAIT IN MUSCULAR DEFICIENCIES

1. The paralytic gait

The essential features in the paralytic gait are as follows:

a) The inability to control pelvic oscillations and to secure the pelvic position against gravity.

b) The inability to stabilize the knee joint during the period of support and to lengthen the extremity against gravity.

c) The inability to develop restraint.

d) The inability to develop propulsion.

ad a) The stabilization of the pelvis and the control of pelvic oscillations. 1)

The gluteus maximus gait. The characteristic feature is the backward thrust of the trunk (Fig. 21). It begins shortly after the heel is set to the ground and it reaches its maximum when the hip is vertical over the ankle joint, the cycle being completed at the end of the standing period. This backward thrust prevents the jackknifing of the hip joint by causing the line of gravity to fall behind the joint center. Gravity acts as an extensory force and is balanced by the tension of the iliofemoral ligament.

This oscillation of the trunk repeats itself with every step and gives the gait an awkward appearance, but there is little difficulty in slow walking. Only in fast walking the steps are shortened to moderate the backward swaying of the trunk. Attempts have been made to substitute the gluteus maximus by the sacrospinalis muscle with very indifferent results, mainly because the substituting muscle lacks lifting height.

2) The gluteus medius gait. In paralysis of this muscle the oscillations of the

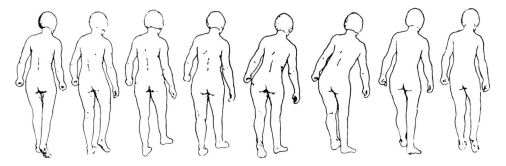

FIG. 22. Gluteus medius gait.

pelvis in the frontal plane are impeded. While the leg swings it assumes an adductory position and this is met by the increased elevation of the pelvis through the action of the contralateral abductors (Fig. 22). Only in this way can the swinging leg clear the ground. This causes a side swaying of the trunk to the sound side. During the supporting period the pelvis drops to the sound side because there is no abductor to secure the rim of the pelvis against the trochanter on the side of paralysis. The center of gravity is now shifted over the affected hip. To minimize the rotation moment at this side or by shifting the center of gravity beyond the hip joint, gravity may act as an abductor supplanting the paralyzed gluteus medius and minimus. The tensor fasciae if acting, is able to secure the pelvic rim but is alone unable to prevent the contralateral pelvic drop. The result is an alternating shift of the body to the sound side during the swinging period and to the affected side during the supporting period of the involved limb. In case of a bilateral paralysis the rhythm is the same but the oscillations are increased and the gait assumes a waddle which is similar to that seen in congenital dislocation of the hip where it is likewise due to insufficiency of the abductors.

The operative attempts to substitute the paralyzed gluteus medius and minimus by other muscles have been much more successful than in paralysis of the gluteus maximus. The best plan is a substitution by the tensor fascia if the latter is active. A simple transposition of the muscle laterally and backward provides it with abductory mobility. A newer, recent method of Mustard[3] of transferring the insertion of the psoas in front or behind the femur into the greater trochanter holds out considerable promise. Less promising is the idea of using a fascial elongation of the vastus lateralis.

b) The instability of the knee joint. 1) Paralysis of the quadriceps. It is well known that people with paralysis of the quadriceps walk well. The stabilization of the knee joint is in danger only during that phase of the supporting period

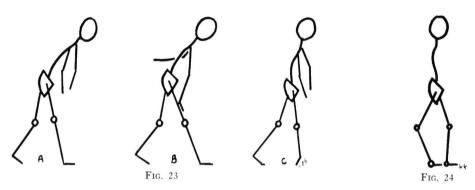

FIG. 23. Paralysis of quadriceps and gluteus maximus. *A*, Line of gravity placed in front of knee joint axis. *B*, Pressing thigh backward with hand. *C*, Locking knee by external rotation of limb and thereby placing knee joint axis in sagittal plane.

FIG. 24. Genu recurvatum with resulting equinus position of foot.

when the standing leg points obliquely forward and the line of gravity falls behind the knee joint axis. The tension of the gluteus maximus combined with that of the soleus, both uniarticular muscles, suffices in ordinary walking to stabilize the knee joint. If any or both of these muscles are paralyzed the patient has several expediencies at his disposal. He can bend his body forward during this phase and so throw the line of gravity forward in front of the knee joint axis; he can place his hands on the thigh and press it backward and let his arm act as quadriceps (Fig. 23); or, he can prevent the jackknifing simply by turning the limb outward so that the knee joint axis runs in the sagittal plane and the joint is locked against flexion. Of these three expediencies the forward bending of the trunk is the one more frequently used.

In the swinging period the lack of the quadriceps prevents the leg from swinging forward promptly so that it lags behind the thigh in the forward pendulum movement, and when the heel is then placed on the floor the knee still shows some degree of flexion. At this moment, however, the other leg is still on the ground carrying out the takeoff and the forward propulsion of the pelvis, assisted by the contralateral extensors of the hip, sees the knee safely over this phase of the standing period.

It is for these reasons that the indications for substituting the missing quadriceps do not always arise. If they do, a judicious selection of the hamstring muscles gives satisfactory results.

2) Paralysis of the hamstrings. Here it is not so much the instability than the increasing deformity of the knee which disturbs the gait. As has been shown before, the hamstrings are active in both periods. In the swing they help flex the knee to shorten the extremity and in the support they are co-stabilizers of the knee joint during the restraint. They again are in action at the moment of propulsion as flexors of the knee.

The greatest difficulty, however, is of static nature, and it lies in a progressing genu recurvatum deformity (Fig. 24). A short leg gait then develops. Furthermore, the marked forward slanting of the leg places the foot in equinus position which makes it difficult to secure enough dorsiflexion in the ankle joint to place the heel on the ground at the phase of restraint. Instead, the whole sole contacts it and the gait becomes sloppy. Efforts have been made to control this static deformity by changing the slant of the tibial condyles. In genu recurvatum the slope is forward and downward. By raising the tibial plateau it can be given a sharp backward slant and the joint be placed in a flexory position.

c) The gait in paralysis of the ankle joint. 1) The insufficiency of restraint. The paralytic dropfoot. Because of its inability to dorsiflex the foot actively the extremity has to be shortened during the swing by raising the knee high on the affected side, the so-called steppage gait.

In the supporting period the lack of dorsiflexion in the ankle joint is more disturbing because the toes now touch the ground first and the forward movement cannot be checked as effectively as it can by heel contact. This sets also a limit to the velocity of the gait. In this respect, the gait resembles that observed in a paralytic genu recurvatum. Operative procedure which restrains the plantar flexion such as the arthrorisis of Lambrinudi or the countersinking of Brewster are able to control this difficulty, at least in part.

2) The insufficiency of the takeoff. The paralytic calcaneus. Here the takeoff is restricted while the restrain is not impaired. The ankle joint is quite stable in walking or in standing, but some difficulty appears in the accelerated walk or in running. The swing is unchanged. In the standing period the heel is set to the ground first and the restraint is normal but as the forward leg gradually goes into backward obliquity the sole does not hug the ground and no takeoff develops. Instead, the heel remains pivoting on the floor. One should expect, as the pelvis on the standing side rotates backward while the other limb swings forward and carries its side of the pelvis with it, that the affected leg would go into inward rotation during the period of support. This, however, is the case only where the calcaneus deformity is so marked that there is no possibility of the sole coming in contact with the ground. In milder cases the patient makes an effort to establish contact with the inner side of the foot. He does this by strong outward rotation in the hip joint and by pronating the foot so the inner border comes in contact with the floor, establishing thereby a sort of takeoff.

There are some very efficient procedures which overcome the calcaneus position and which, at the same time, equalize the lever arms of heel and ball and
redistribute during the supporting period the superincumbent load more evenly
over the ankle joint. We refer here to the astragalectomy of Whitman and the
horizontal transverse section of Davis, two procedures in which the posterior
displacement of the foot against the leg constitutes the principal mechanical
feature.

2. The spastic gait

In the spastic gait we meet with a combination of deficiencies. The alignment
of the trunk over the pelvis is disturbed, the rhythmic shortening and lengthening of the leg is handicapped by the rigidity of the muscles, and for the same
reason the free forward and backward swing of the limb is impeded. The contractural element prevails and it leads to a series of distortions, each of which
gives the gait a characteristic pattern.

A) THE SPASTIC EQUINUS GAIT

The most disturbing feature in this situation is that the patient lacks ground
contact since it is reduced to the balls of the feet. In the standing period this
interferes with both the propulsion and the restraint. At the end of the standing
period the takeoff is produced by first throwing the weight upon the ball of the
foot and then adding a forcible plantar flexion so that it is more a push-off of
the whole ball of the foot than a real takeoff. In the beginning of the standing
period the restraint is received also by the ball of the foot instead of the heel.
The worst feature is the lack of the ground support and the resulting instability
of the whole body. For this reason alone the correction of the equinus deformity
by conservative means or, if these fail, by operative methods is most essential.

B) THE SPASTIC CALCANEUS GAIT

Less common than the spastic equinus is the spastic calcaneus. This deformity offers a more favorable situation. For one thing, the restraint is not
impaired, for reasons already mentioned above. Then, the outward rotation of
the feet, in contrast to the inward rotating equinus gives more ground contact
and this increases the stability. Furthermore, a certain degree of flexion of the
knee accompanies the calcaneus position, and this further increases stability
by lowering the center of gravity of the body. Finally, as in the paralytic calcaneus, the patient is able by strongly pronating the foot to contact the inner
border with the ground and thereby achieve some substitution for the normal
takeoff.

C) THE GAIT IN SPASTIC ADDUCTION AND INWARD ROTATION CONTRACTURE OF THE HIP JOINTS

This contracture produces the greatest difficulty both for standing and walking mainly by making the lateral balance very insecure. The patient is fairly

well able to adjust himself to the forward and backward oscillation of the line of gravity. But it is the regulation of the lateral balance in the frontal plane which becomes so difficult, especially if there is also a spastic equinus. In unilateral paralysis the gait is much less disturbed in the infantile hemiplegic than it is in the adult. Walking on even ground is easier and it can be accomplished with short steps and decreased cadence. The center of gravity is displaced toward the sound side and usually two-thirds or more of the body weight is borne on this side. Since the hip is contracted in adduction it cannot swing forward freely, and the same maneuvers of lifting the pelvis by means of the contralateral abductors is necessary in order to clear the extremity off the ground.

In bilateral cases with both hips rotated inward and adducted, the contraction may reach the stage where the limbs cross or at least where the knees are pressed tightly against each other. If there is an additional spastic equinus the mechanism of walking becomes very difficult. It resenbles that of a bilateral hip contracture gait but it is further aggravated by the flexion of the knees and by the varus position of the feet. One wonders how these patients can walk at all. It seems to be possible only by the rotatory movements carried out by the trunk which are imparted to the pelvis and to the lower extremities and which, finally, resolve themselves in an alternating pivoting about the balls of the feet. The release of these adduction and inward rotations of the hips together with the correction of the equinus deformity of the feet accomplishes a proper realignment and provides a degree of stability which in many cases opens the way for a successful muscle educational program.

d) The Gait in Spastic Flexion Contracture of the Hip

This is an accompanying feature of the adduction-inward rotation contracture; the essential effect is the exaggerated lordosis. If conservative measures to correct this deformity fail, the tenotomy of the iliopsoas, the contracture of which is the main cause of the deformity, will cause the lordosis to disappear and will abolish the flexor spasm (Peterson[4]).

III. THE PROSTHETIC GAIT (SLOCUM[6])
A. AMPUTATION THROUGH THE DISTAL TARSUS

1. Lisfranc's amputation

Slocum finds that the gait is awkward because the absence of the metatarsal heads precludes any takeoff; the patient assumes a calcaneus gait and the step is shortened. During the period of swing the dorsiflexion is missing, if the extensors do not act. However, this does not necessarily occur if the dorsal extensors are joined to the flexors and are not allowed to retract.

After the heel strikes the ground the following plantar flexion becomes more abrupt and the sole drops to the floor rather suddenly and not gradually, as is normal. The principal feature seems to be the deficiency of the dorsiflexors. In the swinging period the hip must rise to give passage to the swinging leg if shortening by dorsiflexion of the ankle joint is lost.

2. Chopart's amputation

We believe that here also some of the difficulties which were mentioned above can be avoided by securing the flexors to the extensor group of the foot and by obtaining in this manner sufficient dorsiflexion for the swinging period without elevating the pelvis on the swinging side. It seems that this can be accomplished to better advantage in Chopart's than in Lisfranc's amputation.

B. THE AMPUTATION OF THE FOOT

1. The Syme amputation (Slocum[6]) (Fig. 25)

With the prosthesis, Slocum found the gait pattern almost normal. The knee is slightly more flexed in the swinging period to give clearance to the foot since

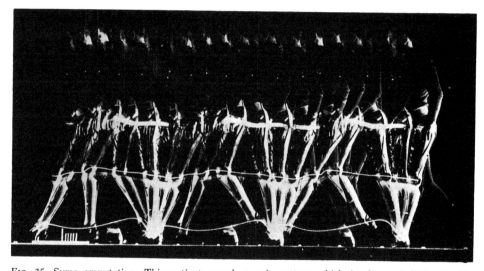

FIG. 25. Syme amputation. This patient reveals a gait pattern which is almost indistinguishable from the normal. The chief point of variance lies in the swing phase where the hip is flexed slightly more than in the normal, to allow clearance of the foot which lacks active dorsiflexion. The pelvic path curve is normal; the knee curve is normal, the knee passing through the biphasic extension characteristic of normal gait since it is extended when the heel strikes the ground, then passes into flexion, and finally into extension again as the limb reaches the perpendicular. Tibial rotation is present with flexion of the knee, as might be anticipated. The path curves of the heel and toe (not shown in this illustration) are normal except that they show a slightly greater height, which is utilized to clear the ground. (Slocum)

there is no active dorsiflexion. This shows also in the path curves of the heel and toe of the prosthesis which are slightly higher.

2. The Pirogoff amputation

The situation is similar to that of the Syme amputation. Without prosthesis, the patient assumes a true heel walk with no takeoff, and propulsion is carried out only through the heel. The standing period is shortened and a limp appears.

C. BELOW-THE-KNEE AMPUTATION (SLOCUM[6]) (FIG. 26)

The spring action of the ankle and the buffer of the metatarsophalangeal joint of the prosthesis permit a sort of takeoff when the patient walks slowly and on even ground. Slocum[6] finds that with a stump of ideal length the amputee walks excellently; the gait varies only slightly from the normal because more flexion in the hip and knee is necessary to clear the ground. Of course, the knee hinges of the prosthesis prevent the normal rotation of the leg, outward on extension and inward on flexion, and as the heel is placed to the floor outward rotation

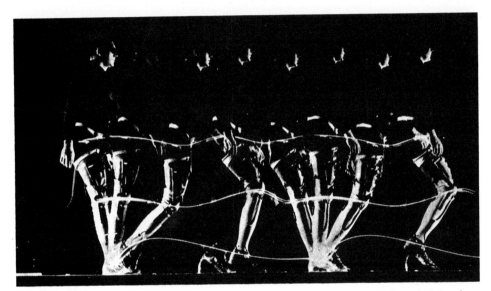

Fig. 26. The below-knee amputation. This well-developed and well-coordinated below-knee amputee, with a stump of ideal length, has excellent gait. It varies from the normal in that the hip flexes slightly more to lift the foot clear of the ground, the thigh does not rotate externally during flexion because of the fixation of the knee hinges, and the foot falls slightly lateral to the line of progression. Because of this last, the wear pattern of the shoe is from the midline of the heel to the midline of the toe. (Slocum)

must come from the hip joint. Inman[2] showed that with a well-fitting prosthesis the gait is practically normal.

The ill-fitting prosthesis has a definite effect on the gait. If the stump fits poorly and slides up or down, then the limb becomes periodically longer and shorter with corresponding effect of increased knee flexion during the swing and a short leg tilt during the standing period. Failure of the spring and buffer mechanism at the ankle and the metatarsophalangeal joint of the prosthesis causes a certain rigidity and lack of propulsion in the second part of the standing period. This means shortening of the step length and an asymmetrical limp.

D. ABOVE-KNEE AMPUTATION (FIG. 27)

Slocum's[6] stroboscopic pictures show a difference in the gait according to the type of weight bearing. In the end-bearing amputation the elevation of the pelvis is greater in the swinging period; the flexion of the hip also increases just before the foot is placed on the ground; the knee joint of the prosthesis does not come to complete extension until the leg has assumed about 20° of backward obliquity. It is the hip joint which is extended suddenly in the support period to secure the knee joint. This extension of the hip has the effect that the stump forces the thigh piece backward by action of the gluteals and so stabilizes the knee joint.

In the ischial-bearing stump (Fig. 28), Slocum[6] finds that the gait of the

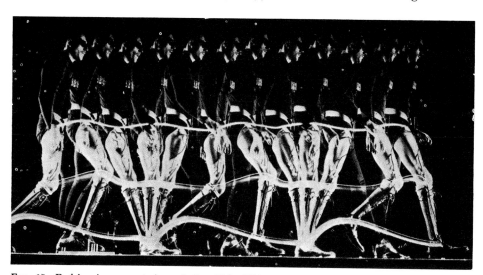

Fig. 27. End-bearing amputation of the thigh. This patient shows variations from the normal pattern of gait typical of this level. A number of important differences are noted: In the phase of swing, the elevation of the pelvis is greater than it is in the phase of support; flexion of the thigh increases throughout to reach its maximum just before the heel is to be placed on the ground; the knee reaches complete extension when the thigh is flexed to about 20 degrees beyond the perpendicular. As the phase of support is entered, the pelvis drops downward and forward and the hip is extended abruptly to maintain the artificial knee joint in extension; because of a slight overswing of the foot which occurs immediately before this, the extension of the hip causes a backward movement of the whole extremity and gives the effect of "digging in the heel" (the term applied to this process by the limb makers).

The prosthesis for amputation at this level has no rigid fixation to the pelvic belt, and the socket has some tendency to drift downward on the stump under the influence of gravity. This makes it more difficult for the amputee to clear the ground with the foot and accounts in some measure for the fact that the extreme rise of the pelvis during the swing phase is found only in the gait of this level. (Slocum)

above-knee amputee varies little from the normal during the supporting period. The limb strikes the ground with the heel first, but with full extension of the knee, the ankle goes into dorsiflexion. In the swinging period there is some difference in respect to the clearing of the toes. This is done either by elevation of the pelvis, which makes the pelvic path curve higher, or else more flexion is given to the hip joint.

The artificial limb in the above-knee amputee does not have the stability to maintain extension in the knee joint during the first phase of the standing

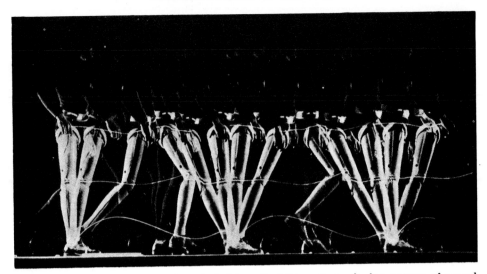

Fig. 28. The ischial-bearing above-knee limb. The gait of this patient, who has a stump of normal length, is similar to that of the normal during the phase of support (except for the lack of biphasic extension), but he utilizes a different technique to bring the foot forward clear of the ground during the phase of swing. Here, instead of elevation of the pelvis, the hip is flexed forward rapidly, then quickly extended, and the whole extremity is lowered to bring the heel in contact with the ground as the body moves forward. The effect on the path curve of the knee is a sharp backward jog immediately after full flexion. This type of gait is characteristic of those who prefer the knee friction to be loose, so that the leg swings more freely about the knee. (Slocum)

period until the line of gravity passes over the knee joint axis. In the second part of its standing period the knee joint should go into flexion, the extremity shortened and the foot prepared for the takeoff. As a result, the supporting period is shortened, particularly in the phase of propulsion (Slocum[6]). Inman[2] finds in his path curve tracings definite anomalies and the expenditure of the gait is increased 27% over that of the normal.

IV. SUMMARY

1. Aberrations of the gait either in the border line or the manifestly pathological cases do not present an entirely new scheme of walking but are merely variations or distortions of the normal pattern.

2. Borderline cases are either physiological exaggerations or suppressions of the normal elements of the gait, or they are due to dynamic deficiences still within physiological limits.

3. The sailor gait consists of a physiological increase of the lateral and horizontal oscillations of the pelvis and the shoulders, associated with increased pelvic inclination; in the processional gait the time of the swing is increased, the restraint is decreased due to the higher position of the hips; in the "goose step" the swing is prolonged and the slamming of the foot interferes with the deploy.

4. An example of exaggeration of restrain is the mincing gait which has a short step length but not the increased cadence natural to it.

5. Borderline cases from dynamic deficiences are the incoordinate gait in which the supporting area is enlarged by spread of the legs; the slouch or malposture gait with decreased amplitude of swing, lagging legs, short steps, decreased cadence, retardation of all phases of the gait, bent knees, forward thrown trunk; and the fatigue gait showing similar features. The latter particularly is marked by the flexed hips and knees which lower the center of gravity, giving the body greater stability.

6. In the short leg walk the "dip" in the frontal plane starts at the moment of planting the heel and continues during the entire period of support. Due to the more violent upward and backward thrust the frontal plane oscillations of the pelvis have greater amplitude on the affected side. The trunk hangs over on this side. While the short leg swings through, the equalizing flexion of the joints of the longer leg is maintained to save the muscular effort of elevating the center of gravity. The patient walks better with the short leg in equinus or with a patten or even a prosthesis, depending on the degree of shortening. In the horizontal plane there is a greater backward thrust as the heel is set to the ground, the hip going into adduction and inward rotation, restrained by the adductors and inward rotators of the other leg. In the sagittal plane the shortening of the long leg avoids the laborious up and down movements of the center of gravity. Early fatigue and exhaustion can best be avoided by equalization of the leg length by mechanical or operative means.

7. The contracted limb is too short for any phase of the gait where periodic lengthening is required for propulsion or restraint. Also, the contracture position may be unsuitable for one or the other phase, be it support, swing, restrain or propulsion.

8. In hip ankylosis the other hip is the center of motion in the frontal and horizontal planes. In the sagittal plane the substitution for for- and backward swing is taken up by the lumbar spine as well as by the sound hip. The best position for the ankylosed hip is neither ab- nor adduction, slight flexion and neutral rotation. Equalization of the limbs is much preferable to trying to lengthen the limb by an ankylosis in abduction. The desirable degree of flexion is a compromise between standing and sitting.

9. Ankylosis of the hip in outward rotation necessitates circumduction of the pelvis around the sound hip with wide amplitude in order to bring the knee joint into the plane of progression. When the affected leg is on the ground the forward swing of the sound leg occurs by pivoting on the ball of the affected side. Adduction contracture combined with outward rotation adds to it the raising of the pelvis of the affected side. Bilateral ankylosis of the hip eliminates the normal pattern. The center of motion is transferred to ball or heel and the momentum for pivoting comes from the trunk and the upper extremities.

10. In flexion contracture of the knee the standing period is affected. Dorsiflexion of the foot is difficult, restraint is interfered with. Forward obliquity is diminished, the upward and backward impact from the floor is weakened, which sets a limit to the velocity. Up to 30° flexion, there is little interference with the gait. In extension contracture the limb is too long and during swing

it must either be circumducted or the pelvis raised.

11. Contracture or ankylosis of the ankle joint in equinus causes loss of deploy but the propulsion is less affected than the restraint. The rigid ankle gives the leg a strong backward thrust stabilizing the knee. In fixed calcaneus deformity the restraint is in full force but the propulsion is impaired. The period of support is shortened, that of the swing is normal.

12. Fixed contracture or ankylosis of subastragalar and midtarsal joints affects more the propulsion than the restraint. Fixed varus or valgus deformity, if there is good weight tolerance, eliminates deploy altogether. However, secondary effects are increased external rotation of the tibia in inveterate club-foot and genu valgum deformity.

13. In the static disabilities of the hip joint such as coxa vara and congenital dislocation the difficulty lies in the inability of the hip muscles to immobilize the pelvis during the supporting period (Trendelenburg's sign). The shortening of the tensor causes increased pelvic inclination. The deficiency results in a waddle. In slow walk support and swing are approximately normal but in fast walk the oscillations of the trunk become very marked and fatigue sets in early.

14. An antalgic gait due to weight-bearing intolerance of the spine is seen in spinal tuberculosis. The gait is guarded and slow. Swinging and supporting periods are reduced. In cervical tuberculosis head and neck are held forward. In dorsal tuberculosis the trunk is in forward flexion; in the lumbar, the spine is hyperextended and the trunk thrown backward.

15. In median sacrolumbar strain the sacrospinalis is contracted to stabilize the forward flexed antalgic position. The gait is slow, the steps short. In lateral sacrolumbar strain the stance is asymmetrical, the body inclining to the side of the lesion and bent forward.

16. In sciatic radiation, whether due to a protruded disc or to nerve impingement in the intervertebral foramen, the posture is asymmetrical, usually inclined forward and to the sound side. This contralateral tilt improves the patency of the intervertebral foramen. Inclination to the affected side, while relaxing the roots, increases at the same time the pressure stress on the disc. The gait is guarded, the cadence low, the steps short.

17. In the coxalgic gait there is a definite reduction of the weight-bearing period. The position in the acute stages is flexion, abduction and outward rotation. The center of gravity is thrown to the affected side by the tilt of the body. The knee is in compensatory flexion, the ankle joint in plantar flexion. The ground is touched with the toes. As the other leg swings forward, the horizontal rotation of the pelvis is enacted by pivoting on the ball of the affected side. When the latter swings forward the center of motion is the sound hip. In the frontal plane the forward swing of the affected limb has to be cleared by elevation of the pelvis. In the psoas contracture gait only the extension is limited and the oscillations of the pelvis are symmetrical.

18. In the gonalgic gait the knee is in midposition of 25° flexion, the ankle in equinus. The gait is asymmetrical because of restriction of the period of support. Fast walking causes a decided limp.

19. In the podalgic gait, so far as the weight tolerance is not entirely abolished, the foot is held in position commensurate with the site of tenderness: equinus in case of a painful heel, calcaneus for a tender callus of the forefoot or anterior arch strain; walk on the lateral border of the foot in strain of the longitudinal arch. If the ankle joint is involved the foot is held in equinus to reduce the shock of restraint. The antalgic limp due to the shortened period of support is marked.

20. The gluteus maximus gait caused by paralysis of this muscle is characterized by the backward thrust of the trunk setting in shortly after the heel is set to the ground. It prevents the hip from jackknifing forward. There is little difficulty in the slow walk but the step becomes short to moderate the backward thrust of the trunk when the walk is accelerated.

21. In paralysis of the gluteus medius the swinging leg goes in adduction necessitating the elevation of the pelvis on the affected side by the sound side abductors. The trunk sways to the sound side. In the supporting phase the pelvis drops to the sound side and the body is now shifted to the affected hip. The result is an alternating shift. In bilateral paralysis the gait is a symmetrical waddle. The best operative substitution is by the tensor, if available.

22. Walking is possible in paralysis of the quadriceps because the gluteus maximus and the soleus stabilize the knee; if both or one of the latter are also paralyzed, jackknifing of the knee can be prevented by bending the body forward or pressing the thigh backward with the hand. Forward swing of the leg is retarded because of loss of quadriceps.

23. Paralysis of the hamstrings does not cause much instability but may lead to genu recurvatum. This, again, may produce dfficulties in dorsiflexing the ankle for the restraint.

24. The paralytic dropfoot is an obstacle to the restraint and during the swing it forces increased raising of the knee, a steppage gait. This sets a limit to the velocity of the gait. Operative procedures to restrain plantar flexion are Lambrinudi's and Brewster's methods.

25. In the paralytic calcaneus the propulsion is impaired as there is no takeoff. The ankle is stable, the swing unchanged, restraint normal. The hip goes in inward rotation only in cases where the deformity is so marked that the sole cannot be brought to touch the ground. Otherwise the sole can be made to contact the ground by strong outward rotation at the hip and pronation of the foot. Efficient operative procedures are Whitman's astragalectomy and the horizontal transverse section of Davis.

26. The most disturbing feature in the spastic equinus is the instability caused by the supporting area being reduced to the ball of the foot. Both propulsion and restraint are impaired. Correction of the equinus position is essential in the interest of stability. Stability is much better in the uncommon spastic calcaneus. The restraint is not affected and the patient is able to accomplish a sort of takeoff by strongly pronating the feet and contacting the floor with their inner border

27. Spastic adduction and inward rotation of the hips causes great difficulties in standing and walking mainly because of inability to maintain lateral balance. In unilateral cases the center of gravity remains on the sound side which carries two-thirds or more of the body weight. In bilateral cases the situation parallels that of bilateral hip ankylosis in adduction. Walking is only possible by alternatingly pivoting about the ball of the foot. An accompanying flexion contracture increases the difficulty because of its effect on the anteroposterior balance. Measures to correct the contractures in all planes must be installed.

28. After Lisfranc's amputation of the forefoot there is often a lack of the dorsiflexors of the foot, which causes the foot to drop into plantar flexion abruptly after the heel touches the ground and which also forces the hip to be elevated during the swing. In Chopart's amputation it seems easier to establish the balance by connecting the extensors with the flexors instead of allowing them to retract.

29. After a Syme amputation the prosthesis provides an almost normal gait. Only the knee may have to be flexed more in the swinging period. The situation is similar in Pirogoff's amputation, only here the limb remains too long for a proper prosthesis. Without it, the gait is pylon type, no takeoff, propulsion entirely through the heel.

30. In below-the-knee amputations the amputee walks excellently with a proper prosthesis which provides a spring action at the ankle and buffer action at the metatarsophalangeal joints. On the other hand, an ill-fitting prosthesis affects the gait by causing the limb to become periodically longer during the swing and shorter during the support period.

31. The above-the-knee amputee shows a difference in the gait according to the type of weight bearing. In the end-bearing stump there is more elevation of the pelvis in the swing and more flexion of the hip. Complete extension of the prosthesis does not occur until at 20° backward obliquity. In ischial weight bearing the gait varies little from the normal. In the swinging period the toes are cleared either by elevation of the pelvis or by increased flexion in the hip. The amputee cannot maintain extension in the knee joint until the line of gravity passes over and in front of the knee joint axis. The supporting period is shortened particularly in the phase of propulsion. The energy expenditure of the gait is increased, according to Inman, 27% over the normal.

BIBLIOGRAPHY

1. CRAMER, K.: Gipsverbandbehandlung bei Ischias. *Ztschr. f. Orthop. Chir., XIV:XLVI:* 685, 1905.
2. INMAN, V. T.: Functional aspect of the abductors of the hip. *J. Bone & Joint Surg., XXIX:3,* 607, July 1944.
3. MUSTARD, W. T.: Iliopsoas transfer for weakness of the hip abductors. *J. Bone & Joint Surg., XXXIV:*647, July 1952.
4. PETERSEN, L. T.: Tenotomy in the treatment of spastic paralysis with special reference to tenotomy of the iliopsoas. *J. Bone & Joint Surg., XXXII A:4,* 875, Oct. 1950.
5. PUTTI, V.: *Lumboartrite e Sciatica Vertebrale.* Bologna, C. Cappelli, 1936.
6. SLOCUM, D. B.: *An Atlas of Amputations.* St. Louis, C. V. Mosby, 1949.

AUTHOR INDEX

SUBJECT INDEX

697

KINESIOLOGY

By

ARTHUR STEINDLER, M.D., (Hon.) F.R.C.S. Eng.,
F.A.C.S., F.I.C.S.

Fifth Printing